EXPERIMENTAL STRESS ANALYSIS

Second Edition

JAMES W. DALLY

Professor of Mechanical Engineering
University of Maryland, College Park

WILLIAM F. RILEY

Professor of Engineering Science and Mechanics
Iowa State University

McGRAW-HILL BOOK COMPANY

New York   St. Louis   San Francisco   Auckland   Bogotá   Düsseldorf
Johannesburg   London   Madrid   Mexico   Montreal   New Delhi   Panama
Paris   São Paulo   Singapore   Sydney   Tokyo   Toronto

EXPERIMENTAL STRESS ANALYSIS

34567890 F G R F G R   83210

This book was set in Times Roman. The editors were B. J. Clark and
Madelaine Eichberg; the cover was designed by Scott Chelius; the production supervisor
was Leroy A. Young. The drawings were done by J & R Services, Inc.
Fairfield Graphics was printer and binder.

Library of Congress Cataloging in Publication Data

Dally, James W
    Experimental stress analysis.

    Includes bibliographies and index.
    1. Strains and stresses. 2. Photoelasticity.
3. Strain gages. I. Riley, William Franklin,
date   joint author. II. Title.
TA407.D32 1978   620.1'123   76-393
ISBN 0-07-015204-7

CONTENTS

PART FOUR OPTICAL METHODS OF STRESS ANALYSIS

PREFACE

This book has been written for upper-division undergraduate students or graduate students beginning to study experimental stress analysis. The material has been divided into four separate parts:

Part 1. Elementary Elasticity, three chapters
Part 2. Brittle-Coating Methods, one chapter
Part 3. Strain-Measurement, six chapters
Part 4. Optical Methods, six chapters

Each part of the book is essentially independent so that the instructor can be quite flexible in selection of the course content. For instance, a two- or three-credit course on strain gages can be offered by using two chapters of Part 1 and all of Part 3. Parts 1 and 4 can be combined to provide a thorough three- or four-credit course on photoelasticity and moiré. Selected chapters from all four parts can be organized to introduce the broader field of experimental stress analysis. A complete detailed treatment of the subject matter covered here supplemented with laboratory exercises on brittle coatings, strain gages, photoelasticity, and moiré will require six- to eight-credit hours

The essential feature of the text is its completeness in introducing the entire range of experimental methods to the student. A reasonably deep coverage is presented of the theory necessary to understand experimental stress analysis and of the four primary methods employed: brittle coatings, strain gages, photoelasticity, and moiré. While the primary emphasis is placed on the theory of experimental stress analysis, the important experimental techniques associated with each of the four major methods are covered in sufficient detail to permit the student to begin laboratory work with a firm understanding of experimental procedures. Many exercises designed to support and extend the treatment and to show the application of the theory have been placed at the end of all of the chapters.

Laboratory exercises have not been included, since the laboratory work will depend so strongly on local conditions such as the equipment and supplies available, the interests of the instructor and students, and the local industrial problems of current interest. It is believed that the instructor is best qualified to specify the associated laboratory exercises on the basis of interest, equipment, supplies, and time available for this important supplement to the course.

This second edition incorporates extensive revisions which reflect the changes in experimental methods that have occurred in the past 13 years. With brittle coatings, material covering the new nontoxic, nonflammable coatings has been introduced. With electrical strain gages, the changes in the text indicate the almost exclusive use of foil gages today, the marked improvements in instrumentation associated with dynamic recording, and the new data acquisition systems which are being employed to process the experimental data. The text, in part, accounts for the very significant advances made in the past decade in optical methods of experimental stress analysis. The basic chapter on properties of light has been completely revised to account for the behavior of coherent light, and the coverage on both moiré and birefringent coatings has been expanded to a complete chapter on each method. In spite of these extensive revisions, space limitations and publishing costs did prevail and it was necessary to delete coverage of several important topics including statistics, holography, failure theories, fracture mechanics, and nondestructive testing.

It is anticipated that the instructor will, in certain instances, treat these topics by using his own lecture notes or recent papers published in the technical journals. While the text is not in any sense a treatise, it does contain most of the fundamental material necessary to present a complete and practical course on the theory of experimental stress analysis.

The material presented here has been assembled by both authors over a period of 25 years. Courses have been developed on Experimental Stress Analysis, Photoelasticity, and Photomechanics at Illinois Institute of Technology, Cornell University, Iowa State University, and the University of Maryland. The material has been shown to be interesting and comprehensible by the students participating in these courses. The mathematics employed in the treatment can easily be understood by senior undergraduates. Cartesian notation and/or vector notation has been used to enhance understanding of the field equations. A great deal of effort was devoted to the selection and preparation of the illustrations employed. These illustrations complement the text and should aid appreciably in presenting the material to the student.

The contribution of many hundreds of investigators working in the field of experimental mechanics should be acknowledged. This book represents a summary of the state of the art in a field which is continually being advanced by the combined efforts of many patient investigators. In particular, we should like to thank Dr. A. J. Durelli, our mentor for many years. Also, we should acknowledge excellent illustrations provided by M. M. Leven, D. Post, C. E. Taylor, and the many suppliers of commercial equipment. Thanks are also due to Professors Robert Mark of Princeton University, and C. W. Smith of Virginia Polytechnic Institute and State University for their careful review of the manuscript.

James W. Dally
William F. Riley

LIST OF SYMBOLS

a	amplitude of a light wave
A	area
b	aperture width
B	strength of a magnetic field
B_L	lateral boundary
c	relative stress optical coefficient
c	velocity of light in a vacuum
c_1, c_2	stress optical coefficients
C	capacitance
C	galvanometer constant
C_c	coating coefficient of sensitivity
C_s	specimen coefficient of sensitivity
C_v	Poisson's ratio mismatch correction factor
d	degree of damping
D	damping coefficient
D	diameter
D	volume dilatation ($\epsilon_1 + \epsilon_2 + \epsilon_3$)
D_0	fluid damping constant
D_0	fog density of a photographic film
e	electron charge
E	electromotive force or voltage
E	exposure
E	magnitude of a light vector
E	modulus of elasticity

E	potential gradient
E^c	modulus of elasticity of a coating
E_m	back electromotive force
E_0	exposure inertia of a photographic film
E^s	modulus of elasticity of a specimen
E^*	modulus of elasticity of a calibrating beam
f	focal length of a lens
f	frequency
f_ϵ	material strain-fringe value
f_σ	material stress-fringe value
F	force
F_{CB}	bending correction factor
F_{CR}	reinforcing correction factor
F_x, F_y, F_z	cartesian components of the body-force intensity
F_r, F_θ, F_z	polar components of the body-force intensity
g	gravitational constant
G	shear modulus of elasticity
G	torsional spring constant
h	thickness
i	current density
I	current
I	intensity of light
I	moment of inertia
I_e	intensity of emerging light
I_g	gage current
I_G	galvanometer current
I_i	intensity of incident light
I_r	intensity of reflected light
I_1, I_2, I_3	first, second, and third invariants of stress
J	polar moment of inertia
J_1, J_2, J_3	first, second, and third invariants of strain
k	dielectric constant
K	brittle-coating strength ratio
K	bulk modulus
K	optical strain coefficient
K	strength of a light source
K_t	transverse-sensitivity factor
K_T	compressibility constant
l	length
l_g	gage length
L	length
$\mathscr{L}$	loss factor
M	magnification factor
n	index of refraction
n	integer

n_0	index of refraction in an unstressed medium
n_1, n_2, n_3	index of refraction along the principal directions
n_1, n_2, n_3	principal directions
N	cycles of relative retardation
N	fringe order
N	number of calibration values
N	number of charge carriers
N	number of cycles
p	pitch of a moire grating
p	pressure
P	force, applied load
P	power
P_D	power density
P_g	power dissipated by a gage
q	resistance ratio
Q	figure of merit
r	resistance ratio
R	radius
R	reflection coefficient
R	resistance
R_b	ballast resistor
R_e	equivalent resistance
R_g	gage resistance
R_G	galvanometer resistance
R_M	measuring-circuit resistance
R_p	parallel resistor
R_s	series resistor
R_x	external resistance
s	distance
s_1, s_2	curvilinear coordinates along an isostatic
S	sensitivity index
S	standard deviation
S_a	axial strain sensitivity of a gage
S_A	strain sensitivity of a material
S_c	circuit sensitivity
S_{CG}	galvanometer-circuit sensitivity
S_g	strain sensitivity of a gage, gage factor
S_{pb}	parallel-balance-circuit sensitivity
S_{sc}	strain sensitivity of a semiconductor material
S_t	gage sensitivity to time
S_t	transverse strain sensitivity of a gage
S_T	gage sensitivity to temperature
S_ϵ	strain sensitivity
S_θ	galvanometer sensitivity
S_σ	stress sensitivity

t	time
T	period
T	temperature
T	torque
T	transmission coefficient
$\mathbf{T}_n$	resultant stress
T_{nx}, T_{ny}, T_{nz}	cartesian components of the resultant stress
T_x, T_y, T_z	cartesian components of the surface tractions
u, v, w	cartesian components of displacement
u_r, u_θ, u_z	polar components of displacement
V	voltage
V	volume
w	width
x, y, z	cartesian coordinates
α	angle of incidence
$\alpha, \beta, \theta, \phi$	angles
α	coefficient of thermal expansion
α_p	polarizing angle
β	angle of reflection
β	coefficient of thermal expansion
γ	angle of refraction
γ	shear strain component
γ	temperature coefficient of resistivity
$\gamma_{r\theta}$	shear strain component in polar coordinates
$\left.\begin{array}{l}\gamma_{xy} = \gamma_{yx}\\ \gamma_{yz} = \gamma_{zy}\\ \gamma_{zx} = \gamma_{xz}\end{array}\right\}$	cartesian shear strain components
δ	displacement
δ	linear phase difference
Δ	relative phase difference
Δ	relative retardation
ΔR_T	resistance change due to temperature
ΔR_ϵ	resistance change due to strain
ϵ	normal strain
ϵ_a	axial strain
ϵ_c	calibration strain
ϵ_n	normal strain component
$\epsilon_r, \epsilon_\theta, \epsilon_z$	normal strain components in polar coordinates
ϵ_t	transverse strain
ϵ_{t*}	threshold strain for a brittle coating under a uniaxial state of stress
$\epsilon_{xx}, \epsilon_{yy}, \epsilon_{zz}$	normal strain components in cartesian coordinates
$\epsilon_1, \epsilon_2, \epsilon_3$	principal normal strains
$\epsilon_1^c, \epsilon_2^c$	principal normal strains in a coating
$\epsilon_1^s, \epsilon_2^s$	principal normal strains in a specimen
η	nonlinear term

η_1, η_2	material viscosities
θ	angular deflection
θ_s	steady-state deflection of a galvanometer
λ	Lame's constant
λ	wavelength
μ	mobility of charge carriers
μ	shear modulus
ν	Poisson's ratio
ν^c	Poisson's ratio of a coating
ν^s	Poisson's ratio of a specimen
ν^*	Poisson's ratio of a calibrating beam
ξ	wave number
π	piezoresistive proportionality constant
ρ	radius of curvature
ρ	resistivity coefficient
ρ	specific resistance
σ	normal stress component
σ_n	normal component of the resultant stress
$\sigma_{rr}, \sigma_{\theta\theta}, \sigma_{zz}$	normal stress components in polar coordinates
σ_{uc}^c	ultimate compressive strength of a brittle coating
σ_{ut}^c	ultimate tensile strength of a brittle coating
$\sigma_{xx}, \sigma_{yy}, \sigma_{zz}$	normal stress components in cartesian coordinates
$\sigma_1, \sigma_2, \sigma_3$	principal normal stresses
σ_1^c, σ_2^c	principal normal stresses in a coating
σ_1^s, σ_2^s	principal normal stresses in a specimen
σ_1', σ_2'	secondary principal stresses
σ^*	normal stress in a calibrating beam
τ	shear stress component
τ_n	shear stress component of the resultant stress
$\tau_{r\theta}$	shear stress component in polar components
$\left.\begin{array}{l}\tau_{xy} = \tau_{yx}\\ \tau_{yz} = \tau_{zy}\\ \tau_{zx} = \tau_{xz}\end{array}\right\}$	shear stress components in cartesian coordinates
ϕ	Airy's stress function
ω	angular frequency
Ω	body-force function

EXPERIMENTAL
STRESS ANALYSIS

PART
ONE

ELEMENTARY ELASTICITY

ONE

STRESS

1.1 INTRODUCTION

An experimental stress analyst must have a thorough understanding of stress, strain, and the laws relating stress to strain. For this reason, Part 1 of this text has been devoted to the elementary concepts of the theory of elasticity. The first chapter deals with stresses produced in a body due to external and body-force loadings. The second chapter deals with deformations and strains produced by the loadings and with relations between the stresses and strains. The third chapter covers plane problems in the theory of elasticity, important since a large part of a first course in experimental stress analysis deals with two-dimensional problems. Also treated is the stress-function approach to the solution of plane problems. Upon completing the subject matter of Part 1 of the text, the student should have a firm understanding of stress and strain and should be able to solve some of the more elementary two-dimensional problems in the theory of elasticity by using the Airy's-stress-function approach.

1.2 DEFINITIONS

Two basic types of force act on a body to produce stresses. Forces of the first type are called *surface forces* for the simple reason that they act on the surfaces of the body. Surface forces are generally exerted when one body comes in contact with another. Forces of the second type are called *body forces* since they act on each element of the body. Body forces are commonly produced by centrifugal, gravitational, or other force fields. The most common body forces are gravitational, being present to some degree in almost all cases. For many practical applications,

3

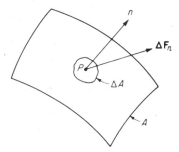

Figure 1.1 Arbitrary surface (either internal or external) showing the resultant of all forces acting over the element of area ΔA.

however, they are so small compared with the surface forces present that they can be neglected without introducing serious error. Body forces are included in the following analysis for the sake of completeness.

Consider an arbitrary internal or external surface, which may be plane or curvilinear, as shown in Fig. 1.1. Over a small area ΔA of this surface in the neighborhood of an arbitrary point P, a system of forces acts which has a resultant represented by the vector $\Delta \mathbf{F}_n$ in the figure. It should be noted that the line of action of the resultant force vector $\Delta \mathbf{F}_n$ does not necessarily coincide with the outer normal n associated with the element of area ΔA. If the resultant force $\Delta \mathbf{F}_n$ is divided by the increment of area ΔA, the average stress which acts over the area is obtained. In the limit as ΔA approaches zero, a quantity defined as the resultant stress $\mathbf{T}_n$ acting at the point P is obtained. This limiting process is illustrated in equation form below.

$$\mathbf{T}_n = \lim_{\Delta A \to 0} \frac{\Delta \mathbf{F}_n}{\Delta A} \tag{1.1}$$

The line of action of this resultant stress $\mathbf{T}_n$ coincides with the line of action of the resultant force $\Delta \mathbf{F}_n$, as illustrated in Fig. 1.2. It is important to note at this point that the resultant stress $\mathbf{T}_n$ is a function of both the position of the point P in the body and the orientation of the plane which is passed through the point and identified by its outer normal n. In a body subjected to an arbitrary system of

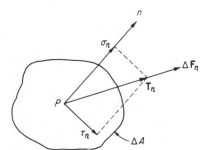

Figure 1.2 Resolution of the resultant stress T_n into its normal and tangential components σ_n and T_n.

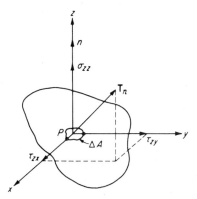

Figure 1.3 Resolution of the resultant stress T_n into its three cartesian components τ_{zx}, τ_{zy}, and σ_{zz}.

loads, both the magnitude and the direction of the resultant stress $\mathbf{T}_n$ at any point P change as the orientation of the plane under consideration is changed.

As illustrated in Fig. 1.2, it is possible to resolve $\mathbf{T}_n$ into two components: one σ_n normal to the surface is known as the resultant normal stress, while the component τ_n is known as the resultant shearing stress.

Cartesian components of stress for any coordinate system can also be obtained from the resultant stress. Consider first a surface whose outer normal is in the positive z direction, as shown in Fig. 1.3. If the resultant stress $\mathbf{T}_n$ associated with this particular surface is resolved into components along the x, y, and z axes, the cartesian stress components τ_{zx}, τ_{zy}, and σ_{zz} are obtained. The components τ_{zx} and τ_{zy} are shearing stresses since they act tangent to the surface under consideration. The component σ_{zz} is a normal stress since it acts normal to the surface.

If the same procedure is followed using surfaces whose outer normals are in the positive x and y directions, two more sets of cartesian components, τ_{xy}, τ_{xz}, σ_{xx}, and τ_{yx}, τ_{yz}, σ_{yy}, respectively, can be obtained. The three different sets of three cartesian components for the three selections of the outer normal are summarized in the array below:

$$\left\|\begin{array}{ccc} \sigma_{xx} & \tau_{xy} & \tau_{xz} \\ \tau_{yx} & \sigma_{yy} & \tau_{yz} \\ \tau_{zx} & \tau_{zy} & \sigma_{zz} \end{array}\right\| \quad \begin{array}{l} \text{outer normal parallel to the } x \text{ axis} \\ \text{outer normal parallel to the } y \text{ axis} \\ \text{outer normal parallel to the } z \text{ axis} \end{array}$$

From this array, it is clear that nine cartesian components of stress exist. These components can be arranged on the faces of a small cubic element, as shown in Fig. 1.4. The sign convention employed in placing the cartesian stress components on the faces of this cube is as follows: if the outer normal defining the cube face is in the direction of increasing x, y, or z, then the associated normal and shear stress components are also in the direction of positive x, y, or z. If the outer normal is in the direction of negative x, y, or z, then the normal and shear stress components are also in the direction of negative x, y, or z. As for subscript convention, the first

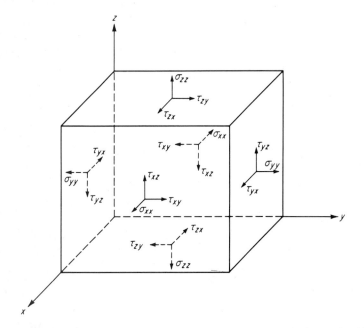

Figure 1.4 Cartesian components of stress acting on the faces of a small cubic element.

subscript refers to the outer normal and defines the plane upon which the stress component acts, whereas the second subscript gives the direction in which the stress acts. Finally, for normal stresses, positive signs indicate tension and negative signs indicate compression.

1.3 STRESS AT A POINT

At a given point of interest within a body, the magnitude and direction of the resultant stress $\mathbf{T}_n$ depend upon the orientation of the plane passed through the point. Thus an infinite number of resultant-stress vectors can be used to represent the resultant stress at each point since an infinite number of planes can be passed through each point. It is easy to show, however, that the magnitude and direction of each of these resultant-stress vectors can be specified in terms of the nine cartesian components of stress acting at the point. This can be seen by considering equilibrium of the elemental tetrahedron shown in Fig. 1.5. In this figure the stresses acting over the four faces of the tetrahedron are represented by their average values. The average value is denoted by placing a $\sim$ sign over the stress symbol. In order for the tetrahedron to be in equilibrium, the following condition must be satisfied. First consider equilibrium in the x direction:

$$\tilde{T}_{nx} A - \tilde{\sigma}_{xx} A \cos (n, x) - \tilde{\tau}_{yx} A \cos (n, y) - \tilde{\tau}_{zx} A \cos (n, z) + \tilde{F}_x \tfrac{1}{3} h A = 0$$

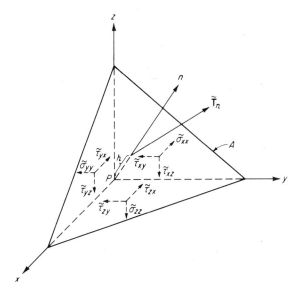

Figure 1.5 Elemental tetrahedron at point P showing the average stresses which act over its four faces.

where h = altitude of tetrahedron

A = area of base of tetrahedron

$\tilde{F}_x$ = average body-force intensity in x direction

$\tilde{T}_{nx}$ = component of resultant stress in x direction

and $A \cos (n, x)$, $A \cos (n, y)$, and $A \cos (n, z)$ are the projections of the area A on the yz, xz, and xy planes, respectively.

By letting the altitude $h \to 0$, after eliminating the common factor A from each term of the expression, it can be seen that the body-force term vanishes, the average stresses become exact stresses at the point P, and the previous expression becomes

$$T_{nx} = \sigma_{xx} \cos (n, x) + \tau_{yx} \cos (n, y) + \tau_{zx} \cos (n, z) \tag{1.2a}$$

Two similar expressions are obtained by considering equilibrium in the y and z directions:

$$T_{ny} = \tau_{xy} \cos (n, x) + \sigma_{yy} \cos (n, y) + \tau_{zy} \cos (n, z) \tag{1.2b}$$

$$T_{nz} = \tau_{xz} \cos (n, x) + \tau_{yz} \cos (n, y) + \sigma_{zz} \cos (n, z) \tag{1.2c}$$

Once the three cartesian components of the resultant stress for a particular plane have been determined by employing Eqs. (1.2), the resultant stress $\mathbf{T}_n$ can be determined by using the expression

$$\mathbf{T}_n = \sqrt{T_{nx}^2 + T_{ny}^2 + T_{nz}^2}$$

The three direction cosines which define the line of action of the resultant stress $\mathbf{T}_n$ are

$$\cos (T_n, x) = \frac{T_{nx}}{|\mathbf{T}_n|} \qquad \cos (T_n, y) = \frac{T_{ny}}{|\mathbf{T}_n|} \qquad \cos (T_n, z) = \frac{T_{nz}}{|\mathbf{T}_n|}$$

The normal stress σ_n and the shearing stress τ_n which act on the plane under consideration can be obtained from the expressions

$$\sigma_n = |\mathbf{T}_n| \cos (T_n, n) \qquad \text{and} \qquad \tau_n = |\mathbf{T}_n| \sin (T_n, n)$$

The angle between the resultant-stress vector $\mathbf{T}_n$ and the normal to the plane n can be determined by using the well-known relationship

$$\cos (T_n, n) = \cos (T_n, x) \cos (n, x) + \cos (T_n, y) \cos (n, y)$$
$$+ \cos (T_n, z) \cos (n, z)$$

It should also be noted that the normal stress σ_n can be determined by considering the projections of T_{nx}, T_{ny}, and T_{nz} onto the normal to the plane under consideration. Thus

$$\sigma_n = T_{nx} \cos (n, x) + T_{ny} \cos (n, y) + T_{nz} \cos (n, z)$$

Once σ_n has been determined, τ_n can easily be found since

$$\tau_n = \sqrt{T_n^2 - \sigma_n^2}$$

1.4 STRESS EQUATIONS OF EQUILIBRIUM

In a body subjected to a general system of body and surface forces, stresses of variable magnitude and direction are produced throughout the body. The distribution of these stresses must be such that the overall equilibrium of the body is maintained; furthermore, equilibrium of each element in the body must be maintained. This section deals with the equilibrium of the individual elements of the body. On the element shown in Fig. 1.6, only the stress and body-force components which act in the x direction are shown. Similar components exist and act in the y and z directions. The stress values shown are average stresses over the faces of an element which is assumed to be very small. A summation of forces in the x direction gives

$$\left(\sigma_{xx} + \frac{\partial \sigma_{xx}}{\partial x} dx - \sigma_{xx}\right) dy\, dz + \left(\tau_{yx} + \frac{\partial \tau_{yx}}{\partial y} dy - \tau_{yx}\right) dx\, dz$$
$$+ \left(\tau_{zx} + \frac{\partial \tau_{zx}}{\partial z} dz - \tau_{zx}\right) dx\, dy + F_x\, dx\, dy\, dz = 0$$

Dividing through by $dx\, dy\, dz$ gives

$$\frac{\partial \sigma_{xx}}{\partial x} + \frac{\partial \tau_{yx}}{\partial y} + \frac{\partial \tau_{zx}}{\partial z} + F_x = 0 \qquad (1.3a)$$

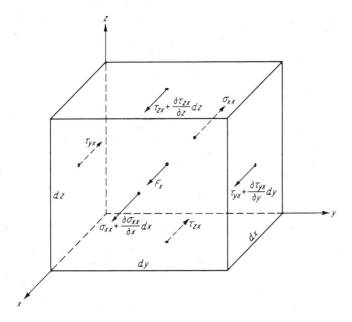

Figure 1.6 Small element removed from a body, showing the stresses acting in the x direction only.

By considering the force and stress components in the y and z directions, it can be established in a similar fashion that

$$\frac{\partial \tau_{xy}}{\partial x} + \frac{\partial \sigma_{yy}}{\partial y} + \frac{\partial \tau_{zy}}{\partial z} + F_y = 0 \qquad (1.3b)$$

$$\frac{\partial \tau_{xz}}{\partial x} + \frac{\partial \tau_{yz}}{\partial y} + \frac{\partial \sigma_{zz}}{\partial z} + F_z = 0 \qquad (1.3c)$$

where F_x, F_y, F_z are body-force intensities (in lb/in^3 or N/m^3) in the x, y, and z directions, respectively.

Equations (1.3) are the well-known stress equations of equilibrium which any theoretically or experimentally obtained stress distribution must satisfy. In obtaining these equations, three of the six equilibrium conditions have been employed. The three remaining conditions can be utilized to establish additional relationships between the stresses.

Consider the element shown in Fig. 1.7. Only those stress components which will produce a moment about the y axis are shown. Since the coordinate system has been selected with its origin at the centroid of the element, the normal stress components and the body forces do not produce any moments.

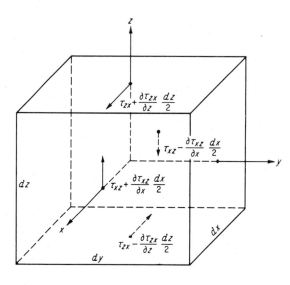

Figure 1.7 Small element removed from a body, showing the stresses which produce a moment about the y axis.

A summation of moments about the y axis gives the following expression:

$$\left(\tau_{zx} + \frac{\partial \tau_{zx}}{\partial z}\frac{dz}{2}\right) dx\, dy\, \frac{dz}{2} + \left(\tau_{zx} - \frac{\partial \tau_{zx}}{\partial z}\frac{dz}{2}\right) dx\, dy\, \frac{dz}{2}$$

$$- \left(\tau_{xz} + \frac{\partial \tau_{xz}}{\partial x}\frac{dx}{2}\right) dy\, dz\, \frac{dx}{2} - \left(\tau_{xz} - \frac{\partial \tau_{xz}}{\partial x}\frac{dx}{2}\right) dy\, dz\, \frac{dx}{2} = 0$$

which reduces to

$$\tau_{zx}\, dx\, dy\, dz - \tau_{xz}\, dx\, dy\, dz = 0$$

Therefore,

$$\tau_{zx} = \tau_{xz} \tag{1.4a}$$

The remaining two equilibrium conditions can be used in a similar manner to establish that

$$\tau_{xy} = \tau_{yx} \tag{1.4b}$$

$$\tau_{yz} = \tau_{zy} \tag{1.4c}$$

The equalities given in Eqs. (1.4) reduce the nine cartesian components of stress to six independent components, which may be expressed in the following array:

$$\begin{array}{ccc} \sigma_{xx} & \tau_{xy} & \tau_{zx} \\ \tau_{xy} & \sigma_{yy} & \tau_{yz} \\ \tau_{zx} & \tau_{yz} & \sigma_{zz} \end{array}$$

1.5 LAWS OF STRESS TRANSFORMATION

It has previously been shown that the resultant-stress vector $\mathbf{T}_n$ acting on an arbitrary plane defined by the outer normal n can be determined by substituting

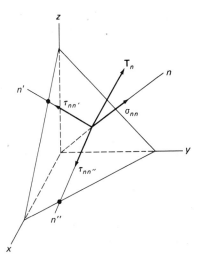

Figure 1.8 Resolution of $\mathbf{T}_n$ into three cartesian components σ_{nn}, $\tau_{nn'}$, and $\tau_{nn''}$.

the six independent cartesian components of stress into Eqs. (1.2). However, it is often desirable to make another transformation, namely, that from the stress components σ_{xx}, σ_{yy}, σ_{zz}, τ_{xy}, τ_{yz}, τ_{zx}, which refer to an $Oxyz$ coordinate system, to the stress components $\sigma_{x'x'}$, $\sigma_{y'y'}$, $\sigma_{z'z'}$, $\tau_{x'y'}$, $\tau_{y'z'}$, $\tau_{z'x'}$, which refer to an $Ox'y'z'$ coordinate system. The transformation equations commonly used to perform this operation will be developed in this section.

Consider an element similar to Fig. 1.5 with an inclined face having outer normal n. Two mutually perpendicular directions n' and n'' can then be denoted in the plane of the inclined face, as shown in Fig. 1.8. The resultant stress $\mathbf{T}_n$ acting on the inclined face can be resolved into components along the directions n, n', and n'' to yield the stresses σ_{nn}, $\tau_{nn'}$, and $\tau_{nn''}$. This resolution of the resultant stress into components can be accomplished most easily by utilizing the cartesian components T_{nx}, T_{ny}, and T_{nz}. Thus

$$\sigma_{nn} = T_{nx} \cos(n, x) + T_{ny} \cos(n, y) + T_{nz} \cos(n, z)$$

$$\tau_{nn'} = T_{nx} \cos(n', x) + T_{ny} \cos(n', y) + T_{nz} \cos(n', z)$$

$$\tau_{nn''} = T_{nx} \cos(n'', x) + T_{ny} \cos(n'', y) + T_{nz} \cos(n'', z)$$

If the results from Eqs. (1.2) and (1.4) are substituted into these expressions, the following important equations are obtained:

$$\sigma_{nn} = \sigma_{xx} \cos^2(n, x) + \sigma_{yy} \cos^2(n, y) + \sigma_{zz} \cos^2(n, z)$$
$$+ 2\tau_{xy} \cos(n, x) \cos(n, y) + 2\tau_{yz} \cos(n, y) \cos(n, z)$$
$$+ 2\tau_{zx} \cos(n, z) \cos(n, x) \tag{1.5a}$$

$$\tau_{nn'} = \sigma_{xx} \cos(n, x) \cos(n', x) + \sigma_{yy} \cos(n, y) \cos(n', y)$$
$$+ \sigma_{zz} \cos(n, z) \cos(n', z)$$
$$+ \tau_{xy}[\cos(n, x) \cos(n', y) + \cos(n, y) \cos(n', x)]$$
$$+ \tau_{yz}[\cos(n, y) \cos(n', z) + \cos(n, z) \cos(n', y)]$$
$$+ \tau_{zx}[\cos(n, z) \cos(n', x) + \cos(n, x) \cos(n', z)] \quad (1.5b)$$
$$\tau_{nn''} = \sigma_{xx} \cos(n, x) \cos(n'', x) + \sigma_{yy} \cos(n, y) \cos(n'', y)$$
$$+ \sigma_{zz} \cos(n, z) \cos(n'', z)$$
$$+ \tau_{xy}[\cos(n, x) \cos(n'', y) + \cos(n, y) \cos(n'', x)]$$
$$+ \tau_{yz}[\cos(n, y) \cos(n'', z) + \cos(n, z) \cos(n'', y)]$$
$$+ \tau_{zx}[\cos(n, z) \cos(n'', x) + \cos(n, x) \cos(n'', z)] \quad (1.5c)$$

Equations (1.5) provide the means for determining normal- and shear-stress components at a point associated with any set of cartesian reference axes provided the stresses associated with one set of axes are known.

Expressions for the stress components $\sigma_{x'x'}$, $\sigma_{y'y'}$, $\sigma_{z'z'}$, $\tau_{x'y'}$, $\tau_{y'z'}$, $\tau_{z'x'}$ can be obtained directly from Eq. (1.5a) or Eq. (1.5b) by employing the following procedure.

In order to determine $\sigma_{x'x'}$, select a plane having an outer normal n coincident with x'. A resultant stress $\mathbf{T}_n = \mathbf{T}_{x'}$ is associated with this plane. The normal stress $\sigma_{x'x'}$ associated with this plane is obtained directly from Eq. (1.5a) by substituting x' for n. Thus

$$\sigma_{x'x'} = \sigma_{xx} \cos^2(x', x) + \sigma_{yy} \cos^2(x', y)$$
$$+ \sigma_{zz} \cos^2(x', z) + 2\tau_{xy} \cos(x', x) \cos(x', y)$$
$$+ 2\tau_{yz} \cos(x', y) \cos(x', z) + 2\tau_{zx} \cos(x', z) \cos(x', x) \quad (1.6a)$$

By selecting n coincident with the y' and z' axes and following the same procedure, expressions for $\sigma_{y'y'}$ and $\sigma_{z'z'}$ can be obtained as follows:

$$\sigma_{y'y'} = \sigma_{yy} \cos^2(y', y) + \sigma_{zz} \cos^2(y', z)$$
$$+ \sigma_{xx} \cos^2(y', x) + 2\tau_{yz} \cos(y', y) \cos(y', z)$$
$$+ 2\tau_{zx} \cos(y', z) \cos(y', x) + 2\tau_{xy} \cos(y', x) \cos(y', y) \quad (1.6b)$$
$$\sigma_{z'z'} = \sigma_{zz} \cos^2(z', z) + \sigma_{xx} \cos^2(z', x)$$
$$+ \sigma_{yy} \cos^2(z', y) + 2\tau_{zx} \cos(z', z) \cos(z', x)$$
$$+ 2\tau_{xy} \cos(z', x) \cos(z', y) + 2\tau_{yz} \cos(z', y) \cos(z', z) \quad (1.6c)$$

The shear-stress component $\tau_{x'y'}$ is obtained by selecting a plane having outer normal n coincident with x' and the in-plane direction n' coincident with y', as

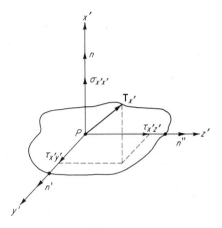

Figure 1.9 Resolution of $T_{x'}$ into three cartesian stress components $\sigma_{x'x'}$, $\tau_{x'y'}$, and $\tau_{x'z'}$.

shown in Fig. 1.9. The shear stress $\tau_{x'y'}$ is then obtained from Eq. (1.5b) by substituting x' for n and y' for n'. Thus

$$
\begin{aligned}
\tau_{x'y'} = {}& \sigma_{xx} \cos (x', x) \cos (y', x) \\
& + \sigma_{yy} \cos (x', y) \cos (y', y) + \sigma_{zz} \cos (x', z) \cos (y', z) \\
& + \tau_{xy}[\cos (x', x) \cos (y', y) + \cos (x', y) \cos (y', x)] \\
& + \tau_{yz}[\cos (x', y) \cos (y', z) + \cos (x', z) \cos (y', y)] \\
& + \tau_{zx}[\cos (x', z) \cos (y', x) + \cos (x', x) \cos (y', z)]
\end{aligned} \tag{1.6d}
$$

By selecting n and n' coincident with the y' and z', and z' and x' axes, additional expressions can be developed for $\tau_{y'z'}$ and $\tau_{z'x'}$, respectively, as follows:

$$
\begin{aligned}
\tau_{y'z'} = {}& \sigma_{yy} \cos (y', y) \cos (z', y) \\
& + \sigma_{zz} \cos (y', z) \cos (z', z) + \sigma_{xx} \cos (y', x) \cos (z', x) \\
& + \tau_{yz}[\cos (y', y) \cos (z', z) + \cos (y', z) \cos (z', y)] \\
& + \tau_{zx}[\cos (y', z) \cos (z', x) + \cos (y', x) \cos (z', z)] \\
& + \tau_{xy}[\cos (y', x) \cos (z', y) + \cos (y', y) \cos (z', x)]
\end{aligned} \tag{1.6e}
$$

$$
\begin{aligned}
\tau_{z'x'} = {}& \sigma_{zz} \cos (z', z) \cos (x', z) \\
& + \sigma_{xx} \cos (z', x) \cos (x', x) + \sigma_{yy} \cos (z', y) \cos (x', y) \\
& + \tau_{zx}[\cos (z', z) \cos (x', x) + \cos (z', x) \cos (x', z)] \\
& + \tau_{xy}[\cos (z', x) \cos (x', y) + \cos (z', y) \cos (x', x)] \\
& + \tau_{yz}[\cos (z', y) \cos (x', z) + \cos (z', z) \cos (x', y)]
\end{aligned} \tag{1.6f}
$$

These six equations permit the six cartesian components of stress relative to the $Oxyz$ coordinate system to be transformed into a different set of six cartesian components of stress relative to an $Ox'y'z'$ coordinate system.

1.6 PRINCIPAL STRESSES

In Sec. 1.2 it was noted that the resultant-stress vector $\mathbf{T}_n$ at a given point P depended upon the choice of the plane upon which the stress acted. If a plane is selected such that $\mathbf{T}_n$ coincides with the outer normal n, as shown in Fig. 1.10, it is clear that the shear stress τ_n vanishes and that $\mathbf{T}_n$, σ_n, and n are coincident.

If n is selected so that it coincides with $\mathbf{T}_n$, then the plane defined by n is known as a principal plane. The direction given by n is a principal direction, and the normal stress acting on this particular plane is a principal stress. In every state of stress there exist at least three principal planes, which are mutually perpendicular, and associated with these principal planes there are at most three distinct principal stresses. These statements can be established by referring to Fig. 1.10 and noting that

$$T_{nx} = \sigma_n \cos(n, x) \qquad T_{ny} = \sigma_n \cos(n, y) \qquad T_{nz} = \sigma_n \cos(n, z) \qquad (a)$$

If Eqs. (1.2) are substituted into Eqs. (a), the following expressions are obtained:

$$\sigma_{xx} \cos(n, x) + \tau_{yx} \cos(n, y) + \tau_{zx} \cos(n, z) = \sigma_n \cos(n, x)$$
$$\tau_{xy} \cos(n, x) + \sigma_{yy} \cos(n, y) + \tau_{zy} \cos(n, z) = \sigma_n \cos(n, y) \qquad (b)$$
$$\tau_{xz} \cos(n, x) + \tau_{yz} \cos(n, y) + \sigma_{zz} \cos(n, z) = \sigma_n \cos(n, z)$$

Rearranging Eqs. (b) gives

$$(\sigma_{xx} - \sigma_n) \cos(n, x) + \tau_{yx} \cos(n, y) + \tau_{zx} \cos(n, z) = 0$$
$$\tau_{xy} \cos(n, x) + (\sigma_{yy} - \sigma_n) \cos(n, y) + \tau_{zy} \cos(n, z) = 0 \qquad (c)$$
$$\tau_{xz} \cos(n, x) + \tau_{yz} \cos(n, y) + (\sigma_{zz} - \sigma_n) \cos(n, z) = 0$$

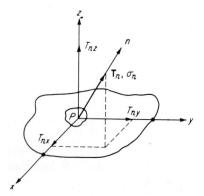

Figure 1.10 Coincidence of T_n with the outer normal n indicates that the shear stresses vanish and that σ_n becomes equal in magnitude to T_n.

Solving for any of the direction cosines, say cos (n, x), by determinants gives

$$\cos(n, x) = \frac{\begin{vmatrix} 0 & \tau_{yx} & \tau_{zx} \\ 0 & \sigma_{yy} - \sigma_n & \tau_{zy} \\ 0 & \tau_{yz} & \sigma_{zz} - \sigma_n \end{vmatrix}}{\begin{vmatrix} \sigma_{xx} - \sigma_n & \tau_{yx} & \tau_{zx} \\ \tau_{xy} & \sigma_{yy} - \sigma_n & \tau_{zy} \\ \tau_{xz} & \tau_{yz} & \sigma_{zz} - \sigma_n \end{vmatrix}} \tag{d}$$

It is clear that nontrivial solutions for the direction cosines of the principal plane will exist only if the determinant in the denominator is zero. Thus

$$\begin{vmatrix} \sigma_{xx} - \sigma_n & \tau_{yx} & \tau_{zx} \\ \tau_{xy} & \sigma_{yy} - \sigma_n & \tau_{zy} \\ \tau_{xz} & \tau_{yz} & \sigma_{zz} - \sigma_n \end{vmatrix} = 0 \tag{e}$$

Expanding the determinant after substituting Eqs. (1.4) gives the following important cubic equation:

$$\sigma_n^3 - (\sigma_{xx} + \sigma_{yy} + \sigma_{zz})\sigma_n^2$$
$$+ (\sigma_{xx}\sigma_{yy} + \sigma_{yy}\sigma_{zz} + \sigma_{zz}\sigma_{xx} - \tau_{xy}^2 - \tau_{yz}^2 - \tau_{zx}^2)\sigma_n$$
$$- (\sigma_{xx}\sigma_{yy}\sigma_{zz} - \sigma_{xx}\tau_{yz}^2 - \sigma_{yy}\tau_{zx}^2 - \sigma_{zz}\tau_{xy}^2 + 2\tau_{xy}\tau_{yz}\tau_{zx}) = 0 \tag{1.7}$$

The roots of this cubic equation are the three principal stresses. By substituting the six cartesian components of stress into this equation, one can solve for σ_n and obtain three real roots. Three possible solutions exist.

1. If σ_1, σ_2, σ_3 are distinct, then n_1, n_2, and n_3 are unique and mutually perpendicular.
2. If $\sigma_1 = \sigma_2 \neq \sigma_3$, then n_3 is unique and every direction perpendicular to n_3 is a principal direction associated with $\sigma_1 = \sigma_2$.
3. If $\sigma_1 = \sigma_2 = \sigma_3$, then a hydrostatic state of stress exists and every direction is a principal direction.

Once the three principal stresses have been established, they can be substituted individually into Eqs. (c) to give three sets of simultaneous equations which together with the relation

$$\cos^2(n, x) + \cos^2(n, y) + \cos^2(n, z) = 1$$

can be solved to give the three sets of direction cosines defining the principal planes. A numerical example of the procedure used in computing principal stresses and directions is given in the exercises at the end of the chapter.

In treating principal stresses it is often useful to order them so that $\sigma_1 > \sigma_2 > \sigma_3$. When the stresses are ordered in this fashion, σ_1 is the normal stress having the largest algebraic value at a given point and σ_3 is the normal stress having the smallest algebraic value. It is important to recall in this ordering process that tensile stresses are considered positive and compressive stresses are considered negative.

Another important concept is that of stress invariants. It was noted in Sec. 1.5 that a state of stress could be described by its six cartesian stress components with respect to either the $Oxyz$ coordinate system or the $Ox'y'z'$ coordinate system. Furthermore, Eqs. (1.6) were established to give the relationship between these two systems. In addition to Eqs. (1.6), three other relations exist which are called the three invariants of stress. To establish these invariants, refer to Eq. (1.7), which is the cubic equation in terms of the principal stresses σ_1, σ_2, and σ_3. By recalling that σ_1, σ_2, and σ_3 are independent of the cartesian coordinate system employed, it is clear that the coefficients of Eq. (1.7) which contain cartesian components of the stresses must also be independent or invariant of the coordinate system. Thus, from Eq. (1.7) it is clear that

$$I_1 = \sigma_{xx} + \sigma_{yy} + \sigma_{zz} = \sigma_{x'x'} + \sigma_{y'y'} + \sigma_{z'z'}$$

$$I_2 = \sigma_{xx}\sigma_{yy} + \sigma_{yy}\sigma_{zz} + \sigma_{zz}\sigma_{xx} - \tau_{xy}^2 - \tau_{yz}^2 - \tau_{zx}^2$$

$$= \sigma_{x'x'}\sigma_{y'y'} + \sigma_{y'y'}\sigma_{z'z'} + \sigma_{z'z'}\sigma_{x'x'} - \tau_{x'y'}^2 - \tau_{y'z'}^2 - \tau_{z'x'}^2 \quad (1.8)$$

$$I_3 = \sigma_{xx}\sigma_{yy}\sigma_{zz} - \sigma_{xx}\tau_{yz}^2 - \sigma_{yy}\tau_{zx}^2 - \sigma_{zz}\tau_{xy}^2 + 2\tau_{xy}\tau_{yz}\tau_{zx}$$

$$= \sigma_{x'x'}\sigma_{y'y'}\sigma_{z'z'} - \sigma_{x'x'}\tau_{y'z'}^2 - \sigma_{y'y'}\tau_{z'x'}^2 - \sigma_{z'z'}\tau_{x'y'}^2 + 2\tau_{x'y'}\tau_{y'z'}\tau_{z'x'}$$

where I_1, I_2, and I_3 are the first, second, and third invariants of stress, respectively. If the $Oxyz$ coordinate system is selected coincident with the principal directions, Eqs. (1.8) reduce to

$$I_1 = \sigma_1 + \sigma_2 + \sigma_3 \qquad I_2 = \sigma_1\sigma_2 + \sigma_2\sigma_3 + \sigma_3\sigma_1 \qquad I_3 = \sigma_1\sigma_2\sigma_3 \quad (1.9)$$

1.7 MAXIMUM SHEAR STRESS

In developing equations for maximum shear stresses, the special case will be considered in which $\tau_{xy} = \tau_{yz} = \tau_{zx} = 0$. No loss in generality is introduced by considering this special case since it involves only a reorientation of the reference axes to coincide with the principal directions. In the following development n_1, n_2, and n_3 will be used to denote the principal directions. In Sec. 1.3 the resultant stress on an oblique plane was given by

$$T_n^2 = T_{nx}^2 + T_{ny}^2 + T_{nz}^2 \tag{a}$$

Substitution of values for T_{nx}, T_{ny}, and T_{nz} from Eqs. (1.2) with principal normal stresses and zero shearing stresses yields

$$T_n^2 = \sigma_1^2 \cos^2 (n, n_1) + \sigma_2^2 \cos^2 (n, n_2) + \sigma_3^2 \cos^2 (n, n_3) \tag{b}$$

Also from Eq. (1.5a)

$$\sigma_n = \sigma_1 \cos^2 (n, n_1) + \sigma_2 \cos^2 (n, n_2) + \sigma_3 \cos^2 (n, n_3) \qquad (c)$$

Since $\tau_n^2 = T_n^2 - \sigma_n^2$, an expression for the shear stress τ_n on the oblique plane is obtained from Eqs. (b) and (c) after substituting $l = \cos (n, n_1)$, $m = \cos (n, n_2)$, and $n = \cos (n, n_3)$ as

$$\tau_n^2 = \sigma_1^2 l^2 + \sigma_2^2 m^2 + \sigma_3^2 n^2 - (\sigma_1 l^2 + \sigma_2 m^2 + \sigma_3 n^2)^2 \qquad (d)$$

The planes on which maximum and minimum shearing stresses occur can be obtained from Eq. (d) by differentiating with respect to the direction cosines l, m, and n. One of the direction cosines, n for example, in Eq. (d) can be eliminated by solving the expression

$$l^2 + m^2 + n^2 = 1 \qquad (e)$$

for l and substituting into Eq. (d). Thus

$$\tau_n^2 = (\sigma_1^2 - \sigma_3^2)l^2 + (\sigma_2^2 - \sigma_3^2)m^2 + \sigma_3^2 - [(\sigma_1 - \sigma_3)l^2 + (\sigma_2 - \sigma_3)m^2 + \sigma_3]^2 \, (f)$$

By taking the partial derivatives of Eq. (f), first with respect to l and then with respect to m, and equating to zero, the following equations are obtained for determining the direction cosines associated with planes having maximum and minimum shearing stresses:

$$l[\tfrac{1}{2}(\sigma_1 - \sigma_3) - (\sigma_1 - \sigma_3)l^2 - (\sigma_2 - \sigma_3)m^2] = 0 \qquad (g)$$

$$m[\tfrac{1}{2}(\sigma_2 - \sigma_3) - (\sigma_1 - \sigma_3)l^2 - (\sigma_2 - \sigma_3)m^2] = 0 \qquad (h)$$

One solution of these equations is obviously $l = m = 0$. Then from Eq. (e), $n = \pm 1$ (a principal plane with zero shear). Solutions different from zero are also possible for this set of equations. Consider first that $m = 0$; then from Eq. (g), $l = \pm(\tfrac{1}{2})^{1/2}$ and from Eq. (e), $n = \pm(\tfrac{1}{2})^{1/2}$. Also if $l = 0$, then from Eq. (h), $m = \pm(\tfrac{1}{2})^{1/2}$ and from Eq. (e), $n = \pm(\tfrac{1}{2})^{1/2}$. Repeating the above procedure by eliminating l and m in turn from Eq. (f) yields other values for the direction cosines which make the shearing stresses maximum or minimum. Substituting the values $l = \pm(\tfrac{1}{2})^{1/2}$ and $n = \pm(\tfrac{1}{2})^{1/2}$ into Eq. (d) yields

$$\tau_n^2 = \tfrac{1}{2}\sigma_1^2 + 0 + \tfrac{1}{2}\sigma_3^2 - (\tfrac{1}{2}\sigma_1 + 0 + \tfrac{1}{2}\sigma_3)^2$$

from which

$$\tau_n = \tfrac{1}{2}(\sigma_1 - \sigma_3)$$

Similarly, using the other values for the direction cosines which make the shearing stresses maximum gives

$$\tau_n = \tfrac{1}{2}(\sigma_1 - \sigma_2) \quad \text{and} \quad \tau_n = \tfrac{1}{2}(\sigma_2 - \sigma_3)$$

Of these three possible results, the largest magnitude will be obtained from $\sigma_1 - \sigma_3$ if the principal stresses are ordered such that $\sigma_1 \geq \sigma_2 \geq \sigma_3$. Thus

$$\tau_{max} = \tfrac{1}{2}(\sigma_{max} - \sigma_{min}) = \tfrac{1}{2}(\sigma_1 - \sigma_3) \qquad (1.10)$$

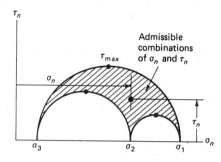

Figure 1.11 Mohr's Circle for the three-dimensional state of stress.

A useful aid for visualizing the complete state of stress at a point is the three-dimensional Mohr's circle shown in Fig. 1.11. This representation, which is similar to the familiar two-dimensional Mohr's circle, shows the three principal stresses, the maximum shearing stresses, and the range of values within which the normal- and shear-stress components must lie for a given state of stress.

1.8 THE TWO-DIMENSIONAL STATE OF STRESS

For two-dimensional stress fields where $\sigma_{zz} = \tau_{zx} = \tau_{yz} = 0$, z' is coincident with z, and θ is the angle between x and x', Eqs. (1.6a) to (1.6f) reduce to

$$\sigma_{x'x'} = \sigma_{xx} \cos^2 \theta + \sigma_{yy} \sin^2 \theta + 2\tau_{xy} \sin \theta \cos \theta$$

$$= \frac{\sigma_{xx} + \sigma_{yy}}{2} + \frac{\sigma_{xx} - \sigma_{yy}}{2} \cos 2\theta + \tau_{xy} \sin 2\theta \qquad (1.11a)$$

$$\sigma_{y'y'} = \sigma_{yy} \cos^2 \theta + \sigma_{xx} \sin^2 \theta - 2\tau_{xy} \sin \theta \cos \theta$$

$$= \frac{\sigma_{yy} + \sigma_{xx}}{2} + \frac{\sigma_{yy} - \sigma_{xx}}{2} \cos 2\theta - \tau_{xy} \sin 2\theta \qquad (1.11b)$$

$$\tau_{x'y'} = \sigma_{yy} \cos \theta \sin \theta - \sigma_{xx} \cos \theta \sin \theta + \tau_{xy}(\cos^2 \theta - \sin^2 \theta)$$

$$= \frac{\sigma_{yy} - \sigma_{xx}}{2} \sin 2\theta + \tau_{xy} \cos 2\theta \qquad (1.11c)$$

$$\sigma_{z'z'} = \tau_{z'x'} = \tau_{y'z'} = 0 \qquad (1.11d)$$

The relationships between stress components given in Eqs. (1.11) can be graphically represented by using Mohr's circle of stress, as indicated in Fig. 1.12. In this diagram, normal-stress components σ are plotted horizontally, while shear-stress components τ are plotted vertically. Tensile stresses are plotted to the right of the τ axis. Compressive stresses are plotted to the left. Shear-stress components which tend to produce a clockwise rotation of a small element surrounding the point are plotted above the σ axis. Those tending to produce a counterclockwise

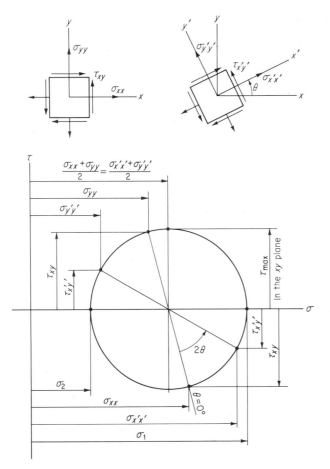

Figure 1.12 Mohr's circle of stress.

rotation are plotted below. When plotted in this manner, the stress components associated with each plane through the point are represented by a point on the circle. The diagram thus gives an excellent visual picture of the state of stress at a point. Mohr's circle and Eqs. (1.11) are often used in experimental stress-analysis work when stress components are transformed from one coordinate system to another. These relationships will be used frequently in later sections of this text, where strain gages and photoelasticity methods of analysis are discussed. Since two-dimensional stress systems are often considered in subsequent chapters, it will be useful to consider the principal stresses which occur in a two-dimensional stress system. If a coordinate system is chosen so that $\sigma_{zz} = \tau_{zx} = \tau_{yz} = 0$, then a state of plane stress exists and Eq. (1.7) reduces to

$$\sigma_n[\sigma_n^2 - (\sigma_{xx} + \sigma_{yy})\sigma_n + (\sigma_{xx}\sigma_{yy} - \tau_{xy}^2)] = 0 \qquad (a)$$

Solving this equation for the three principal stresses yields

$$\sigma_1, \sigma_2 = \frac{\sigma_{xx} + \sigma_{yy}}{2} \pm \sqrt{\left(\frac{\sigma_{xx} - \sigma_{yy}}{2}\right)^2 + \tau_{xy}^2} \qquad \sigma_3 = 0 \qquad (1.12)$$

The two direction cosines which define the two principal planes can be determined from Eq. (1.11c), which gives $\tau_{x'y'}$ in terms of $\sigma_{xx}, \sigma_{yy}, \tau_{xy}$ and the angle θ between x and x'. If x' and y' are selected so that $x' = n_1$ and $y' = n_2$, then $\tau_{x'y'}$ must vanish since no shearing stresses can exist on principal planes. Thus the following equation can be written:

$$\frac{\sigma_{yy} - \sigma_{xx}}{2} \sin 2(n_1, x) + \tau_{xy} \cos 2(n_1, x) = 0 \qquad (1.13)$$

Dividing through by $\cos 2(n_1, x)$ and simplifying gives

$$\tan 2(n_1, x) = \frac{2\tau_{xy}}{\sigma_{xx} - \sigma_{yy}} \qquad (1.14a)$$

and hence

$$\cos 2(n_1, x) = \frac{\sigma_{xx} - \sigma_{yy}}{\sqrt{(\sigma_{xx} - \sigma_{yy})^2 + 4\tau_{xy}^2}} \qquad (1.14b)$$

$$\sin 2(n_1, x) = \frac{2\tau_{xy}}{\sqrt{(\sigma_{xx} - \sigma_{yy})^2 + 4\tau_{xy}^2}} \qquad (1.14c)$$

Equations (1.14) are used in solving for the direction of n_1 if the cartesian stress components $\tau_{xy}, \sigma_{xx}, \sigma_{yy}$ are known.

1.9 STRESSES RELATIVE TO A PRINCIPAL COORDINATE SYSTEM

If the coordinate system $Oxyz$ is selected to coincide with the three principal directions n_1, n_2, n_3, then $\sigma_1 = \sigma_{xx}, \sigma_2 = \sigma_{yy}, \sigma_3 = \sigma_{zz}$, and $\tau_{xy} = \tau_{yz} = \tau_{zx} = 0$. This reduces the six components of stress to three, which permits a considerable simplification in some of the previous results. Equations (1.2) become

$$T_{nx} = \sigma_1 \cos(n, x) \qquad T_{ny} = \sigma_2 \cos(n, y) \qquad T_{nz} = \sigma_3 \cos(n, z) \quad (1.15)$$

and Equations (1.6) reduce to

$$\sigma_{x'x'} = \sigma_1 \cos^2(x', x) + \sigma_2 \cos^2(x', y) + \sigma_3 \cos^2(x', z)$$
$$\sigma_{y'y'} = \sigma_1 \cos^2(y', x) + \sigma_2 \cos^2(y', y) + \sigma_3 \cos^2(y', z)$$
$$\sigma_{z'z'} = \sigma_1 \cos^2(z', x) + \sigma_2 \cos^2(z', y) + \sigma_3 \cos^2(z', z)$$
$$\tau_{x'y'} = \sigma_1 \cos(x', x) \cos(y', x) + \sigma_2 \cos(x', y) \cos(y', y)$$
$$+ \sigma_3 \cos(x', z) \cos(y', z) \qquad (1.16)$$

$$\tau_{y'z'} = \sigma_1 \cos(y', x) \cos(z', x) + \sigma_2 \cos(y', y) \cos(z', y)$$
$$+ \sigma_3 \cos(y', z) \cos(z', z)$$
$$\tau_{z'x'} = \sigma_1 \cos(z', x) \cos(x', x) + \sigma_2 \cos(z', y) \cos(x', y)$$
$$+ \sigma_3 \cos(z', z) \cos(x', z)$$

Often experimental methods yield principal stresses directly, and in these cases Eqs. (1.16) are frequently used to obtain the stresses acting on other planes.

1.10 SPECIAL STATES OF STRESS

Two states of stress occur so frequently in practice that they have been classified. They are the state of pure shearing stress and the hydrostatic state of stress. Both are defined below.

1. A state of pure shear stress exists if one particular set of axes $Oxyz$ can be found such that $\sigma_{xx} = \sigma_{yy} = \sigma_{zz} = 0$. It can be shown that this particular set of axes $Oxyz$ exists if and only if the first invariant of stress $I_1 = 0$. The proof of this condition is beyond the scope of this text. Two of the infinite number of arrays which represent a state of pure shearing stress are given below.

$$\begin{Vmatrix} 0 & \tau_{xy} & \tau_{xz} \\ \tau_{xy} & 0 & \tau_{yz} \\ \tau_{xz} & \tau_{yz} & 0 \end{Vmatrix} \quad \text{or} \quad \begin{Vmatrix} \sigma_{xx} & \tau_{xy} & \tau_{xz} \\ \tau_{xy} & \sigma_{yy} = -\sigma_{xx} & \tau_{yz} \\ \tau_{xz} & \tau_{yz} & 0 \end{Vmatrix}$$

Pure shear   Can be converted to the form shown on the left by a suitable rotation of the coordinate system

2. A state of stress is said to be hydrostatic if $\sigma_{xx} = \sigma_{yy} = \sigma_{zz} = -p$ and all the shearing stresses vanish. In photoelastic work a hydrostatic state of stress is often called an isotropic state of stress. The stress array for this case is

$$\begin{Vmatrix} -p & 0 & 0 \\ 0 & -p & 0 \\ 0 & 0 & -p \end{Vmatrix}$$

One particularly important property of these two states of stress is that they can be combined to form a general state of stress. Of more importance, however, is the fact that any state of stress can be separated into a state of pure shear plus a hydrostatic state of stress. This is easily seen from the three arrays shown below:

$$
\begin{Vmatrix} \sigma_{xx} & \tau_{xy} & \tau_{xz} \\ \tau_{xy} & \sigma_{yy} & \tau_{yz} \\ \tau_{xz} & \tau_{yz} & \sigma_{zz} \end{Vmatrix} = \begin{Vmatrix} -p & 0 & 0 \\ 0 & -p & 0 \\ 0 & 0 & -p \end{Vmatrix}
$$

$$
+ \begin{Vmatrix} \sigma_{xx} + p & \tau_{xy} & \tau_{xz} \\ \tau_{xy} & \sigma_{yy} + p & \tau_{yz} \\ \tau_{xz} & \tau_{yz} & \sigma_{zz} + p \end{Vmatrix} \tag{1.17}
$$

General state of stress = hydrostatic state of stress
+ state of pure shearing stress

It is immediately clear that the array on the left represents a general state of stress and that the center array represents a hydrostatic state of stress; however, the right-hand array represents a state of pure shear if and only if its first stress invariant is zero. This fact implies that

$$
(\sigma_{xx} + p) + (\sigma_{yy} + p) + (\sigma_{zz} + p) = 0
$$

Hence
$$
p = -\tfrac{1}{3}(\sigma_{xx} + \sigma_{yy} + \sigma_{zz}) \tag{1.18}
$$

If the p represented in the hydrostatic state of stress satisfies Eq. (1.18), then the separation of the state of stress given in Eq. (1.17) is valid. In the study of plasticity, the effect of the hydrostatic stresses is usually neglected; consequently, the principle illustrated above is quite important.

EXERCISES

1.1 At a point in a stressed body, the cartesian components of stress are $\sigma_{xx} = 60$ MPa, $\sigma_{yy} = -30$ MPa, $\sigma_{zz} = 30$ MPa, $\tau_{xy} = 40$ MPa, $\tau_{yz} = \tau_{zx} = 0$. Determine the normal and shear stresses on a plane whose outer normal has the direction cosines

$$\cos{(n, x)} = \tfrac{6}{11} \qquad \cos{(n, y)} = \tfrac{6}{11} \qquad \cos{(n, z)} = \tfrac{7}{11}$$

1.2 At a point in a stressed body, the cartesian components of stress are $\sigma_{xx} = 70$ MPa, $\sigma_{yy} = 60$ MPa, $\sigma_{zz} = 50$ MPa, $\tau_{xy} = 20$ MPa, $\tau_{yz} = -20$ MPa, $\tau_{zx} = 0$. Determine the normal and shear stresses on a plane whose outer normal has the direction cosines

$$\cos{(n, x)} = \tfrac{12}{25} \qquad \cos{(n, y)} = \tfrac{15}{25} \qquad \cos{(n, z)} = \tfrac{16}{25}$$

1.3 At a point in a stressed body, the cartesian components of stress are $\sigma_{xx} = 40$ MPa, $\sigma_{yy} = 60$ MPa, $\sigma_{zz} = 40$ MPa, $\tau_{xy} = 80$ MPa, $\tau_{yz} = 50$ MPa, $\tau_{zx} = 60$ MPa. Determine (a) the normal and shear stresses on a plane whose outer normal has the direction cosines

$$\cos{(n, x)} = \tfrac{4}{9} \qquad \cos{(n, y)} = \tfrac{4}{9} \qquad \cos{(n, z)} = \tfrac{7}{9}$$

and (b) the angle between T_n and the outer normal n.

1.4 At a point in a stressed body, the cartesian components of stress are $\sigma_{xx} = 60$ MPa, $\sigma_{yy} = 40$ MPa, $\sigma_{zz} = 20$ MPa, $\tau_{xy} = 40$ MPa, $\tau_{yz} = 20$ MPa, $\tau_{zx} = 30$ MPa. Determine (a) the normal and shear stresses on a plane whose outer normal has the direction cosines

$$\cos(n, x) = \tfrac{1}{3} \qquad \cos(n, y) = \tfrac{2}{3} \qquad \cos(n, z) = \tfrac{2}{3}$$

and (b) the angle between T_n and the outer normal n.

1.5 Determine the normal and shear stresses on a plane whose outer normal makes equal angles with the x, y, and z axes if the cartesian components of stress at the point are

$$\sigma_{xx} = \sigma_{yy} = \sigma_{zz} = 0 \qquad \tau_{xy} = 75 \text{ MPa} \qquad \tau_{yz} = 0 \qquad \tau_{zx} = 100 \text{ MPa}$$

1.6 The following stress distribution has been determined for a machine component:

$$\sigma_{xx} = 3x^2 - 3y^2 - z \qquad \sigma_{yy} = 3y^2 \qquad \sigma_{zz} = 3x + y - z + \tfrac{5}{4}$$

$$\tau_{xy} = z - 6xy - \tfrac{3}{4} \qquad \tau_{yz} = 0 \qquad \tau_{zx} = x + y - \tfrac{3}{2}$$

Is equilibrium satisfied in the absence of body forces?

1.7 If the state of stress at any point in a body is given by the equations

$$\sigma_{xx} = ax + by + cz \qquad \sigma_{yy} = dx^2 + ey^2 + fz^2 \qquad \sigma_{zz} = gx^3 + hy^3 + iz^3$$

$$\tau_{xy} = k \qquad \tau_{yz} = ly + mz \qquad \tau_{zx} = nx^2 + pz^2$$

what equations must the body-force intensities F_x, F_y, F_z satisfy?

1.8 At a point in a stressed body, the cartesian components of stress are $\sigma_{xx} = 90$ MPa, $\sigma_{yy} = 60$ MPa, $\sigma_{zz} = 30$ MPa, $\tau_{xy} = 30$ MPa, $\tau_{yz} = 30$ MPa, $\tau_{zx} = 60$ MPa. Transform this set of cartesian stress components into a new set of cartesian stress components relative to an $Ox'y'z'$ set of coordinates where the $Ox'y'z'$ axes are defined as:

θ	Case 1	Case 2	Case 3	Case 4
$x - x'$	$\pi/4$	$\pi/2$	0	$\pi/2$
$y - y'$	$\pi/4$	$\pi/2$	$\pi/2$	0
$z - z'$	0	0	$\pi/2$	$\pi/2$

1.9 At a point in a stressed body, the cartesian components of stress are $\sigma_{xx} = 70$ MPa, $\sigma_{yy} = 60$ MPa, $\sigma_{zz} = 50$ MPa, $\tau_{xy} = 20$ MPa, $\tau_{yz} = -20$ MPa, $\tau_{zx} = 0$. Transform this set of cartesian stress components into a new set of cartesian stress components relative to an $Ox'y'z'$ set of coordinates where the $Ox'y'z'$ axes are defined by the following direction cosines:

	x	y	z
x'	$\tfrac{2}{3}$	$\tfrac{2}{3}$	$-\tfrac{1}{3}$
y'	$-\tfrac{2}{3}$	$\tfrac{1}{3}$	$-\tfrac{2}{3}$
z'	$-\tfrac{1}{3}$	$\tfrac{2}{3}$	$\tfrac{2}{3}$

1.10 For the state of stress at the point of Exercise 1.1, determine the principal stresses and the maximum shear stress at the point.

1.11 For the state of stress at the point of Exercise 1.2, determine the principal stresses and the maximum shear stress at the point.

1.12 For the state of stress at the point of Exercise 1.3, determine the principal stresses and the maximum shear stress at the point.

1.13 For the state of stress at the point of Exercise 1.4, determine the principal stresses and the maximum shear stress at the point.

1.14 Determine the principal stresses and the maximum shear stress at the point $x = \frac{1}{2}, y = 1, z = \frac{3}{4}$ for the stress distribution given in Exercise 1.6.

1.15 At a point in a stressed body, the cartesian components of stress are $\sigma_{xx} = 50$ MPa, $\sigma_{yy} = 50$ MPa, $\sigma_{zz} = 50$ MPa, $\tau_{xy} = 100$ MPa, $\tau_{yz} = 0$, $\tau_{zx} = 50$ MPa. Determine (a) the principal stresses and the maximum shear stress at the point and (b) the orientation of the plane on which the maximum tensile stress acts.

1.16 At a point in a stressed body, the cartesian components of stress are $\sigma_{xx} = \sigma_{yy} = \sigma_{zz} = 0$, $\tau_{xy} = 75$ MPa, $\tau_{yz} = 0$, $\tau_{zx} = 100$ MPa. Determine (a) the principal stresses and the maximum shear stress at the point and (b) the orientation of the plane on which the maximum tensile stress acts.

1.17 At a point in a stressed body, the cartesian components of stress are $\sigma_{xx} = \sigma_{yy} = \sigma_{zz} = 25$ MPa, $\tau_{xy} = 100$ MPa, $\tau_{yz} = 0$, $\tau_{zx} = 75$ MPa. Determine the principal stresses and the associated principal directions. Check on the invariance of I_1, I_2, and I_3.

1.18 At a point in a stressed body, the cartesian components of stress are $\sigma_{xx} = \sigma_{yy} = \sigma_{zz} = 0$, $\tau_{xy} = \tau_{yz} = \tau_{zx} = 100$ MPa. Determine the principal stresses and the associated principal directions. Check on the invariance of I_1, I_2, and I_3.

1.19 A machine component is subjected to loads which produce the following stress field in a region where an oilhole must be drilled: $\sigma_{xx} = 100$ MPa, $\sigma_{yy} = -50$ MPa, $\sigma_{zz} = 50$ MPa, $\tau_{xy} = 50$ MPa, $\tau_{yz} = \tau_{zx} = 0$. To minimize the effects of stress concentrations, the hole must be drilled along a line parallel to the direction of the maximum tensile stress in the region. Determine the direction cosines associated with the centerline of the hole with respect to the reference $Oxyz$ coordinate system.

1.20 A two-dimensional state of stress ($\sigma_{zz} = \tau_{zx} = \tau_{zy} = 0$) exists at a point on the free surface of a machine component. The remaining cartesian components of stress are $\sigma_{xx} = 100$ MPa, $\sigma_{yy} = -80$ MPa, $\tau_{xy} = -40$ MPa. Determine (a) the principal stresses and their associated directions at the point and (b) the maximum shear stress at the point.

1.21 A two-dimensional state of stress ($\sigma_{zz} = \tau_{zx} = \tau_{zy} = 0$) exists at a point on the surface of a loaded member. Determine the principal stresses and the maximum shear stress at the point if the remaining cartesian components of stress are $\sigma_{xx} = 90$ MPa, $\sigma_{yy} = 60$ MPa, $\tau_{xy} = 40$ MPa.

1.22 A two-dimensional state of stress ($\sigma_{zz} = \tau_{zx} = \tau_{zy} = 0$) exists at a point on the surface of a loaded member. The remaining cartesian components of stress are $\sigma_{xx} = 100$ MPa, $\sigma_{yy} = 70$ MPa, $\tau_{xy} = 20$ MPa. Determine the principal stresses and the maximum shear stress at the point.

1.23 A two-dimensional state of stress ($\sigma_{zz} = \tau_{zx} = \tau_{zy} = 0$) exists at a point on the surface of a loaded member. The remaining cartesian components of stress are $\sigma_{xx} = 90$ MPa, $\sigma_{yy} = 40$ MPa, $\tau_{xy} = 60$ MPa. Determine the principal stresses and the maximum shear stress at the point.

1.24 Solve Exercise 1.22 by means of Mohr's circle.

1.25 Solve Exercise 1.23 by means of Mohr's circle.

1.26 At the point of Exercise 1.22, determine the normal and shear stresses on a plane whose outer normal has the direction cosines

$$\cos(n, x) = \tfrac{3}{5} \qquad \cos(n, y) = \tfrac{4}{5} \qquad \cos(n, z) = 0$$

1.27 At the point of Exercise 1.23, determine the normal and shear stresses on a plane whose outer normal has the direction cosines

$$\cos(n, x) = \tfrac{1}{3} \qquad \cos(n, y) = \tfrac{2}{3} \qquad \cos(n, z) = \tfrac{2}{3}$$

1.28 There is a crack in a plate of steel which makes the material in that area weak in tension and shear. The plate must be used for a member which will be loaded to produce the following state of stress in the plane of the plate: $\sigma_{xx} = 100$ MPa, $\sigma_{yy} = -60$ MPa, $\tau_{xy} = 20$ MPa. How should the x and y axes be oriented with respect to the crack in order to minimize the effect of the crack?

1.29 At a point in a metal machine part the principal stresses are $\sigma_1 = 150$ MPa, $\sigma_2 = 100$ MPa, $\sigma_3 = 50$ MPa. Determine the normal and shear stresses on a plane whose outer normal has the direction cosines

$$\cos(n, n_1) = \frac{\sqrt{3}}{2} \qquad \cos(n, n_2) = 0 \qquad \cos(n, n_3) = \frac{1}{2}$$

1.30 If the three principal stresses relative to the $Oxyz$ reference system are $\sigma_1 = \sigma_{xx} = 100$ MPa, $\sigma_2 = \sigma_{yy} = 80$ MPa, $\sigma_3 = \sigma_{zz} = -20$ MPa, determine the six cartesian components of stress relative to the $Ox'y'z'$ reference system where $Ox'y'z'$ is defined as:

θ	Case 1	Case 2	Case 3	Case 4
$x - x'$	$\pi/4$	$\pi/2$	0	$\pi/4$
$y - y'$	$\pi/4$	$\pi/2$	$\pi/4$	0
$z - z'$	0	0	$\pi/4$	$\pi/4$

1.31 Resolve the general state of stress given in Exercise 1.1 into a hydrostatic state of stress and a state of pure shearing stress.

1.32 Resolve the general state of stress given in Exercise 1.2 into a hydrostatic state of stress and a state of pure shearing stress.

1.33 Resolve the general state of stress given in Exercise 1.4 into a hydrostatic state of stress and a state of pure shearing stress.

1.34 Resolve the two-dimensional state of stress given in Exercise 1.21 into a hydrostatic state of stress and a state of pure shearing stress.

1.35 Determine the octahedral normal and shearing stresses associated with the principal stresses σ_1, σ_2, and σ_3. Octahedral normal and shearing stresses occur on planes whose outer normal makes equal angles with the principal directions n_1, n_2, and n_3.

REFERENCES

1. Boresi, A. P., and P. P. Lynn: "Elasticity in Engineering Mechanics," chap. 3, Prentice-Hall, Inc., Englewood Cliffs, N.J., 1974.
2. Chou, P. C., and N. J. Pagano: "Elasticity," chap. 1, D. Van Nostrand Company, Inc., Princeton, N.J., 1967.
3. Durelli, A. J., E. A. Phillips, and C. H. Tsao: "Introduction to the Theoretical and Experimental Analysis of Stress and Strain," chap. 1, McGraw-Hill Book Company, New York, 1958.
4. Love, A. E. H.: "A Treatise on the Mathematical Theory of Elasticity," chap. 2, Dover Publications, Inc., New York, 1944.
5. Sechler, E. E.: "Elasticity in Engineering," chaps. 2 and 3, John Wiley & Sons, Inc., New York, 1952.
6. Sokolnikoff, I. S.: "Mathematical Theory of Elasticity," 2d ed., chap. 2, McGraw-Hill Book Company, New York, 1956.
7. Southwell, R. V.: "An Introduction to the Theory of Elasticity," chap. 8, Oxford University Press, Fair Lawn, N.J., 1953.
8. Timoshenko, S. P., and J. N. Goodier: "Theory of Elasticity," 2d ed., chaps. 1, 8, and 9, McGraw-Hill Book Company, New York, 1951.

TWO

STRAIN AND THE STRESS-STRAIN RELATIONS

2.1 INTRODUCTION

In the preceding chapter the state of stress which develops at an arbitrary point within a body as a result of surface- or body-force loadings was discussed. The relationships obtained were based on the conditions of equilibrium, and since no assumptions were made regarding body deformations or physical properties of the material of which the body was composed, the results are valid for any material and for any amount of body deformation. In this chapter the subject of body deformation and associated strain will be discussed. Since strain is a pure geometric quantity, no restrictions on body material will be required. However, in order to obtain linear equations relating displacement to strain, restrictions must be placed on the allowable deformations. In a later section, when the stress-strain relations are developed, the elastic constants of the body material must be considered.

2.2 DEFINITIONS OF DISPLACEMENT AND STRAIN

If a given body is subjected to a system of forces, individual points of the body will, in general, move. This movement of an arbitrary point is a vector quantity known as a *displacement*. If the various points in the body undergo different movements, each can be represented by its own unique displacement vector. Each vector can be resolved into components parallel to a set of cartesian coordinate axes such that u, v, and w are the displacement components in the x, y, and z directions, respectively.

Motion of the body may be considered as the sum of two parts:

1. A translation and/or rotation of the body as a whole
2. The movement of the points of the body relative to each other

The translation or rotation of the body as a whole is known as *rigid-body motion*. This type of motion is applicable to either the idealized rigid body or the real deformable body. The movement of the points of the body relative to each other is known as a *deformation* and is obviously a property of real bodies only. Rigid-body motions can be large or small. Deformations, in general, are small except when rubberlike materials or specialized structures such as long, slender beams are involved.

Strain is a geometric quantity which depends on the relative movements of two or three points in the body and therefore is related only to the deformation displacements. Since rigid-body displacements do not produce strains, they will be neglected in all further developments in this chapter. In the preceding chapter two types of stress were discussed: normal stress and shear stress. This same classification will be used for strains. A normal strain is defined as the change in length of a line segment between two points divided by the original length of the line segment. A shearing strain is defined as the angular change between two line segments which were originally perpendicular. The relationships between strains and displacements can be determined by considering the deformation of an arbitrary cube in a body as a system of loads is applied. This deformation is illustrated in Fig. 2.1, in which a general point P is moved through a distance u in the x direction, v in the y direction, and w in the z direction. The other corners of the cube are also displaced and, in general, they will be displaced by amounts which differ from those at point P. For example the displacements u^*, v^*, and w^* associated with point Q can be expressed in terms of the displacements u, v, and w at point P by means of a Taylor-series expansion. Thus

$$u^* = u + \frac{\partial u}{\partial x}\Delta x + \frac{\partial u}{\partial y}\Delta y + \frac{\partial u}{\partial z}\Delta z + \cdots$$

$$v^* = v + \frac{\partial v}{\partial x}\Delta x + \frac{\partial v}{\partial y}\Delta y + \frac{\partial v}{\partial z}\Delta z + \cdots \tag{2.1}$$

$$w^* = w + \frac{\partial w}{\partial x}\Delta x + \frac{\partial w}{\partial y}\Delta y + \frac{\partial w}{\partial z}\Delta z + \cdots$$

The terms shown in the above expressions are the only significant terms if it is assumed that the cube is sufficiently small for higher-order terms such as $(\Delta x)^2$, $(\Delta y)^2$, $(\Delta z)^2$, ... to be neglected. Under these conditions, planes will remain plane and straight lines will remain straight lines in the deformed cube, as shown in Fig. 2.1.

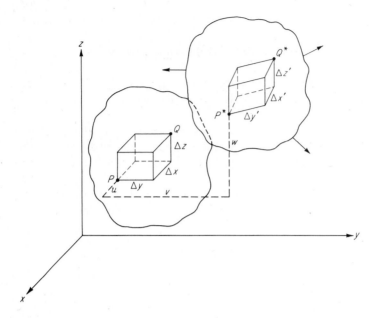

Figure 2.1 The distortion of an arbitrary cube in a body due to the application of a system of forces.

The average normal strain along an arbitrary line segment was previously defined as the change in length of the line segment divided by its original length. This normal strain can be expressed in terms of the displacements experienced by points at the ends of the segment. For example, consider the line PQ originally oriented parallel to the x axis, as shown in Fig. 2.2. Since y and z are constant along PQ, Eqs. (2.1) yield the following displacements for point Q if the displacements for point P are u, v, and w:

$$u^* = u + \frac{\partial u}{\partial x}\Delta x \qquad v^* = v + \frac{\partial v}{\partial x}\Delta x \qquad w^* = w + \frac{\partial w}{\partial x}\Delta x$$

From the definition of normal strain,

$$\epsilon_{xx} = \frac{\Delta x' - \Delta x}{\Delta x} \tag{a}$$

which is equivalent to

$$\Delta x' = (1 + \epsilon_{xx})\,\Delta x \tag{b}$$

As shown in Fig. 2.2, the deformed length $\Delta x'$ can be expressed in terms of the displacement gradients as

$$(\Delta x')^2 = \left[\left(1 + \frac{\partial u}{\partial x}\right)\Delta x\right]^2 + \left(\frac{\partial v}{\partial x}\Delta x\right)^2 + \left(\frac{\partial w}{\partial x}\Delta x\right)^2 \tag{c}$$

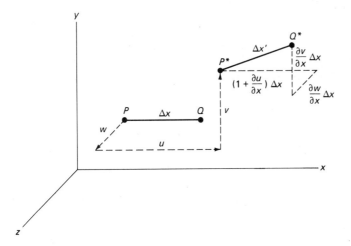

Figure 2.2 Displacement gradients associated with the normal strain ϵ_{xx}.

Squaring Eq. (*b*) and substituting Eq. (*c*) yields

$$(1 + \epsilon_{xx})^2 (\Delta x)^2 = \left[1 + 2\frac{\partial u}{\partial x} + \left(\frac{\partial u}{\partial x}\right)^2 + \left(\frac{\partial v}{\partial x}\right)^2 + \left(\frac{\partial w}{\partial x}\right)^2 \right](\Delta x)^2$$

or

$$\epsilon_{xx} = \sqrt{1 + 2\frac{\partial u}{\partial x} + \left(\frac{\partial u}{\partial x}\right)^2 + \left(\frac{\partial v}{\partial x}\right)^2 + \left(\frac{\partial w}{\partial x}\right)^2} - 1 \qquad (2.2a)$$

In a similar manner considering line segments originally oriented parallel to the *y* and *z* axes leads to

$$\epsilon_{yy} = \sqrt{1 + 2\frac{\partial v}{\partial y} + \left(\frac{\partial v}{\partial y}\right)^2 + \left(\frac{\partial w}{\partial y}\right)^2 + \left(\frac{\partial u}{\partial y}\right)^2} - 1 \qquad (2.2b)$$

$$\epsilon_{zz} = \sqrt{1 + 2\frac{\partial w}{\partial z} + \left(\frac{\partial w}{\partial z}\right)^2 + \left(\frac{\partial u}{\partial z}\right)^2 + \left(\frac{\partial v}{\partial z}\right)^2} - 1 \qquad (2.2c)$$

The shear-strain components can also be related to the displacements by considering the changes in right angle experienced by the edges of the cube during deformation. For example, consider lines *PQ* and *PR*, as shown in Fig. 2.3. The angle θ^* between *P*Q** and *P*R** in the deformed state can be expressed in terms of the displacement gradients since the cosine of the angle between any two intersecting lines in space is the sum of the pairwise products of the direction cosines of the lines with respect to the same set of reference axes. Thus

$$\cos\theta^* = \left[\left(1 + \frac{\partial u}{\partial x}\right)\frac{\Delta x}{\Delta x'} \right]\left(\frac{\partial u}{\partial y}\frac{\Delta y}{\Delta y'}\right) + \left(\frac{\partial v}{\partial x}\frac{\Delta x}{\Delta x'}\right)\left[\left(1 + \frac{\partial v}{\partial y}\right)\frac{\Delta y}{\Delta y'}\right]$$

$$+ \left(\frac{\partial w}{\partial x}\frac{\Delta x}{\Delta x'}\right)\left(\frac{\partial w}{\partial y}\frac{\Delta y}{\Delta y'}\right) \qquad (d)$$

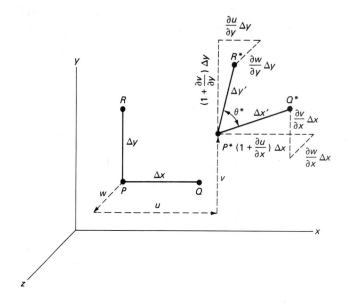

Figure 2.3 Displacement gradients associated with the shear strain γ_{xy}.

From the definition of shear strain

$$\gamma_{xy} = \left(\frac{\pi}{2} - \theta^*\right) \tag{e}$$

therefore

$$\sin \gamma_{xy} = \sin\left(\frac{\pi}{2} - \theta^*\right) = \cos \theta^* \tag{f}$$

Substituting Eq. (d) into Eq. (f) and simplifying yields

$$\sin \gamma_{xy} = \left[\left(1 + \frac{\partial u}{\partial x}\right)\frac{\partial u}{\partial y} + \left(1 + \frac{\partial v}{\partial x}\right)\frac{\partial v}{\partial y} + \frac{\partial w}{\partial x}\frac{\partial w}{\partial y}\right]\left(\frac{\Delta x \, \Delta y}{\Delta x' \, \Delta y'}\right)$$

From Eq. (b)

$$\Delta x' = (1 + \epsilon_{xx})\,\Delta x \qquad \text{and} \qquad \Delta y' = (1 + \epsilon_{yy})\,\Delta y$$

therefore

$$\gamma_{xy} = \arcsin \frac{\dfrac{\partial u}{\partial y} + \dfrac{\partial v}{\partial x} + \dfrac{\partial u}{\partial x}\dfrac{\partial u}{\partial y} + \dfrac{\partial v}{\partial x}\dfrac{\partial v}{\partial y} + \dfrac{\partial w}{\partial x}\dfrac{\partial w}{\partial y}}{(1 + \epsilon_{xx})(1 + \epsilon_{yy})} \tag{2.3a}$$

In a similar manner by considering two line segments originally oriented parallel to the y and z axes and the z and x axes

$$\gamma_{yz} = \arcsin \frac{\dfrac{\partial v}{\partial z} + \dfrac{\partial w}{\partial y} + \dfrac{\partial v}{\partial y}\dfrac{\partial v}{\partial z} + \dfrac{\partial w}{\partial y}\dfrac{\partial w}{\partial z} + \dfrac{\partial u}{\partial y}\dfrac{\partial u}{\partial z}}{(1 + \epsilon_{yy})(1 + \epsilon_{zz})} \tag{2.3b}$$

$$\gamma_{zx} = \arcsin \frac{\dfrac{\partial w}{\partial x} + \dfrac{\partial u}{\partial z} + \dfrac{\partial w}{\partial z}\dfrac{\partial w}{\partial x} + \dfrac{\partial u}{\partial z}\dfrac{\partial u}{\partial x} + \dfrac{\partial v}{\partial z}\dfrac{\partial v}{\partial x}}{(1 + \epsilon_{zz})(1 + \epsilon_{xx})} \tag{2.3c}$$

Equations (2.2) and (2.3) represent a common engineering description of strain in terms of positions of points in a body before and after deformation. In the development of these equations, no limitations were imposed on the magnitudes of the strains. One restriction was introduced, however, when the higher-order terms in the Taylor-series expansion for displacement were neglected. This restriction has the effect of limiting the length of the line segment (gage length) used for strain determinations unless displacement gradients ($\partial u/\partial x$, $\partial u/\partial y$, ...) in the region of interest are essentially constant. If displacement gradients change rapidly with position in the region of interest, very short gage lengths will be required for accurate strain measurements.

In a wide variety of engineering problems, the displacements and strains produced by the applied loads are very small. Under these conditions, it can be assumed that products and squares of displacement gradients will be small with respect to the displacement gradients and therefore can be neglected. With this assumption Eqs. (2.2) and (2.3) reduce to the strain-displacement equations frequently encountered in the theory of elasticity. The reduced form of the equations is

$$\epsilon_{xx} = \frac{\partial u}{\partial x} \tag{2.4a}$$

$$\epsilon_{yy} = \frac{\partial v}{\partial y} \tag{2.4b}$$

$$\epsilon_{zz} = \frac{\partial w}{\partial z} \tag{2.4c}$$

$$\gamma_{xy} = \frac{\partial v}{\partial x} + \frac{\partial u}{\partial y} \tag{2.4d}$$

$$\gamma_{yz} = \frac{\partial w}{\partial y} + \frac{\partial v}{\partial z} \tag{2.4e}$$

$$\gamma_{zx} = \frac{\partial u}{\partial z} + \frac{\partial w}{\partial x} \tag{2.4f}$$

Equations (2.4) indicate that it is a simple matter to convert a displacement field into a strain field. However, as will be emphasized later, an entire displacement field is rarely determined experimentally. Usually, strains are determined at a number of small areas on the surface of the body through the use of strain gages. In certain problems, however, the displacement field can be computed analytically, and in these instances Eqs. (2.4) become very important.

2.3 STRAIN EQUATIONS OF TRANSFORMATION

Now that the normal and shearing strains in the x, y, z directions have been determined, consider the normal strain in an arbitrary direction. Refer to Fig. 2.4 and consider the elongation of the diagonal PQ. By definition the strain along PQ is

$$\epsilon_{PQ} = \frac{P^*Q^* - PQ}{PQ} \tag{a}$$

From geometric considerations, as illustrated in Fig. 2.4,

$$(PQ)^2 = (\Delta x)^2 + (\Delta y)^2 + (\Delta z)^2 \tag{b}$$

$$(P^*Q^*)^2 = (\Delta x^*)^2 + (\Delta y^*)^2 + (\Delta z^*)^2 \tag{c}$$

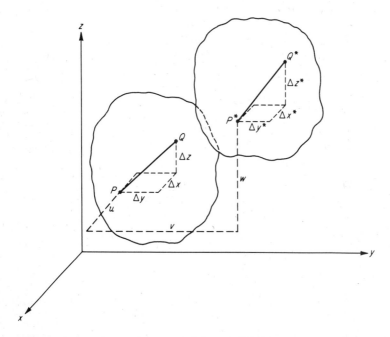

Figure 2.4 Displacements of points P and Q in a body which result from the application of a system of loads.

In general, the component Δx^* will have a different length than the component Δx because of the deformation of the body in the x direction. From Fig. 2.4 it can easily be seen that

$$\Delta x^* = \left(1 + \frac{\partial u}{\partial x}\right)\Delta x + \frac{\partial u}{\partial y}\Delta y + \frac{\partial u}{\partial z}\Delta z$$

$$\Delta y^* = \frac{\partial v}{\partial x}\Delta x + \left(1 + \frac{\partial v}{\partial y}\right)\Delta y + \frac{\partial v}{\partial z}\Delta z \qquad (d)$$

$$\Delta z^* = \frac{\partial w}{\partial x}\Delta x + \frac{\partial w}{\partial y}\Delta y + \left(1 + \frac{\partial w}{\partial z}\right)\Delta z$$

If Eqs. (d) are substituted into Eq. (c), the length of the deformed line segment P^*Q^* can be computed. In the substitution, since the deformations are extremely small, the products and squares of derivatives can be neglected. Thus

$$(P^*Q^*)^2 = \left(1 + 2\frac{\partial u}{\partial x}\right)(\Delta x)^2 + \left(1 + 2\frac{\partial v}{\partial y}\right)(\Delta y)^2$$

$$+ \left(1 + 2\frac{\partial w}{\partial z}\right)(\Delta z)^2 + 2\left(\frac{\partial u}{\partial y} + \frac{\partial v}{\partial x}\right)\Delta x\,\Delta y$$

$$+ 2\left(\frac{\partial v}{\partial z} + \frac{\partial w}{\partial y}\right)\Delta y\,\Delta z + 2\left(\frac{\partial w}{\partial x} + \frac{\partial u}{\partial z}\right)\Delta z\,\Delta x \qquad (e)$$

Equation (a) can be rearranged in the following form:

$$\epsilon_{PQ} = \frac{P^*Q^*}{PQ} - 1$$

or

$$(\epsilon_{PQ} + 1)^2 = \left(\frac{P^*Q^*}{PQ}\right)^2$$

If Eqs. (b) and (e) are substituted into this rearranged form of Eq. (a), the following equation can be obtained after some rearrangement of terms:

$$(\epsilon_{PQ} + 1)^2 = \cos^2(x, PQ) + \cos^2(y, PQ) + \cos^2(z, PQ)$$

$$+ 2\frac{\partial u}{\partial x}\cos^2(x, PQ) + 2\frac{\partial v}{\partial y}\cos^2(y, PQ) + 2\frac{\partial w}{\partial z}\cos^2(z, PQ)$$

$$+ 2\left(\frac{\partial u}{\partial y} + \frac{\partial v}{\partial x}\right)\cos(x, PQ)\cos(y, PQ)$$

$$+ 2\left(\frac{\partial v}{\partial z} + \frac{\partial w}{\partial y}\right)\cos(y, PQ)\cos(z, PQ)$$

$$+ 2\left(\frac{\partial w}{\partial x} + \frac{\partial u}{\partial z}\right)\cos(z, PQ)\cos(x, PQ) \qquad (f)$$

If the left side of Eq. (f) is expanded, the ϵ_{PQ}^2 term can be neglected since it is of the same order of magnitude as the products and squares of displacement derivatives which were neglected in a previous step of this development. Recall also that

$$\cos^2 (x, PQ) + \cos^2 (y, PQ) + \cos^2 (z, PQ) = 1$$

Thus the basic equation for the strain along an arbitrary line segment is

$$
\epsilon_{PQ} = \frac{\partial u}{\partial x} \cos^2 (x, PQ) + \frac{\partial v}{\partial y} \cos^2 (y, PQ) + \frac{\partial w}{\partial z} \cos^2 (z, PQ)
$$

$$
+ \left(\frac{\partial u}{\partial y} + \frac{\partial v}{\partial x} \right) \cos (x, PQ) \cos (y, PQ)
$$

$$
+ \left(\frac{\partial v}{\partial z} + \frac{\partial w}{\partial y} \right) \cos (y, PQ) \cos (z, PQ)
$$

$$
+ \left(\frac{\partial w}{\partial x} + \frac{\partial u}{\partial z} \right) \cos (z, PQ) \cos (x, PQ) \tag{2.5a}
$$

Equations (2.4) and (2.5a) can be used to determine $\epsilon_{x'x'}$ by choosing the direction of PQ parallel to the x' axis. Then

$$
\epsilon_{x'x'} = \epsilon_{xx} \cos^2 (x, x') + \epsilon_{yy} \cos^2 (y, x')
$$

$$
+ \epsilon_{zz} \cos^2 (z, x') + \gamma_{xy} \cos (x, x') \cos (y, x')
$$

$$
+ \gamma_{yz} \cos (y, x') \cos (z, x') + \gamma_{zx} \cos (z, x') \cos (x, x') \tag{2.6a}
$$

In a similar manner $\epsilon_{y'y'}$ and $\epsilon_{z'z'}$ can be determined by choosing the direction of PQ parallel to the y' and z' axes, respectively:

$$
\epsilon_{y'y'} = \epsilon_{yy} \cos^2 (y, y') + \epsilon_{zz} \cos^2 (z, y')
$$

$$
+ \epsilon_{xx} \cos^2 (x, y') + \gamma_{yz} \cos (y, y') \cos (z, y')
$$

$$
+ \gamma_{zx} \cos (z, y') \cos (x, y') + \gamma_{xy} \cos (x, y') \cos (y, y') \tag{2.6b}
$$

$$
\epsilon_{z'z'} = \epsilon_{zz} \cos^2 (z, z') + \epsilon_{xx} \cos^2 (x, z')
$$

$$
+ \epsilon_{yy} \cos^2 (y, z') + \gamma_{zx} \cos (z, z') \cos (x, z')
$$

$$
+ \gamma_{xy} \cos (x, z') \cos (y, z') + \gamma_{yz} \cos (y, z') \cos (z, z') \tag{2.6c}
$$

A similar but somewhat more involved derivation can be used to establish the shearing strains. Consider the angular change in an arbitrary right angle formed by two line segments PQ_1 and PQ_2. The shearing strain γ_{PQ_1, PQ_2} can be shown

[3, p. 44]† to be given by

$$\gamma_{PQ_1, PQ_2} = 2\epsilon_{xx} \cos(x, PQ_1) \cos(x, PQ_2)$$
$$+ 2\epsilon_{yy} \cos(y, PQ_1) \cos(y, PQ_2)$$
$$+ 2\epsilon_{zz} \cos(z, PQ_1) \cos(z, PQ_2)$$
$$+ \gamma_{xy}[\cos(x, PQ_1) \cos(y, PQ_2) + \cos(x, PQ_2) \cos(y, PQ_1)]$$
$$+ \gamma_{yz}[\cos(y, PQ_1) \cos(z, PQ_2) + \cos(y, PQ_2) \cos(z, PQ_1)]$$
$$+ \gamma_{zx}[\cos(z, PQ_1) \cos(x, PQ_2) + \cos(z, PQ_2) \cos(x, PQ_1)]$$

$$(2.5b)$$

By choosing PQ_1 parallel to x' and PQ_2 parallel to y', an expression for $\gamma_{x'y'}$ is obtained as follows:

$$\gamma_{x'y'} = 2\epsilon_{xx} \cos(x, x') \cos(x, y')$$
$$+ 2\epsilon_{yy} \cos(y, x') \cos(y, y')$$
$$+ 2\epsilon_{zz} \cos(z, x') \cos(z, y')$$
$$+ \gamma_{xy}[\cos(x, x') \cos(y, y') + \cos(x, y') \cos(y, x')]$$
$$+ \gamma_{yz}[\cos(y, x') \cos(z, y') + \cos(y, y') \cos(z, x')]$$
$$+ \gamma_{zx}[\cos(z, x') \cos(x, y') + \cos(z, y') \cos(x, x')] \qquad (2.6d)$$

Similarly

$$\gamma_{y'z'} = 2\epsilon_{yy} \cos(y, y') \cos(y, z')$$
$$+ 2\epsilon_{zz} \cos(z, y') \cos(z, z')$$
$$+ 2\epsilon_{xx} \cos(x, y') \cos(x, z')$$
$$+ \gamma_{yz}[\cos(y, y') \cos(z, z') + \cos(y, z') \cos(z, y')]$$
$$+ \gamma_{zx}[\cos(z, y') \cos(x, z') + \cos(z, z') \cos(x, y')]$$
$$+ \gamma_{xy}[\cos(x, y') \cos(y, z') + \cos(x, z') \cos(y, y')] \qquad (2.6e)$$

$$\gamma_{z'x'} = 2\epsilon_{zz} \cos(z, z') \cos(z, x')$$
$$+ 2\epsilon_{xx} \cos(x, z') \cos(x, x')$$
$$+ 2\epsilon_{yy} \cos(y, z') \cos(y, x')$$
$$+ \gamma_{zx}[\cos(z, z') \cos(x, x') + \cos(z, x') \cos(x, z')]$$
$$+ \gamma_{xy}[\cos(x, z') \cos(y, x') + \cos(x, x') \cos(y, z')]$$
$$+ \gamma_{yz}[\cos(y, z') \cos(z, x') + \cos(y, x') \cos(z, z')] \qquad (2.6f)$$

---

† Numbers in brackets refer to numbered references at the end of the chapter.

Equations (2.6a) to (2.6f) are the strain equations of transformation and can be used to transform the six cartesian components of strain ϵ_{xx}, ϵ_{yy}, ϵ_{zz}, γ_{xy}, γ_{yz}, γ_{zx} relative to the $Oxyz$ reference system to six other cartesian components of strain relative to the $Ox'y'z'$ reference system.

A comparison of Eqs. (2.6) with the stress equations of transformation [Eq. (1.6)] shows remarkable similarities:

$$\sigma_{xx} \leftrightarrow \epsilon_{xx} \qquad 2\tau_{xy} \leftrightarrow \gamma_{xy}$$
$$\sigma_{yy} \leftrightarrow \epsilon_{yy} \qquad 2\tau_{yz} \leftrightarrow \gamma_{yz} \qquad (2.7)$$
$$\sigma_{zz} \leftrightarrow \epsilon_{zz} \qquad 2\tau_{zx} \leftrightarrow \gamma_{zx}$$

Here the symbol $\leftrightarrow$ indicates an interchange. This interchange is important since many of the derivations given in the preceding chapter for stresses can be converted directly into strains. Some of these conversions are indicated in the next section.

2.4 PRINCIPAL STRAINS

From the similarity between the laws of stress and strain transformation it can be concluded that there exist at most three distinct principal strains with their three associated principal directions. By substituting the conversions indicated by Eqs. (2.7) into Eq. (1.7), the cubic equation whose roots give the principal strains is obtained:

$$\epsilon_n^3 - (\epsilon_{xx} + \epsilon_{yy} + \epsilon_{zz})\epsilon_n^2$$
$$+ \left(\epsilon_{xx}\epsilon_{yy} + \epsilon_{yy}\epsilon_{zz} + \epsilon_{zz}\epsilon_{xx} - \frac{\gamma_{xy}^2}{4} - \frac{\gamma_{yz}^2}{4} - \frac{\gamma_{zx}^2}{4}\right)\epsilon_n$$
$$- \left(\epsilon_{xx}\epsilon_{yy}\epsilon_{zz} - \epsilon_{xx}\frac{\gamma_{yz}^2}{4} - \epsilon_{yy}\frac{\gamma_{zx}^2}{4} - \epsilon_{zz}\frac{\gamma_{xy}^2}{4} + \frac{\gamma_{xy}\gamma_{yz}\gamma_{zx}}{4}\right) = 0 \qquad (2.8)$$

As with principal stresses, three situations exist:

$$\epsilon_1 \neq \epsilon_2 \neq \epsilon_3 \qquad \epsilon_1 = \epsilon_2 \neq \epsilon_3 \qquad \epsilon_1 = \epsilon_2 = \epsilon_3$$

The significance of these three cases is determined from the discussion in Sec. 1.6, page 14.

Similarly, there are three strain invariants which are analogous to the three stress invariants. By substituting Eqs. (2.7) into Eqs. (1.8), the following expressions are obtained for the strain invariants:

$$J_1 = \epsilon_{xx} + \epsilon_{yy} + \epsilon_{zz}$$
$$J_2 = \epsilon_{xx}\epsilon_{yy} + \epsilon_{yy}\epsilon_{zz} + \epsilon_{zz}\epsilon_{xx} - \frac{\gamma_{xy}^2}{4} - \frac{\gamma_{yz}^2}{4} - \frac{\gamma_{zx}^2}{4} \qquad (2.9)$$
$$J_3 = \epsilon_{xx}\epsilon_{yy}\epsilon_{zz} - \frac{\epsilon_{xx}\gamma_{yz}^2}{4} - \frac{\epsilon_{yy}\gamma_{zx}^2}{4} - \frac{\epsilon_{zz}\gamma_{xy}^2}{4} + \frac{\gamma_{xy}\gamma_{yz}\gamma_{zx}}{4}$$

It is clear that other equations derived in Chap. 1 for stresses could easily be converted into equations in terms of strains. A few more will be covered in the exercises at the end of the chapter, and others will be converted as the need arises.

2.5 COMPATIBILITY

From a given displacement field, i.e., three equations expressing u, v, and w as functions of x, y, and z, a unique strain field can be determined by using Eqs. (2.4). However, an arbitrary strain field may yield an impossible displacement field, i.e., one in which the body might contain voids after deformation. A valid displacement field can be ensured only if the body under consideration is simply connected and if the strain field satisfies a set of equations known as the *compatibility relations*. The six equations of compatibility which must be satisfied are

$$\frac{\partial^2 \gamma_{xy}}{\partial x \, \partial y} = \frac{\partial^2 \epsilon_{xx}}{\partial y^2} + \frac{\partial^2 \epsilon_{yy}}{\partial x^2} \tag{2.10a}$$

$$\frac{\partial^2 \gamma_{yz}}{\partial y \, \partial z} = \frac{\partial^2 \epsilon_{yy}}{\partial z^2} + \frac{\partial^2 \epsilon_{zz}}{\partial y^2} \tag{2.10b}$$

$$\frac{\partial^2 \gamma_{zx}}{\partial z \, \partial x} = \frac{\partial^2 \epsilon_{zz}}{\partial x^2} + \frac{\partial^2 \epsilon_{xx}}{\partial z^2} \tag{2.10c}$$

$$2\frac{\partial^2 \epsilon_{xx}}{\partial y \, \partial z} = \frac{\partial}{\partial x}\left(-\frac{\partial \gamma_{yz}}{\partial x} + \frac{\partial \gamma_{zx}}{\partial y} + \frac{\partial \gamma_{xy}}{\partial z}\right) \tag{2.10d}$$

$$2\frac{\partial^2 \epsilon_{yy}}{\partial z \, \partial x} = \frac{\partial}{\partial y}\left(\frac{\partial \gamma_{yz}}{\partial x} - \frac{\partial \gamma_{zx}}{\partial y} + \frac{\partial \gamma_{xy}}{\partial z}\right) \tag{2.10e}$$

$$2\frac{\partial^2 \epsilon_{zz}}{\partial x \, \partial y} = \frac{\partial}{\partial z}\left(\frac{\partial \gamma_{yz}}{\partial x} + \frac{\partial \gamma_{zx}}{\partial y} - \frac{\partial \gamma_{xy}}{\partial z}\right) \tag{2.10f}$$

In order to derive Eq. (2.10a), begin by recalling

$$\gamma_{xy} = \frac{\partial u}{\partial y} + \frac{\partial v}{\partial x} \tag{a}$$

Differentiating γ_{xy} once with respect to x and then again with respect to y gives

$$\frac{\partial^2 \gamma_{xy}}{\partial x \, \partial y} = \frac{\partial^3 u}{\partial x \, \partial y^2} + \frac{\partial^3 v}{\partial x^2 \, \partial y} \tag{b}$$

Note that

$$\frac{\partial^2 \epsilon_{xx}}{\partial y^2} = \frac{\partial^3 u}{\partial x \, \partial y^2} \quad \text{and} \quad \frac{\partial^2 \epsilon_{yy}}{\partial x^2} = \frac{\partial^3 v}{\partial x^2 \, \partial y} \tag{c}$$

Substituting Eqs. (c) into Eq. (b) gives

$$\frac{\partial^2 \gamma_{xy}}{\partial x \, \partial y} = \frac{\partial^2 \epsilon_{xx}}{\partial y^2} + \frac{\partial^2 \epsilon_{yy}}{\partial x^2} \tag{d}$$

which establishes Eq. (2.10a), and by the same methods Eqs. (2.10b) and (2.10c) could be verified. The proof of Eq. (2.10d) is obtained by considering four identities:

$$\frac{\partial^2 \epsilon_{xx}}{\partial y\, \partial z} = \frac{\partial^3 u}{\partial x\, \partial y\, \partial z} \tag{e}$$

$$\frac{\partial^2 \gamma_{xy}}{\partial x\, \partial z} = \frac{\partial^3 u}{\partial x\, \partial y\, \partial z} + \frac{\partial^3 v}{\partial x^2\, \partial z} \tag{f}$$

$$\frac{\partial^2 \gamma_{zx}}{\partial x\, \partial y} = \frac{\partial^3 w}{\partial x^2\, \partial y} + \frac{\partial^3 u}{\partial x\, \partial y\, \partial z} \tag{g}$$

$$\frac{\partial^2 \gamma_{yz}}{\partial x^2} = \frac{\partial^3 w}{\partial x^2\, \partial y} + \frac{\partial^3 v}{\partial x^2\, \partial z} \tag{h}$$

Now by forming

$$2(e) = (f) + (g) - (h) \tag{i}$$

thus obtaining

$$2\frac{\partial^2 \epsilon_{xx}}{\partial y\, \partial z} = \frac{\partial}{\partial x}\left(\frac{\partial \gamma_{xy}}{\partial z} + \frac{\partial \gamma_{zx}}{\partial y} - \frac{\partial \gamma_{yz}}{\partial x}\right) \tag{j}$$

Eq. (2.10d) is verified. The remaining two compatibility relations can be established in an identical manner.

In order to gain a better physical understanding of the compatibility relations, consider a two-dimensional body made up of a large number of small, square elements. When the body is loaded, the elements deform. By measuring angle changes and length changes, the strains which develop in each element can be determined. This procedure is accomplished theoretically by differentiating the displacement field. Consider now the inverse problem. Suppose a large number of small, deformed elements are given which must be fitted together to form a body free of voids and discontinuities. If and only if each element is properly strained can the body be reassembled without voids. The deformed elements correspond to the case of the prescribed strain field. The check to determine whether the elements are all properly strained and hence compatible with each other represents the compatibility relations. If these relations are satisfied, the elements will fit together properly, thus guaranteeing a satisfactory displacement field.

2.6 EXAMPLE OF A DISPLACEMENT FIELD COMPUTED FROM A STRAIN FIELD

If a circular shaft of radius a is loaded in torsion, the following strain field is produced:

$$\gamma_{zx} = -ay \qquad \gamma_{yz} = ax \qquad \epsilon_{xx} = \epsilon_{yy} = \epsilon_{zz} = \gamma_{xy} = 0 \tag{a}$$

where the z axis of the reference system is coincident with the centerline of the shaft. The first step in solving for the displacement field is to check the compatibility conditions:

1. The body must be simply connected, a condition which is obviously satisfied in this case.
2. The strain relations given in Eqs. (a) must satisfy all the compatibility relations given in Eqs. (2.10). It is clear that this linear system of strains does satisfy this requirement.

Next substitute Eqs. (a) into Eqs. (2.4) and integrate:

$$\epsilon_{xx} = \frac{\partial u}{\partial x} = 0 \qquad u = f(y, z)$$

$$\epsilon_{yy} = \frac{\partial v}{\partial y} = 0 \qquad v = g(x, z)$$

$$\epsilon_{zz} = \frac{\partial w}{\partial z} = 0 \qquad w = h(x, y) \tag{b}$$

$$\gamma_{xy} = \frac{\partial u}{\partial y} + \frac{\partial v}{\partial x} = 0 = \frac{\partial f(y, z)}{\partial y} + \frac{\partial g(x, z)}{\partial x}$$

The last of Eqs. (b) can be satisfied only if both right-hand terms are functions of z alone; hence

$$\frac{\partial f(y, z)}{\partial y} = -\frac{\partial g(x, z)}{\partial x} = F(z) \tag{c}$$

Integrating Eq. (c) gives

$$f(z) = yF(z) + C_1 = u \qquad g(z) = -xF(z) + C_2 = v \tag{d}$$

Recall the value of γ_{yz} from Eqs. (a), the definition of γ_{yz} from Eq. (2.4e), and the functional relation for w from the third of Eqs. (b) and form

$$\gamma_{yz} = \frac{\partial w}{\partial y} + \frac{\partial v}{\partial z} = ax = \frac{\partial h(x, y)}{\partial y} - x\frac{dF(z)}{dz} \tag{e}$$

Equation (e) can be satisfied if and only if both the right-hand terms are functions of x alone; hence

$$\frac{\partial h(x)}{\partial y} = H(x) \qquad \frac{dF(z)}{dz} = C_3 \tag{f}$$

Substituting Eqs. (f) into Eq. (e) gives

$$H(x) - C_3 x = ax$$

$$H(x) = (a + C_3)x \tag{g}$$

Integrating the second of Eqs. (f) yields

$$F(z) = C_3 z + C_4 \qquad (h)$$

Substituting Eq. (h) into Eqs. (d) gives

$$u = y(C_3 z + C_4) + C_1 \qquad v = -x(C_3 z + C_4) + C_2 \qquad (i)$$

By Eqs. (b), (f), and (g),

$$w = h(x, y) = \int H(x) \, \partial y = \int (a + C_3) x \, \partial y = (a + C_3) xy + C_5 \qquad (j)$$

Thus far five constants of integration have been introduced, and five of the six cartesian components of strain which were given have been used. By employing the last strain relation, the arbitrary constant C_3 can be evaluated. Recall from Eqs. (a):

$$\gamma_{zx} = \frac{\partial w}{\partial x} + \frac{\partial u}{\partial z} = (a + C_3) y + C_3 y = -ay \qquad C_3 = -a \qquad (k)$$

Substituting Eqs. (k) into Eqs. (i) and (j) gives

$$u = -ayz + C_4 y + C_1 \qquad v = axz - C_4 x + C_2 \qquad w = C_5 \qquad (l)$$

The constants C_1, C_2, and C_5 indicate rigid-body translation of the shaft. The constant C_4 indicates rigid-body rotation of the shaft. It is clear upon differentiation that $C_1, C_2, C_3,$ and C_4 do not enter into the strains produced in the shaft and hence are not a part of the displacement due to deformation. The deformation displacements are given by

$$u = -ayz \qquad v = axz \qquad w = 0 \qquad (m)$$

As indicated by this simple example, the process by which the displacement field is calculated from a given strain field is quite lengthy. On the other hand, it is a very simple matter to go from a complex displacement field to a strain field.

2.7 VOLUME DILATATION

Consider a small, rectangular element in a deformed body which has its edges oriented along the principal axes. The length of each side of the block may have changed; however, the element will not be distorted since there are no shearing strains acting on the faces. The change in volume of such an element divided by the initial volume is, by definition, the volume dilatation D, that is,

$$D = \frac{V^* - V}{V}$$

where V is the initial volume, equal to the product of the three sides of the element, a_1, a_2, a_3, before deformation and V^* is the final volume after straining, equal to the product of the three sides a_1^*, a_2^*, a_3^*, after deformation. Since

$$a_1^* = a_1(1 + \epsilon_1) \qquad a_2^* = a_2(1 + \epsilon_2) \qquad a_3^* = a_3(1 + \epsilon_3)$$

it follows that

$$D = \frac{a_1 a_2 a_3 (1 + \epsilon_1)(1 + \epsilon_2)(1 + \epsilon_3) - a_1 a_2 a_3}{a_1 a_2 a_3}$$

If the higher-order strain terms are neglected,

$$D = \epsilon_1 + \epsilon_2 + \epsilon_3 = J_1 \tag{2.11}$$

Equation (2.11) indicates that the volume dilatation D is equal to the first invariant of strain. Since the first invariant of strain is independent of the coordinate system being used, the volume dilatation of an element is independent of the reference frame forming its sides. Volume dilatation is thus a coordinate-independent concept.

2.8 STRESS-STRAIN RELATIONS

Thus far stress and strain have been discussed individually, and no assumptions have been required regarding the behavior of the material except that it was a continuous medium.† In this section, stress will be related to strain; therefore, certain restrictive assumptions regarding the body material must be introduced. The first of these assumptions regards linearity of the stress versus strain in the body. With a linear stress-strain relationship it is possible to write the general stress-strain expressions as follows:

$$\sigma_{xx} = K_{11}\epsilon_{xx} + K_{12}\epsilon_{yy} + K_{13}\epsilon_{zz} + K_{14}\gamma_{xy} + K_{15}\gamma_{yz} + K_{16}\gamma_{zx}$$

$$\sigma_{yy} = K_{21}\epsilon_{xx} + K_{22}\epsilon_{yy} + K_{23}\epsilon_{zz} + K_{24}\gamma_{xy} + K_{25}\gamma_{yz} + K_{26}\gamma_{zx}$$

$$\sigma_{zz} = K_{31}\epsilon_{xx} + K_{32}\epsilon_{yy} + K_{33}\epsilon_{zz} + K_{34}\gamma_{xy} + K_{35}\gamma_{yz} + K_{36}\gamma_{zx} \tag{2.12}$$

$$\tau_{xy} = K_{41}\epsilon_{xx} + K_{42}\epsilon_{yy} + K_{43}\epsilon_{zz} + K_{44}\gamma_{zy} + K_{45}\gamma_{yz} + K_{46}\gamma_{zx}$$

$$\tau_{yz} = K_{51}\epsilon_{xx} + K_{52}\epsilon_{yy} + K_{53}\epsilon_{zz} + K_{54}\gamma_{xy} + K_{55}\gamma_{yz} + K_{56}\gamma_{zx}$$

$$\tau_{zx} = K_{61}\epsilon_{xx} + K_{62}\epsilon_{yy} + K_{63}\epsilon_{zz} + K_{64}\gamma_{xy} + K_{65}\gamma_{yz} + K_{66}\gamma_{zx}$$

where K_{11} to K_{66} are the coefficients of elasticity of the material and are independent of the magnitudes of both the stress and the strain, provided the elastic limit of the material is not exceeded. If the elastic limit is exceeded, the linear relationship between stress and strain no longer holds, and Eqs. (2.12) are not valid.

There are 36 coefficients of elasticity in Eqs. (2.12); however, they are not all independent. By strain energy considerations, which are beyond the scope of this

---

† Actually most metals are not strictly continuous since they are composed of a large number of rather small grains. However, the grains are in almost all cases small enough in comparison with the size of the body for the body to behave as if it were a continuous medium.

book, the number of independent coefficients of elasticity can be reduced to 21. This reduction is quite significant; however, even with 21 constants, Eqs. (2.12) may be considered rather long and involved. By assuming that the material is isotropic, i.e., that the elastic constants are the same in all directions and hence independent of the choice of a coordinate system, the 21 coefficients of elasticity reduce to two constants. The stress-strain relationships then reduce to

$$\sigma_{xx} = \lambda J_1 + 2\mu\epsilon_{xx} \qquad \sigma_{yy} = \lambda J_1 + 2\mu\epsilon_{yy} \qquad \sigma_{zz} = \lambda J_1 + 2\mu\epsilon_{zz}$$

$$\tau_{xy} = \mu\gamma_{xy} \qquad \tau_{yz} = \mu\gamma_{yz} \qquad \tau_{zx} = \mu\gamma_{zx} \tag{2.13}$$

where J_1 = first invariant of strain $(\epsilon_{xx} + \epsilon_{yy} + \epsilon_{zz})$
     λ = Lamé's constant
     μ = shear modulus

Equations (2.13) can be solved to give the strains as a function of stress:

$$\epsilon_{xx} = \frac{\lambda + \mu}{\mu(3\lambda + 2\mu)}\sigma_{xx} - \frac{\lambda}{2\mu(3\lambda + 2\mu)}(\sigma_{yy} + \sigma_{zz})$$

$$\epsilon_{yy} = \frac{\lambda + \mu}{\mu(3\lambda + 2\mu)}\sigma_{yy} - \frac{\lambda}{2\mu(3\lambda + 2\mu)}(\sigma_{xx} + \sigma_{zz})$$

$$\epsilon_{zz} = \frac{\lambda + \mu}{\mu(3\lambda + 2\mu)}\sigma_{zz} - \frac{\lambda}{2\mu(3\lambda + 2\mu)}(\sigma_{yy} + \sigma_{xx})$$

$$\gamma_{xy} = \frac{1}{\mu}\tau_{xy} \qquad \gamma_{yz} = \frac{1}{\mu}\tau_{yz} \qquad \gamma_{zx} = \frac{1}{\mu}\tau_{zx} \tag{2.14}$$

The elastic coefficients μ and λ shown in Eqs. (2.13) and (2.14) arise from a mathematical treatment of the general linear stress-strain relations. In experimental work, Lamé's constant λ is rarely used since it has no physical significance; however, as will be shown later, the shear modulus has physical significance and can easily be measured.

Consider a two-dimensional case of pure shear where

$$\sigma_{xx} = \sigma_{yy} = \sigma_{zz} = \tau_{zx} = \tau_{yz} = 0 \qquad \tau_{xy} = \text{applied shearing stress}$$

From Eqs. (2.14),

$$\mu = \frac{\tau_{xy}}{\gamma_{xy}} \tag{2.15a}$$

Hence, the shear modulus μ is the ratio of the shearing stress to the shearing strain in a two-dimensional state of pure shear.

In a conventional tension test which is often used to determine the mechanical properties of materials, a long, slender bar is subjected to a state of uniaxial stress in, say, the x direction. In this instance

$$\sigma_{yy} = \sigma_{zz} = \tau_{xy} = \tau_{yz} = \tau_{zx} = 0 \qquad \sigma_{xx} = \text{applied normal stress}$$

From Eqs. (2.14),

$$\epsilon_{xx} = \frac{\lambda + \mu}{\mu(3\lambda + 2\mu)} \sigma_{xx} \tag{a}$$

$$\epsilon_{yy} = \epsilon_{zz} = -\frac{\lambda}{2\mu(3\lambda + 2\mu)} \sigma_{xx} \tag{b}$$

In elementary strength-of-materials texts, the stress-strain relations for the case of uniaxial stress are often written

$$\epsilon_{xx} = \frac{1}{E} \sigma_{xx} \tag{c}$$

$$\epsilon_{yy} = \epsilon_{zz} = -\frac{v}{E} \sigma_{xx} \tag{d}$$

By equating the coefficients in Eqs. (a) and (b) to those in Eqs. (c) and (d),

$$E = \frac{\mu(3\lambda + 2\mu)}{\lambda + \mu} \tag{2.15b}$$

$$v = \frac{\lambda}{2(\lambda + \mu)} \tag{2.15c}$$

where E is the modulus of elasticity and v is Poisson's ratio, defined as

$$v = -\frac{\epsilon_{yy}}{\epsilon_{xx}} \tag{2.15d}$$

Equations (2.15b) and (2.15c) indicate the conversion from Lamé's constant λ and the shear modulus μ to the more commonly used modulus of elasticity E and Poisson's ratio v.

To establish the definition and physical significance of a fifth elastic constant, consider a state of hydrostatic stress where

$$\sigma_{xx} = \sigma_{yy} = \sigma_{zz} = -p \qquad \tau_{xy} = \tau_{yz} = \tau_{zx} = 0$$

where p is the uniform pressure acting on the body.

Adding together the first three of Eqs. (2.13) gives

$$-3p = (3\lambda + 2\mu)J_1$$

or

$$p = -\frac{3\lambda + 2\mu}{3} J_1 = -KJ_1 = -KD$$

Thus

$$K = \frac{3\lambda + 2\mu}{3} = -\frac{p}{D} \tag{2.15e}$$

The constant K is known as the bulk modulus and is the ratio of the applied hydrostatic pressure to the volume dilatation.

Table 2.1 Relationships between the elastic constants

	λ equals	μ equals	E equals	ν equals	K equals
λ, μ			$\dfrac{\mu(3\lambda + 2\mu)}{\lambda + \mu}$	$\dfrac{\lambda}{2(\lambda + \mu)}$	$\dfrac{3\lambda + 2\mu}{3}$
λ, E		$\dfrac{A\dagger + (E - 3\lambda)}{4}$		$\dfrac{A\dagger - (E + \lambda)}{4\lambda}$	$\dfrac{A\dagger + (3\lambda + E)}{6}$
λ, ν		$\dfrac{\lambda(1 - 2\nu)}{2\nu}$	$\dfrac{\lambda(1 + \nu)(1 - 2\nu)}{\nu}$		$\dfrac{\lambda(1 + \nu)}{3\nu}$
λ, K		$\dfrac{3(K - \lambda)}{2}$	$\dfrac{9K(K - \lambda)}{3K - \lambda}$	$\dfrac{\lambda}{3K - \lambda}$	
μ, E	$\dfrac{\mu(2\mu - E)}{E - 3\mu}$			$\dfrac{E - 2\mu}{2\mu}$	$\dfrac{\mu E}{3(3\mu - E)}$
μ, ν	$\dfrac{2\mu\nu}{1 - 2\nu}$		$2\mu(1 + \nu)$		$\dfrac{2\mu(1 + \nu)}{3(1 - 2\nu)}$
μ, K	$\dfrac{3K - 2\mu}{3}$		$\dfrac{9K\mu}{3K + \mu}$	$\dfrac{3K - 2\mu}{2(3K + \mu)}$	
E, ν	$\dfrac{\nu E}{(1 + \nu)(1 - 2\nu)}$	$\dfrac{E}{2(1 + \nu)}$			$\dfrac{E}{3(1 - 2\nu)}$
K, E	$\dfrac{3K(3K - E)}{9K - E}$	$\dfrac{3EK}{9K - E}$		$\dfrac{3K - E}{6K}$	
ν, K	$\dfrac{3K\nu}{1 + \nu}$	$\dfrac{3K(1 - 2\nu)}{2(1 + \nu)}$	$3K(1 - 2\nu)$		

† $A = \sqrt{E^2 + 2\lambda E + 9\lambda^2}$.

Five elastic constants λ, μ, E, ν, and K have been discussed. The constant λ has no physical significance and is employed because it simplifies, mathematically speaking, the stress-strain relations. The constant μ has both mathematical and physical significance. It is used extensively in torsional problems. The constants E and ν are the most widely recognized of the five constants considered and are used in almost all areas of stress analysis. The rather specialized bulk modulus K is used primarily for computing volume changes in a given body subjected to hydrostatic pressure. As indicated previously, there are two and only two independent elastic constants. The five constants discussed are related to each other as shown in Table 2.1.

Since the constants E and ν will be used almost exclusively throughout the remainder of this text, Eqs. (2.15b) and (2.15c) have been substituted into Eqs. (2.13) and (2.14) to obtain expressions for strain in terms of stress and the constants

$$\epsilon_{xx} = \frac{1}{E}[\sigma_{xx} - v(\sigma_{yy} + \sigma_{zz})]$$

$$\epsilon_{yy} = \frac{1}{E}[\sigma_{yy} - v(\sigma_{xx} + \sigma_{zz})]$$

$$\epsilon_{zz} = \frac{1}{E}[\sigma_{zz} - v(\sigma_{yy} + \sigma_{xx})]$$

$$\gamma_{xy} = \frac{2(1+v)}{E}\tau_{xy} \qquad \gamma_{yz} = \frac{2(1+v)}{E}\tau_{yz} \qquad \gamma_{zx} = \frac{2(1+v)}{E}\tau_{zx} \qquad (2.16)$$

and for stress in terms of strain and the constants

$$\sigma_{xx} = \frac{E}{(1+v)(1-2v)}[(1-v)\epsilon_{xx} + v(\epsilon_{yy} + \epsilon_{zz})]$$

$$\sigma_{yy} = \frac{E}{(1+v)(1-2v)}[(1-v)\epsilon_{yy} + v(\epsilon_{xx} + \epsilon_{zz})]$$

$$\sigma_{zz} = \frac{E}{(1+v)(1-2v)}[(1-v)\epsilon_{zz} + v(\epsilon_{xx} + \epsilon_{yy})]$$

$$\tau_{xy} = \frac{E}{2(1+v)}\gamma_{xy} \qquad \tau_{yz} = \frac{E}{2(1+v)}\gamma_{yz} \qquad \tau_{zx} = \frac{E}{2(1+v)}\gamma_{zx} \qquad (2.17)$$

2.9 STRAIN-TRANSFORMATION EQUATIONS AND STRESS-STRAIN RELATIONS FOR A TWO-DIMENSIONAL STATE OF STRESS

Simplified forms of the strain-transformation equations and the stress-strain relations, which will be extremely useful in later chapters when brittle coating and electrical-resistance strain-gage analyses are discussed, are the equations applicable to the strain field associated with a two-dimensional state of stress $(\sigma_{zz} = \tau_{zx} = \tau_{zy} = 0)$.

The strain-transformation equations can be obtained from Eqs. (2.6) by selecting z' coincident with z and noting from Eqs. (2.16) that $\gamma_{zx} = \gamma_{yz} = 0$. The notation can also be simplified by denoting the angle between x' and x as θ. The equations obtained are

$$\epsilon_{x'x'} = \epsilon_{xx}\cos^2\theta + \epsilon_{yy}\sin^2\theta + \gamma_{xy}\sin\theta\cos\theta$$

$$\epsilon_{y'y'} = \epsilon_{yy}\cos^2\theta + \epsilon_{xx}\sin^2\theta - \gamma_{xy}\sin\theta\cos\theta$$

$$\gamma_{x'y'} = 2(\epsilon_{yy} - \epsilon_{xx})\sin\theta\cos\theta + \gamma_{xy}(\cos^2\theta - \sin^2\theta) \qquad (2.18)$$

$$\epsilon_{z'z'} = \epsilon_{zz} \qquad \gamma_{yz'} = \gamma_{z'x'} = 0$$

The stress-strain relations for a two-dimensional state of stress are obtained by substituting $\sigma_{zz} = \tau_{zx} = \tau_{yz} = 0$ into Eqs. (2.16). Thus

$$\epsilon_{xx} = \frac{1}{E}(\sigma_{xx} - v\sigma_{yy}) \qquad \epsilon_{yy} = \frac{1}{E}(\sigma_{yy} - v\sigma_{xx}) \qquad \epsilon_{zz} = -\frac{v}{E}(\sigma_{xx} + \sigma_{yy})$$

$$\gamma_{xy} = \frac{2(1 + v)}{E}\tau_{xy} \qquad \gamma_{yz} = \gamma_{zx} = 0 \tag{2.19}$$

In a similar manner the equations for stress in terms of strain for the two-dimensional state of stress are obtained from Eqs. (2.17). Thus

$$\sigma_{xx} = \frac{E}{1 - v^2}(\epsilon_{xx} + v\epsilon_{yy}) \qquad \sigma_{yy} = \frac{E}{1 - v^2}(\epsilon_{yy} + v\epsilon_{xx})$$

$$\sigma_{zz} = \tau_{zx} = \tau_{yz} = 0 \qquad \tau_{xy} = \frac{E}{2(1 + v)}\gamma_{xy} \tag{2.20}$$

One additional relationship which relates the strain ϵ_{zz} to the measured strains ϵ_{xx} and ϵ_{yy} in experimental analyses is obtained from Eq. (2.17) by substituting $\sigma_{zz} = 0$. Thus

$$\epsilon_{zz} = -\frac{v}{1 - v}(\epsilon_{xx} + \epsilon_{yy}) \tag{2.21}$$

This equation can be used to establish the magnitude of the third principal strain associated with a two-dimensional state of stress. This information is useful for maximum shear-strain determinations.

EXERCISES

2.1 Given the displacement field

$$u = (3x^4 + 2x^2y^2 + x + y + z^3 + 3)(10^{-3})$$
$$v = (3xy + y^3 + y^2z + z^2 + 1)(10^{-3})$$
$$w = x^2 + xy + yz + zx + y^2 + z^2 + 2)(10^{-3})$$

Compute the associated strains at point (1, 1, 1). Compare the results obtained by using Eqs. (2.2) and (2.3) with those obtained by using Eqs. (2.4).

2.2 Given the displacement field

$$u = (x^2 + y^4 + 2y^2z + yz)(10^{-3})$$
$$v = (xy + xz + 3x^2z)(10^{-3})$$
$$w = (y^4 + 4y^3 + 2z^2)(10^{-3})$$

Compute the associated strains at point (2, 2, 2). Compare the results obtained by using Eqs. (2.2) and (2.3) with those obtained by using Eqs. (2.4).

2.3 Transform the set of cartesian strain components

$$\epsilon_{xx} = 600 \times 10^{-6} \qquad \epsilon_{yy} = 400 \times 10^{-6} \qquad \epsilon_{zz} = 200 \times 10^{-6}$$

$$\gamma_{xy} = 400 \times 10^{-6} \qquad \gamma_{yz} = 200 \times 10^{-6} \qquad \gamma_{zx} = 300 \times 10^{-6}$$

into a new set of cartesian strain components relative to an $Ox'y'z'$ set of coordinates, where the $Ox'y'z'$ axes are defined as:

θ	Case 1	Case 2	Case 3	Case 4
$x - x'$	$\pi/4$	$\pi/2$	0	$\pi/2$
$y - y'$	$\pi/4$	$\pi/2$	$\pi/2$	0
$z - z'$	0	0	$\pi/2$	$\pi/2$

2.4 At a point in a stressed body, the cartesian components of strain are

$$\epsilon_{xx} = 600 \times 10^{-6} \qquad \epsilon_{yy} = 900 \times 10^{-6} \qquad \epsilon_{zz} = 600 \times 10^{-6}$$

$$\gamma_{xy} = 1200 \times 10^{-6} \qquad \gamma_{yz} = 750 \times 10^{-6} \qquad \gamma_{zx} = 900 \times 10^{-6}$$

Transform this set of cartesian strain components into a new set of cartesian strain components relative to an $Ox'y'z'$ set of coordinates where the $Ox'y'z'$ axes are defined by the following direction cosines:

	x	y	z
x'	$\frac{2}{3}$	$\frac{2}{3}$	$-\frac{1}{3}$
y'	$-\frac{2}{3}$	$\frac{1}{3}$	$-\frac{2}{3}$
z'	$-\frac{1}{3}$	$\frac{2}{3}$	$\frac{2}{3}$

2.5 At a point in a stressed body, the cartesian components of strain are

$$\epsilon_{xx} = 900 \times 10^{-6} \qquad \epsilon_{yy} = 600 \times 10^{-6} \qquad \epsilon_{zz} = 300 \times 10^{-6}$$

$$\gamma_{xy} = 300 \times 10^{-6} \qquad \gamma_{yz} = 300 \times 10^{-6} \qquad \gamma_{zx} = 600 \times 10^{-6}$$

Transform this set of cartesian strain components into a new set of cartesian strain components relative to an $Ox'y'z'$ set of coordinates where the $Ox'y'z'$ axes are defined by the following direction cosines:

	x	y	z
x'	$\frac{1}{9}$	$-\frac{8}{9}$	$\frac{4}{9}$
y'	$\frac{4}{9}$	$\frac{4}{9}$	$\frac{7}{9}$
z'	$-\frac{8}{9}$	$\frac{1}{9}$	$\frac{4}{9}$

2.6 At a point in a stressed body, the cartesian components of strain are

$$\epsilon_{xx} = 990 \times 10^{-6} \qquad \epsilon_{yy} = 825 \times 10^{-6} \qquad \epsilon_{zz} = 550 \times 10^{-6}$$

$$\gamma_{xy} = 330 \times 10^{-6} \qquad \gamma_{yz} = 660 \times 10^{-6} \qquad \gamma_{zx} = 275 \times 10^{-6}$$

Transform this set of cartesian strain components into a new set of cartesian strain components relative to an $Ox'y'z'$ set of coordinates where the $Ox'y'z'$ axes are defined by the following direction cosines:

	x	y	z
x'	$\frac{9}{11}$	$\frac{2}{11}$	$-\frac{6}{11}$
y'	$\frac{6}{11}$	$-\frac{6}{11}$	$\frac{7}{11}$
z'	$\frac{2}{11}$	$\frac{9}{11}$	$\frac{6}{11}$

2.7 Determine the three principal strains and the maximum shearing strain at the point having the cartesian strain components given in Exercise 2.3. Check the three strain invariants.

2.8 Determine the three principal strains and the maximum shearing strain at the point having the cartesian strain components given in Exercise 2.4. Check the three strain invariants.

2.9 Determine the three principal strains and the principal strain directions at the point having the cartesian strain components given in Exercise 2.5.

2.10 Determine the three principal strains and the principal-strain directions at the point having the cartesian strain components given in Exercise 2.6.

2.11 Determine whether the following strain fields are compatible:

$$\epsilon_{xx} = 2x^2 + 3y^2 + z + 1 \qquad \epsilon_{xx} = 3y^2 + xy$$

$$\epsilon_{yy} = 2y^2 + x^2 + 3z + 2 \qquad \epsilon_{yy} = 2y + 4z + 3$$

$$\epsilon_{zz} = 3x + 2y + z^2 + 1 \qquad \epsilon_{zz} = 3zx + 2xy + 3yz + 2$$

$$\gamma_{xy} = 8xy \qquad\qquad\qquad \gamma_{xy} = 6xy$$

$$\gamma_{yz} = 0 \qquad\qquad\qquad\quad \gamma_{yz} = 2x$$

$$\gamma_{zx} = 0 \qquad\qquad\qquad\quad \gamma_{zx} = 2y$$

2.12 Determine whether the following strain fields are compatible:

$$\epsilon_{xx} = 3x^2 + 4xy - 8y^2 \qquad \epsilon_{xx} = 12x^2 - 12y^2 - 4z$$

$$\epsilon_{yy} = 2x^2 + xy + 3y^2 \qquad \epsilon_{yy} = 12y^2 - 12x^2 + 4z$$

$$\epsilon_{zz} = 0 \qquad\qquad\qquad\quad \epsilon_{zz} = 12x + 4y - z + 5$$

$$\gamma_{xy} = -x^2 - 12xy - 4y^2 \qquad \gamma_{xy} = 4z - 48xy - 3$$

$$\gamma_{yz} = 0 \qquad\qquad\qquad\quad \gamma_{yz} = 0$$

$$\gamma_{zx} = 0 \qquad\qquad\qquad\quad \gamma_{zx} = 4x + 4y - 6$$

2.13 Given the strain field

$$\epsilon_{xx} = ay \qquad \epsilon_{yy} = by \qquad \epsilon_{zz} = by$$

$$\gamma_{xy} = 0 \qquad \gamma_{yz} = 0 \qquad \gamma_{zx} = 0$$

Compute the displacement fields. What physical problem does this strain field represent?

2.14 Determine the volume dilatation for the strain field in Exercise 2.5.

2.15 Determine the volume dilatation for the strain field in Exercise 2.6.

2.16 Determine the volume dilation at point $(2, 1, 2)$ of the displacement field given in Exercise 2.2.

2.17 Determine λ, μ, and K for steel, brass, aluminum, and plastic CR-39, using the following values for E and v:

Material	E, GPa	v
Steel	200	0.29
Brass	110	0.33
Aluminum	70	0.33
CR-39	2.4	0.44

2.18 A cube of steel ($E = 200$ GPa and $v = 0.30$) is loaded with a uniformly distributed pressure of 200 MPa on the four faces having outward normals in the x and y directions. Rigid constraints limit the total deformation of the cube in the z direction to 0.05 mm. Determine the normal stress, if any, which develops in the z direction. The length of a side of the cube is 250 mm.

2.19 Determine the change in volume of a 25-mm cube of aluminum when dropped a distance of 8 km to the ocean floor.

2.20 Determine the stresses at a point in a steel ($E = 200$ GPa, $v = 0.30$) machine component if the cartesian components of strain at the point are as listed in Exercise 2.3.

2.21 Determine the stresses at a point in an aluminum ($E = 70$ GPa, $v = 0.33$) machine component if the cartesian components of strain at the point are as listed in Exercise 2.4.

2.22 The cartesian components of stress at a point in a steel machine part are

$$\sigma_{xx} = 200 \text{ MPa} \qquad \sigma_{yy} = 70 \text{ MPa} \qquad \sigma_{zz} = 140 \text{ MPa}$$

$$\tau_{xy} = 100 \text{ MPa} \qquad \tau_{yz} = 50 \text{ MPa} \qquad \tau_{zx} = 60 \text{ MPa}$$

Determine the principal strains at the point if $E = 200$ GPa and $v = 0.30$.

2.23 At a point on the free surface of an alloy-steel ($E = 200$ GPa, $v = 0.30$) machine part normal strains of 1000×10^{-6}, 2000×10^{-6}, and 1200×10^{-6} were measured at respective angles of 0, 60, and 120° with respect to the x axis. Design considerations limit the maximum normal stress to 510 MPa, the maximum shearing stress to 275 MPa, the maximum normal strain to 2200×10^{-6}, and the maximum shear strain to 2500×10^{-6}. What is your evaluation of the design?

2.24 A thick-walled cylindrical pressure vessel will be used to store gas under a pressure of 100 MPa. During initial pressurization of the vessel, axial and hoop components of strain were measured on the inside and outside surfaces. On the inside surface the axial strain was 500×10^{-6} and the hoop strain was 750×10^{-6}. On the outside surface the axial strain was 500×10^{-6}, and the hoop strain was 100×10^{-6}. Determine the axial and hoop components of stress associated with these strains if $E = 200$ GPa and $v = 0.30$.

2.25 The cartesian components of stress at a point in a steel ($E = 200$ GPa, $v = 0.30$) machine part are as follows:

$$\sigma_{xx} = 140 \text{ MPa} \qquad \sigma_{yy} = -60 \text{ MPa} \qquad \sigma_{zz} = 70 \text{ MPa}$$

$$\tau_{xy} = 140 \text{ MPa} \qquad \tau_{yz} = 0 \qquad \tau_{zx} = 0$$

Determine the three principal strains, the principal-strain directions, and the maximum shearing strain.

2.26 Mohr's circle for stress and Mohr's circle for strain are convenient graphical methods for visualizing three-dimensional states of stress and strain at points in a stressed body. ($E = 200$ GPa, $v = 0.30$.) At a particular point in a body the three principal stresses are

$$\sigma_1 = 110 \text{ MPa} \qquad \sigma_2 = 50 \text{ MPa} \qquad \sigma_3 = -30 \text{ MPa}$$

(a) Sketch the three-dimensional Mohr's circle for stress at the point.

(b) Sketch the three-dimensional Mohr's circle for strain at the point.

(c) On a plane through the point the shearing stress $\tau = 70$ MPa. What normal stress must exist on this plane?

(d) On another plane through the point the shearing stress $\tau = 50$ MPa. Within what range of values must the normal stress associated with this plane fall?

(e) Along a line through the point (say the x axis), the normal strain is zero. What range of values may the shearing strain γ assume for different orientations of the x axis?

2.27 A thin rubber membrane is stretched in such a manner that the following uniform strain field is produced:

$$\epsilon_{xx} = 6000 \times 10^{-6} \qquad \epsilon_{yy} = -5000 \times 10^{-6} \qquad \gamma_{xy} = 3000 \times 10^{-6}$$

A rectangle is drawn on the membrane before stretching. How should the rectangle be oriented if the angles are to remain 90° during stretching?

2.28 A thin rectangular aluminum ($E = 70$ GPa, $v = \frac{1}{3}$) plate 75 by 100 mm is acted upon by a two-dimensional stress distribution which produces the following uniform strains in the plate:

$$\epsilon_{xx} = 2500 \times 10^{-6} \qquad \epsilon_{yy} = -500 \times 10^{-6} \qquad \gamma_{xy} = 1500 \times 10^{-6}$$

(a) Determine the changes in length of the diagonals of the plate.

(b) Determine the maximum shearing strain in the plate. Indicate on a sketch two of the initially perpendicular lines in the plate associated with this maximum shearing strain.

REFERENCES

1. Boresi, A. P., and P. P. Lynn: "Elasticity in Engineering Mechanics," chap. 2, Prentice-Hall, Inc., Englewood Cliffs, N.J., 1974.
2. Chou, P. C., and N. J. Pagano: "Elasticity," chaps. 2 and 3, D. Van Nostrand Company, Inc., Princeton, N.J., 1967.
3. Durelli, A. J., E. A. Phillips, and C. H. Tsao: "Introduction to the Theoretical and Experimental Analysis of Stress and Strain," chaps. 2 and 4, McGraw-Hill Book Company, New York, 1958.
4. Love, A. E. H.: "A Treatise on the Mathematical Theory of Elasticity," chaps. 1 and 3, Dover Publications, Inc., New York, 1944.
5. Sechler, E. E.: "Elasticity in Engineering," chaps. 4 and 5, John Wiley & Sons, Inc., New York, 1952.
6. Sokolnikoff, I. S.: "Mathematical Theory of Elasticity," 2d ed., chaps. 1 and 3, McGraw-Hill Book Company, New York, 1956.
7. Southwell, R. V.: "An Introduction to the Theory of Elasticity," chaps. 9 and 10, Oxford University Press, Fair Lawn, N.J., 1953.
8. Timoshenko, S. P., and J. N. Goodier: "Theory of Elasticity," 2d ed., chaps. 1, 8, and 9, McGraw-Hill Book Company, New York, 1951.

THREE

BASIC EQUATIONS AND
PLANE ELASTICITY THEORY

3.1 FORMULATION OF THE PROBLEM

In the general three-dimensional elasticity problem there are 15 unknown quantities which must be determined at every point in the body, namely, the 6 cartesian components of stress, the 6 cartesian components of strain, and the 3 components of displacement. Attempts can be made to obtain a solution to a given problem after the following quantities have been adequately defined:

1. The geometry of the body
2. The boundary conditions
3. The body-force field as a function of position
4. The elastic constants

In order to solve for the above-mentioned 15 unknown quantities, 15 independent equations are required. Three are provided by the stress equations of equilibrium [Eqs. (1.3)], six are provided by the strain-displacement relations [Eqs. (2.4)], and the remaining six can be obtained from the stress-strain expressions [Eqs. (2.16)].

A solution to an elasticity problem, in addition to satisfying these 15 equations, must also satisfy the boundary conditions. In other words, the stresses acting over the surface of the body must produce tractions which are equivalent to the loads being applied to the body. Boundary conditions are often classified to define the four different types of boundary-value problem listed below:

Type 1. If the displacements are prescribed over the entire boundary, the problem is classified as a type 1 boundary-value problem. As an example, consider a long, slender rod which is given an axial displacement, say, u and transverse displacements v and w. In this instance displacements are prescribed over the entire boundary of the rod.

Type 2. The most frequently encountered boundary-value problem is the type where normal and shearing forces are given over the entire surface. For instance, a sphere subjected to a uniform hydrostatic pressure has zero shearing stress and a normal stress equal to $-p$ on the surface and hence is a type 2 boundary-value problem.

Type 3. This is a mixed boundary-value problem where the normal and shearing forces are given over a portion of the boundary and the displacements are given over the remainder of the body. To illustrate this type of problem, consider the shrinking of a sleeve over a shaft. In the shrinking process a radial displacement is given to the sleeve at the interface between the shaft and the sleeve. On all other surfaces of the sleeve, both the normal and the shearing components of stress are zero.

Type 4. This type of boundary-value problem is the most general of the four considered. Over a portion of the surface, displacements are prescribed. Over a second portion of the surface, normal and shearing stresses are prescribed. Over a third portion of the surface, the normal component of displacement and the shearing component of stress are prescribed. Over a fourth portion of the surface, the shearing component of displacement and the normal component of stress are prescribed. Obviously, the first three types of problem can be regarded as special cases of this general fourth type.

One of the most difficult problems encountered in any experimental study is the design and construction of the loading fixture for applying the required displacements or tractions to the model being studied. The classifications given previously should be kept in mind when one designs the fixture. In general, it has been found that tractions cannot be adequately simulated by applying a displacement field to the model and vice versa. Type 1 and type 2 boundary-value problems are usually the easiest to approach experimentally. In general, type 3 and type 4 problems offer more difficulties in properly loading the model.

3.2 FIELD EQUATIONS

Thus far in the development four sets of field equations have been discussed, namely, the stress equations of equilibrium, the strain-displacement relations, the stress-strain expressions, and the equations of compatibility. Quite often two or more of these sets of equations can be combined to give a new set which may be more applicable to a specific problem. As an example, consider the six stress-displacement equations which can be obtained from the six stress-strain relations and the six strain-displacement equations by substituting Eqs. (2.4) into Eqs. (2.16).

$$\frac{\partial u}{\partial x} = \frac{1}{E}[\sigma_{xx} - v(\sigma_{yy} + \sigma_{zz})]$$

$$\frac{\partial v}{\partial y} = \frac{1}{E}[\sigma_{yy} - v(\sigma_{zz} + \sigma_{xx})]$$

$$\frac{\partial w}{\partial z} = \frac{1}{E}[\sigma_{zz} - v(\sigma_{xx} + \sigma_{yy})]$$

$$\frac{\partial u}{\partial y} + \frac{\partial v}{\partial x} = \frac{1}{\mu}\tau_{xy} \qquad \frac{\partial v}{\partial z} + \frac{\partial w}{\partial y} = \frac{1}{\mu}\tau_{yz} \qquad \frac{\partial w}{\partial x} + \frac{\partial u}{\partial z} = \frac{1}{\mu}\tau_{zx} \qquad (3.1)$$

It is interesting to note that the set of equations consisting of the stress equations of equilibrium [Eqs. (1.3)] and the stress-displacement relations [Eqs. (3.1)] are expressed as nine equations in terms of nine unknowns. The reduction in the number of unknowns from 15 to 9 was made possible by eliminating the strains.

The problem can be reduced further (from nine to three unknowns) if the stress equations of equilibrium [Eqs. (1.3)] are combined with the stress-displacement equations (3.1). The displacement equations of equilibrium obtained can be written as follows:

$$\nabla^2 u + \frac{1}{1-2v}\frac{\partial}{\partial x}\left(\frac{\partial u}{\partial x} + \frac{\partial v}{\partial y} + \frac{\partial w}{\partial z}\right) + \frac{1}{\mu}F_x = 0$$

$$\nabla^2 v + \frac{1}{1-2v}\frac{\partial}{\partial y}\left(\frac{\partial u}{\partial x} + \frac{\partial v}{\partial y} + \frac{\partial w}{\partial z}\right) + \frac{1}{\mu}F_y = 0 \qquad (3.2)$$

$$\nabla^2 w + \frac{1}{1-2v}\frac{\partial}{\partial z}\left(\frac{\partial u}{\partial x} + \frac{\partial v}{\partial y} + \frac{\partial w}{\partial z}\right) + \frac{1}{\mu}F_z = 0$$

where ∇^2 is the operator $\partial^2/\partial x^2 + \partial^2/\partial y^2 + \partial^2/\partial z^2$.

It is clear that a solution of the displacement equations of equilibrium will yield the three displacements u, v, w. Once the displacements are known, the six strains and the six stresses can easily be obtained by using Eqs. (2.4) to obtain the strains and Eqs. (2.16) to obtain the stresses.

Analytical solutions for three-dimensional elasticity problems are quite difficult to obtain, and the number of problems which have been solved in an exact fashion to date is surprisingly small. The most successful approach to date has been through the use of the Boussinesq-Popkovich stress functions, which are defined so as to satisfy Eq. (3.2). The development of this approach is somewhat involved and is therefore beyond the scope and objectives of this elementary treatment of the theory of elasticity. The interested student should consult the selected references at the end of the chapter for a detailed development of the Boussinesq-Popkovich stress-function approach.

Before this section is completed, the stress equations of compatibility will be developed since they are the basis for an important theorem regarding the dependence of stresses on the elastic constants. If the stress-strain relations [Eqs. (2.16)], the stress equations of equilibrium [Eqs. (1.3)], and the strain compatibility equations [Eqs. (2.10)] are combined, the six stress equations of compatibility are obtained as follows:

$$\nabla^2 \sigma_{xx} + \frac{1}{1+v} \frac{\partial^2}{\partial x^2} I_1 = -\frac{v}{1-v} \left(\frac{\partial F_x}{\partial x} + \frac{\partial F_y}{\partial y} + \frac{\partial F_z}{\partial z} \right) - 2 \frac{\partial F_x}{\partial x}$$

$$\nabla^2 \sigma_{yy} + \frac{1}{1+v} \frac{\partial^2}{\partial y^2} I_1 = -\frac{v}{1-v} \left(\frac{\partial F_x}{\partial x} + \frac{\partial F_y}{\partial y} + \frac{\partial F_z}{\partial z} \right) - 2 \frac{\partial F_y}{\partial y}$$

$$\nabla^2 \sigma_{zz} + \frac{1}{1+v} \frac{\partial^2}{\partial z^2} I_1 = -\frac{v}{1-v} \left(\frac{\partial F_x}{\partial x} + \frac{\partial F_y}{\partial y} + \frac{\partial F_z}{\partial z} \right) - 2 \frac{\partial F_z}{\partial z} \qquad (3.3)$$

$$\nabla^2 \tau_{xy} + \frac{1}{1+v} \frac{\partial^2}{\partial x\, \partial y} I_1 = -\left(\frac{\partial F_x}{\partial y} + \frac{\partial F_y}{\partial x} \right)$$

$$\nabla^2 \tau_{yz} + \frac{1}{1+v} \frac{\partial^2}{\partial y\, \partial z} I_1 = -\left(\frac{\partial F_y}{\partial z} + \frac{\partial F_z}{\partial y} \right)$$

$$\nabla^2 \tau_{zx} + \frac{1}{1+v} \frac{\partial^2}{\partial z\, \partial x} I_1 = -\left(\frac{\partial F_x}{\partial z} + \frac{\partial F_z}{\partial x} \right)$$

where I_1 is the first invariant of stress $\sigma_{xx} + \sigma_{yy} + \sigma_{zz}$ and F_x, F_y, F_z are the body-force intensities in the x, y, z directions, respectively.

If this system of six equations is solved for the six cartesian stress components, and if the boundary conditions are satisfied, the problem can be considered solved. Of great importance to the experimentalist is the appearance of elastic constants in Eqs. (3.3). Recall that equations of stress equilibrium did not contain elastic constants. Since only Poisson's ratio v appears in Eqs. (3.3), it follows that the stresses are independent of the modulus of elasticity E of the model material and can at most depend upon Poisson's ratio alone. Of course, this is true only for a simply connected body since the strain compatibility equations are valid only for this condition.

This independence of the stresses on the elastic modulus is very important in three-dimensional photoelasticity, where a low-modulus plastic model is used to simulate a metal prototype. Only the difference in Poisson's ratio between the model and the prototype is a source of error. The very large difference between the moduli of elasticity of the model and the prototype does not produce any significant errors in the determination of stresses using a three-dimensional photoelastic approach, provided the strains induced in the photoelastic model remain sufficiently small.

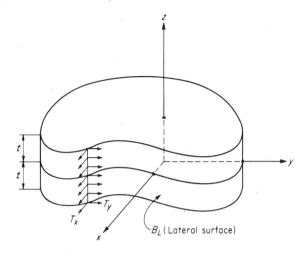

Figure 3.1 A body which may be considered for the plane-elasticity approach is bounded on the top and bottom by two parallel planes and is bounded laterally by any surface which is normal to the top and bottom planes.

3.3 THE PLANE ELASTIC PROBLEM

In the theory of elasticity there exists a special class of problems, known as *plane problems*, which can be solved more readily than the general three-dimensional problem since certain simplifying assumptions can be made in their treatment. The geometry of the body and the nature of the loading on the boundaries which permit a problem to be classified as a plane problem are as follows:

By definition a plane body consists of a region of uniform thickness bounded by two parallel planes and by any closed lateral surface B_L, as indicated by Fig. 3.1. Although the thickness of the body must be uniform, it need not be limited. It may be very thick or very thin; in fact, these two extremes represent the most desirable cases for this approach, as will be pointed out later.
In addition to the restrictions on the geometry of the body, the following restrictions are imposed on the loads applied to the plane body.

1. Body forces, if they exist, cannot vary through the thickness of the region, that is, $F_x = F_x(x, y)$ and $F_y = F_y(x, y)$. Furthermore, the body force in the z direction must equal zero.
2. The surface tractions or loads on the lateral boundary B_L must be in the plane of the model and must be uniformly distributed across the thickness, i.e., constant in the z direction. Hence, $T_x = T_x(x, y)$, $T_y = T_y(x, y)$, and $T_z = 0$.
3. No loads can be applied on the parallel planes bounding the top and bottom surfaces, that is, $\mathbf{T}_n = 0$ on $z = \pm t$.

   Once the geometry and loading have been defined, stresses can be determined by using either the plane-strain or the plane-stress approach. Usually the plane-strain approach is used when the body is very thick relative to its lateral dimensions. The plane-stress approach is employed when the body is relatively thin in relation to its lateral dimensions.

3.4 THE PLANE-STRAIN APPROACH

If it is assumed that the strains in the body are plane, i.e., the strains in the x and y directions are functions of x and y alone, and also that the strains in the z directions are equal to zero, the strain-displacement relation [Eqs. (2.4)] can be simplified as follows:

$$\epsilon_{xx} = \frac{\partial u}{\partial x} \qquad \epsilon_{yy} = \frac{\partial v}{\partial y} \qquad \epsilon_{zz} = \frac{\partial w}{\partial z} = 0$$

$$\gamma_{xy} = \frac{\partial u}{\partial y} + \frac{\partial v}{\partial x} \qquad \gamma_{yz} = \frac{\partial v}{\partial z} + \frac{\partial w}{\partial y} = 0 \qquad \gamma_{zx} = \frac{\partial w}{\partial x} + \frac{\partial u}{\partial z} = 0 \qquad (3.4)$$

Similarly, if Eqs. (3.4) are substituted into Eqs. (2.13), a reduced form of the stress-strain relations for the case of plane strain is obtained:

$$\sigma_{xx} = \lambda J_1 + 2\mu\epsilon_{xx} \qquad \sigma_{yy} = \lambda J_1 + 2\mu\epsilon_{yy} \qquad \sigma_{zz} = \lambda J_1$$

$$\tau_{xy} = \mu\gamma_{xy} \qquad \tau_{yz} = \tau_{zx} = 0 \qquad (3.5)$$

where $J_1 = \epsilon_{xx} + \epsilon_{yy}$. In addition, the stress equations of equilibrium [Eqs. (1.3)] reduce to

$$\frac{\partial \sigma_{xx}}{\partial x} + \frac{\partial \tau_{xy}}{\partial y} + F_x = 0 \qquad \frac{\partial \tau_{xy}}{\partial x} + \frac{\partial \sigma_{yy}}{\partial y} + F_y = 0 \qquad (3.6)$$

Any solution for a plane-strain problem must satisfy Eqs. (3.4) to (3.6) in addition to the boundary conditions on the lateral boundary B_L and the bounding planes. The boundary conditions on B_L can be expressed in terms of the stresses by referring to Eqs. (1.2), which give the x, y, and z components of the resultant-stress vector in terms of the cartesian components of stress. Thus, on B_L the following relations must be satisfied:

$$T_{nx} = \sigma_{xx} \cos{(n, x)} + \tau_{xy} \cos{(n, y)}$$

$$T_{ny} = \tau_{xy} \cos{(n, x)} + \sigma_{yy} \cos{(n, y)} \qquad (3.7)$$

$$T_{nz} = 0$$

where T_{nx}, T_{ny}, T_{nz} are the x, y, z components of the stresses applied to the body on surface B_L. Finally, on the two parallel bounding planes,

$$\mathbf{T}_n = 0 \qquad (3.8)$$

i.e., no tractions are applied to these surfaces; hence, τ_{yz}, τ_{zx}, σ_{zz} must be zero on these surfaces.

It is clear from Eqs. (3.5) that σ_{zz} will be equal to zero, as demanded by Eq. (3.8), only when the dilatation J_1 is equal to zero. In most problems J_1 will not be equal to zero; therefore, the solution will not be exact since the boundary conditions on the parallel planes are violated. In many problems this violation of the boundary conditions can be cleared by superimposing an equal and opposite distribution of σ_{zz} (residual solution) onto the original solution.

It is possible to obtain an exact solution to the residual problem only when σ_{zz} is a linear function of x and y. When σ_{zz} is nonlinear, an approximate solution based on Saint-Venant's principle† is often utilized. When the nonlinear distribution of σ_{zz} on the parallel boundaries is replaced by a linear distribution which is statically equivalent, the solution will be valid only in regions well removed from the parallel bounding planes. Thus, it is clear that the plane-strain approach is necessarily limited to the central regions of bodies such as shafts or dams which are very long, i.e., thick, relative to their lateral dimensions. In the central region of such a long body, the stresses σ_{xx}, σ_{yy}, and τ_{xy} can be found from the solution of the original problem since the superposition of the residual solution onto the original problem does not influence these stresses but only serves to make σ_{zz} vanish.

In this section the plane-strain approach has been discussed without indicating a method for solving for σ_{xx}, σ_{yy}, and τ_{xy}. This problem will be treated later in this chapter when the Airy's-stress-function approach is discussed. In this plane-strain section it is important for the student to understand the plane-strain assumption, why it usually leads to a violation of the boundary conditions on the two parallel planes, and finally how these undesired stresses can be removed from the planes by superimposing a statically equivalent linear stress system. Also quite important is Saint-Venant's principle, since an experimentalist in simulating loads often relies on this principle to permit simplification in the design of the loading fixtures.

3.5 PLANE STRESS

In the preceding section it was noted that the plane-strain method is limited to very long or thick bodies. In those cases where the body thickness is small relative to its lateral dimensions, it is advantageous to assume that

$$\sigma_{zz} = \tau_{yz} = \tau_{zx} = 0 \tag{3.9}$$

throughout the thickness of the plate. With this assumption the stress equations of equilibrium again reduce to

$$\frac{\partial \sigma_{xx}}{\partial x} + \frac{\partial \tau_{xy}}{\partial y} + F_x = 0 \qquad \frac{\partial \tau_{xy}}{\partial x} + \frac{\partial \sigma_{yy}}{\partial y} + F_y = 0 \tag{3.10}$$

† Saint-Venant's principle states that a system of forces acting over a small region of the boundary can be replaced by a statically equivalent system of forces without introducing appreciable changes in the distribution of stresses in regions well removed from the area of load application.

and the stress-strain relations [Eqs. (2.13)] become

$$\sigma_{xx} = \lambda J_1 + 2\mu\epsilon_{xx} \qquad \sigma_{yy} = \lambda J_1 + 2\mu\epsilon_{yy} \qquad \sigma_{zz} = \lambda J_1 + 2\mu\epsilon_{zz} = 0$$

$$\tau_{xy} = \mu\gamma_{xy} \qquad \tau_{yz} = \mu\gamma_{yz} = 0 \qquad \tau_{zx} = \mu\gamma_{zx} = 0 \tag{3.11}$$

From the third of Eqs. (3.11) the following relationship can be obtained:

$$\epsilon_{zz} = -\frac{\lambda}{\lambda + 2\mu}(\epsilon_{xx} + \epsilon_{yy}) \tag{a}$$

With this value of ϵ_{zz} the first strain invariant J_1 becomes

$$J_1 = \frac{2\mu}{\lambda + 2\mu}(\epsilon_{xx} + \epsilon_{yy}) \tag{b}$$

Substituting the value for J_1 given in Eq. (b) into Eqs. (3.11) yields

$$\sigma_{xx} = \frac{2\lambda\mu}{\lambda + 2\mu}(\epsilon_{xx} + \epsilon_{yy}) + 2\mu\epsilon_{xx}$$

$$\sigma_{yy} = \frac{2\lambda\mu}{\lambda + 2\mu}(\epsilon_{xx} + \epsilon_{yy}) + 2\mu\epsilon_{yy} \tag{3.12}$$

$$\tau_{xy} = \mu\gamma_{xy} \qquad \sigma_{zz} = \tau_{yz} = \tau_{zx} = 0$$

Unfortunately, in the general case σ_{xx}, σ_{yy}, and τ_{xy} are not independent of z, and thus the boundary conditions imposed on the boundary B_L cannot be rigorously satisfied. To overcome this difficulty, average stresses and displacements over the thickness are commonly used. If the body is relatively thin, these averages closely approximate the true boundary conditions on B_L. Average values for the stresses and displacements over the thickness of the body are obtained as follows:

$$\tilde{\sigma}_{xx} = \frac{1}{2t}\int_{-t}^{t} \sigma_{xx}\, dz \qquad \tilde{\sigma}_{yy} = \frac{1}{2t}\int_{-t}^{t} \sigma_{yy}\, dz \qquad \tilde{\tau}_{xy} = \frac{1}{2t}\int_{-t}^{t} \tau_{xy}\, dz$$

$$\tilde{u} = \frac{1}{2t}\int_{-t}^{t} u\, dz \qquad \tilde{v} = \frac{1}{2t}\int_{-t}^{t} v\, dz \tag{3.13}$$

The symbol $\sim$ over the stresses and displacements indicates average values. Substituting the average values of the stresses into Eqs. (1.2) gives the boundary conditions which must be satisfied on B_L:

$$T_{nx} = \tilde{\sigma}_{xx} \cos(n, x) + \tilde{\tau}_{xy} \cos(n, y)$$

$$T_{ny} = \tilde{\tau}_{xy} \cos(n, x) + \tilde{\sigma}_{yy} \cos(n, y) \tag{3.14}$$

If the equations which the plane-strain and the plane-stress solutions must satisfy are compared, it can be observed that they are identical except for the comparison between Eqs. (3.5) and (3.11). An examination of a typical equation from each of these sets,

$$\sigma_{xx} = \begin{cases} \lambda(\epsilon_{xx} + \epsilon_{yy}) + 2\mu\epsilon_{xx} & \text{plane strain} \\[2mm] \dfrac{2\lambda\mu}{\lambda + 2\mu}(\epsilon_{xx} + \epsilon_{yy}) + 2\mu\epsilon_{xx} & \text{plane stress} \end{cases}$$

indicates that they are identical except for the coefficients of the $\epsilon_{xx} + \epsilon_{yy}$ term. Since all other equations for the plane-stress and plane-strain solutions are identical, results from plane strain can be transformed into plane stress by letting

$$\lambda \rightarrow \frac{2\lambda\mu}{\lambda + 2\mu}$$

which is equivalent to letting

$$\frac{v}{1 - v} \rightarrow v \tag{3.15}$$

In a similar manner a plane-stress solution can be transformed into a plane-strain solution by letting

$$\frac{2\lambda\mu}{\lambda + 2\mu} \rightarrow \lambda$$

or

$$v \rightarrow \frac{v}{1 - v} \tag{3.16}$$

In the plane-stress approach it is generally assumed that

$$\sigma_{zz} = \tau_{yz} = \tau_{zx} = 0$$

and the unknown stresses σ_{xx}, σ_{yy}, and τ_{xy} will have a z dependence. As a result of this z dependence, the boundary conditions on B_L are violated. This difficulty can be eliminated and an approximate solution to the problem can be obtained by using average values for the stresses and displacements. Finally, it was shown that plane-stress and plane-strain solutions can be transformed from one case into the other by a simple replacement involving Poisson's ratio, as indicated in Eqs. (3.15) and (3.16).

3.6 AIRY'S STRESS FUNCTION

In the plane problem three unknowns σ_{xx}, σ_{yy}, and τ_{xy} must be determined which will satisfy the required field equations and boundary conditions. The most convenient sets of field equations to use in this determination are the two equations of equilibrium and one stress equation of compatibility.

The equilibrium equations in two dimensions are

$$\frac{\partial \sigma_{xx}}{\partial x} + \frac{\partial \tau_{xy}}{\partial y} + F_x = 0 \tag{3.17a}$$

$$\frac{\partial \tau_{xy}}{\partial x} + \frac{\partial \sigma_{yy}}{\partial y} + F_y = 0 \tag{3.17b}$$

The stress compatibility equation for the case of plane strain is

$$\nabla^2(\sigma_{xx} + \sigma_{yy}) = -\frac{2(\lambda + \mu)}{\lambda + 2\mu}\left(\frac{\partial F_x}{\partial x} + \frac{\partial F_y}{\partial y}\right) \tag{3.17c}$$

Suppose the body-force field is defined by $\Omega(x, y)$ so that the body-force intensities are given by

$$F_x = -\frac{\partial \Omega}{\partial x} \qquad F_y = -\frac{\partial \Omega}{\partial y} \tag{3.18}$$

Then by substituting Eqs. (3.18) into Eqs. (3.17) and noting that $2(\lambda + \mu)/(\lambda + 2\mu) = 1/(1 - v)$, it is apparent that

$$\frac{\partial \sigma_{xx}}{\partial x} + \frac{\partial \tau_{xy}}{\partial y} = \frac{\partial \Omega}{\partial x} \qquad \frac{\partial \tau_{xy}}{\partial x} + \frac{\partial \sigma_{yy}}{\partial y} = \frac{\partial \Omega}{\partial y} \tag{3.19}$$

$$\nabla^2\left(\sigma_{xx} + \sigma_{yy} - \frac{\Omega}{1 - v}\right) = 0$$

Equations (3.19) represent the three field equations which σ_{xx}, σ_{yy}, and τ_{xy} must satisfy.

Assume that the stresses can be represented by a stress function ϕ such that

$$\sigma_{xx} = \frac{\partial^2 \phi}{\partial y^2} + \Omega \qquad \sigma_{yy} = \frac{\partial^2 \phi}{\partial x^2} + \Omega \qquad \tau_{xy} = -\frac{\partial^2 \phi}{\partial x \, \partial y} \tag{3.20}$$

If Eqs. (3.20) are substituted into Eqs. (3.19), it can be seen that the two equations of equilibrium are exactly satisfied, and the last of Eqs. (3.19) gives

$$\nabla^4 \phi = -\frac{1 - 2v}{1 - v}\nabla^2 \Omega \tag{3.21}$$

Thus, equilibrium and compatibility are immediately satisfied if ϕ satisfies Eq. (3.21). The expression ϕ is known as Airy's stress function. If Eq. (3.21) is solved for ϕ, an expression containing x, y, and a number of constants will be obtained. The constants are evaluated from the boundary conditions given in Eqs. (3.17), and the stresses are computed from ϕ according to Eqs. (3.20). Of course, evaluation of ϕ from Eq. (3.21) produces stresses for the plane-strain case.

Stresses for the plane-stress case can be obtained by letting $v/(1 - v) \to v$, as indicated in Eq. (3.15). This substitution leads to

$$\nabla^4 \phi = -(1 - v) \nabla^2 \Omega \qquad (3.22)$$

which is valid for plane-stress problems.

It is important to note that if the body-force intensities are zero or constant, such as those encountered in a gravitational field, then

$$\nabla^2 \Omega = 0$$

and Eqs. (3.21) and (3.22) both become

$$\nabla^4 \phi = 0 \qquad (3.23a)$$

This is a biharmonic equation, which can also be written in the form

$$\frac{\partial^4 \phi}{\partial x^4} + 2 \frac{\partial^4 \phi}{\partial x^2 \, \partial y^2} + \frac{\partial^4 \phi}{\partial y^4} = 0 \qquad (3.23b)$$

Examination of this equation shows that ϕ and thus σ_{xx}, σ_{yy}, and τ_{xy} are independent of the elastic constants. This consideration is very important in two-dimensional photoelasticity since it indicates that the stresses obtained from a plastic model are identical to those in a metal prototype if the model is simply connected and subjected to a zero or a uniform body-force field. Differences in the values of the modulus of elasticity and Poisson's ratio between model and prototype do not influence the results for the stresses. There are exceptions to the simply connected restriction, however, which will be covered in a later chapter on photoelasticity.

3.7 AIRY'S STRESS FUNCTION IN CARTESIAN COORDINATES

Any Airy's stress function used in the solution of a plane problem must satisfy Eqs. (3.23a) and (3.23b) and provide stresses via Eqs. (3.20) which satisfy the defined boundary conditions. Some Airy's stress functions commonly used are polynomials in x and y. In this section, polynomials from the first to the fifth degree will be considered.

A. Airy's Stress Function in Terms of a First-Degree Polynomial

$$\phi_1 = a_1 x + b_1 y$$

It is clear from Eqs. (3.20) that

$$\sigma_{xx} = \sigma_{yy} = \tau_{xy} = 0 \qquad (3.24)$$

and that Eqs. (3.23a) and (3.23b) are satisfied. This function is suitable only for indicating a stress-free field and therefore is of little use in the solution of any problem.

B. Airy's Stress Function in Terms of a Second-Degree Polynomial

$$\phi_2 = a_2 x^2 + b_2 xy + c_2 y^2$$

From Eqs. (3.20) the stresses are

$$\sigma_{xx} = 2c_2 \qquad \sigma_{yy} = 2a_2 \qquad \tau_{xy} = -b_2 \tag{3.25}$$

Note that Eqs. (3.23a) and (3.23b) are satisfied and that the stress function ϕ_2 gives a uniform stress field over the entire body which is independent of x and y.

C. Airy's Stress Function in Terms of a Third-Degree Polynomial

$$\phi_3 = a_3 x^3 + b_3 x^2 y + c_3 xy^2 + d_3 y^3$$

Again, by use of Eqs. (3.20) the stresses are given by

$$\sigma_{xx} = 2c_3 x + 6d_3 y \qquad \sigma_{yy} = 6a_3 x + 2b_3 y \qquad \tau_{xy} = -2b_3 x - 2c_3 y \tag{3.26}$$

Equations (3.23a) and (3.23b) are satisfied unconditionally, and the stress function ϕ_3 provides a linearly varying stress field over the body.

D. Airy's Stress Function in Terms of a Fourth-Degree Polynomial

$$\phi_4 = a_4 x^4 + b_4 x^3 y + c_4 x^2 y^2 + d_4 xy^3 + e_4 y^4$$

From Eqs. (3.20) it is apparent that

$$\sigma_{xx} = 2c_4 x^2 + 6d_4 xy + 12e_4 y^2$$
$$\sigma_{yy} = 12a_4 x^2 + 6b_4 xy + 2c_4 y^2 \tag{3.27}$$
$$\tau_{xy} = -3b_4 x^2 - 4c_4 xy - 3d_4 y^2$$

When ϕ_4 is substituted into Eq. (3.23b), it should be noted that it is not unconditionally satisfied. In order for $\nabla^4 \phi = 0$ it is necessary that

$$e_4 = -\left(a_4 + \frac{c_4}{3}\right)$$

Substituting this equation into the relations for the stresses gives

$$\sigma_{xx} = 2c_4 x^2 + 6d_4 xy - 12a_4 y^2 - 4c_4 y^2$$

and σ_{yy} and τ_{xy} are unchanged. Thus, ϕ_4 yields a stress field which is a second-degree polynomial in x and y.

E. Airy's Stress Function in Terms of a Fifth-Degree Polynomial

$$\phi_5 = a_5 x^5 + b_5 x^4 y + c_5 x^3 y^2 + d_5 x^2 y^3 + e_5 xy^4 + f_5 y^5$$

Employing Eqs. (3.20) to solve for the stresses gives

$$\sigma_{xx} = 2c_5 x^3 + 6d_5 x^2 y + 12e_5 xy^2 + 20f_5 y^3$$
$$\sigma_{yy} = 20a_5 x^3 + 12b_5 x^2 y + 6c_5 xy^2 + 2d_5 y^3$$
$$\tau_{xy} = -4b_5 x^3 - 6c_5 x^2 y - 6d_5 xy^2 - 4e_5 y^3$$

Again, note that ϕ_5 must be subjected to certain conditions involving the constants e_5 and f_5. For Eqs. (3.23a) and (3.23b) to be satisfied, these conditions are

$$e_5 = -(5a_5 + c_5) \qquad f_5 = -\tfrac{1}{5}(b_5 + d_5)$$

Subject to the restrictive conditions listed above, the cartesian stress components become

$$\sigma_{xx} = 2c_5 x^3 + 6d_5 x^2 y - 12(5a_5 + c_5)xy^2 - 4(b_5 + d_5)y^3$$
$$\sigma_{yy} = 20a_5 x^3 + 12b_5 x^2 y + 6c_5 xy^2 + 2d_5 y^3 \qquad (3.28)$$
$$\tau_{xy} = -4b_5 x^3 - 6c_5 x^2 y - 6d_5 xy^2 + 4(5a_5 + c_5)y^3$$

Thus, it is clear that ϕ_5 yields a stress field which is a third-degree polynomial in x and y. It is possible to continue this procedure to ϕ_6, ϕ_7, etc., as long as Eqs. (3.23a) and (3.23b) are satisfied. It is also possible to add together two or more stress functions to form another, for example, $\phi^* = \phi_2 + \phi_3$. Thus, by simply adding terms or by eliminating terms from the stress function it is theoretically possible to build up any stress field that can be expressed as a function of x and y.

3.8 EXAMPLE PROBLEM

Airy's stress function expressed in cartesian coordinates can be employed to solve a particular class of two-dimensional problems where the boundaries of the body can be adequately represented by the cartesian reference frame. As an example, consider the simply supported beam with uniform loads shown in Fig. 3.2. An examination of the loading conditions indicates that

$$\sigma_{yy} = \begin{cases} \tau_{xy} = 0 & \text{at } y = \dfrac{-h}{2} \\[2mm] -q & \text{at } y = \dfrac{+h}{2} \end{cases} \qquad (a)$$

$$\tau_{xy} = 0 \qquad \text{at } y = \dfrac{+h}{2} \qquad (b)$$

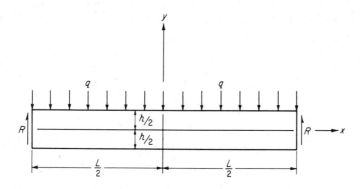

Figure 3.2 Simply supported beam of length L, height h, and unit depth subjected to a uniformly distributed load.

Also at $x = \pm L/2$

$$\int_{-h/2}^{h/2} \tau_{xy}\, dy = R = \frac{qL}{2} \qquad (c)$$

$$\int_{-h/2}^{h/2} \sigma_{xx}\, dy = 0 \qquad (d)$$

$$\int_{-h/2}^{h/2} \sigma_{xx} y\, dy = 0 \qquad (e)$$

Note that the bending moment (and consequently σ_{xx}) is a maximum at position $x = 0$ and decreases with a change in x in either the positive or the negative direction. This is possible only if the stress function contains even functions of x. Note also that σ_{yy} varies from zero at $y = -h/2$ to a maximum value of $-q$ at $y = +h/2$; thus the stress function must contain odd functions of y. From the stress functions listed in Sec. 3.7, the following even and odd functions can be selected to form a new stress function ϕ which satisfies the previously listed conditions.

$$\phi = a_2 x^2 + b_3 x^2 y + d_3 y^3 + a_4 x^4 + b_5 x^4 y + d_5 x^2 y^3 + f_5 y^5 \qquad (f)$$

This stress function ϕ must satisfy the equation $\nabla^4 \phi = 0$; hence

$$a_4 = 0 \qquad f_5 = -\tfrac{1}{5}(b_5 + d_5) \qquad (g)$$

From Eqs. (3.20) the cartesian stress components are

$$\sigma_{xx} = 6d_3 y + 6d_5 x^2 y - 4(b_5 + d_5)y^3$$
$$\sigma_{yy} = 2a_2 + 2b_3 y + 12b_5 x^2 y + 2d_5 y^3 \qquad (h)$$
$$\tau_{xy} = -2b_3 x - 4b_5 x^3 - 6d_5 xy^2$$

Examination of the boundary conditions shown in Eq. (a) indicates that σ_{yy} must be independent of x; hence the coefficient $b_5 = 0$. Consequently, Eqs. (h) reduce to

$$\sigma_{xx} = 6d_3 y + 6d_5 x^2 y - 4d_5 y^3$$
$$\sigma_{yy} = 2a_2 + 2b_3 y + 2d_5 y^3 \qquad (i)$$
$$\tau_{xy} = -2b_3 x - 6d_5 xy^2$$

The problem can be solved if the coefficients a_2, b_3, d_3, and d_5 can be selected so that the boundary conditions given in Eqs. (a) to (e) are satisfied. From Eqs. (a)

$$\sigma_{yy} = 0 = 2a_2 + 2b_3\left(-\frac{h}{2}\right) + 2d_5\left(-\frac{h}{2}\right)^3 \qquad a_2 - \frac{b_3 h}{2} - \frac{d_5 h^3}{8} = 0 \qquad (j)$$

and from Eqs. (b)

$$\sigma_{yy} = -q = 2a_2 + 2b_3\frac{h}{2} + 2d_5\left(\frac{h}{2}\right)^3 \qquad a_2 + \frac{b_3 h}{2} + \frac{d_5 h^3}{8} = -\frac{q}{2} \qquad (k)$$

Adding Eqs. (j) and (k) gives

$$a_2 = -\frac{q}{4} \qquad (l)$$

From Eqs. (a) and (b)

$$\tau_{xy} = 0 = -2x\left[b_3 + 3d_5\left(\pm\frac{h}{2}\right)^2\right] \qquad b_3 = -\frac{3}{4}h^2 d_5 \qquad (m)$$

Substituting Eqs. (m) into Eqs. (j),

$$\frac{d_5 h^3}{8} - \frac{3h^3 d_5}{8} = -\frac{q}{4} \qquad d_5 = \frac{q}{h^3} \qquad (n)$$

and

$$b_3 = -\frac{3}{4}\frac{q}{h} \qquad (o)$$

With the values of a_2, b_3, and d_5 given by Eqs. (l), (o), and (n), respectively, Eqs. (c) and (d) are identically satisfied. Equation (e) can be used to solve for the remaining unknown d_3.

$$\int_{-h/2}^{h/2}\left(6d_3 y^2 + \frac{3}{2}\frac{q}{h^3}L^2 y^2 - 4\frac{qy^4}{L^3}\right)dy = 0$$

$$\left[2d_3 y^3 + \frac{qL^2 y^3}{2h^3} - \frac{4qy^5}{5h^3}\right]_{-h/2}^{+h/2} = 0 \qquad (p)$$

Solving Eq. (p) for d_3 gives

$$d_3 = \frac{q}{240I}(2h^2 - 5L^2) \qquad (q)$$

where $I = h^3/12$ is the moment of inertia of the unit-width beam. Substituting Eqs. (q), (o), (n), and (l) into Eqs. (i) gives the final equations for the cartesian components of stress:

$$\sigma_{xx} = \frac{q}{8I}(4x^2 - L^2)y + \frac{q}{60I}(3h^2y - 20y^3)$$

$$\sigma_{yy} = \frac{q}{24I}(4y^3 - 3h^2y - h^3) \qquad (r)$$

$$\tau_{xy} = \frac{qx}{8I}(h^2 - 4y^2)$$

The conventional strength-of-materials solution for this problem, namely, that $\sigma_{xx} = My/I$, gives

$$\sigma_{xx} = \frac{q}{8I}(4x^2 - L^2)y \qquad (s)$$

which is identical with the first term of the relation given for σ_{xx} in Eqs. (r). The second term, $(q/60I)(3h^2y - 20y^3)$, is a correction term for the strength-of-materials solution. In the strength-of-materials approach, recall that it is assumed that plane sections remain plane after bending. This is not exactly true, and as a consequence the solution obtained lacks the correction term shown above. It is clear that the correction term is small when $L \gg h$, and the strength-of-materials solution will be sufficiently accurate.

This simple example illustrates how elementary elasticity theory can be employed to extend the student's understanding of the distribution of stresses in simple two-dimensional problems. Other examples are included in the exercises at the end of this chapter.

3.9 TWO-DIMENSIONAL PROBLEMS IN POLAR COORDINATES

In Sec. 3.6 the Airy's-stress-function approach to the solution of two-dimensional elasticity problems in cartesian coordinates was developed. This method was then applied to solve an elementary problem which was well suited to the cartesian reference frame. In many problems, however, the geometry of the body does not lend itself to the use of a cartesian coordinate system, and it is more expeditious to work with a different system. A large class of problems (such as circular rings, curved beams, and half-planes) can be solved by employing a commonly used system, the polar coordinate system. In any elasticity problem the proper choice of the coordinate system is extremely important since this choice establishes the complexity of the mathematical expressions employed to satisfy the field equations and the boundary conditions.

In order to solve two-dimensional elasticity problems by employing a polar-coordinate reference frame, the equations of equilibrium, the definition of Airy's stress function, and one of the stress equations of compatibility must be reestablished in terms of polar coordinates. On the following pages the equations of equilibrium will be derived by considering a polar element instead of a cartesian element. The equations for the polar components of stress in terms of Airy's stress function as well as the stress equation of compatibility will be transformed from cartesian to polar coordinates. Finally, a set of stress functions is developed which satisfies the stress equation of compatibility.

The stress equations of equilibrium in polar coordinates can be derived from the free-body diagram of the polar element shown in Fig. 3.3. The element is assumed to be very small. The average values of the normal and shearing stresses which act on surface 1 are denoted by σ_{rr} and $\tau_{r\theta}$, respectively. Since the stresses may vary as a function of r, values of the normal and shearing stresses on surface 3 are given by $\sigma_{rr} + (\partial \sigma_{rr}/\partial r)\, dr$ and $\tau_{r\theta} + (\partial \tau_{r\theta}/\partial r)\, dr$. Similarly, the average values of the normal and shearing stresses which act on surface 2 are given by $\sigma_{\theta\theta}$ and $\tau_{r\theta}$. Since the stresses may also vary as a function of θ, values of the normal and shearing stresses on surface 4 are $\sigma_{\theta\theta} + (\partial \sigma_{\theta\theta}/\partial \theta)\, d\theta$ and $\tau_{r\theta} + (\partial \tau_{r\theta}/\partial \theta)\, d\theta$.

For a polar element of unit thickness to be in a state of equilibrium the sum of all forces in the radial r and tangential θ directions must equal zero. Summing forces first in the radial direction and considering the body-force intensity F_r gives the equation of equilibrium

$$\left(\sigma_{rr} + \frac{\partial \sigma_{rr}}{\partial r}\, dr\right)(r + dr)\, d\theta - \sigma_{rr} r\, d\theta - \left[\sigma_{\theta\theta}\, dr + \left(\sigma_{\theta\theta} + \frac{\partial \sigma_{\theta\theta}}{\partial \theta}\, d\theta\right) dr\right]\frac{d\theta}{2}$$

$$+ \left(\tau_{r\theta} + \frac{\partial \tau_{r\theta}}{\partial \theta}\, d\theta - \tau_{r\theta}\right) dr + F_r r\, d\theta\, dr = 0 \qquad (a)$$

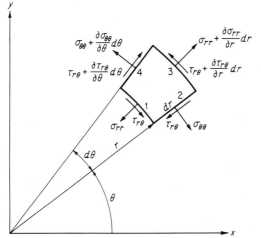

Figure 3.3 Polar element of unit depth showing the stresses acting on the four faces.

Dividing Eq. (*a*) by *dr dθ* and simplifying gives

$$\frac{\partial \sigma_{rr}}{\partial r} dr - \frac{\partial \sigma_{\theta\theta}}{\partial \theta} \frac{d\theta}{2} + \sigma_{rr} + \frac{\partial \sigma_{rr}}{\partial r} r - \sigma_{\theta\theta} + \frac{\partial \tau_{r\theta}}{\partial \theta} + F_r r = 0 \tag{b}$$

If the element is made infinitely small by permitting *dr* and *dθ* each to approach zero, the first two terms in Eq. (*b*) also approach zero and the expression can be rewritten as

$$\frac{\partial \sigma_{rr}}{\partial r} + \frac{1}{r} \frac{\partial \tau_{r\theta}}{\partial \theta} + \frac{1}{r} (\sigma_{rr} - \sigma_{\theta\theta}) + F_r = 0 \tag{3.29a}$$

The equation of equilibrium in the tangential direction can be derived in the same manner if the forces acting in the *θ* direction on the polar element are summed and set equal to zero. Hence

$$\frac{1}{r} \frac{\partial \sigma_{\theta\theta}}{\partial \theta} + \frac{\partial \tau_{r\theta}}{\partial r} + \frac{2\tau_{r\theta}}{r} + F_\theta = 0 \tag{3.29b}$$

Equations (3.29*a*) and (3.29*b*) represent the equations of equilibrium in polar coordinates. They are analogous to the equations of equilibrium in cartesian coordinates presented in Eqs. (3.17*a*) and (3.17*b*). Any solution to an elasticity problem must satisfy these field equations.

3.10 TRANSFORMATION OF THE EQUATION $\nabla^4\phi = 0$ INTO POLAR COORDINATES

In the coverage of Airy's stress function given in Sec. 3.6 it was shown that the stress function *φ* had to satisfy the biharmonic equation $\nabla^4\phi = 0$, provided the body forces are zero or constants. In polar coordinates the stress function must satisfy this same equation; however, the definition of the ∇^4 operator must be modified to suit the polar-coordinate system. This modification may be accomplished by transforming the ∇^4 operator from the cartesian system to the polar system.

In transforming from cartesian coordinates to polar coordinates, recall that

$$r^2 = x^2 + y^2 \qquad \theta = \arctan \frac{y}{x} \tag{3.30}$$

where *r* and *θ* are defined in Fig. 3.3.

Differentiating Eqs. (3.30) gives

$$\frac{\partial r}{\partial x} = \frac{x}{r} = \cos\theta \qquad \frac{\partial r}{\partial y} = \frac{y}{r} = \sin\theta$$

$$\frac{\partial \theta}{\partial x} = -\frac{y}{r^2} = -\frac{\sin\theta}{r} \qquad \frac{\partial \theta}{\partial y} = \frac{x}{r^2} = \frac{\cos\theta}{r} \tag{3.31}$$

The form of the ∇^4 operator in cartesian coordinates is

$$\nabla^4\phi = \left(\frac{\partial^2}{\partial x^2} + \frac{\partial^2}{\partial y^2}\right)\left(\frac{\partial^2\phi}{\partial x^2} + \frac{\partial^2\phi}{\partial y^2}\right)$$

Individual elements of this expression can be transformed by employing Eqs. (3.30) and (3.31) as follows. If it is assumed that ϕ is a function of r and θ,

$$\frac{\partial\phi}{\partial x} = \frac{\partial\phi}{\partial r}\frac{\partial r}{\partial x} + \frac{\partial\phi}{\partial\theta}\frac{\partial\theta}{\partial x} \tag{a}$$

$$\frac{\partial^2\phi}{\partial x^2} = \frac{\partial\phi}{\partial r}\frac{\partial^2 r}{\partial x^2} + \left(\frac{\partial r}{\partial x}\right)^2\frac{\partial^2\phi}{\partial r^2} + 2\frac{\partial^2\phi}{\partial r\,\partial\theta}\frac{\partial r}{\partial x}\frac{\partial\theta}{\partial x} + \frac{\partial\phi}{\partial\theta}\frac{\partial^2\theta}{\partial x^2} + \left(\frac{\partial\theta}{\partial x}\right)^2\frac{\partial^2\phi}{\partial\theta^2} \tag{b}$$

Substituting the equalities given in Eqs. (3.31) into Eq. (b) yields

$$\frac{\partial^2\phi}{\partial x^2} = \frac{\sin^2\theta}{r}\frac{\partial\phi}{\partial r} + \cos^2\theta\frac{\partial^2\phi}{\partial r^2} - \frac{\sin 2\theta}{r}\frac{\partial^2\phi}{\partial r\,\partial\theta}$$

$$+ \frac{\sin 2\theta}{r^2}\frac{\partial\phi}{\partial\theta} + \frac{\sin^2\theta}{r^2}\frac{\partial^2\phi}{\partial\theta^2} \tag{3.32a}$$

Following the same procedure makes it clear that

$$\frac{\partial^2\phi}{\partial y^2} = \frac{\cos^2\theta}{r}\frac{\partial\phi}{\partial r} + \sin^2\theta\frac{\partial^2\phi}{\partial r^2} + \frac{\sin 2\theta}{r}\frac{\partial^2\phi}{\partial r\,\partial\theta}$$

$$- \frac{\sin 2\theta}{r^2}\frac{\partial\phi}{\partial\theta} + \frac{\cos^2\theta}{r^2}\frac{\partial^2\phi}{\partial\theta^2} \tag{3.32b}$$

$$\frac{\partial^2\phi}{\partial x\,\partial y} = -\frac{\sin\theta\cos\theta}{r}\frac{\partial\phi}{\partial r} + \sin\theta\cos\theta\frac{\partial^2\phi}{\partial r^2} + \frac{\cos 2\theta}{r}\frac{\partial^2\phi}{\partial r\,\partial\theta}$$

$$- \frac{\cos 2\theta}{r^2}\frac{\partial\phi}{\partial\theta} - \frac{\sin\theta\cos\theta}{r^2}\frac{\partial^2\phi}{\partial\theta^2} \tag{3.32c}$$

Adding Eqs. (3.32a) and (3.32b) gives

$$\frac{\partial^2\phi}{\partial x^2} + \frac{\partial^2\phi}{\partial y^2} = \frac{\partial^2\phi}{\partial r^2} + \frac{1}{r}\frac{\partial\phi}{\partial r} + \frac{1}{r^2}\frac{\partial^2\phi}{\partial\theta^2} \tag{3.33}$$

Furthermore, it is easily seen that

$$\nabla^4\phi = \left(\frac{\partial^2}{\partial x^2} + \frac{\partial^2}{\partial y^2}\right)\left(\frac{\partial^2\phi}{\partial x^2} + \frac{\partial^2\phi}{\partial y^2}\right)$$

$$= \left(\frac{\partial^2}{\partial r^2} + \frac{1}{r}\frac{\partial}{\partial r} + \frac{1}{r^2}\frac{\partial^2}{\partial\theta^2}\right)\left(\frac{\partial^2\phi}{\partial r^2} + \frac{1}{r}\frac{\partial\phi}{\partial r} + \frac{1}{r^2}\frac{\partial^2\phi}{\partial\theta^2}\right) = 0 \tag{3.34}$$

Equation (3.34) is the stress equation of compatibility in terms of Airy's stress function referred to a polar coordinate system.

3.11 POLAR COMPONENTS OF STRESS IN TERMS OF AIRY'S STRESS FUNCTION

By referring to the two-dimensional equations of stress transformation [Eqs. (1.11)], expressions can be obtained which relate the polar stress components σ_{rr}, $\sigma_{\theta\theta}$, and $\tau_{r\theta}$ to the cartesian stress components σ_{xx}, σ_{yy}, and τ_{xy} as follows:

$$\sigma_{rr} = \sigma_{xx} \cos^2 \theta + \sigma_{yy} \sin^2 \theta + \tau_{xy} \sin 2\theta$$

$$\sigma_{\theta\theta} = \sigma_{yy} \cos^2 \theta + \sigma_{xx} \sin^2 \theta - \tau_{xy} \sin 2\theta \qquad (3.35)$$

$$\tau_{r\theta} = (\sigma_{yy} - \sigma_{xx}) \sin \theta \cos \theta + \tau_{xy} \cos 2\theta$$

If Eqs. (3.20) are substituted into Eqs. (3.35) and Ω set equal to zero (which is equivalent to setting both F_x and F_y equal to zero), then

$$\sigma_{rr} = \frac{\partial^2 \phi}{\partial y^2} \cos^2 \theta + \frac{\partial^2 \phi}{\partial x^2} \sin^2 \theta - \frac{\partial^2 \phi}{\partial x\, \partial y} \sin 2\theta$$

$$\sigma_{\theta\theta} = \frac{\partial^2 \phi}{\partial x^2} \cos^2 \theta + \frac{\partial^2 \phi}{\partial y^2} \sin^2 \theta + \frac{\partial^2 \phi}{\partial x\, \partial y} \sin 2\theta \qquad (3.36)$$

$$\tau_{r\theta} = \left(\frac{\partial^2 \phi}{\partial x^2} - \frac{\partial^2 \phi}{\partial y^2} \right) \sin \theta \cos \theta - \frac{\partial^2 \phi}{\partial x\, \partial y} \cos 2\theta$$

If the results from Eqs. (3.32a) to (3.32c) are substituted into Eqs. (3.36), the polar components of stress in terms of Airy's stress function are obtained:

$$\sigma_{rr} = \frac{1}{r} \frac{\partial \phi}{\partial r} + \frac{1}{r^2} \frac{\partial^2 \phi}{\partial \theta^2} \qquad \sigma_{\theta\theta} = \frac{\partial^2 \phi}{\partial r^2}$$

$$\tau_{r\theta} = \frac{1}{r^2} \frac{\partial \phi}{\partial \theta} - \frac{1}{r} \frac{\partial^2 \phi}{\partial r\, \partial \theta} \qquad (3.37)$$

When Airy's stress function ϕ in polar coordinates has been established, these relations can be employed to determine the stress field as a function of r and θ.

3.12 FORMS OF AIRY'S STRESS FUNCTION IN POLAR COORDINATES

The equation $\nabla^4 \phi = 0$ is a fourth-order biharmonic partial differential equation which can be reduced to an ordinary fourth-order differential equation by using a separation-of-variables technique, where

$$\phi^{(n)} = R_n(r) \begin{vmatrix} \cos n\theta \\ \sin n\theta \end{vmatrix}$$

The resulting differential equation is an Euler type which yields four different stress functions upon solution. These stress functions are tabulated, together with the stress and displacement distributions which they provide, on the following pages.

One of the stress functions obtained can be expressed in the following form:

$$\phi^{(0)}(r) = a_0 + b_0 \ln r + c_0 r^2 + d_0 r^2 \ln r \qquad (3.38a)$$

By using Eqs. (3.37), the stresses associated with this particular stress function can be expressed as

$$\sigma_{rr} = \frac{b_0}{r^2} + 2c_0 + d_0(1 + 2 \ln r)$$

$$\sigma_{\theta\theta} = -\frac{b_0}{r^2} + 2c_0 + d_0(3 + 2 \ln r) \qquad \tau_{r\theta} = 0 \qquad (3.38b)$$

The displacements associated with this function can be determined by integrating the stress displacement relations, giving

$$u_r = \frac{1}{E}\left[-(1 + v)\frac{b_0}{r} + 2(1 - v)c_0 r + 2(1 - v)d_0 r \ln r - (1 + v)d_0 r \right]$$

$$+ \alpha_2 \cos \theta + \alpha_3 \sin \theta$$

$$u_\theta = \frac{1}{E}(4d_0 r\theta) - \alpha_1 r - \alpha_2 \sin \theta + \alpha_3 \cos \theta \qquad (3.38c)$$

where u_r and u_θ are the displacements in the radial and circumferential directions, respectively. The terms containing α_1, α_2, and α_3 are associated with rigid-body displacements.

It should be noted that the stresses in this solution are independent of θ; hence, the stress function $\phi^{(0)}$ should be employed to solve problems which have rotational symmetry.

One of the other stress functions and the stresses and displacements associated with it can be expressed as follows:

$$\phi^{(1)} = \left(a_1 r + \frac{b_1}{r} + c_1 r^3 + d_1 r \ln r \right)\begin{Bmatrix} \sin \theta \\ \cos \theta \end{Bmatrix}$$

$$\sigma_{rr} = \left(-\frac{2b_1}{r^3} + 2c_1 r + \frac{d_1}{r} \right)\begin{Bmatrix} \sin \theta \\ \cos \theta \end{Bmatrix}$$

$$\sigma_{\theta\theta} = \left(\frac{2b_1}{r^3} + 6c_1 r + \frac{d_1}{r} \right)\begin{Bmatrix} \sin \theta \\ \cos \theta \end{Bmatrix}$$

$$\tau_{r\theta} = \left(-\frac{2b_1}{r^3} + 2c_1 r + \frac{d_1}{r} \right)\begin{Bmatrix} -\cos \theta \\ \sin \theta \end{Bmatrix}$$

$$u_r = \frac{1}{E}\left\{\left[(1+v)\frac{b_1}{r^2} + (1-3v)c_1 r^2 - (1+v)d_1\right.\right. \tag{3.39}$$

$$+ (1-v)d_1 \ln r\Big]\begin{Bmatrix}\sin\theta\\\cos\theta\end{Bmatrix} - (2d_1\theta)\begin{Bmatrix}\cos\theta\\-\sin\theta\end{Bmatrix}\right\} + \alpha_2\cos\theta + \alpha_3\sin\theta$$

$$u_\theta = \frac{1}{E}\left\{\left[-(1+v)\frac{b_1}{r^2} - (5+v)c_1 r^2 + (1-v)d_1 \ln r\right]\begin{Bmatrix}\cos\theta\\-\sin\theta\end{Bmatrix}\right.$$

$$+ (2d_1\theta)\begin{Bmatrix}\sin\theta\\\cos\theta\end{Bmatrix}\right\} - \alpha_1 r - \alpha_2\sin\theta + \alpha_3\cos\theta$$

The third stress function of interest and its associated stresses and displacements are as follows:

$$\phi^{(n)} = (a_n r^n + b_n r^{-n} + c_n r^{2+n} + d_n r^{2-n})\begin{Bmatrix}\sin n\theta\\\cos n\theta\end{Bmatrix}$$

$$\sigma_{rr} = [a_n(n-n^2)r^{n-2} - b_n(n+n^2)r^{-n-2} + c_n(2+n-n^2)r^n$$

$$+ d_n(2-n-n^2)r^{-n}]\begin{Bmatrix}\sin n\theta\\\cos n\theta\end{Bmatrix}$$

$$\sigma_{\theta\theta} = [a_n(n^2-n)r^{n-2} + b_n(n^2+n)r^{-n-2} + c_n(2+3n+n^2)r^n$$

$$+ d_n(2-3n+n^2)r^{-n}]\begin{Bmatrix}\sin n\theta\\\cos n\theta\end{Bmatrix}$$

$$\tau_{r\theta} = [a_n(n^2-n)r^{n-2} - b_n(n+n^2)r^{-n-2} + c_n(n+n^2)r^n$$

$$+ d_n(n-n^2)r^{-n}]\begin{Bmatrix}-\cos n\theta\\\sin n\theta\end{Bmatrix} \tag{3.40}$$

$$u_r = \frac{1}{E}\{-a_n(1+v)nr^{n-1} + b_n(1+v)nr^{-n-1}$$

$$+ c_n[4-(1+v)(2+n)]r^{n+1}$$

$$+ d_n[4-(1+v)(2-n)]r^{-n+1}\}\begin{Bmatrix}\sin n\theta\\\cos n\theta\end{Bmatrix} + \alpha_2\cos\theta + \alpha_3\sin\theta$$

$$u_\theta = \frac{1}{E}\{-a_n(1+v)nr^{n-1} - b_n(1+v)nr^{-n-1}$$

$$- c_n[4+(1+v)n]r^{n+1} + d_n[4-(1+v)n]r^{-n+1}\}\begin{Bmatrix}\cos n\theta\\-\sin n\theta\end{Bmatrix}$$

$$- \alpha_1 r - \alpha_2\sin\theta + \alpha_3\cos\theta$$

For the stress function $\phi^{(n)}$ the value of n can be greater than or equal to 2 (that is, $n \geq 2$).

The fourth stress function of interest and the associated stresses and displacements are expressed as follows:

$$\phi^{(*)} = a_* \theta + b_* r^2 \theta + c_* r\theta \sin\theta + d_* r\theta \cos\theta$$

$$\sigma_{rr} = 2b_* \theta + 2c_* \frac{\cos\theta}{r} - 2d_* \frac{\sin\theta}{r}$$

$$\sigma_{\theta\theta} = 2b_* \theta$$

$$\tau_{r\theta} = \frac{a_*}{r^2} - b_* \tag{3.41}$$

$$u_r = \frac{1}{E}[2(1-v)b_* r\theta + (1-v)c_* \theta \sin\theta + 2c_* \ln r \cos\theta$$

$$+ (1-v)d_* \theta \cos\theta - 2d_* \ln r \sin\theta] + \alpha_2 \cos\theta + \alpha_3 \sin\theta$$

$$u_\theta = \frac{1}{E}\Bigg[-(1+v)\frac{a_*}{r} + (3-v)b_* r - 4b_* r \ln r$$

$$- (1+v)c_* \sin\theta - 2c_* \ln r \sin\theta + (1-v)c_* \theta \cos\theta$$

$$- (1+v)d_* \cos\theta - 2d_* \ln r \cos\theta - (1-v)d_* \theta \sin\theta\Bigg]$$

$$- \alpha_1 r - \alpha_2 \sin\theta + \alpha_3 \cos\theta$$

In the example problems which follow, the stress functions previously listed will be employed to determine the stresses and displacements for problems which lend themselves to polar coordinates. As the selection of the stress function is often the most difficult phase of the problem, particular emphasis will be placed on the reasoning behind the selection.

3.13 STRESSES AND DISPLACEMENTS IN A CIRCULAR CYLINDER SUBJECTED TO INTERNAL AND EXTERNAL PRESSURE

Consider the long hollow cylinder shown in Fig. 3.4, which is subjected to an internal pressure p_i and an external pressure p_o. The inside and outside radii of the cylinder are denoted as a and b, respectively.

As stated previously, the first step in the solution of an elasticity problem after the geometry of the body has been defined is to establish the boundary conditions. For the problem under consideration these conditions can be listed as follows:

$$\sigma_{rr} = -p_i \qquad \tau_{r\theta} = 0 \qquad \text{at } r = a$$

$$\sigma_{rr} = -p_o \qquad \tau_{r\theta} = 0 \qquad \text{at } r = b \tag{a}$$

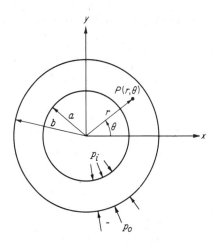

Figure 3.4 Circular cylinder subjected to internal and external pressures.

An examination of the boundary conditions indicates that they are independent of θ; hence the four stress functions $\phi^{(0)}$, $\phi^{(1)}$, $\phi^{(n)}$, and $\phi^{(*)}$ should be inspected to determine which will provide a stress field independent of θ. The stress function $\phi^{(0)}$ given in Eq. (3.38a) yields stresses which satisfy this requirement, as shown below:

$$\sigma_{rr} = \frac{b_0}{r^2} + 2c_0 + d_0(1 + 2 \ln r)$$

$$\sigma_{\theta\theta} = -\frac{b_0}{r^2} + 2c_0 + d_0(3 + 2 \ln r) \qquad (b)$$

$$\tau_{r\theta} = 0$$

Equations (b) will provide the desired solution to the problem if the constants b_0, c_0, and d_0 can be determined so that the boundary conditions given in Eqs. (a) are satisfied.

An examination of Eqs. (b) indicates that the condition $\tau_{r\theta} = 0$ throughout the body satisfies part of the boundary conditions. From symmetry considerations it is also obvious that both u_r and u_θ must be independent of θ. This condition can be satisfied only if $d_0 = 0$ in Eqs. (3.38c). The two remaining constants b_0 and c_0 can be evaluated by using the remaining boundary conditions in Eqs. (a):

$$\sigma_{rr} = -p_i = \frac{b_0}{a^2} + 2c_0 \qquad \sigma_{rr} = -p_o = \frac{b_0}{b^2} + 2c_0 \qquad (c)$$

Solving Eqs. (c) for b_0 and c_0 yields

$$b_0 = \frac{a^2 b^2 (p_o - p_i)}{b^2 - a^2} \qquad c_0 = \frac{a^2 p_i - b^2 p_o}{2(b^2 - a^2)}$$

These values when substituted into Eqs. (3.38) provide the required solution.

$$\sigma_{rr} = \frac{a^2b^2(p_o - p_i)}{(b^2 - a^2)r^2} + \frac{a^2p_i - b^2p_o}{b^2 - a^2}$$

$$\sigma_{\theta\theta} = -\frac{a^2b^2(p_o - p_i)}{(b^2 - a^2)r^2} + \frac{a^2p_i - b^2p_o}{b^2 - a^2}$$

$$\tau_{r\theta} = 0 \tag{3.42}$$

$$u_r = \frac{1}{E}\left[-(1 + v)\frac{a^2b^2(p_o - p_i)}{(b^2 - a^2)r} + (1 - v)\frac{a^2p_i - b^2p_o}{b^2 - a^2}r\right]$$

$$u_\theta = 0$$

Equations (3.42) give the stresses and displacements at a point $P(r, \theta)$ in the cylinder if the two pressures, the radii, and the elastic constants are known. Three special cases of this problem are of interest.

Case 1: External pressure equals zero Setting $p_o = 0$ in Eqs. (3.42) leads to

$$\sigma_{rr} = \frac{a^2p_i}{b^2 - a^2}\left(1 - \frac{b^2}{r^2}\right) \qquad \sigma_{\theta\theta} = \frac{a^2p_i}{b^2 - a^2}\left(1 + \frac{b^2}{r^2}\right) \qquad \tau_{r\theta} = 0$$

$$u_r = \frac{a^2p_i}{Er(b^2 - a^2)}[(1 + v)b^2 + (1 - v)r^2] \qquad u_\theta = 0 \tag{3.43}$$

This special case is often encountered when dealing with stresses in piping systems or pressure vessels.

Case 2: Internal pressure equals zero Setting $p_i = 0$ in Eqs. (3.42) leads to

$$\sigma_{rr} = \frac{b^2p_o}{b^2 - a^2}\left(\frac{a^2}{r^2} - 1\right) \qquad \sigma_{\theta\theta} = -\frac{b^2p_o}{b^2 - a^2}\left(\frac{a^2}{r^2} + 1\right) \qquad \tau_{r\theta} = 0$$

$$u_r = -\frac{b^2p_o}{Er(b^2 - a^2)}[(1 + v)a^2 + (1 - v)r^2] \qquad u_\theta = 0 \tag{3.44}$$

When external pressure is applied to a cylindrical shell, the problem of buckling should also be considered.

Case 3: External pressure on a solid circular cylinder When one sets $a = 0$ in Eqs. (3.44), the hole in the cylinder vanishes and the stresses become

$$\sigma_{rr} = \sigma_{\theta\theta} = -p_o \qquad \tau_{r\theta} = 0$$

$$u_r = -\frac{1 - v}{E}p_o r \qquad u_\theta = 0 \tag{3.45}$$

3.14 STRESS DISTRIBUTION IN A THIN, INFINITE PLATE WITH A CIRCULAR HOLE SUBJECTED TO UNIAXIAL TENSILE LOADS

A thin plate of infinite length and width with a circular hole is shown in Fig. 3.5. The plate is subjected to a uniform tensile-type load which produces a uniform stress σ_0 in the y direction at $r = \infty$. The distribution of the stresses about the hole, along the x axis, and along the y axis can be determined by using the Airy's-stress-function approach.

The boundary conditions which must be satisfied are

$$\sigma_{rr} = \tau_{r\theta} = 0 \qquad \text{at } r = a \qquad (a)$$

$$\sigma_{yy} = \sigma_0 \qquad \text{at } r \to \infty$$

$$\sigma_{xx} = \tau_{xy} = 0 \qquad \text{at } r \to \infty \qquad (b)$$

Selection of a stress function for this particular problem is difficult since none of the four functions previously tabulated is satisfactory. In order to overcome this difficulty, a method of superposition is commonly used which employs two different stress functions. The first function is selected such that the stresses associated with it satisfy the boundary conditions at $r \to \infty$ but in general violate the conditions on the boundary of the hole. The second stress function must have associated stresses which cancel the stresses on the boundary of the hole without influencing the stresses at $r \to \infty$. An illustration of this superposition process is presented in Fig. 3.6.

The boundary conditions at $r \to \infty$ can be satisfied by the uniform stress field associated with the stress function ϕ_2 in Eqs. (3.25). For the case of uniaxial

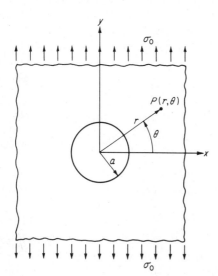

Figure 3.5 Thin, infinite plate with a circular hole subjected to a uniaxial tensile stress σ_0.

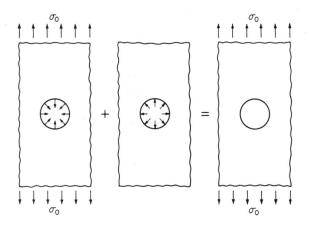

Figure 3.6 The method of superposition.

tension in the y direction, ϕ_2 reduces to

$$\phi_2 = a_2 x^2 = \frac{\sigma_0 x^2}{2} \tag{c}$$

The stresses throughout the plate for a plate without a hole are

$$\sigma_{yy} = \sigma_0 \qquad \sigma_{xx} = \tau_{xy} = 0 \tag{d}$$

If an imaginary hole of radius a is cut into the plate, the stresses σ_{rr}, $\sigma_{\theta\theta}$, and $\tau_{r\theta}$ on the boundary of the imaginary hole can be computed from Eqs. (3.31) as follows:

$$\sigma_{rr}^I = \sigma_0 \sin^2 \theta = \frac{\sigma_0}{2}(1 - \cos 2\theta)$$

$$\sigma_{\theta\theta}^I = \sigma_0 \cos^2 \theta = \frac{\sigma_0}{2}(1 + \cos 2\theta) \tag{e}$$

$$\tau_{r\theta}^I = \sigma_0 \sin \theta \cos \theta = \frac{\sigma_0}{2} \sin 2\theta$$

In the original problem the boundary conditions at $r = a$ were

$$\sigma_{rr} = \tau_{r\theta} = 0$$

The boundary conditions to be satisfied by the stresses associated with the second stress function are therefore

$$\sigma_{rr} = -\sigma_0 \sin^2 \theta = -\frac{\sigma_0}{2}(1 - \cos 2\theta) \qquad \text{at } r = a$$

$$\sigma_{rr} = \tau_{r\theta} = \sigma_{\theta\theta} = 0 \qquad \text{at } r \to \infty \tag{f}$$

$$\tau_{r\theta} = -\sigma_0 \sin \theta \cos \theta = -\frac{\sigma_0}{2} \sin 2\theta \qquad \text{at } r = a$$

From Eqs. (f) it is apparent that the stresses σ'_{rr} and $\tau_{r\theta}$ are functions of $\sin 2\theta$ and $\cos 2\theta$, which suggests $\phi^{(2)}$ given by Eqs. (3.40) as a possible stress function. Inspection of Eqs. (3.40) indicates, however, that this function can satisfy the boundary conditions only for $\tau_{r\theta}$. From Eqs. (3.38), however, it can be seen that the stresses associated with $\phi^{(0)}$ can satisfy the boundary conditions for σ_{rr} without influencing $\tau_{r\theta}$. Thus, the stress function $\phi^{(0)} + \phi^{(2)}$ may be applicable. From Eqs. (3.38) and (3.40),

$$\phi^{(0)} + \phi^{(2)} = a_0 + b_0 \ln r + c_0 r^2 + d_0 r^2 \ln r$$

$$+ (a_2 r^2 + b_2 r^{-2} + c_2 r^4 + d_2) \cos 2\theta \tag{g}$$

$$\sigma_{rr} = \frac{b_0}{r^2} + 2c_0 + d_0(1 + 2 \ln r)$$

$$- \left(2a_2 + \frac{6b_2}{r^4} + \frac{4d_2}{r^2}\right) \cos 2\theta \tag{h}$$

$$\sigma_{\theta\theta} = -\frac{b_0}{r^2} + 2c_0 + d_0(3 + 2 \ln r)$$

$$+ \left(2a_2 + \frac{6b_2}{r^4} + 10c_2 r^2\right) \cos 2\theta \tag{i}$$

$$\tau_{r\theta} = \left(2a_2 - \frac{6b_2}{r^4} + 6c_2 r^2 - \frac{2d_2}{r^2}\right) \sin 2\theta \tag{j}$$

Equations (h) to (j) contain seven unknowns: $b_0, c_0, d_0, a_2, b_2, c_2,$ and d_2. Since $\sigma_{\theta\theta} = \sigma_{rr} = \tau_{r\theta} = 0$ as $r \to \infty$,

$$c_0 = d_0 = a_2 = c_2 = 0 \tag{k}$$

and Eqs. (h) to (j) reduce to

$$\sigma_{rr} = \frac{1}{r^2}\left[b_0 - \left(\frac{6b_2}{r^2} + 4d_2\right) \cos 2\theta\right] \tag{l}$$

$$\sigma_{\theta\theta} = \frac{1}{r^2}\left(-b_0 + \frac{6b_2}{r^2} \cos 2\theta\right) \tag{m}$$

$$\tau_{r\theta} = -\frac{1}{r^2}\left[\left(\frac{6b_2}{r^2} + 2d_2\right) \sin 2\theta\right] \tag{n}$$

From the boundary conditions at $r = a$,

$$\tau_{r\theta} = -\frac{1}{a^2}\left[\left(\frac{6b_2}{a^2} + 2d_2\right) \sin 2\theta\right] = -\frac{\sigma_0}{2} \sin 2\theta \tag{o}$$

$$\sigma_{rr} = \frac{1}{a^2}\left[b_0 - \left(\frac{6b_2}{a^2} + 4d_2\right) \cos 2\theta\right] = -\frac{\sigma_0}{2}(1 - \cos 2\theta) \tag{p}$$

Solving Eqs. (*o*) and (*p*) for the coefficients gives

$$b_0 = -\frac{\sigma_0 a^2}{2} \qquad b_2 = \frac{\sigma_0 a^4}{4} \qquad d_2 = -\frac{\sigma_0 a^2}{2} \tag{q}$$

Substituting Eqs. (*q*) into Eqs. (*l*) to (*n*) gives

$$\sigma_{rr}^{\text{II}} = -\frac{\sigma_0 a^2}{2r^2}\left[1 + \left(\frac{3a^2}{r^2} - 4\right)\cos 2\theta\right]$$

$$\sigma_{\theta\theta}^{\text{II}} = \frac{\sigma_0 a^2}{2r^2}\left(1 + \frac{3a^2}{r^2}\cos 2\theta\right)$$

$$\tau_{r\theta}^{\text{II}} = -\frac{\sigma_0 a^2}{2r^2}\left[\left(\frac{3a^2}{r^2} - 2\right)\sin 2\theta\right]$$

The required solution for the original problem is obtained by superposition as follows:

$$\sigma_{rr} = \sigma_{rr}^{\text{I}} + \sigma_{rr}^{\text{II}} = \frac{\sigma_0}{2}\left\{\left(1 - \frac{a^2}{r^2}\right)\left[1 + \left(\frac{3a^2}{r^2} - 1\right)\cos 2\theta\right]\right\}$$

$$\sigma_{\theta\theta} = \sigma_{\theta\theta}^{\text{I}} + \sigma_{\theta\theta}^{\text{II}} = \frac{\sigma_0}{2}\left[\left(1 + \frac{a^2}{r^2}\right) + \left(1 + \frac{3a^4}{r^4}\right)\cos 2\theta\right] \tag{3.46}$$

$$\tau_{r\theta} = \tau_{r\theta}^{\text{I}} + \tau_{r\theta}^{\text{II}} = \frac{\sigma_0}{2}\left[\left(1 + \frac{3a^2}{r^2}\right)\left(1 - \frac{a^2}{r^2}\right)\sin 2\theta\right]$$

Equations (3.46) give the polar components of stress at any point in the body defined by r, θ. Through the use of Eqs. (3.46) the stresses along the x axis, along the y axis, and about the boundary of the hole can easily be computed.

The stresses along the x axis can be obtained by setting $\theta = 0$ and $r = x$ in Eqs. (3.46):

$$\sigma_{rr} = \sigma_{xx} = \frac{\sigma_0}{2}\left(1 - \frac{a^2}{x^2}\right)\frac{3a^2}{x^2}$$

$$\sigma_{\theta\theta} = \sigma_{yy} = \frac{\sigma_0}{2}\left(2 + \frac{a^2}{x^2} + \frac{3a^4}{x^4}\right) \tag{3.47}$$

$$\tau_{r\theta} = \tau_{xy} = 0$$

The distribution of σ_{xx}/σ_0 and σ_{yy}/σ_0 is plotted as a function of position along the x axis in Fig. 3.7. An examination of this figure clearly indicates that the presence of the hole in the infinite plate under uniaxial tension increases the σ_{yy} stress by a factor of 3. This factor is often called a *stress concentration factor*. In a later chapter it will be shown how photoelasticity can be effectively employed to determine stress concentration factors.

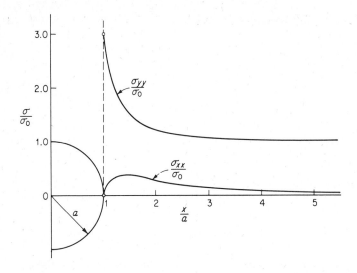

Figure 3.7 Distribution of σ_{xx}/σ_0 and σ_{yy}/σ_0 along the x axis.

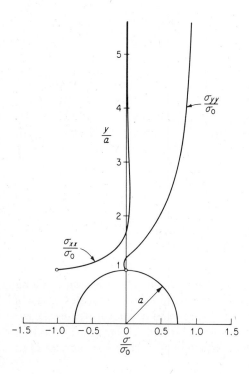

Figure 3.8 Distribution of σ_{xx}/σ_0 and σ_{yy}/σ_0 along the y axis.

In a similar manner the stresses along the y axis can be obtained by setting $\theta = \pi/2$ and $r = y$ in Eqs. (3.46):

$$\sigma_{rr} = \sigma_{yy} = \frac{\sigma_0}{2}\left(2 - \frac{5a^2}{y^2} + \frac{3a^4}{y^4}\right)$$

$$\sigma_{\theta\theta} = \sigma_{xx} = \frac{\sigma_0}{2}\left(\frac{a^2}{y^2} - \frac{3a^4}{y^4}\right) \tag{3.48}$$

$$\tau_{r\theta} = \tau_{xy} = 0$$

A distribution of σ_{xx}/σ_0 and σ_{yy}/σ_0 is plotted as a function of position along the y axis in Fig. 3.8. In this figure it can be noted that $\sigma_{xx}/\sigma_0 = -1$ at the boundary of the hole; thus the influence of the hole not only produces a concentration of the stresses but in this case also produces a change in the sign of the stresses.

The distribution of $\sigma_{\theta\theta}$ about the boundary of the hole is obtained by setting $r = a$ into Eqs. (3.46):

$$\sigma_{rr} = \tau_{r\theta} = 0 \qquad \sigma_{\theta\theta} = \sigma_0(1 + 2\cos 2\theta) \tag{3.49}$$

The distribution of $\sigma_{\theta\theta}/\sigma_0$ about the boundary of the hole is shown in Fig. 3.9. The maximum $\sigma_{\theta\theta}/\sigma_0$ occurs at the x axis ($\sigma_{\theta\theta}/\sigma_0 = 3$), and the minimum occurs at the y axis ($\sigma_{\theta\theta}/\sigma_0 = -1$). At the point defined by $\theta = 60°$ on the boundary of the hole, all stresses are zero. This type of point is commonly referred to as a singular point.

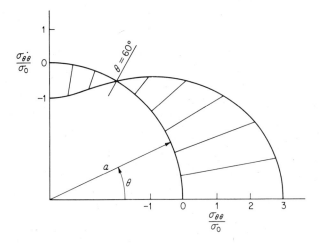

Figure 3.9 Distribution of $\sigma_{\theta\theta}/\sigma_0$ about the boundary of the hole.

EXERCISES

3.1 Determine the stress-strain and the strain-stress relations from Eqs. (2.16) and (2.17) with the assumptions of the plane-stress problem.

3.2 Determine the stress-strain and the strain-stress relations from Eqs. (2.16) and (2.17) with the assumption of the plane-strain problem.

3.3 Verify Eq. (3.17c).

3.4 In Eqs. (3.18), the body-force intensities were defined by

$$F_x = -\frac{\partial \Omega}{\partial x} \quad \text{and} \quad F_y = -\frac{\partial \Omega}{\partial y}$$

Determine Ω for a gravitational field of intensity q in the x direction.

3.5 Verify Eqs. (3.25).

3.6 Establish Eqs. (3.29).

3.7 Determine the stresses in the uniformly loaded cantilever beam shown in Fig. E3.7, using the Airy's-stress-function approach. Assume a unit thickness.

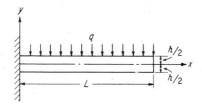

Figure E3.7

3.8 For the built-in triangular plate shown in Fig. E3.8, verify that the applicable stress function is

$$\phi = \frac{q \cot \alpha}{2(1 - \alpha \cot \alpha)}\left[-x^2 \tan \alpha + xy + (x^2 + y^2)\left(\alpha - \arctan\frac{y}{x}\right)\right]$$

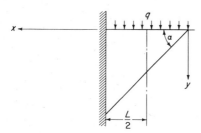

Figure E3.8

For the particular case of $\alpha = 45°$ and a plate of unit thickness, determine the normal stress distribution along the line $x = L/2$ and compare the solution with the results determined by using elementary beam theory.

3.9 Show that

$$\frac{\partial^2 \phi}{\partial y^2} = \frac{\partial \phi}{\partial r}\frac{\partial^2 r}{\partial y^2} + \frac{\partial^2 \phi}{\partial r^2}\left(\frac{\partial r}{\partial y}\right)^2 + 2\frac{\partial^2 \phi}{\partial r \partial \theta}\frac{\partial r}{\partial y}\frac{\partial \theta}{\partial y} + \frac{\partial \phi}{\partial \theta}\frac{\partial^2 \theta}{\partial y^2} + \frac{\partial^2 \phi}{\partial \theta^2}\left(\frac{\partial \theta}{\partial y}\right)^2$$

and verify Eq. (3.32b).

3.10 Show that

$$\frac{\partial^2 \phi}{\partial x \, \partial y} = \frac{\partial \phi}{\partial r}\frac{\partial^2 r}{\partial x \, \partial y} + \frac{\partial^2 \phi}{\partial r^2}\frac{\partial r}{\partial x}\frac{\partial r}{\partial y} + \frac{\partial^2 \phi}{\partial r \, \partial \theta}\frac{\partial r}{\partial x}\frac{\partial \theta}{\partial y}$$

$$+ \frac{\partial \phi}{\partial \theta}\frac{\partial^2 \theta}{\partial x \, \partial y} + \frac{\partial^2 \phi}{\partial \theta \, \partial r}\frac{\partial \theta}{\partial x}\frac{\partial r}{\partial y} + \frac{\partial^2 \phi}{\partial \theta^2}\frac{\partial \theta}{\partial x}\frac{\partial \theta}{\partial y}$$

and verify Eq. (3.32c).

3.11 Verify Eqs. (3.37).

3.12 Determine the polar components of the stresses σ_{rr}, $\sigma_{\theta\theta}$ in a thick-walled cylindrical vessel whose inside diameter is 1200 mm and outside diameter is 1500 mm if the vessel is subjected to an internal pressure of 35 MPa. Determine the radial displacement of the outside diameter if the vessel is made from steel, that is, $E = 200$ GPa and $v = 0.29$.

3.13 Show how the radial displacement of the inside and outside surfaces of a cylindrical pressure vessel subjected to internal pressure can be used to determine E and v.

3.14 Determine the stresses and displacements in the curved beam shown in Fig. E3.14 when subjected to the moment load M.

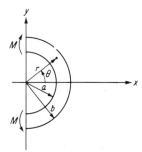

Figure E3.14

3.15 In Exercise 3.14, let a increase while holding $b - a$ constant and compare the values of the maximum stress $\sigma_{\theta\theta}$ from curved-beam theory and from straight-beam theory. Draw conclusions regarding the influence of the radius of curvature on $\sigma_{\theta\theta}$.

3.16 Determine the stresses in a semi-infinite plate due to a normal load acting on its edge as shown in Fig. E3.16. *Hint:* Try the stress function ϕ^*.

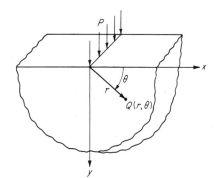

Figure E3.16

3.17 Determine the stresses in a semi-infinite plate due to a shear load acting on its edge as shown in Fig. E3.17. *Hint:* Use the stress function ϕ^*.

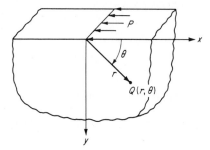

Figure 3.17

3.18 Determine the stresses in a semi-infinite plate due to an inclined load acting on its edge as shown in Fig. E3.18. *Hint:* Employ the solutions to Exercises 3.16 and 3.17 and superimpose solutions.

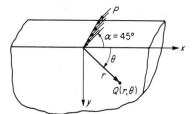

Figure E3.18

3.19 Determine the stresses in a semi-infinite body due to a distributed normal load (Fig. E3.19).

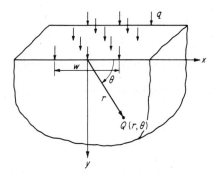

Figure E3.19

3.20 A steel ring is shrunk into another steel ring, as shown in Fig. E3.20. Determine the maximum

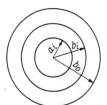

Figure E3.20

interference possible without yielding one of the rings if the yield strength of both rings is 825 MPa, $E = 200$ GPa, $v = 0.30$, and

	Case 1	Case 2	Case 3	Case 4
a_i, mm	100	100	100	0
b_i, mm	125	115	150	100
b_o, mm	150	200	175	150

3.21 Discuss the possibility of fabricating gun tubes by shrink fitting two or more long cylinders together to form the tube.

REFERENCES

1. Boresi, A. P., and P. P. Lynn: "Elasticity in Engineering Mechanics," chaps. 4–6, Prentice-Hall, Inc., Englewood Cliffs, N.J., 1974.
2. Chou, P. C., and N. J. Pagano: "Elasticity," chaps. 4 and 5, D. Van Nostrand Company, Inc., Princeton, N.J., 1967.
3. Durelli, A. J., E. A. Phillips, and C. H. Tsao: "Introduction to the Theoretical and Experimental Analysis of Stress and Strain," chaps. 6 and 9, McGraw-Hill Book Company, New York, 1958.
4. Sechler, E. E.: "Elasticity in Engineering," chaps. 6 and 8, John Wiley & Sons, Inc., New York, 1952.
5. Timoshenko, S. P., and J. N. Goodier: "Theory of Elasticity," 2d ed., chap. 4, McGraw-Hill Book Company, New York, 1951.
6. Kirsch, G.: Die Theorie der Elasticität und die Bedürfnisse der Festigkeitlehre, *Z. Ver. Dtsch. Ing.*, vol. 32, pp. 797–807, 1898.
7. Sternberg, E., and M. Sadowsky: Three-dimensional Solution for the Stress Concentration around a Circular Hole in a Plate of Arbitrary Thickness, *J. Appl. Mech.*, vol. 16, pp. 27–38, 1949.
8. Howland, R. C. J.: On the Stresses in the Neighborhood of a Circular Hole in a Strip under Tension, *Trans. R. Soc.*, vol. A229, pp. 49–86, 1929.
9. Inglis, C. E.: Stresses in a Plate Due to the Presence of Cracks and Sharp Corners, *Proc. Inst. Nav. Arch.*, vol. 55, part 1, pp. 219–230, 1913.
10. Greenspan, M.: Effect of a Small Hole on the Stresses in a Uniformly Loaded Plate, *Q. Appl. Math.*, vol. 2, pp. 60–71, 1944.
11. Jeffrey, G. B.: Plane Stress and Plane Strain in Bi-polar Co-ordinates, *Phil. Trans. R. Soc.*, vol. A-221, pp. 265–293, 1920.
12. Mindlin, R. D.: Stress Distribution around a Hole Near the Edge of a Plate under Tension, *Proc. SESA*, vol. V, no. 2, pp. 56–68, 1948.
13. Ling, C. B.: Stresses in a Notched Strip under Tension, *J. Appl. Mech.*, vol. 14, pp. 275–280, 1947.

PART
TWO

BRITTLE-COATING METHODS

FOUR

THEORY OF THE BRITTLE-COATING METHOD

4.1 INTRODUCTION [1–9]

The brittle-coating method of experimental stress analysis provides a simple and direct approach for solving a large class of industrial-design problems where extreme accuracy is not required. The method is based on the perfect adhesion of a coating with brittle characteristics to the component being studied. When loads are applied to the component, the strains which develop are transmitted to the coating. Here they produce a system of coating stresses which ultimately causes the coating to fail by cracking. Since the coating is designed to fail at a low stress level, the component is not overstressed and the method is classified as nondestructive.

The first brittle coatings to be employed were those which naturally formed on structural members, such as mill scale on hot-rolled steel and oxides on heated surfaces. These brittle coatings failed by flaking or cracking when the base material yielded under load and excessive strains were produced. To improve the visibility of the crack patterns, the structural members were often coated with whitewash. The cracking and flaking of the coating produced dark lines which were readily visible against the white background. The first artificially produced brittle coating was a mixture of shellac and alcohol, which Sauerwald and Wieland employed in 1925 to indicate regions of plastic strain. This shellac and alcohol mixture was an improvement over the whitewash technique, but the strains required for the coating to fail were still in the yield region for the base material, and the method could not be classified as nondestructive. In order to reduce the strain sensitivity, other coatings were developed and introduced by

Dietrich and Lehr in Germany in 1932, by Portevin and Cymboliste in France in 1934, and finally in the United States by Ellis in 1937.

Ellis, working with the Magnaflux Corporation in Chicago, made his resin-based coating commercially available in 1941. The formulation of the coating is relatively simple, with three basic constituents: (1) zinc resinate base, (2) carbon disulfide as a solvent, and (3) dibutyl phthalate as a plasticizer to vary the degree of brittleness of the coating. This resin-based coating, known as Stresscoat, has been widely used in the United States, Europe, and Asia.

In recent years both Magnaflux Corporation and Photolastic, Inc. have introduced new resin-based brittle coatings. In these coatings the solvent carbon disulfide has been replaced with methylene chloride to provide a nonflammable, low-toxicity product which is much safer to employ in typical industrial laboratories. The plasticizer in these coatings has also been changed in order to reduce the typical threshold strain to about 500 μin/in. (μm/m). These commercial developments were preceded by formulation of a new nonflammable, low-toxicity coating called Straintec by Racine, Taracks, and Killinger of the General Motors Truck and Coach Division.

Ceramic-based brittle coatings, introduced by Singdale in 1954, are porcelain enamels mixed with lead borosilicate. A similar ceramic-based coating has been marketed under the trade name All-Temp by the Magnaflux Corp. These ceramic-based brittle coatings extend the applicability of the brittle-coating method to temperatures approaching 700°F (370°C). Unfortunately, the coatings must be fired onto the component at temperatures ranging from 950 to 1100°F (510 to 595°C), a procedure that severely limits their use.

A glass lacquer was developed by Hickson at the Royal Aircraft Establishment in England in 1965. This coating consists of a mixture of lithium hydroxide, boric acid, and water. The coating dries at room temperature and is similar in many respects to the resin-based brittle coatings.

The primary advantages of the brittle-coating method are:

1. It provides nearly whole-field data for both magnitude and direction of principal stresses and does not require a tedious point-per-point analysis.
2. It does not require the construction of a model and can usually be applied to a prototype of the actual machine component being studied.
3. Since it can be applied to the actual machine component in operation, it is not necessary to simulate or even to know the value of the loads acting on the component.
4. The data obtained are converted to stresses in a simple fashion without an elaborate or complex mathematical analysis.

The main disadvantage of the method is the variable precision of the stress distributions which are indicated by the coating. The precision or accuracy of the measurements is strongly dependent on temperature and humidity variations during the test and load-time history. The accuracy also depends to a smaller degree on the thickness of the coating and the biaxiality of the stress field being measured. In order to minimize errors, the experimentalist should understand the behavior of the coating as completely as possible. If proper precautions are taken during the preparations and actual testing, the accuracy of the determinations from the brittle-coating method is usually sufficient for stress analysis related to programs of the industrial-development type.

4.2 COATING STRESSES [10, 11]

The brittle coating is usually air-sprayed onto the component being studied until a layer of coating from 0.002 to 0.008 in (0.05 to 0.20 mm) thick has accumulated. After the coating has dried at room temperature or cured at an elevated temperature in an oven, the component is subjected to a system of loads which causes the coating to fail by cracking. The stresses in the coating which produce the crack pattern are related to the stresses in the specimen, as will be shown later.

Since the coating is usually very thin relative to the thickness of the specimen, it is valid to assume that the surface strains occurring on the specimen are transmitted to the coating without magnification or attenuation and, moreover, that these coating strains are uniform through the thickness of the coating. Most of the investigators working with brittle coatings assume that the coating fails when it is strained to a certain critical value called the *threshold strain*. The value of the threshold strain ϵ_{t*} is determined by calibrating the coating in a manner which will be discussed in detail later. The principal stress in the specimen, perpendicular to the crack in the coating, is computed by using the equation

$$\sigma_1^s = E^s \epsilon_{t*} \tag{4.1}$$

where σ_1^s = principal stress in specimen at location of coating crack
E^s = modulus of elasticity of specimen material
ϵ_{t*} = threshold strain required to crack coating when specimen is subjected to uniaxial state of stress

This equation for computing the principal stresses is extremely simple and is widely accepted as being sufficiently accurate for brittle-coating work; however, it does have one important shortcoming. This formulation neglects the influence on the failure of the coating of a biaxial stress system which usually exists in actual practice.

To examine the implications of the biaxial stress system, consider a surface element which includes both the coating and the specimen, as shown in Fig. 4.1. Since the coating is thin relative to the thickness of the specimen, the strains developed at the surface of the specimen are transmitted without change to the coating; thus

$$\epsilon_{xx}^c = \epsilon_{xx}^s \qquad \epsilon_{yy}^c = \epsilon_{yy}^s \tag{4.2}$$

Also it is reasonable to assume that

$$\sigma_{zz}^c = \sigma_{zz}^s = 0 \tag{4.3}$$

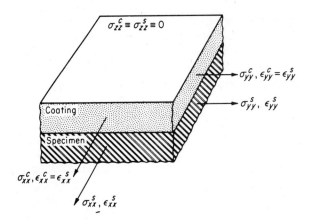

$$\sigma^c_{zz} \equiv \sigma^s_{zz} \equiv 0$$

$$\sigma^c_{yy}, \epsilon^c_{yy} = \epsilon^s_{yy}$$

$$\sigma^s_{yy}, \epsilon^s_{yy}$$

Coating

Specimen

$$\sigma^c_{xx}, \epsilon^c_{xx} = \epsilon^s_{xx}$$

$$\sigma^s_{xx}, \epsilon^s_{xx}$$

Figure 4.1 Elemental section of coating and specimen showing stresses and strains.

Within the limits of the previous assumptions it is clear that a state of plane stress exists in both the coating and the boundary layer of the specimen; therefore, the strains may be expressed in terms of the stresses by employing Eqs. (2.19). Hence

$$\epsilon^s_{xx} = \frac{1}{E^s}(\sigma^s_{xx} - v^s\sigma^s_{yy}) \qquad \epsilon^s_{yy} = \frac{1}{E^s}(\sigma^s_{yy} - v^s\sigma^s_{xx})$$

$$\epsilon^c_{xx} = \frac{1}{E^c}(\sigma^c_{xx} - v^c\sigma^c_{yy}) \qquad \epsilon^c_{yy} = \frac{1}{E^c}(\sigma^c_{yy} - v^c\sigma^c_{xx}) \qquad (4.4)$$

Substituting Eqs. (4.4) into Eqs. (4.2) yields

$$\frac{1}{E^s}(\sigma^s_{xx} - v^s\sigma^s_{yy}) = \frac{1}{E^c}(\sigma^c_{xx} - v^c\sigma^c_{yy})$$

$$\frac{1}{E^s}(\sigma^s_{yy} - v^s\sigma^s_{xx}) = \frac{1}{E^c}(\sigma^c_{yy} - v^c\sigma^c_{xx}) \qquad (a)$$

These equations can be solved for σ^c_{xx} and σ^c_{yy} as follows:

$$\sigma^c_{xx} = \frac{E^c}{E^s(1 - v^{c2})}[(1 - v^cv^s)\sigma^s_{xx} + (v^c - v^s)\sigma^s_{yy}]$$

$$\sigma^c_{yy} = \frac{E^c}{E^s(1 - v^{c2})}[(1 - v^cv^s)\sigma^s_{yy} + (v^c - v^s)\sigma^s_{xx}] \qquad (4.5)$$

These equations represent the plane state of stress produced in the coating by the plane state of stress acting on the surface of a specimen.

A brittle coating is usually calibrated by applying it to a calibrating beam and subjecting the beam to a known deflection. If the beam is considered as the specimen, the specimen stresses are given by

$$\sigma^s_{xx} = E^s\epsilon^s_{xx} \qquad \sigma^s_{yy} \approx 0 \qquad (4.6)$$

The procedure followed in calibrating is to set

$$\epsilon_{xx}^s = \epsilon_{t*}$$ (4.7)

to obtain the threshold strain for the coating. Substituting Eqs. (4.6) and (4.7) into Eqs. (4.5) gives the coating stress for the calibration beam:

$$\sigma_{xx}^c = \frac{E^c}{1 - v^{c2}}(1 - v^c v^*)\epsilon_{t*} = \frac{E^c}{E^*(1 - v^{c2})}(1 - v^c v^*)\sigma_{xx}^*$$

$$\sigma_{yy}^c = \frac{E^c}{1 - v^{c2}}(v^c - v^*)\epsilon_{t*} = \frac{E^c}{E^*(1 - v^{c2})}(v^c - v^*)\sigma_{xx}^*$$ (4.8)

where $v^* =$ Poisson's ratio for calibrating beam
$\quad\ \sigma^* =$ uniaxial stress in calibrating beam required to crack coating
$\quad\ E^* =$ modulus of elasticity of calibrating beam

From Eqs. (4.8) it is apparent that a uniaxial state of stress in the calibrating beam produces a biaxial state of stress in the coating. The biaxial stress system results from the mismatch in Poisson's ratio between the specimen and the coating. Since v^c is about 0.42, this mismatch is appreciable when brittle coatings are employed on metals where v^s ranges between 0.29 and 0.33.

4.3 FAILURE THEORIES [12, 13]

In order to predict stresses in the specimen from observations of the crack pattern in the coating, some law of failure for the coating must be utilized. Three laws of failure—maximum-normal-strain theory, maximum-normal-stress theory, and the Mohr theory—have in the past been considered to represent the behavior of the coating. The maximum-normal-strain and the maximum-normal-stress theories are not verified experimentally. The Mohr theory of failure overcomes many of the shortcomings of the other two theories; however, the analysis involved is somewhat more complex.

The Mohr theory of failure has been employed frequently in the past to predict failure stresses in materials such as cast iron, concrete, rocks, and ceramics, in which the ultimate compressive strength is much greater than the ultimate tensile strength. Since it is well known that the compressive strength of a brittle coating greatly exceeds its tensile strength, the Mohr theory of failure appears to be an ideal choice for the theory of failure which is necessary to relate coating failure to specimen stresses.

In this application of the Mohr theory of failure, three important cases arise:

Case 1: $\qquad\qquad\qquad\qquad \sigma_{xx}^c \geq \sigma_{yy}^c \geq 0$

Case 2: $\qquad\qquad\qquad\qquad \sigma_{xx}^c \geq 0 \geq \sigma_{yy}^c$ (4.9)

Case 3: $\qquad\qquad\qquad\qquad 0 \geq \sigma_{xx}^c \geq \sigma_{yy}^c$

where the coordinate system Oxy is considered to be principal.

The state of stress in case 1 is biaxial tension, and the Mohr theory of failure coincides with the maximum-normal-stress theory of failure. For case 1, failure of the coating occurs when

FoR $\quad \sigma_{xx}^c \geq \sigma_{yy}^c \geq 0$

$$\frac{\sigma_{xx}^c}{\sigma_{ut}^c} \geq 1 \qquad (4.10a)$$

The state of stress in case 2 is also biaxial, with σ_{xx}^c tensile and σ_{yy}^c compressive. In this case the Mohr theory gives the following failure relationship:

FoR $\quad \sigma_{xx}^c \geq 0 \geq \sigma_{yy}^c$

$$\frac{\sigma_{xx}^c}{\sigma_{ut}^c} - \frac{\sigma_{yy}^c}{\sigma_{uc}^c} \geq 1 \qquad (4.10b)$$

where σ_{ut}^c is the ultimate tensile strength of the coating and σ_{uc}^c is the ultimate compressive strength of the coating (a positive number). The state of stress associated with case 3 is biaxial compression, with both σ_{xx}^c and σ_{yy}^c compressive. The failure equation for this state of stress is

FoR $\quad 0 \geq \sigma_{xx}^c \geq \sigma_{yy}^c$

$$-\frac{\sigma_{yy}^c}{\sigma_{uc}^c} \geq 1 \qquad (4.10c)$$

Assume in this formulation of the Mohr theory that

$$\sigma_{uc}^c = K\sigma_{ut}^c \qquad (4.11)$$

where K is the strength ratio equal to $\sigma_{uc}^c/\sigma_{ut}^c$ for the brittle coating.

The coating stresses given in the three failure equations must be converted to specimen stresses in order to determine how failure of the coating is related to the specimen stresses. By substituting Eqs. (4.3), (4.5), (4.8), and (4.11) into Eqs. (4.10), the failure conditions for the coating based upon specimen stress are obtained as follows:

$$\frac{\sigma_{xx}^s}{\sigma_{xx}^*} + \frac{v^c - v^s}{1 - v^c v^s} \frac{\sigma_{yy}^s}{\sigma_{xx}^*} \geq 1 \qquad (4.12a)$$

if $\sigma_{xx}^c \geq \sigma_{yy}^c \geq 0$ (case 1);

$$\left[1 - \frac{v^c - v^s}{K(1 - v^c v^s)}\right]\frac{\sigma_{xx}^s}{\sigma_{xx}^*} + \left(\frac{v^c - v^s}{1 - v^c v^s} - \frac{1}{K}\right)\frac{\sigma_{yy}^s}{\sigma_{xx}^*} \geq 1 \qquad (4.12b)$$

if $\sigma_{xx}^c \geq 0 \geq \sigma_{yy}^c$ (case 2);

$$-\left(\frac{\sigma_{yy}^s}{\sigma_{xx}^*} + \frac{v^c - v^s}{1 - v^c v^s}\frac{\sigma_{xx}^s}{\sigma_{xx}^*}\right) \geq K \qquad (4.12c)$$

if $0 \geq \sigma_{xx}^c \geq \sigma_{yy}^c$ (case 3).

Equations (4.12) relate the specimen stresses to the coating failure when the specimen is subjected to any biaxial stress field. For a better visualization of this failure criterion, these three equations have been plotted in Fig. 4.2. This figure shows that the three cases given above are separated by lines which represent

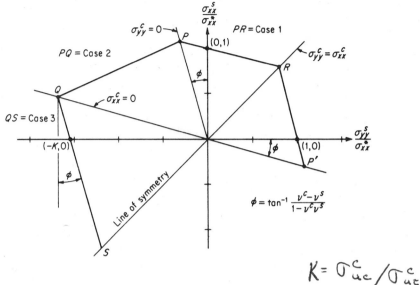

Figure 4.2 Failure chart for the Mohr theory of failure.

states of coating stress where $\sigma^c_{xx} = 0, \sigma^c_{yy} = 0, \sigma^c_{xx} = \sigma^c_{yy}$. From Eqs. (4.5) it can be directly established that

$$\frac{\sigma^s_{yy}}{\sigma^*_{xx}} = -\frac{v^c - v^s}{1 - v^c v^s} \frac{\sigma^s_{xx}}{\sigma^*_{xx}} \qquad \text{for } \sigma^c_{yy} = 0 \qquad (4.13a)$$

$$\frac{\sigma^s_{xx}}{\sigma^*_{xx}} = -\frac{v^c - v^s}{1 - v^c v^s} \frac{\sigma^s_{yy}}{\sigma^*_{xx}} \qquad \text{for } \sigma^c_{xx} = 0 \qquad (4.13b)$$

and finally

$$\frac{\sigma^s_{xx}}{\sigma^*_{xx}} = \frac{\sigma^s_{yy}}{\sigma^*_{xx}} \qquad \text{for } \sigma^c_{yy} = \sigma^c_{xx} \qquad (4.13c)$$

The failure line associated with case 1 is independent of the strength ratio K and is determined by the ratio $(v^c - v^s)/(1 - v^c v^s)$. On the other hand, the failure lines associated with cases 2 and 3 depend upon both K and the ratio $(v^c - v^s)/(1 - v^c v^s)$. A typical failure diagram for a resin-based brittle coating $(v^c = 0.42)$ employed on an aluminum specimen $(v^s = 0.33)$ is presented in Fig. 4.3. Failure lines are shown for $K = 2, 4, 6$, and 8. Experimental data on the failure of brittle coatings under biaxial stress states are limited; however, a few points obtained in tests conducted by Durelli indicate that a value of 4 for K is a close approximation for the ST1200-1210 type of Stresscoat.

If the value $K = 4$ is considered to be representative for resin-based brittle coatings, it is possible to vary v^s to determine the influence of the biaxiality of the stress field for specimens fabricated from glass, steel, aluminum, and plastic. The results presented in Fig. 4.4 show that when $\sigma^s_{yy} > 0$, the effect of biaxiality is relatively small, the value of $\sigma^s_{xx}/\sigma^*_{xx}$ always exceeding 0.85. However, when

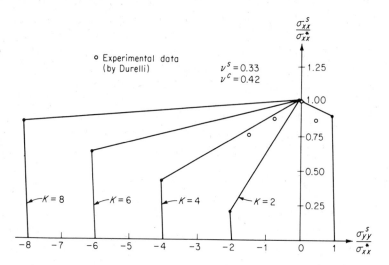

Figure 4.3 Failure curves for various values of K with $v^c = 0.42$ and $v^s = 0.33$.

$\sigma_{yy}^s < 0$, the effect of biaxiality can be quite large, particularly for specimens with high Poisson's ratio. In these instances, the simple failure relation given in Eq. (4.1) can produce large (although conservative) errors.

In view of the fact that the law of failure is the very foundation of the brittle-coating method, it is clear that more research work should be performed in order to firmly establish the proper theory and to determine correction factors for Eq. (4.1). It is believed that the Mohr theory of failure shows a great deal of

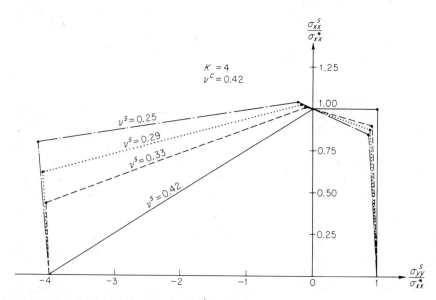

Figure 4.4 Failure curves for $K = 4$ and various Poisson's ratios for the specimen.

promise in this regard, but it must be substantiated by more experimental evidence.

Other failure theories have been examined to account for the effects of biaxiality. These include the maximum-normal-strain theory of failure and the maximum-normal-stress theory of failure; however, experimental evidence shows that failure of the ST1200-1210 series of Stresscoat does not correspond to either of these theories. For this reason, the derivation of the failure equations for these two theories is deferred to the exercises at the end of this chapter.

4.4 BRITTLE-COATING CRACK PATTERNS [14, 15]

The manner in which a brittle coating fails by cracking depends entirely upon the state of stress in the specimen to which it adheres. The failure behavior of the coating is determined by the magnitudes of σ_1 and σ_2, the principal stresses in the coating. Consider the following special cases when the specimen is subjected to direct loading.

Case 1: $\sigma_1 > 0$, $\sigma_2 < 0$, $\sigma_3 = 0$ In this case, only one set of cracks forms, and they are perpendicular to σ_1 (see Fig. 4.5). These cracks indicate the principal stress direction of σ_2, and, as a consequence, the cracks represent stress trajectories, or isostatics.

Case 2: $\sigma_1 > \sigma_2 > 0$, $\sigma_3 = 0$ In this example, two families of cracks can form. The first set of cracks due to σ_1 forms perpendicular to σ_1 and parallel to σ_2 (see Fig. 4.6). When the stress level of σ_2 becomes sufficiently high, a second family of cracks will form perpendicular to σ_2 and parallel to σ_1. When both families of cracks are produced, a complete set of isostatics, or stress trajectories, is obtained.

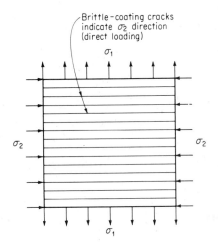

Figure 4.5 Brittle-coating crack pattern produced by a state of stress where $\sigma_1 > 0$, $\sigma_2 < 0$, $\sigma_3 = 0$.

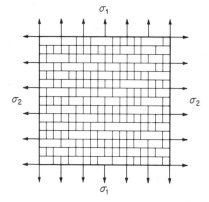

Figure 4.6 Brittle-coating crack pattern produced by a state of stress where $\sigma_1 > \sigma_2 > 0$, $\sigma_3 = 0$.

Crack patterns of this type are often encountered in testing the cylindrical portion of pressure vessels, where σ_1 (the hoop stress) is twice as large as σ_2 (the axial stress). If the pressure is increased slowly, the first cracks will appear in the coating along the axis of the cylinder and will be due to the hoop stress. Later, after the hydrostatic pressure acting on the cylindrical vessel has more than doubled, a second set of cracks will form in the hoop direction because of the axial stresses. An example of a brittle-coating crack formation on a cylindrical pressure vessel is shown in Fig. 4.7. In the central portion of the cylinder, at the bottom of the

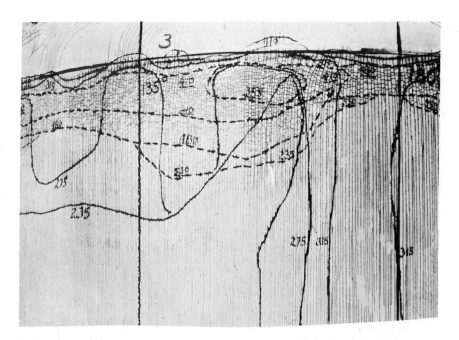

Figure 4.7 Enlarged view of the brittle-coating crack pattern over a portion of a cylindrical pressure vessel.

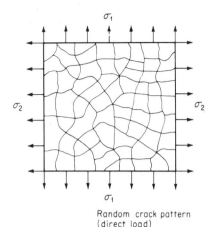

Random crack pattern
(direct load)

Figure 4.8 Brittle-coating crack pattern produced by a state of stress where $\sigma_1 = \sigma_2 > 0$, $\sigma_3 = 0$.

figure, only the cracks due to σ_1 (the hoop stress) have formed. At the top of the photograph, near the weld which joins the cylinder to the hemispherical head, σ_2 (the axial stress) is sufficient to induce the second set of cracks in the coating.

Case 3: $\sigma_1 = \sigma_2 > 0$, $\sigma_3 = 0$ When $\sigma_1 = \sigma_2$ at any point on the body, the stress system is said to be isotropic and every direction is a principal direction. If the values of σ_1 and σ_2 are sufficiently high, the coating will fail; however, the crack pattern produced will be random in character. Crack patterns of this form are often referred to as *craze patterns* since the cracks have no preferential direction. A schematic illustration of this type of crack pattern is given in Fig. 4.8.

Crack patterns of this type occur when pressure vessels with hemispherical head closures are tested, since $\sigma_1 = \sigma_2$. An illustration of a typical pattern obtained from the hemispherical head of a pressure vessel is shown in Fig. 4.9. Cases where σ_1 and σ_2 are either compressive or too small to produce crack patterns will be covered in subsequent sections.

Case 4: $\sigma_2 < \sigma_1 < 0$, $\sigma_3 = 0$ In this instance, the coating is subjected to a state of biaxial compressive stress and will not crack under direct load. However, if the compressive stresses are sufficiently large to satisfy Eq. (4.12c), the coating will fail by flaking from the surface of the specimen. Observations of regions where brittle coatings have failed by flaking are not common in elastic analyses, where loads are limited to maintain stresses in the specimen below the yield strength of the specimen material. However, when such loads are increased and regions of the specimen experience plastic flow, flaking is commonly observed, particularly in the local neighborhood of concentrated loads.

In many practical problems which are approached by the brittle-coating method, a particular region is often highly stressed and the remainder of the body is not stressed to a level sufficient to produce coating response. These low-stressed regions are often as important to an engineer seeking weight reduction as the

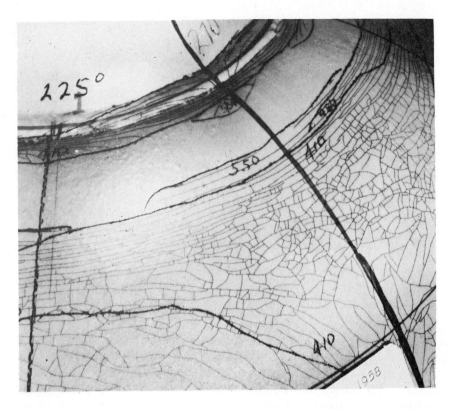

Figure 4.9 Brittle-coating crack pattern on a hemispherical head closure. (The random directional property which cracks exhibit indicates nearly equal stresses in the circumferential and meridional directions.)

highly stressed region, since material can be removed from the low-stressed regions without endangering the service life of the component. The only information imparted to the investigator by a coating which will not crack is that the stress is lower than σ_{xx}^*; that is, the lowest specimen stress required to crack the coating. It is usually necessary to employ strain gages to obtain specific values for the stresses in regions where the coating does not naturally respond. However, with some inducement, the brittle coating can provide information regarding the directions of the stresses in these regions. When the principal stress directions are known, the number of strain gages required for principal stress determinations at a point is reduced from three to two.

It is possible to obtain coating cracks in low-stressed regions by employing refrigeration techniques with brittle coating. First, the specimen is loaded until the stress in the critical region is just below the yield stress; then, the coating is subjected to a rapid temperature drop while under load. The rapid temperature drop produces a state of hydrostatic tension in the coating which is superimposed upon the existing stresses in the coating due to the load. The combined load and thermal stresses are usually sufficient to produce coating failure and a crack pattern. The direction of the resulting cracks is coincident with one of the principal stresses due to the load since the isotropic thermal stresses have no preferential directions. This technique is simple to apply, and the results are accurate provided the coating stresses due to the load are sufficiently large.

In order to reduce the temperature of the coating, i.e., refrigeration, one of two techniques is commonly employed. In the first method ice water is sponged over the area of the coating which has not previously responded. This method is not always successful since it does not produce sufficiently high thermal stresses to give total coating response. A larger temperature change can be obtained by passing a stream of compressed air through a box of dry ice before it is directed onto the surface of the coated model. The stream of very cold air can be accurately directed, and the resulting crack patterns can be closely controlled.

In many applications of the brittle-coating method, situations are encountered where one or possibly both of the principal stresses are compressive. In these instances it is not possible to obtain, by direct loading procedures, the crack patterns associated with the compressive stresses, since a brittle coating does not crack as a result of compressive stress. In order to circumvent this difficulty, a relaxation technique is employed which is exactly opposite to the direct-loading procedure. That is, a load is applied to the coated specimen before it has had the opportunity to dry. The load is maintained on the coated specimen until drying is complete. Under these conditions the coating is stress-free (neglecting residual stresses, which are always present in the coating), while the specimen is highly stressed in compression. If a portion of the load is relieved, the specimen will stretch, since it was previously compressed, and tensile stresses will develop in the coating which are sufficiently high to indicate regions of compressive specimen stress. Thus, relaxation of the load on the specimen reverses the coating stresses and permits a reasonably accurate determination of the compressive stresses occurring in the specimen.

4.5 CRACK DETECTION [16]

When a load is applied to a specimen treated with a brittle coating, cracks will form in the coating. If the coating is properly applied, these cracks will remain open after the load is removed. To employ the coating for either directional information or stress-level data accurately, it is necessary to locate all the cracks and to record the value of the load at which they occur. The complete test procedure will be covered in detail later; however, it is convenient at this time to discuss crack-detection techniques.

The cracks produced in brittle coatings are very fine. They are probably V-shaped with a depth equal to the coating thickness and a width estimated to be less than one-twentieth of the coating thickness. In order to observe these fine cracks visually, a focused light source must be directed at oblique incidence to the surface and normal to the crack. Often the location, direction, and the load at which the first crack will appear are not known. Consequently, a great deal of time is expended in a brittle-coating analysis in examining the coating for crack location.

A method of electrified-particle inspection which greatly reduces the experience needed in locating the crack patterns has been developed by Magnaflux Corporation under the trade name Statiflux. Water containing wetting agents to reduce its surface tension is applied to the coating to be examined. If any cracks are present, the "wet water" flows into the cracks, fills them, and makes electrical contact with the metal specimen. The surface of the specimen is then rubbed dry with a facial tissue. This drying process removes the water from the surface of the coating but does not remove the water from the cracks. Next, a talcum powder, negatively charged, is sprayed through a special gun onto the specimen. The negatively charged particles of powder are attracted electrically to the grounded water contained in the cracks. The powder forms small, white mounds over the cracks and provides an excellent means of locating the crack pattern. A typical

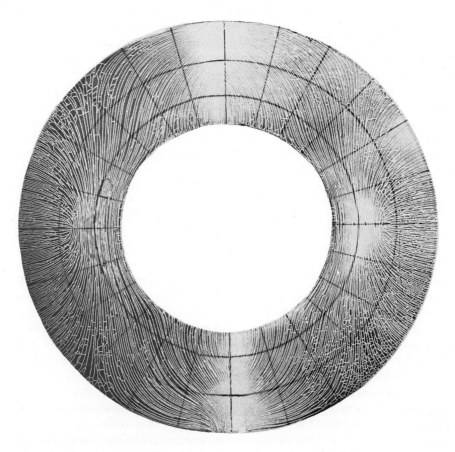

Figure 4.10 Statiflux method of crack detection. (*Courtesy of A. J. Durelli.*)

example of a Statiflux crack pattern is illustrated in Fig. 4.10. The method also can be used very successfully as an aid to the photographer in recording the crack patterns.

The prime disadvantage of the Statiflux method of crack detection is that the "wet water" must repeatedly be applied and removed from the coating. Any deviation from a uniform application of the water over the complete surface of the coating will probably result in a temperature change due to differentials in evaporation rates. As will be pointed out in Sec. 4.7, a temperature change is to be avoided in any quantitative brittle-coating analysis.

Magnaflux also markets a red dye etchant, which can be used with some of the resin-based coatings to increase the visibility of the crack patterns for photographic purposes. The dye etchant is a mixture containing turpentine, machine oil, and red dye. The etchant is applied to the surface of a cracked brittle coating for approximately 1 min. During this time the etchant begins to attack the coating in the neighborhood of the cracks, thus making them wider. If the etchant is left in contact with the coating too long, it will attack the surface of the coating. After the etchant is wiped from the surface of the coating, the coating is cleaned with an etchant emulsifier (soap and water). The dye which has penetrated the cracks is not removed during the cleaning process; thus, the cracks appear as fine red lines on a yellow background. The brittle-coating patterns shown in Figs. 4.7 and 4.9 were dye-etched before the photographs were taken.

It is not advisable to use dye etchant as a means of crack detection during the actual test, since the oils employed in the composition of the etchant influence the behavior of the coating and inconsistent results are obtained.

4.6 CERAMIC-BASED BRITTLE COATINGS [17–19]

Two basic types of brittle coating are commonly employed in stress-analysis problems. The first are resin-based coatings marketed by Magnaflux Corp. under the trade name Stresscoat and by Photolastic, Inc. under the trade name Tens-lac. The second type is a ceramic-based coating marketed by Magnaflux Corp. under the trade name All-Temp. All-Temp consists of finely ground ceramic particles suspended in a volatile carrier. The coating is sprayed onto the specimen, air-dried to a soft powder, and fired at high temperature until the ceramic particles melt and coalesce to form a glaze.

Although use of the resin-based brittle coatings is much more common (because they are easier to apply and to cure), the ceramic-based coatings have several advantages. Since the coating is essentially a porcelain-glass, its strain sensitivity is not markedly affected by changes in test temperature, humidity, or loading time. The threshold strain of ceramic coatings is also independent of coating thickness. For these reasons, it appears that higher accuracies should be possible with ceramic coatings than with resin-based coatings.

Another major advantage of ceramic coatings is their ability to perform at elevated temperatures. The threshold strain of All-Temp coatings employed on

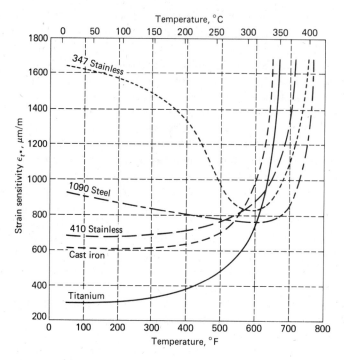

Figure 4.11 Effect of temperature on the strain sensitivity of All-Temp Coatings. (*Magnaflux Corporation.*)

347 and 410 stainless steels, 1090 steel, cast iron, and titanium is shown as a function of temperature in Fig. 4.11. These results show relatively small changes in the strain sensitivity of the coating with temperature over the temperature range from 75 to 550°F (24 to 288°C), indicating that small variations in temperature during the test period would not produce serious errors in the analysis. At about 550°F (288°C) the porcelain coating begins to soften, and the strain sensitivity increases rapidly with further increases in temperature. It appears then that these coatings can be used up to temperatures of 550 to 600°F (288 to 316°C) before they become too insensitive to respond to elastic strains. This is a significant improvement in the upper limit of the temperature range since resin-based coatings can be used only at temperatures less than 100°F (38°C).

The ceramic coating is chemically inert and as such can be employed in hostile environments which would destroy resin-based coatings. Ceramic coatings can be exposed to water under high-velocity flow conditions including cavitation. The coating is LOX-compatible and can be used at cryogenic temperatures provided the coating is cooled slowly from room temperature to the test temperature. The coating can be subjected to oil- and petroleum-based solvents without loss of sensitivity.

Ceramic coatings have one serious disadvantage which greatly limits their use: they are difficult to apply by spraying, and they must be fired at temperatures ranging from 950 to 1100°F (510 to 595°C). Unfortunately, this firing temperature is so high that components fabricated from aluminum, magnesium, plastics, and many heat-treated steels will be damaged by the firing process. If the component is large, a furnace capable of up to 1100°F (595°C) may not be available. The firing temperature is also rather critical. A 25°F (14°C) overtemperature will produce overfired coatings, while a 25°F (14°C) undertemperature will produce only a partially cured coating.

Another disadvantage of ceramic coatings is associated with crack detection since cracks developed in the coating are usually not visible without enhancement. An electrified-particle crack-detecting system or a dye penetrant is commonly used to locate cracks between each loading increment during a test.

The All-Temp coatings available from Magnaflux Corp. are designed to match the thermal coefficient of expansion of the metal used in fabricating the component. As coatings with different coefficients of expansion are selected, the strain sensitivity of the coating can be controlled. If a coating with a coefficient of expansion greater than that of the metal is selected, less strain is required to fail the coating because cooling from the firing temperature introduces an isotropic state of residual tensile stress in the coating. Conversely, coatings selected with a smaller coefficient of expansion than that of the specimen will require very large strains to produce cracking since the cooling process will introduce an isotropic state of residual compressive stress in the coating.

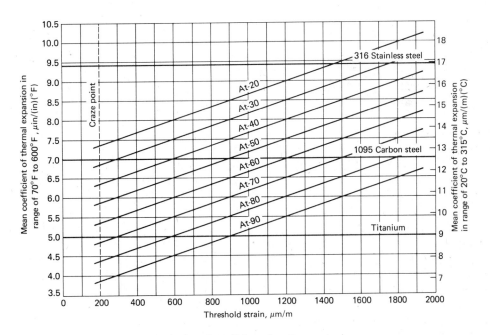

Figure 4.12 All-Temp Coating selection chart. (*Magnaflux Corporation.*)

Eight different grades of All-Temp (AT-20 to AT-90) are available with a wide range of coefficients of expansion. Selection of a particular grade depends upon the coefficient of expansion of the specimen being studied and the threshold strain specified for the test. The coating selection chart presented in Fig. 4.12 is used to specify the particular grade of All-Temp to be employed. For example, with a component fabricated from 1095 carbon steel and a threshold strain of 600 μin/in (μm/m) the coating AT-40 would be selected.

4.7 RESIN-BASED BRITTLE COATINGS [8, 16, 20–24]

There are three different resin-based coatings commercially available in the United States today. Magnaflux markets two different coatings under the Stresscoat trade name, and Photolastic markets a coating known as Tens-lac. The exact composition of these coatings is proprietary; however, the constituents in the non-flammable coatings are probably similar to those found in Straintec. This coating consists of zinc resinate and calcium resinate dissolved in the solvent methylene chloride with oleic acid and dibutyl phthalate used as the plasticizer.

Stresscoat is also available with carbon disulfide as the solvent; while this solvent has many disadvantages, e.g., an autoignition point of 212°F (100°C) and a low threshold limit value of 20 ppm on the toxicity scale, the solvent is very volatile and dries in short periods (12 to 16 h) at room temperature.

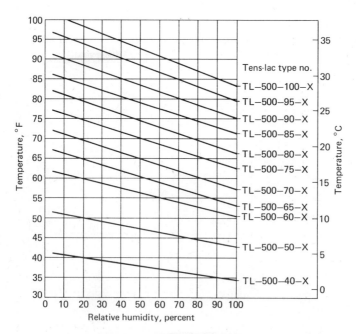

Figure 4.13 Coating selection chart of Tens-lac brittle coating. (*Photolastic, Inc.*)

The sensitivity of resin-based coatings is influenced by atmospheric conditions. Since the thermal coefficient of expansion of the resin base is an order of magnitude greater than the expansion coefficients of typical metals, the coating is very sensitive to small changes in temperature. Relative humidity, in addition to temperature, also affects the behavior of these coatings. Relative humidity is especially important during the cure period when the solvent is diffusing out of the coating.

To account for varying but predictable temperature and humidity conditions, resin-based coatings are available in several different grades. The particular grade is selected to give a specified threshold strain [usually 500 μin/in (μm/m)] for some specified conditions of temperature and humidity. The selection chart for the Tens-lac brittle coating presented in Fig. 4.13 shows that 11 different grades are available. If the temperature and relative humidity for both the cure and testing periods are 70°F (21°C) and 50 percent, respectively, the coating identified as TL-500-75-X should be selected to give a threshold strain of 500 μin/in (μm/m). The different grades of these coatings are essentially the same except for the amount of plasticizer added to the resin-solvent base. For high test temperatures 90 to 100°F (32 to 38°C) little plasticizer is used; however, as the test temperature is decreased, the amount of plasticizer added to the coating is increased.

Variations in the test temperature during the test period, when the brittle-coating crack patterns are being developed, are a major source of error in any attempt to predict the magnitude of specimen stresses. The influence of temperature on a resin-based brittle coating is shown graphically in Fig. 4.14. It is unlikely that the strength of the coating varies appreciably with temperature over this range; hence, the marked change in the strain required to crack the coating is probably due to the change in the coating's residual stresses when it expands or contracts relative to the base material during a temperature change.

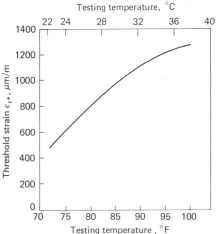

Figure 4.14 Influence of testing temperature on threshold strain for a resin-based brittle coating.

A change of 3 Fahrenheit degrees (1.7 Celsius degrees) in the testing temperature will produce a change of approximately 100 μin/in (μm/m) in the threshold strain. If the coating sensitivity was about 500 μin/in (μm/m) initially, the error produced by a temperature shift of 3 Fahrenheit degrees (1.7 Celsius degrees) during the test would be about 20 percent.

It is obvious that for consistently good results, a brittle-coating laboratory must have at least a temperature-controlled atmosphere. If the component for some reason must be tested in the field or factory, where temperature control is impossible, there are corrective measures which can be taken to minimize the errors due to temperature changes. (1) The weather bureau should be consulted to find when the temperature will be the most stable. Usually on a summer day the temperature increases rapidly during the morning and more slowly in the early afternoon, finally peaking later in the afternoon and then slowly decreasing until sunset. Temperature changes are minimized by conducting the brittle-coating analysis at this peak period in the late afternoon. (2) The test should be conducted as rapidly as possible to shorten the time interval over which the temperature has an opportunity to change. (3) There should be a continuous calibration of the coating so that the threshold strain can be established as a function of time and temperature during the test.

The coatings are sprayed onto the components being analyzed by using aerosol cans or compressed-air spray guns. For resin-based coatings with carbon disulfide as the solvent, the recommended coating thickness is from 0.006 to 0.008 in (0.15 to 0.20 mm). For the nonflammable brittle coatings, where methylene chloride is used as the solvent, the recommended range of thickness is only $\underline{\text{0.002 to 0.004 in (0.05 to 0.10 mm)}}$. At the recommended thicknesses the coatings require approximately 24 h to cure at room temperature.

TENS · LAC

Variations in thickness of the coating from point to point over the surface of the component being analyzed can be a serious source of error. Since the solvent in the coating is eliminated by a diffusion process, the time for a coating to dry depends strongly on its thickness. Thick regions of the coating usually are not as completely dry as thinner regions, and as a result, the coating exhibits a sensitivity which varies from point to point. One method used to eliminate the effects of partially cured coatings is to heat them in an air-circulating oven for 12 to 16 h before testing. These heat-cured coatings become completely dry, and variations in strain sensitivity due to variations in concentration of solvent remaining in the coating are eliminated.

It is important not to overcure coatings at elevated temperatures in excess of 100°F (38°C). Coatings exposed to high temperatures during curing begin to absorb moisture from the air as soon as they are removed from the oven. Since the strain sensitivity of the coating is influenced by the moisture absorbed from the air, the thinner coatings, which require less time to diffuse in higher concentrations of water, become softened before the thicker coatings; consequently, they are less sensitive.

It is evident that room-temperature curing produces a coating with a threshold strain that increases with increasing thickness, while high-temperature

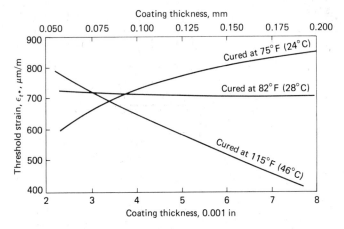

Figure 4.15 Sensitivity trends with respect to thickness for resin-based brittle coatings.

curing produces a coating with a threshold strain that decreases with increasing thickness. These sensitivity trends are illustrated in Fig. 4.15. By curing resin-based coatings at moderate temperatures between 80 and 85°F (27 to 29°C) for 20 h, it is possible to strike a balance between the effects of solvent diffusing out of the coating and water diffusing into the coating and obtain a coating with a threshold strain that is relatively insensitive to changes in coating thickness. The independence of threshold strain on coating thickness with proper curing is illustrated in Fig. 4.15.

In spite of the fact that resin-based coatings are called brittle coatings, they are viscoelastic materials with properties that vary as a function of time. Recommended procedure for determining the threshold strain with a calibration beam involves deflecting the beam in about 1 s and holding the beam in the deflected position for approximately 15 s. The position of the last crack is marked before unloading the beam. Since the threshold strain is a function of this load-time relation, the time required to load a component under study is extremely important. If the loading of the component being tested cannot be accomplished in the same manner, the calibration value of the threshold strain for the coating must be corrected.

Suppose, for example, that the load must be applied slowly to the component. Viscoelastic behavior of the coating during the loading interval results in a relaxation of the stresses in the coating before failure of the coating by cracking. The overall effect of this stress relaxation is to increase the strain required to crack the coating. The effects of loading times longer than 1 s can be taken into account by increasing the value of the threshold strain by the proper amount. Corrected values of threshold strain as a function of loading time are presented in Fig. 4.16 for the Stresscoat ST-10 to ST-100 series of resin-based coatings.

To illustrate the use of the correction chart shown in Fig. 4.16, consider a coating with a 1-s calibration value of threshold strain equal to 500 μin/in (μm/m) which is to be used on a component that requires 80 s to load. In this instance, the

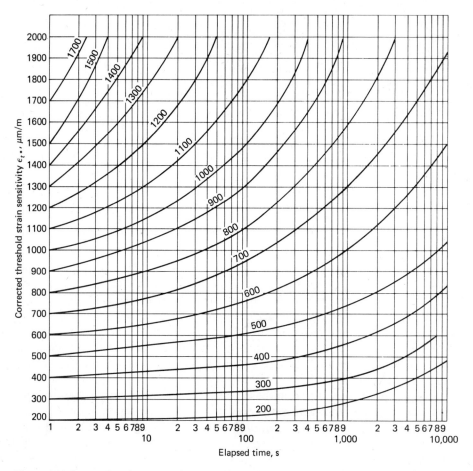

Figure 4.16 Stress-relaxation correction chart for stresscoat series ST-10 to ST-100. (*Magnaflux Corporation.*)

corrected value of threshold strain is 600 μin/in (μm/m), which is 20 percent higher than the original calibration.

Resin-based coatings can also be employed for dynamic stress analyses; however, two precautions must be taken: (1) the coatings must be more sensitive than those normally employed in static applications to ensure that the cracks remain open and visible after the dynamic load is removed; (2) the calibration value of threshold strain must be corrected to account for the effect of duration of load. A combination of strain and time under strain is required for the coating to react by cracking. If the duration of the load decreases, the strain required to crack the coating increases. The influence of load duration on threshold strain for the Stresscoat 1200 to 1210 series of resins is shown in Fig. 4.17. These data can be used to provide approximate corrections for the threshold strain.

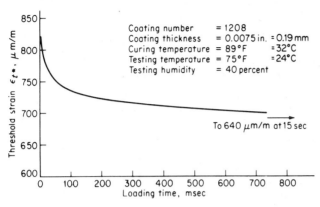

Figure 4.17 Influence of load duration on the threshold strain of Stresscoat.

Resin-based brittle coatings are inexpensive, easy to apply, and provide a simple but direct method for determining the magnitudes and directions of the principal stresses in actual prototypes. The amount of error involved in predicting the magnitudes of the principal stresses depends entirely on the ability of the stress analyst to control the effects of temperature, relative humidity, coating thickness, curing time and temperature, and the load-time function. If control of these parameters can be exercised, and if the effects of biaxiality of the stress field are accounted for in the interpretation of the data, the accuracy can approach ± 10 percent of the maximum stress levels, which is well within the accuracy requirements for most industrial applications.

4.8 TEST PROCEDURES FOR BRITTLE-COATING ANALYSES [16, 25–27]

The equipment required for a brittle-coating analysis can range from a small kit containing a few aerosol cans of coatings, precleaner, and undercoat to a much more complete kit containing a wide assortment of accessories, as illustrated in Fig. 4.18. If aerosol cans of undercoating and brittle coating are employed, the accessories can be kept to a minimum, but they should include at least the following items:

1. Focused light for visual inspection of the coating
2. Beam-bending device and scale for calibration
3. Calibration beams (12 or more)
4. Temperature- and humidity-measuring instruments

Careful preparation of the surface of the component to be tested is essential

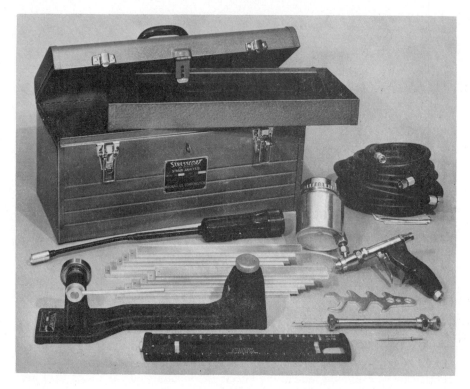

Figure 4.18 A complete kit of brittle coating accessories. (*Magnaflux Corp.*)

for a successful brittle-coating analysis. All paint or other coatings must be completely removed. The surface should then be lightly sanded or sand-blasted and degreased with one or more solvents such as acetone, Freon, or methylene chloride. Next an aluminum undercoating should be sprayed onto the component to provide a uniform reflecting background to increase the visibility of the crack patterns. It is often advantageous to draw a reference grid onto the aluminum undercoating for convenience in locating the crack patterns relative to various regions of the component.

Selection of the proper coating to be sprayed is easily made with the help of coating-selection charts like those presented in Figs. 4.12 and 4.13, which are provided by the manufacturer of the coating. The usual practice is to select a coating with a threshold strain of approximately 500 μin/in (μm/m) which is above the craze limit of the coatings (300 μin/in or μm/m) but well below the yield or fracture strains associated with most engineering materials. The ability to select the proper grade of a coating depends on the accuracy achieved in predicting the weather conditions 1 day ahead or on the ability to control the temperature in the laboratory if the test is conducted indoors.

In an air-conditioned laboratory (even one without humidity control), precise adjustment of the threshold strain can be achieved quite easily. After a few months of brittle-coating work, a logbook of threshold strains for various coating numbers and curing cycles should be established and used to select coatings for future tests. If for some reason the proper value of the threshold strain is not obtained, the temperature of the laboratory can be adjusted [3 Fahrenheit degrees or 1.7 Celsius degrees per 100 μin/in (μm/m)] to give the precise value specified for the test.

Curing the coating should be accomplished in an oven with a uniform temperature distribution if possible. Coatings with carbon disulfide as the solvent should be applied to a thickness of 0.006 to 0.008 in (0.15 to 0.20 mm) and cured for 16 h at 85°F (29°C). The nonflammable coatings with methylene chloride as the solvent should be applied to a thickness of only 0.002 to 0.004 in (0.05 to 0.10 mm) since this solvent requires more time to diffuse out of the coating. Again a curing temperature of 85°F (29°C) for 16 h is recommended to minimize the influence of variations in coating thickness on the threshold strain.

If the component is too large for an available oven, it is often possible to increase the temperature of the room in which the component is housed to 85°F (29°C) during the curing cycle.

Once the coating has been selected, the surface of the component prepared, and the coating sprayed to the proper thickness and cured with the recommended cycle, the actual brittle-coating test can be initiated. The component must be handled carefully, to avoid damage to the coating, during placement in the device which will apply the loads. Loads are commonly applied with universal testing machines, hydraulic cylinders, dead weights, or some form of pressure. The essential element is for it to be possible to control the load and apply it relatively quickly in increments. The first increment of load should be limited to a value which will produce cracking of the coating only over a small area in the region of highest stress. The time to apply the load should be carefully recorded since this time will be needed to correct the calibration value of threshold strain for time of loading effects. The load increment should be maintained on the component for approximately 15 s to permit the coating crack pattern to develop fully. The load should then be removed from the component. After unloading, the entire surface of the coating should be examined carefully for coating cracks. As indicated later, when analysis of data is discussed, the ability to identify the load level at which the first crack appears significantly affects maximum-stress determinations. After each examination, the component can be loaded to a level approximately 20 percent higher than the previous level and the entire process repeated. Care must be taken not to load more frequently than once every 5 min to allow the coating to recover from stresses produced by relaxation of the coating under the load. The crack patterns located after each loading cycle are normally encircled with a line (*isoentatic line*) and marked with a number corresponding to the load which produced the pattern (see Fig. 4.19). The isoentatic lines which encircle the crack pattern separate the cracked and uncracked zones of the coating. The loadings can be continued until the highly stressed regions approach either the yield or fracture strains.

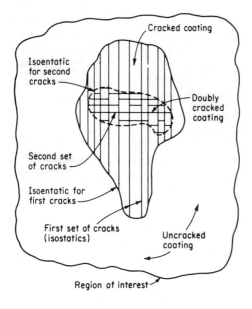

Cracked coating

Isoentatic for second cracks

Doubly cracked coating

Second set of cracks

Isoentatic for first cracks

First set of cracks (isostatics)

Uncracked coating

Region of interest

Figure 4.19 Method of constructing isoentatic lines on a field of brittle coating.

The isoentatic lines constructed after a given load are loci of points of approximately constant principal stress σ_1. In the example shown in Fig. 4.19, the coating cracked a second time under direct load, giving a second family of isostatics and, of course, a second isoentatic family. This second pattern of cracks, which is superimposed on the first pattern, has been encircled with a dashed line, which is a locus of points along which the secondary principal stress σ_2 is approximately constant.

4.9 CALIBRATION PROCEDURES [28]

The isoentatics represent lines of constant stress and are analogous to contour lines on a topographic map. In order to determine the stress associated with each line, it is necessary to calibrate the coating. This calibration is obtained by deflecting the specially prepared beams in the bending fixture shown in Fig. 4.20. This calibration procedure is repeated 10 to 15 times during the course of the experiment so that a number of values of ϵ_{t*} are obtained. These are then evaluated statistically. The mean value of the threshold strain used to compute the stresses associated with each isoentatic is obtained from the relation

$$\bar{x} = \frac{1}{N} \sum_{i=1}^{N} x_i \qquad (4.14)$$

where $\bar{x}$ = estimated mean value of threshold strain
x_i = ith value of threshold strain
N = total number of calibration values used in Eq. (4.14)

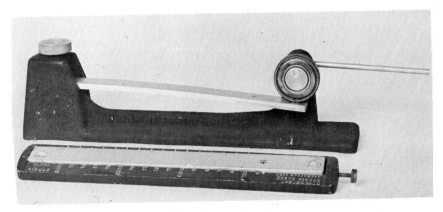

Figure 4.20 Calibration fixture, beams, and strain-measuring scale.

If the estimated mean of ϵ_{t*} is employed in the analysis instead of the results from a single calibration beam, a much more accurate estimate of the true threshold strain is obtained.

By employing multiple calibration techniques, an estimate of the standard deviation of the threshold strain can be computed and predictions on the accuracy of the brittle-coating determinations can be made.

An estimate of the standard deviation of the threshold strain is computed from

$$S_x = \sqrt{\frac{1}{N-1} \sum_{i=1}^{N} (x_i - \bar{x})^2} = \sqrt{\frac{1}{N-1} \left[\sum_{i=1}^{N} x_i^2 - \frac{1}{N} \left(\sum_{i=1}^{N} x_i \right)^2 \right]} \quad (4.15)$$

The true threshold strain x_{true} lies within the limits

$$x_{\text{true}} = \begin{cases} \bar{x} \pm S_x & 68\% \text{ of the time} \\ \bar{x} \pm 2S_x & 95\% \text{ of the time} \end{cases} \quad (4.16)$$

Thus the error introduced by the variable threshold strain is

$$\frac{\pm S_x}{\bar{x}} \quad 68\% \text{ of the time}$$

$$\pm \frac{2S_x}{\bar{x}} \quad 95\% \text{ of the time} \quad (4.17)$$

Typical values of the standard deviation S_x are in the range from 30 to 50 $\mu\text{in/in}$ ($\mu\text{m/m}$) for a typical brittle coating test conducted under laboratory conditions with temperature control to $\pm 1°F$ ($\frac{5}{9}°C$). For a coating with a mean value $\bar{x} = 500$ $\mu\text{in/in}$ ($\mu\text{m/m}$) for the threshold strain and $S_x = 50$ $\mu\text{in/in}$ ($\mu\text{m/m}$), the error associated with the brittle-coating determination will be within ± 10 percent 68 percent of the time and within ± 20 percent 95 percent of the time. This is the calibration error due to the inherent variation in the strength of the coating from point to point. This variation should not be considered the sole source of error in the analysis.

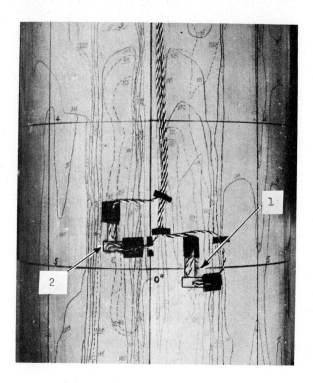

Figure 4.21 Strain gages mounted on the pressure vessel to calibrate the coating.

In calibration of the coating it is often impossible to duplicate the load-time function applied to the component. Loading with manually operated testing machines often requires 10 s or longer, while deflection of the calibration beam can be effected in approximately 1 s. Corrections for this difference in time to apply the load are made by using data similar to those presented in Fig. 4.16. It is often advantageous to use an independent method to check the accuracy of the brittle-coating calibration. Strain gages can be used to accomplish this independent check, as illustrated in Fig. 4.21. Two strain gages were mounted on an isoentatic line, the first perpendicular to the isostatics and the second parallel to the isostatics. The vessel was loaded to the pressure which formed the isoentatic, and the strain-gage readings were taken. These readings for the strains were employed with the well-known stress-strain relations to compute the values for the two principal stresses:

$$\sigma_1^s = \frac{E^s}{1 - v^{s2}} \left(\epsilon_1^s + v^s \epsilon_2^s \right) \qquad \sigma_2^s = \frac{E^s}{1 - v^{s2}} \left(\epsilon_2^s + v^s \epsilon_1^s \right) \tag{4.18}$$

where σ_1^s, σ_2^s = two principal stresses in specimen
E^s = modulus of elasticity of specimen
$\epsilon_1^s, \epsilon_2^s$ = two principal strains
v^s = Poisson's ratio for specimen material

Once the principal stresses σ_1^s and σ_2^s were determined, the threshold strains were computed from Eq. (4.1), which gave

$$\epsilon_{t*1} = \frac{\sigma_1^s}{E^s} \quad \text{and} \quad \epsilon_{t*2} = \frac{\sigma_2^s}{E^s} \tag{4.19}$$

where ϵ_{t*1} and ϵ_{t*2} are the threshold strains for the first and second families of isoentatics, respectively.

Another advantage of utilizing strain gages on the actual component is that the calibration tends to account for the effect of the biaxial stress field on the threshold strain. Also, when a double set of cracks is obtained (one before the other), it is essential that the value of ϵ_{t*2} be calibrated in this manner since it is always greater than ϵ_{t*1}.

4.10 ANALYSIS OF BRITTLE-COATING DATA [29, 30]

If the stresses in the component are linear with respect to the load, the principal stresses associated with a given isoentatic can be determined from the following modifications of Eq. (4.1):

$$\sigma_1^s \Big|_i = E^s \epsilon_{t*1} \frac{p_s}{p_i} \qquad \sigma_2^s \Big|_i = E^s \epsilon_{t*2} \frac{p_s}{p_i} \tag{4.20}$$

where $\sigma_1^s \big|_i$, $\sigma_2^s \big|_i$ = principal stresses in specimen associated with ith isoentatic
p_s = standard reference pressure or load to which all stresses are related; a value of 100 lb/in² was arbitrarily selected as the reference pressure in the following example

Table 4.1 Values of σ_1^s and σ_2^s associated with the isoentatics on the pressure vessel†

Isoentatic number or pressure p_i, lb/in²	Maximum principal stress σ_1^s, lb/in²	Minimum principal stress σ_2^s, lb/in²
145	14,200	
172	11,900	
195	10,500	
235	8,750	
275	7,460	
315	6,520	
355	5,790	
410	5,010	5520
480	4,280	4720
550	3,740	4120

† $\epsilon_{t*1} = 685 \ \mu\text{in/in}$ and $\epsilon_{t*2} = 775 \ \mu\text{in/in}$

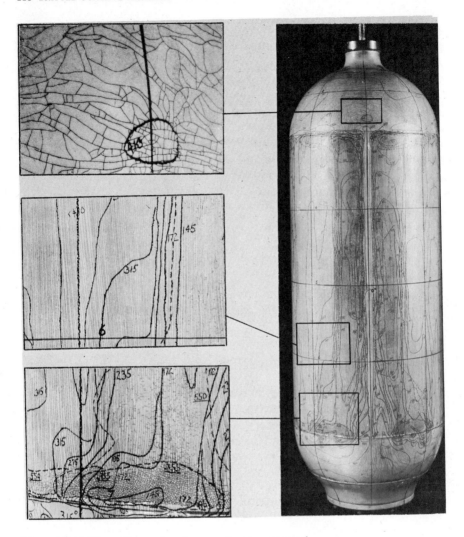

Figure 4.22 Brittle-coating crack patterns on the pressure vessel.

p_i = pressure which caused crack pattern to extend to isoentatic numbered i

and E^s and ϵ_{t*1}, ϵ_{t*2} have been defined previously.

As an example, consider the brittle coating pattern for the thin-walled pressure vessel shown in Fig. 4.22. Values of σ_1^s and σ_2^s computed for each isoentatic by using Eqs. (4.20) are shown in Table 4.1.

By comparing the brittle-coating patterns shown in Fig. 4.22 with the values shown in Table 4.1 it is possible to obtain the magnitudes of the principal stresses over nearly the entire surface of the pressure vessel. The values for the stresses are assumed to be linearly proportional to the pressure. This assumption is justified over most of the surface of the vessel; however, at the longitudinal weld the cylinder is somewhat flattened on each side of the weld. At these imperfections the stresses will not be linear with respect to the pressure, because as the pressure increases, the flattening of the shell becomes less pronounced and the effects of the imperfections on the stress level decrease. The results of the brittle-coating analysis showed clearly that the stresses in the cylindrical pressure vessel were appreciably different from those theoretically predicted by using elementary shell theory. The local imperfections adjacent to the weld produced significant stress concentrations. Bending stresses induced by some of these imperfections are about equal to the membrane stresses and effectively increase the value of the stresses by a factor of 2.

It should be noted that the brittle coating responded over nearly the entire outer surface of the vessel, giving information regarding the stresses near the longitudinal weld, the girth weld, and the forward head, as well as the aft head. Large amounts of data were obtained in a very short time. It was necessary to employ a few strain gages at selected points to complete the stress analysis. It should be emphasized that the strain gages were used to supplement the brittle-coating data, for often the reverse situation is true, namely, that the brittle-coating data are used to supplement the strain-gage data.

EXERCISES

4.1 List the advantages of the brittle-coating method.

4.2 What are the primary disadvantages of the brittle-coating method?

4.3 Determine the stresses in a brittle coating applied to a component fabricated from steel ($E = 200$ GPa, $v = 0.30$) when the specimen stresses are $\sigma_1^s = 210$ MPa and $\sigma_2^s = -140$ MPa (*a*) for a resin-based coating with $E^c = 1.40$ GPa and $v^c = 0.42$ and (*b*) for a ceramic-based coating with $E^c = 70$ GPa and $v^c = 0.25$.

4.4 If the threshold strain in a brittle coating is 500 μm/m, what is the corresponding state of stress in the coating during calibration (*a*) for a resin-based coating with the properties given in Exercise 4.3 and (*b*) for a ceramic-based coating with the properties given in Exercise 4.3?

4.5 Describe qualitatively how the state of stress in the coating on a calibration beam changes after the coating fails by cracking. Extend this description to include the dependence of crack density on the magnitude of the stress applied to the coating after it first fails. Also, describe how the coating thickness influences the crack density. *Note:* Crack density is the number of cracks per unit length.

4.6 The maximum-normal-strain theory of failure for brittle coatings indicates that cracks will form whenever

$$\epsilon_1^c \geq \epsilon_{t*} \qquad \epsilon_2^c \geq \epsilon_{t*} \qquad \epsilon_3^c \geq \epsilon_{t*}$$

Show that the failure equations for the coating in terms of the stresses applied to the specimen are

$$\frac{\sigma_1^s}{\sigma_1^*} - \frac{v^s \sigma_2^s}{\sigma_1^*} \geq 1 \qquad\qquad (4.21a)$$

$$\frac{\sigma_2^s}{\sigma_1^*} - \frac{v^s \sigma_1^s}{\sigma_1^*} \geq 1 \qquad\qquad (4.21b)$$

$$-\frac{v^c(1 - v^s)}{1 - v^c}\left(\frac{\sigma_1^s}{\sigma_1^*} + \frac{\sigma_2^s}{\sigma_1^*}\right) \geq 1 \qquad\qquad (4.21c)$$

4.7 Using the results of Eqs. (4.21), construct the failure diagram for the maximum-normal-strain theory of failure.

4.8 Failure of the coating according to Eqs. (4.21a) and (4.21b) results in an orthogonal set of coating cracks as illustrated in Fig. 4.6. Describe the failure exhibited by the coating when the conditions of Eq. (4.21c) are satisfied.

4.9 The maximum-normal-stress theory of failure for brittle coatings indicates that cracks will form whenever

$$\sigma_1^c \geq \sigma_{ut} \qquad \sigma_2^c \geq \sigma_{ut}$$

where σ_{ut} is the ultimate tensile strength of the coating. Show that the failure equations for the coating in terms of the stresses applied to the specimen are

$$\frac{\sigma_1^s}{\sigma_1^*} + \frac{(v^c - v^s)\sigma_2^s}{(1 - v^c v^s)\sigma_1^*} \geq 1 \qquad \frac{\sigma_2^s}{\sigma_1^*} + \frac{(v^c - v^s)\sigma_1^s}{(1 - v^c v^s)\sigma_1^*} \geq 1 \tag{4.22}$$

4.10 Using the results of Eqs. (4.22), construct the failure diagram for the maximum-normal-stress theory of failure.

4.11 Describe why the condition $\sigma_3^c \geq \sigma_{ut}$ was not considered in Exercise 4.9.

4.12 Using a resin-based coating with the properties given in Exercise 4.3 on a component fabricated from aluminum ($E = 70$ GPa, $v = 0.03$), specify the value of K (the strength ratio of the coating) which will minimize the effect of σ_2^s on the failure of the coating. Assume the coating follows Mohr's theory of failure.

4.13 Repeat Exercise 4.12 for a component fabricated from (a) glass with $v^s = 0.25$, (b) steel with $v^s = 0.29$, and (c) plastic with $v^s = 0.42$.

4.14 Sketch the brittle-coating pattern which would be observed on the following components:
(a) A tensile specimen
(b) A circular shaft subjected to pure torsion
(c) A circular disk subjected to diametrical compression
(d) A circular ring subjected to diametrical compression
(e) A cylindrical thin-walled vessel subjected to internal pressure
(f) A spherical thin-walled vessel subjected to internal pressure

4.15 What is the significance of uncracked areas of the brittle-coating field at the conclusion of a brittle-coating test? Are these regions of any importance with respect to the economic aspects of a given design? What are some procedures which can be employed to gain further information regarding the stresses in these regions?

4.16 For a resin-based brittle coating compute the maximum possible crack width with a coating thickness of 0.004 in (0.10 mm). Assume the cracks have a uniform spacing of 0.060 in (1.52 mm).

4.17 What are the consequences of being unable to locate a cracked region of coating during a brittle-coating test? Can dye etchant or the electrified-particle method of crack detection be employed during the test with (a) resin-based coatings and (b) ceramic-based coatings?

4.18 Why are ceramic-based brittle coatings less sensitive to temperature changes than resin-based coatings? What is the anticipated change in threshold strain of a resin-based coating for a 5°F (2.8°C) change in temperature? What change in strain sensitivity would be encountered with All-Temp on 410 stainless steel as the temperature is increased from 100 to 400°F (38 to 204°C)?

4.19 Select the proper grade of All-Temp for a strain sensitivity of 500 μin/in (μm/m) on the following materials:

(a) 1040 carbon steel   (b) Titanium
(c) Class 60 gray cast iron   (d) 301 stainless steel
(e) 316 stainless steel   (f) Monel
(g) Hastelloy A   (h) Inconel X

4.20 Select the proper grade of Tens-lac resin-based brittle coating for curing and testing under the following atmospheric conditions:

Temperature, °C	30	25	20	10
Relative humidity, %	60	70	40	30

4.21 What is the effect of poor thickness control during application of a resin-based brittle coating? What are normal variations in the thickness of a coating applied to a component with an irregular shape? How can the influence of these variations in the thickness of the coating be minimized?

4.22 Discuss temperature control to minimize errors in a brittle-coating test (with a resin-based coating) when the test is conducted in the field.

4.23 Why is the thickness of a nonflammable type of resin-based coating limited to 0.002 to 0.004 in (0.05 to 0.10 mm) when the thickness of a resin-based coating containing carbon disulfide as the solvent can be two or three times greater?

4.24 Stresscoat ST-75 with a strain sensitivity of 700 μm/m is employed in a test of a structure loaded with sand bags. If 12 min is required to position all the bags, compute the corrected strain sensitivity and comment on the advisability of the test procedure. If the loading is accomplished with hydraulic rams, the loading time (with manual control) can be reduced to 10 s. Determine the corrected sensitivity in this instance.

4.25 Comment on the advisability of using servo-controlled loading methods for brittle-coating testing.

4.26 Prepare test specifications, which include surface preparation, coating selection, coating application procedures, coating curing cycle, loading schedule, and coating inspection and marking procedures for a brittle-coating test of the following components:
   (*a*) An internal combustion engine head
   (*b*) A large T-joint on a gas transmission line
   (*c*) A crane hook
   (*d*) A passenger-car wheel
   (*e*) A pillow block bearing housing
   (*f*) A structural beam
   (*g*) A pressure vessel with pipe supports

4.27 The isoentatic pattern shown in Fig. E4.27 was produced during a direct loading test of an

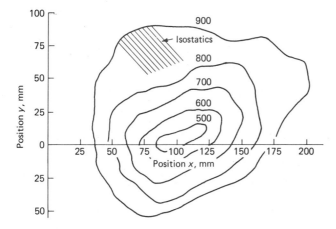

Figure E4.27

aluminum ($E = 70$ GPa, $v = 0.33$) component. Load levels of 500, 600, 700, 800, and 900 N were used during the test. Using only this isoentatic pattern and a corrected calibration value of threshold strain of 570 μm/m, estimate the distribution of σ_1^s along the lines:

(a) $y = 0$ (b) $y = 25$ mm   (c) $y = -25$ mm
(d) $x = 75$ mm   (e) $x = 100$ mm   (f) $x = 125$ mm

(g) What is your estimate of the maximum value of σ_1^s?

4.28 The same aluminum component described in Exercise 4.27 was retested using the relaxation loading method to obtain the isostatic and isoentatic patterns associated with σ_2^s shown in Fig. E4.28. The corrected calibration value of threshold strain was 620 μm/m. Estimate the distribution of σ_2^s along the lines:

(a) $y = 0$ (b) $y = 25$ mm   (c) $y = -25$ mm
(d) $x = 75$ mm   (e) $x = 100$ mm   (f) $x = 125$ mm

(g) What is your estimate of the maximum value of σ_2^s?

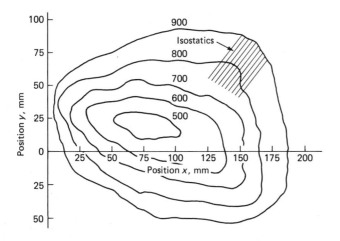

Figure E4.28

4.29 Use the results from Exercise 4.28 together with the Mohr theory of failure to improve your estimate of σ_1^s as determined in Exercise 4.27. For the Mohr theory of failure assume that the coating has a strength ratio $K = 4$. Improve the estimate along the lines:

(a) $y = 0$ (b) $y = 25$ mm   (c) $y = -25$ mm
(d) $x = 75$ mm   (e) $x = 100$ mm   (f) $x = 125$ mm

4.30 A steel machine component is being studied using brittle-coating methods. At a critical point on the specimen, cracks begin to appear when the load is raised to 5 kN. The corrected calibration threshold strain for the coating at this time is 500 μm/m. The load is raised in increments, and at a level of 20 kN cracks perpendicular to the first set begin to appear at the same point. The corrected calibration threshold strain for the second family of cracks is 600 μm/m. Determine the two principal stresses in the specimen at a load level of 5 kN by assuming that the coating cracks according to the maximum-normal-stress theory of failure. Assume that the coating is a resin-based coating with the properties given in Exercise 4.3.

REFERENCES

1. Dietrich, O., and E. Lehr: Das Dehnungslinienverfahren, *Z. Ver. Dtsch. Ing.*, vol. 76, pp. 973–982, 1932.
2. Portevin, A., and M. Cymboliste: Procédé d'étude de la distribution des efforts élastiques dans les pièces métalliques, *Rev. Metal.*, vol. 31, pp. 147–158, 1934.
3. Sauerwald, F., and H. Wieland: Über die Kerbschlagprobe nach Schule-Moser, *Z. Metalkd.*, vol. 17, pp. 358–364, 392–399, 1925.
4. Ellis, G.: "Strain Indicating Lacquers," master's thesis, Massachusetts Institute of Technology, Department of Aeronautical Engineering, 1937.
5. Ellis, G.: Practical Strain Analysis by Use of Brittle Coatings, *Proc. SESA*, vol. I, no. 1, pp. 46–53, 1943.
6. Singdale, F. N.: Improved Brittle Coatings for Use under Widely Varying Temperature Conditions, *Proc. SESA*, vol. XI, no. 2, pp. 173–178, 1954.
7. Durelli, A. J., E. Phillips, and C. H. Tsao: "Introduction to the Theoretical and Experimental Analysis of Stress and Strain," McGraw-Hill Book Company, New York, 1958.
8. Racine, F. L., H. G. Taracks, and J. J. Killinger: Development of a Non-flammable Resin System for Engineering Stress-Strain Applications, *Gen. Mot. Eng. J.*, 4th quarter, 1965, pp. 8–12.
9. Hickson, V. M.: Some New Techniques in Strain Measurement: A Survey, in chap. 11, O. C. Zienkiewicz and G. S. Holister (eds.), "Stress Analysis," John Wiley & Sons, Inc., New York, 1965.
10. Durelli, A. J., and C. H. Tsao: Use of Brittle Coating Data in Stress Analysis, *Proc. SESA*, vol. XI, no. 1, pp. 181–196, 1953.
11. Stokey, W. F.: Elastic and Creep Properties of Stresscoat, *Proc. SESA*, vol. X, no. 1, pp. 179–186, 1952.
12. Durelli, A. J., R. H. Jacobson, and S. Okubo: Further Study of Properties of Stresscoat, *Proc. SESA*, vol. XIII, no. 1, pp. 35–53, 1955.
13. Durelli, A. J., and T. N. DeWolf: Law of Failure of Stresscoat, *Proc. SESA*, vol. VI, no. 2, pp. 68–83, 1949.
14. Durelli, A. J.: Experimental Determination of Isostatic Lines, *J. Appl. Mech.*, vol. 64, pp. A155–160, 1942.
15. Durelli, A. J.: What Kind of Information Does Brittle Coating Give? *Prod. Eng.*, June–July, 1948.
16. Staats, H.: Principles of Stresscoat, *Magnaflux Corp. Bull.*, 2d. ed., 1967.
17. Staats, H. N., and S. J. Baranowski: Calibrated Porcelain Enamel Coatings, *Am. Ceram. Soc. Bull.*, vol. 35, no. 4, 1956.
18. Rollins, C. T., and E. K. Lynn: Experience with Ceramic Coatings, *Exp. Mech.*, vol. 8, no. 3, 1968.
19. Singdale, F. N.: Method of Determining Strain Values in Rigid Articles, U.S. Patent 2,724,964, Nov. 29, 1955.
20. Ellis, G.: Resinous Composition for Determining the Strain Concentration in Rigid Articles, U.S. Patent 2,428,559, Oct. 7, 1947.
21. Tens-lac Brittle Lacquer, *Bull.* TL-205, Photolastic, Inc.
22. Dally, J. W., and A. J. Durelli: Variables Affecting Brittle Coating in Stress Analysis, *Prod. Eng., Des. Dig. Issue*, 1959.
23. Durelli, A. J., and J. W. Dally: Some Properties of Stresscoat under Dynamic Loading, *Proc. SESA*, vol. XV, no. 1, pp. 57–64, 1957.
24. Murthy, P. N.: Theoretical Investigation of Creep and Crack Density Studies in Stresscoat, *Proc. SESA*, vol. XV, no. 1, pp. 57–64, 1957.
25. Dally, J. W., and A. J. Durelli: Stress Analysis of a Reactor Head Closure, *Proc. SESA*, vol. XVII, no. 2, pp. 71–87, 1959.
26. DeForest, A. V., G. Ellis, and F. B. Stern, Jr.: Brittle Coatings for Quantitative Strain Measurements, *J. Appl. Mech.*, vol. 64, pp. 184–188, 1942.
27. Durelli, A. J., J. W. Dally, and S. Morse: Experimental Study of Large-Diameter Thin-walled Pressure Vessels, *Exp. Mech.*, vol. 1, no. 2, pp. 33–42, 1961.

28. Durelli, A. J., S. Okubo, and R. H. Jacobson: Study of Some Properties of Stresscoat, *Proc. SESA*, vol. XII, no. 2, pp. 55–76, 1955.
29. Durelli, A. J., E. A. Phillips, and C. H. Tsao: "Introduction to the Theoretical and Experimental Analysis of Stress and Strain," McGraw-Hill Book Company, New York, 1958.
30. DeForest, A. V., and F. B. Stern, Jr.: Stresscoat and Wire Strain Gage Indications of Residual Stresses, *Proc. SESA*, vol. II, no. 1, pp. 161–169, 1944.

THREE

STRAIN-MEASUREMENT METHODS AND RELATED INSTRUMENTATION

INTRODUCTION TO STRAIN MEASUREMENTS

5.1 DEFINITION OF STRAIN AND ITS RELATION TO EXPERIMENTAL DETERMINATIONS

A state of strain may be characterized by its six cartesian strain components or, equally well, by its three principal strain components with the three associated principal directions. The six cartesian components of strain are defined in terms of the displacement field by the following set of equations when the strains are small (normally the case for elastic analyses):

$$\epsilon_{xx} = \frac{\partial u}{\partial x} \qquad \epsilon_{yy} = \frac{\partial v}{\partial y} \qquad \epsilon_{zz} = \frac{\partial w}{\partial z}$$

$$\gamma_{xy} = \frac{\partial v}{\partial x} + \frac{\partial u}{\partial y} \qquad \gamma_{yz} = \frac{\partial w}{\partial y} + \frac{\partial v}{\partial z} \qquad \gamma_{zx} = \frac{\partial u}{\partial z} + \frac{\partial w}{\partial x} \qquad (2.4)$$

The ϵ components are normal strains, and ϵ_{xx}, for instance, is defined as the change in length of a line segment parallel to the x axis divided by its original length. The γ components are shearing strains, and γ_{xy}, for instance, is defined as the change in the right angle formed by line segments parallel to the x and y axes.

For the most part, strain-gage applications are confined to the free surfaces of a body. The two-dimensional state of stress existing on this surface can be expressed in terms of three cartesian strains ϵ_{xx}, ϵ_{yy}, and γ_{xy}. Thus, if the two displacements u and v can be established over the surface of the body, the strains can be determined directly from Eqs. (2.4). In certain isolated cases, the most appropriate approach for the determination of the stress and strain field is the determination of the displacement field. As an example, consider the very simple problem of a transversely loaded beam. The deflections of the beam $w(x)$ along its longitudinal axis can be accurately determined with relatively simple experimental techniques. The strains and stresses are related to the deflection $w(x)$ of the beam by

$$\epsilon_{xx} = \frac{z}{\rho} = z\frac{d^2w}{dx^2} \qquad \sigma_{xx} = E\epsilon_{xx} = Ez\frac{d^2w}{dx^2} \tag{5.1}$$

where ρ is the radius of curvature of the beam and z is the distance from the neutral axis of the beam to the point of interest

Measurement of the transverse displacements of plates can also be accomplished with relative ease, and stresses and strains computed by employing equations similar to (5.1). In the case of a more general body, however, the displacement field cannot readily be measured. Also, the conversion from displacements to strains requires a determination (by differentiation) of the gradients of experimentally determined displacements at many points on the surface of the specimen. Since the displacements are often difficult to obtain and the differentiation process is subject to large errors, it is advisable to employ a strain gage of one form or another to measure the surface strains directly.

Examination of Eqs. (2.4) shows that the strains ϵ_{xx}, ϵ_{yy}, and γ_{xy} are really the slopes of the displacement surfaces u and v. Moreover, these strains are not, in general, uniform; instead, they vary from point to point. The slopes of the displacement surfaces cannot be established unless the in-plane displacements u and v can be accurately established. Since the in-plane displacements are often exceedingly small in comparison with the transverse (out-of-plane) displacements mentioned previously, their direct measurement over the entire surface of a body is exceedingly difficult. To circumvent this difficulty, one displacement component is usually measured over a small portion of the body along a short line segment, as illustrated in Fig. 5.1. This displacement measurement is converted to strain by the relationship

$$\epsilon_{xx} = \frac{l_x - l_0}{l_0} = \frac{\Delta u}{\Delta x} \tag{5.2}$$

where $\Delta u = l_x - l_0$ is the displacement in the x direction over the length of the line segment $l_0 = \Delta x$. Strain measured in this manner is not exact since the determination is made over some finite length l_0 and not at a point, as the definition for strain ϵ requires. The error involved in this approach depends upon the strain gradient and the length of the line segment l_0. If the strain determination is considered to represent the strain which occurs at the center of the line segment l_0, that is, point x_1, the error involved for various strain gradients is

Case 1, strain constant: $\epsilon_{xx} = k_1$ (no error is induced)
Case 2, strain linear: $\epsilon_{xx} = k_1 x + k_2$ (no error is induced)
Case 3, strain quadratic: $\epsilon_{xx} = k_1 x^2 + k_2 x + k_3$

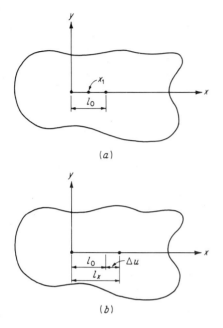

(a)

(b)

Figure 5.1 Strain measurement over a short line segment of length l_0: (a) before deformation; (b) after deformation.

In case 3, an error is involved since the strain at the midpoint x_1 is not equal to the average strain over the gage length l_0. The average strain over the gage length l_0 can be computed as

$$\epsilon_{av} = \frac{\int_0^{l_0} (k_1 x^2 + k_2 x + k_3)\, dx}{l_0} = \frac{k_1 l_0^2}{3} + \frac{k_2 l_0}{2} + k_3 \tag{a}$$

and the strain at the midpoint $x_1 = l_0/2$ is given by

$$\epsilon_{xx}\Big|_{x_1} = \frac{k_1 l_0^2}{4} + \frac{k_2 l_0}{2} + k_3 \tag{b}$$

The difference between the average and midpoint strains represents the error involved and is given by

$$\text{Error} = \frac{k_1 l_0^2}{12} \tag{5.3}$$

In this example the error involved depends upon the values of k_1 and l_0. If the gradient is sharp, the value of k_1 will be significant and the error induced will be large unless l_0 is reduced to an absolute minimum. Other examples corresponding to cubic and quartic strain distributions can also be analyzed; however, the fact that an error is induced is established by considering any strain distribution other than a linear one.

In view of the error introduced by the length of the line segment in certain strain fields, great effort has been expended in reducing the gage length l_0. Two

Figure 5.2 Berry-type strain gage.

factors complicate these efforts. (1) Mechanical difficulties are encountered when l_0 is reduced. Regardless of the type of gage employed to measure the strain, it must have a certain finite size and a certain number of parts. As the size is reduced, the parts become smaller and the dimensional tolerances required on each become prohibitive. (2) The strain to be measured is a very small quantity. Suppose, for example, that strain determinations are to be made with an accuracy of ± 1 μin/in (μm/m) over a gage length of 0.1 in (2.5 mm). The strain gage must measure the corresponding displacement to an accuracy of $\pm 1 \times 10^{-6} \times 0.1 = \pm 1 \times 10^{-7}$ in (2.5 nm) or one-ten-millionth of an inch. These size and accuracy requirements place heavy demands on the talents of investigators in the area of strain-gage development.

The smallest gage developed and sold commercially to date is the electrical-resistance type. This gage is prepared from an ultrathin alloy foil which is photoetched to produce the intricate grid construction with a gage length of only 0.008 in (0.2 mm). On the other hand, mechanical strain gages are still employed in civil engineering structural applications where the gage length $l_0 = 8$ in or 200 mm (Berry strain gage). These Berry gages are rugged, simple to use, and sufficiently accurate in structural applications where the strain distribution is approximately linear over the 8-in (200-mm) gage length. A Berry-type strain gage is shown in Fig. 5.2.

5.2 PROPERTIES OF STRAIN-GAGE SYSTEMS

Historically, the development of strain gages has followed many different paths, and gages have been developed which are based on mechanical, optical, electrical, acoustical, and pneumatic principles. No single gage system, regardless of the principle upon which it is based, has all the properties required of an optimum gage. Thus a need exists for a wide variety of gaging systems to meet the requirements of a wide range of different engineering problems involving strain measurement.

Some of the optimum characteristics commonly used to judge the adequacy of a strain-gage system for a particular application are the following:

1. The calibration constant for the gage should be stable; it should not vary with either time or temperature.
2. The gage should be able to measure strains with an accuracy of ± 1 μin/in (μm/m) over a strain range of 10 percent.
3. The gage size, i.e., the gage length l_0 and width w_0, should be small so that strain at a point is adequately approximated.
4. The response of the gage, largely controlled by its inertia, should be sufficient to permit recording of dynamic strains.
5. The gage system should permit on-location or remote readout.
6. The output from the gage during the readout period should be independent of temperature and other environmental parameters.
7. The gage and the associated auxiliary equipment should be economically feasible.
8. The gage system should not involve overcomplex installation and operational techniques.
9. The gage should exhibit a linear response to strain.
10. The gage should be suitable for use as the sensing element in other transducer systems where an unknown quantity such as pressure is measured in terms of strain.

No single strain-gage system satisfies all these optimum characteristics. The strain-gage system for a particular application can be selected after proper consideration is given to each of these characteristics in terms of the requirements of the measurement to be made. Over the last 50 years a large number of systems, with wide variations in design, have been conceived, developed, and marketed; however, each system has four basic characteristics which deserve additional consideration. These are the gage length l_0, the gage sensitivity, the range of strain, and the accuracy or precision of the readout.

Strains cannot be measured at a point with any type of gage, and, as a consequence, nonlinear strain fields cannot be measured without some degree of error being introduced. In these cases, the error will definitely depend on the gage length l_0 in the manner described in Sec. 5.1 and may also depend on the gage width w_0. The gage size for a mechanical strain gage is characterized by the distance between the two knife-edges in contact with the specimen (the gage length l_0) and by the width of the movable knife-edge (the gage width w_0). The gage length of the metal-film resistance strain gage is determined by the length of the strain portion of the grid, and the width w_0 is determined by the width of the grid. In selecting a gage for a given application, gage length is one of the most important considerations.

A second basic characteristic of a strain gage is its sensitivity. Sensitivity is the smallest value of strain which can be read on the scale associated with the strain gage. The term sensitivity should not be mistaken for accuracy or precision, since very large values of magnification can be designed into a gage to increase its sensitivity; but friction, wear, and deflection introduce large errors which limit the accuracy. In certain applications gages can be employed with sensitivities of less than 1 μin/in (μm/m) if proper procedures are established. In other applications, where sensitivity is not important, 50 to 100 μin/in (μm/m) is often quite sufficient. The choice of a gage is dependent upon the degree of sensitivity required, and quite often the selection of a gage with a very high sensitivity when it is not really necessary needlessly increases the complexity of the measuring method.

A third basic characteristic of a strain gage is its range. Range represents the maximum strain which can be recorded without resetting or replacing the strain gage. The range and sensitivity are interrelated since very sensitive gages respond to small strains with appreciable indicator deflections and the range is usually limited to the full-scale deflection of the indicator. Often it is necessary to compromise between the range and sensitivity characteristics of a gage to obtain reasonable performance for both these categories.

The final basic consideration is the accuracy or precision. As was previously pointed out, sensitivity does not ensure accuracy. Usually the very sensitive instruments are quite prone to errors unless they are employed with the utmost care. In a mechanical strain gage, inaccuracies may result from lost motion such as backlash in a gear train, friction, temperature changes, wear in the mechanism, slippage, or flexure or deflection of the components. On all strain gages there is a readout error whether the gage is manually recorded or the output is read on a digital printer.

5.3 TYPES OF STRAIN GAGE

The problem encountered in measuring strain is to determine the motion between two points some distance l_0 apart. The physical principles which have been employed to accomplish this task are very numerous, and a complete survey will not be attempted; however, a few of the more applicable methods will be covered briefly. The principles employed in strain-gage construction can be used as the basis for classifying the gages into the following four groups:

1. Mechanical
2. Optical
3. Electrical
4. Acoustical

A. Mechanical Strain Gages [1–5]

The mechanical gage to be considered here is the Huggenberger tensometer, which is the most popular and one of the most accurate mechanical strain gages in use today. The gage shown in Figs. 5.3 and 5.4 is based entirely on mechanical principles. The displacement of the movable knife-edge is multiplied by a set of compound levers until a magnification of displacement up to 2000 is obtained.

The operation of the Huggenberger tensometer is illustrated in Fig. 5.4. The frame C supports the two knife-edges A and B as well as the entire lever system and the indicating scale Z. The knife edge A is fixed rigidly to the frame C, whereas the knife-edge B rotates about a fixed point on the frame a distance v_1 from the specimen. This knife-edge thus serves as the first lever in the system, and since it is also an integral part of the arm H, the displacement Δl is transmitted to point M, where it has been magnified to a distance Δs. The value of Δs is given by the simple lever rule as

$$\Delta s = \frac{v_2}{v_1} \Delta l$$

Figure 5.3 The Huggenberger tensometer.

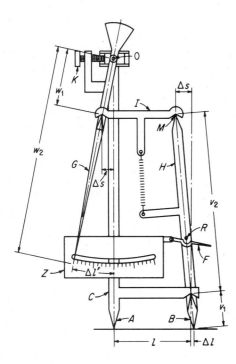

Figure 5.4 The mechanical-lever system employed to magnify displacements in the Huggenberger tensometer.

The motion at point M, that is, Δs, is transmitted through the yoke I to the lever G. The lever G, which rotates about the point 0, serves as the pointer for the strain gage. The pointer moves over a distance $\Delta l'$, which is related to Δs and Δl by the expression

$$\Delta l' = \frac{w_2}{w_1} \Delta s = \frac{w_2 v_2}{w_1 v_1} \Delta l \tag{5.4}$$

The magnification factor is thus $w_2 v_2 / w_1 v_1$, which in the five models commercially available ranges between 300 and 2000.

The Huggenberger tensometer can be reset during a series of strain determinations by rotating the screw which moves the pivot point 0 without disturbing the rest of the linkage system. The gage can be locked when not in use through the pin R and lever F to avoid unnecessary wear on the components. Specifications for the type A gage, one of the most popular Huggenberger models, are listed below:

Multiplication approximately 1200 (exact calibration value provided with the instrument)
Gage length: 1 in convertible to $\frac{1}{2}$ in
Scale 38 divisions: 0.05 in each division
Strain range: 0.004 in without resetting
Dimensions: $6\frac{1}{2}$ in high, $2\frac{1}{16}$ in wide, $\frac{5}{8}$ in deep
Weight: $2\frac{1}{2}$ oz
Accuracy: ± 10 μin/in

The gage is mounted by clamps, springs, or screw pressure applied to the frame C in sufficient magnitude to set the knife-edges into the specimen. This is perhaps the most objectionable feature in the application of the Huggenberger gage. The knife-edges must be set well enough to avoid slippage of either knife-edge yet not tight enough to damage the gage or the specimen. This clamping process is further complicated by the great height of the gage, which makes it quite unstable in mounting. However, when proper fixturing is available and the operator has developed some skill, the mounting procedure can be accomplished in a reasonable time.

The instrument is definitely limited to static measurements of strain since its size and inertia rule out any chance of a reasonable frequency response, which is required in dynamic applications. Care should also be exercised in keeping the instrument in calibration. The knife-edge B has a tendency to wear, thus shortening the distance v_1 and changing the multiplication factor.

In comparison with other mechanical gages the Huggenberger and the Johansson Mikrokator are the most accurate, sensitive, and reliable of the entire line. However, since the advent of the electrical-resistance strain gage, the general use of mechanical strain gages has declined appreciably until today they are used only in special applications.

Another strain gage based on mechanical principles is the scratch gage (Fig. 5.5), consisting of two base plates L and S, which can be welded, bolted, or adhesively bonded to the component being evaluated. When the component is subjected to loads, the base plates displace relative to each other and cause the stylus D to move relative to the target plate T. The stylus records the displacement Δu by scratching the target plate.

The target plate is rotated by a wire drive system, which is activated by the relative displacements between the base plates. For instance, when the gage is subjected to compression and plates L and S are brought closer together, wire B, which is guided and supported by tube BT, causes the target to rotate counterclockwise. This combined action—the direct movement of the stylus together with the rotation of the target—produces a slanted line on the target plate and effects separation of the individual displacements associated with different levels of load.

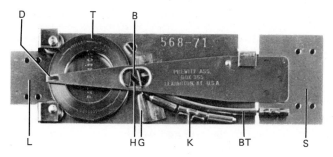

Figure 5.5 The scratch strain-gage system. (*Prewitt Associates.*)

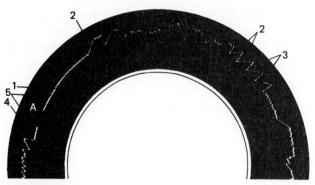

1. Beginning of recording.
2. Compression movement, target rotates.
3. Tension movement, target stationary.
4. End of self-recorder data.
5. Base line arcs.

Figure 5.6 Enlargement of a target ring with a scratch record. (*Prewitt Associates.*)

An enlargement of a typical scratch record on a target plate is shown in Fig. 5.6.

The target plates are removed from the gage after completion of a test, and the scratch patterns are analyzed with an optical comparator or with a special microscope designed and calibrated for use with the target plates. The width of a scratch on a target plate is approximately 0.0005 in (0.0125 mm) and a movement between plates of 0.004 in (0.10 mm) can be distinguished and measured with reasonable accuracy.

The sensitivity and accuracy of a strain measurement depends upon the gage length and the gradient of the strain field. Gages are available in sizes from 3 to 100 in (75 to 2500 mm); thus, precision measurements (sensitivity between 50 and 100 μin/in or μm/m) can be made if the strain field is either constant or linear so that the longer gage lengths can be employed. With shorter gage lengths the sensitivity is significantly reduced.

The target rotates a small amount with each cycle of strain above some minimum threshold level; thus, continuous records of strain cycles above the threshold value can be obtained over an extended period of time (during a typical flight of an aircraft for example). The life of a given target depends on the magnitudes of the relative displacement between base plates; however, 1000 cycles of strain above the threshold value can be used as a rough guide for the life. Since the target plates can be changed, the life of a gage installation can be extended by periodic replacement of the plates.

The gages have a limited capability for the measurement of dynamic strains and can withstand accelerations of approximately $33g$ without significant effect on the accuracy of the measurement. Due to the size and weight of the mechanical components, the frequency response is limited, and most suitable applications involve measurement of quasi-static strains.

B. Optical Strain Gages [6–10]

In the first edition of this text, the Tuckerman optical strain gage, marketed by the American Instrument Company, was described. This gage is similar in some ways to the Huggenberger type of mechanical gage in that it incorporates knife-edge attachment to the specimen. The chief difference between the two gages is in the lever system, where the optical gage substitutes light rays for the mechanical levers. This substitution decreases the size and inertia of the device appreciably and permits the optical gage to be employed at low frequencies in dynamic applications. Since 1965, considerable research effort has been devoted to the area of optical methods of experimental stress analysis. The availability of gas and ruby lasers as monochromatic, collimated, and coherent light sources has led to several new developments in strain gages. Two of these developments—the diffraction strain gage and the interferometric strain gage—are described to indicate the capabilities of optical gages which use coherent light.

The diffraction strain gage is quite simple in construction; it consists of two blades that are bonded or welded to the component, as illustrated in Fig. 5.7. The two blades are separated by a distance b to form a narrow aperture and are fixed to the specimen along the edges (see Fig. 5.7) to give a gage length l. A beam of collimated monochromatic light from a helium-neon laser is directed onto the aperture to produce a diffraction pattern that can be observed as a line of dots on a screen a distance R from the aperture. An example of a diffraction pattern is illustrated in Fig. 5.8.

When the distance R to the screen is very large compared with the aperture width b, the distribution of the intensity I of light in the diffraction pattern (see Sec. 11.5) is

$$I = A_0^2 \frac{\sin^2 \beta}{\beta^2} \tag{5.5}$$

where A_0 is the amplitude of the light on the centerline of the pattern ($\theta = 0$) and

$$\beta = \frac{\pi b}{\lambda} \sin \theta \tag{5.6}$$

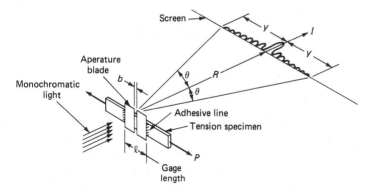

Figure 5.7 Arrangement of the diffraction-type strain gage. (*After T. R. Pryor and W. P. T. North.*)

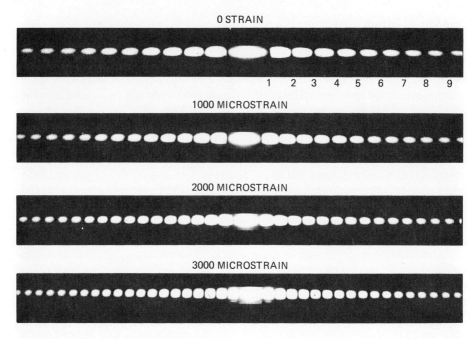

Figure 5.8 Diffractograms showing changes in the diffraction pattern with increasing strain. (*T. R. Pryor and W. P. T. North.*)

where θ is defined in Fig. 5.7 and λ is the wavelength of the light. If the analysis of the diffraction pattern is limited to short distances y from the centerline of the system, $\sin \theta$ is small enough to be represented by y/R and Eq. (5.6) becomes

$$\beta = \frac{\pi b}{\lambda} \frac{y}{R} \tag{5.7}$$

The intensity I vanishes according to Eq. (5.5) when $\sin \beta = 0$ or when $\beta = n\pi$, where $n = 1, 2, 3, \dots$. By considering those points in the diffraction pattern where $I = 0$, it is possible to obtain a relationship between their location in the pattern and the aperture width b. Thus

$$b = \frac{\lambda R n}{y} \tag{5.8}$$

where n is the order of extinction in the diffraction pattern at the point having y as its position.

As the specimen is strained, the deformation results in a change in the aperture width $\Delta b = \epsilon l$ and a corresponding change in the diffraction pattern as indicated in Fig. 5.7. The magnitude of this strain ϵ can be determined from Eq. (5.8) and measurements from the two diffraction patterns. As an example, consider the diffraction pattern after deformation, where

$$b + \Delta b = \frac{\lambda R n^*}{y_1} \qquad (a)$$

and the diffraction pattern before deformation, where

$$b = \frac{\lambda R n^*}{y_0} \qquad (b)$$

Subtracting Eq. (b) from Eq. (a) and simplifying gives the average strain ϵ over the gage length l as

$$\epsilon = \frac{\Delta b}{l} = \frac{\lambda R n^*}{l} \frac{y_0 - y_1}{y_0 y_1} \qquad (5.9)$$

In practice, the order of extinction n^* is selected as high as possible consistent with the optical quality of the diffraction pattern. When the higher orders of extinction are used, the distance y can be measured with sufficient accuracy with calipers and an engineering scale and elaborate measuring devices can be avoided.

The diffraction strain gage is extremely simple to install and use provided the component can be observed during the test. The method has many advantages for strain measurement at high temperatures since it is automatically temperature-compensated if the blades are constructed of the same material as the specimen.

A second optical method of strain measurement utilizes the interference patterns produced when coherent, monochromatic light from a source such as a helium-neon laser is reflected from two shallow V-grooves ruled on a highly polished portion of the specimen surface. The V-grooves are usually cut with a diamond to a depth of approximately 0.000040 in (0.001 mm) and are spaced approximately 0.005 in (0.125 mm) apart. An interference pattern from a pair of grooves having this depth and spacing and a groove angle of 110° is shown in Fig. 5.9.

When the grooves which serve as the reflective surfaces are small enough to cause the light to diffract, and when the grooves are close enough together to permit the diffracted light rays to superimpose and produce an interference pattern, the intensity of light in the pattern is given by the expression

$$I = 4A_0^2 \frac{\sin^2 \beta}{\beta^2} \cos^2 \phi \qquad (5.10)$$

where $\beta = \dfrac{\pi b}{\lambda} \sin \theta \qquad \phi = \dfrac{\pi d}{\lambda} \sin \theta$

and b = width of groove
d = width between grooves
θ = angle from central maximum, as previously defined

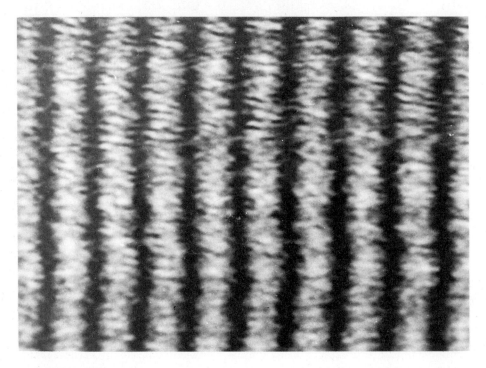

Figure 5.9 Interference fringe pattern produced by reflected light from two V-shaped grooves. (*Courtesy of W. N. Sharpe, Jr.*)

As the light is reflected from the sides of the V grooves, two different interference patterns are formed, as indicated in Fig. 5.10. In an actual experimental situation, the fringe patterns are observed on screens located approximately 8 in (200 mm) from the grooves.

The intensity in the interference pattern goes to zero and a dark fringe is produced whenever $\beta = n\pi$, with $n = 1, 2, 3, \ldots$, or when $\phi = (m + \frac{1}{2})\pi$, with $m = 0, 1, 2, \ldots$. When the specimen is strained, both the distance d between

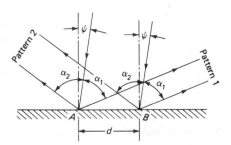

Figure 5.10 Schematic diagram showing the light rays which form the two interference patterns.

grooves and the width b of the grooves change. These effects produce shifts in the fringes of the two interference patterns which can be related to the average strain between the two grooves. The proof is beyond the scope of this presentation; however, it is shown in Ref. 10 that

$$\epsilon = \frac{(\Delta N_1 - \Delta N_2)\lambda}{2d \sin \alpha} \tag{5.11}$$

where ΔN_1 and ΔN_2 are the changes in fringe order in the two patterns produced by the strain and α is the angle between the incident light beam and the diffracted rays which produce the interference pattern.

The interferometric strain gage offers a method for measuring strain without actual use of a gage, thus eliminating any reinforcing effects or bonding difficulties. Since no contact is made, the method can be employed on rotating parts or in hostile environments. Again temperature compensation is automatic, and the method can be employed at very high temperatures.

C. Electrical Strain Gages

During the past 30 years, electrical strain gages have become so widely accepted that they now dominate the entire strain-gage field except for a few special applications. The most important electrical strain gage is the resistance type, which will be covered in much greater detail in subsequent chapters. Two less commonly employed electrical strain gages, the capacitance type and the inductance type, will be introduced in this section. Although these gages have limited use in conventional stress analysis, they are often employed in transducer applications and on occasion find special application in measuring strain.

The capacitance strain gage [11] The capacitance of the parallel-plate capacitor illustrated in Fig. 5.11 can be computed from the relation

$$C = \begin{cases} 0.225 \dfrac{kA}{h} & \dfrac{A}{h} \text{ in inches} \\[3mm] 8.86 \times 10^{-3} \dfrac{kA}{h} & \dfrac{A}{h} \text{ in millimeters} \end{cases} \tag{5.12}$$

where C = capacitance, pF
k = dielectric constant of the medium between two plates
A = cross-sectional area of the plates
h = distance between two parallel plates

The flat-plate capacitor can be employed as a strain or displacement gage in one of three possible ways: (1) by changing the gap h between the plates; (2) by moving one plate in a transverse direction with respect to the other, thereby changing the area A between the two plates; and (3) by moving a body with a dielectric constant higher than air between the two plates.

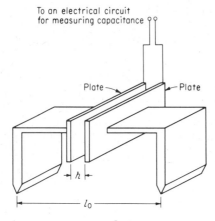

To an electrical circuit
for measuring capacitance

Plate

Plate

h

l_0

Figure 5.11 Schematic illustration of a capacitor strain gage with a variable air gap.

A capacitor-type strain gage with a variable air gap is shown schematically in Fig. 5.11. The change in capacitance for a small change in the air gap h can be obtained by differentiating Eq. (5.12) with respect to h to obtain

$$\frac{dC}{dh} = 0.225kA\left(-\frac{1}{h^2}\right) = -\frac{C}{h} \tag{5.13}$$

When the gage illustrated in Fig. 5.11 is mounted to a specimen which in turn is loaded, the gage length will change by an amount Δl and the air gap h will change by $\Delta h = \Delta l$. Hence, the strain ϵ produces a change in capacitance ΔC which is given by Eq. (5.13) as

$$\epsilon = \frac{\Delta l}{l_0} = \frac{\Delta h}{l_0} = -\frac{h}{l_0}\frac{\Delta C}{C}$$

or

$$\frac{\Delta C}{C} = -\frac{l_0}{h}\epsilon \tag{5.14}$$

For a capacitor gage with $l_0 = 1$ in (25 mm), $h = 0.010$ in (0.25 mm), and $\epsilon = 1\ \mu\text{in/in}$ (μm/m), the value obtained for $\Delta C/C$ is 10^{-4}. Electric circuits (a DeSauty ac bridge) can be employed to accurately measure capacitance changes as small as $\Delta C = C \times 10^{-4}$ for both static applications and low-frequency dynamic applications. Thus, the sensitivity and accuracy of the capacitor gage are quite sufficient for application to the general problem of determining strain distributions. The primary disadvantage of the capacitor gage is its relatively large size and its mechanical attachment through knife-edges. The application of the capacitor principle to other transducer systems, however, does show promise since it can be employed at elevated temperatures and the system involves very low operating forces.

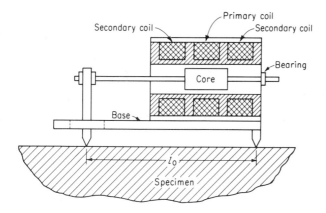

Figure 5.12 Schematic illustration of a linear differential transformer employed as a strain transducer.

The inductance strain gage [12–13] Of the many types of inductance measuring systems which could be employed to measure strain, the differential-transformer system will be considered here. The linear differential transformer is an excellent device for converting mechanical displacement into an electrical signal. It can be employed in a large variety of transducers, including strain, displacement, pressure, acceleration, force, and temperature. A schematic illustration of a linear differential transformer employed as a strain-gage transducer is shown in Fig. 5.12. Mechanical knife-edges are displaced over the gage length by the strain induced in the specimen. This displacement is transmitted to the core, which moves relative to the coils, and an electrical output is produced across the coils.

A linear differential transformer has three coils, a primary coil and two secondary coils on either side of the primary. A core of magnetic material supported on a shaft of nonmagnetic material is positioned in the center of the coils, as shown in the circuit drawing in Fig. 5.13. As the core moves within the coils, it varies the mutual inductance between the primary and each secondary winding, with one secondary becoming more tightly coupled to the primary and the other secondary becoming more loosely coupled. The two secondary coils are wired in series opposition, and consequently the output voltage E_0 is the difference between the voltages developed in each secondary, that is, $E_0 = E_1 - E_2$. In a symmetrically constructed transformer a null output should occur when the core is at the center

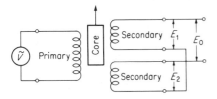

Figure 5.13 Schematic diagram of the linear-differential-transformer circuit.

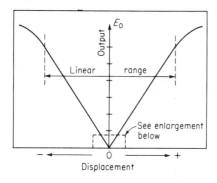

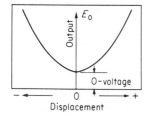

Figure 5.14 Output voltage as a function of core position in a linear differential transformer.

point between the two secondary coils. Movement of the core in one direction off the null position will induce an unbalance in E_1 and E_2, and some net E_0 will be indicated. Displacement in the opposite direction will also produce an unbalance, with a resulting E_0 which is 180° out of phase with the first output. If the direction of the displacement must be known, the phase angle must be determined.

The above description refers to an ideal differential transformer in which the three coils are purely inductive and perfectly symmetrical, thus giving a perfect null voltage when the core is in the zero position. In reality, the differential transformer is not perfect and the output voltage goes to a minimum rather than a null at the zero position of the core. The output voltage for a typical differential transformer as a function of core position is given in Fig. 5.14.

The factor which is most detrimental to the formation of the null voltage is the capacitive coupling between the primary and secondary coils. This capacitive coupling will not produce opposing voltages across the two windings; hence E_0 will not go to null even when the inductively coupled voltages cancel out. The zero voltage is normally less than 1 percent of the maximum voltage; and unless extreme sensitivities are required, the error it induces is quite small.

The sensitivity of commercial differential transformers is of the order of 0.5 to 4 V/in (0.02 to 0.16 V/mm) of displacement per volt of excitation applied to the primary coil. The primary-coil voltage varies from 5 to 18 V at frequencies from 60 Hz to 2.4 kHz. At rated excitation voltages the sensitivities obtainable range from 4 to 65 V/in (0.16 to 2.6 V/mm) of travel.

The range of the differential transformer varies with the design of the unit. With commercially available transformers it is possible to obtain ranges which vary from 0.020 to 2 in (0.5 to 50 mm). The greater the range of a particular transformer, the lower the sensitivity.

The errors of the linear differential transformer amount, in general, to about 0.5 percent of the maximum linear output. The deviation from linearity of the commercial units is also of the order 0.5 percent of the specified range. The dynamic response is limited by the mass of the core and supporting mechanical assembly. It is also limited electrically by the frequency of the applied ac voltage. This is a carrier frequency, which should be at least 10 times the maximum frequency being measured.

The linear differential transformer requires a very small driving force (a fraction of a gram) to move the core. The operation of the differential transformer can be severely affected, however, by the presence of metal masses or by stray magnetic fields. A magnetic shield is employed around the coil holder to minimize these effects.

In recent years the linear differential transformer has been supplied with a dc to ac power source and with a demodulator to condition the output voltage. When packaged as a single unit, the device is known as a DCDT. It is small, rugged, and extremely useful in the laboratory. It can be used for many applications since it can be powered with an ordinary 6-V dry cell and can be read out on a digital multimeter. Both these items are readily available in most laboratories. The use of the linear differential transformer or the DCDT for strain-gage applications has been limited because of the mechanical-attachment problem; however, it is one of the best displacement transducers available for general laboratory use.

D. Acoustical Strain Gages [14–16]

Acoustical strain gages have been employed in a variety of forms in several countries since the late 1920s. To date they have been largely supplemented by the electrical-resistance strain gage. However, they are unique among all forms of strain gages in view of their long-term stability and freedom from drift over extended time periods. The acoustical gage described here, due to R. S. Jerrett, was developed in 1944. The strain-measuring system is based on the use of two acoustic gages referenced as a test gage and a reference gage. The significant parts of a gage are shown schematically in Fig. 5.15.

In the figure it can be seen that the gage has the commonly employed knife-edge mounting provision. One knife-edge is mounted to the main body, while the other knife-edge is mounted in a bearing suspension and is free to elongate with the specimen. The gage length l_0 is 3 in (76 mm). One end of a steel wire is attached to the movable knife-edge while the other end of the wire passes through a small hole in the fixed knife-edge and is attached to a tension screw. The movable knife-edge is connected to a second tension screw by a leaf spring. This design permits the initial tension in the wire to be applied without the transmission of load to the knife-edges. The wire passes between the pole pieces of two small electromagnets. One of these magnets is used to keep the wire vibrating at its natural frequency; the other is employed to pick up the frequency of the system.

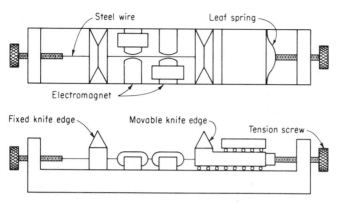

Figure 5.15 Schematic drawing showing the operation of the Jerrett acoustical strain gage.

Electrically both magnets operate together in that the signal from the pickup magnet is amplified and fed back into the driving magnet to keep the string excited in its natural frequency.

The reference gage is very similar to the test gage except that the knife-edges are removed and a micrometer is used to tension the wire. A helical spring is employed in series with the wire to give larger rotations of the micrometer head for small changes of stress in the wire.

To operate the system, the test gage is mounted and adjusted and the reference gage is placed near it to attempt to compensate temperature effects. Both the gages are energized, and each wire will emit a musical note. If the frequency of vibrations from the two gages is not the same, beats will occur. The micrometer setting is varied on the reference gage until the beat frequency decreases to zero. The reading of the micrometer is then taken and the strain is applied to the test gage. The change in tension in the wire of the test gage produces a change in frequencies, and it is necessary to adjust the reference gage until the beats are eliminated. This new micrometer reading is proportional to the strain.

If the test gage is located in a remote position and the beat signals from the test and reference gages cannot be developed, it is possible to balance the two gages by using an oscilloscope. The voltage output from the pickup coils of each gage are displayed while operating the oscilloscope in the xy mode. The resulting Lissajous figure provides the readout which permits adjustment of the reference gage to the frequency of the test gage.

The natural frequency of a wire held between two fixed points is given by

$$f = \frac{1}{2L}\sqrt{\frac{g\sigma}{w}} \qquad (5.15)$$

where L = length of wire between supports
g = gravitational constant
σ = stress in wire
w = density of wire material
f = frequency

In terms of strain in the wire the frequency is governed by the following equation, which comes directly from Eq. (5.15):

$$f = \frac{1}{2L}\sqrt{\frac{gE\epsilon}{w}} \tag{5.16}$$

where E is the modulus of elasticity.

The sensitivity of this instrument is very high, with possible determinations of displacements of the order of 0.1 μin (2.5 nm). The range is limited, in general, to about one-thousandth of the wire length before over- or understressing of the sensing wire becomes critical. The gage is temperature-sensitive unless the thermal coefficients of expansion of the base and wire are closely matched over the temperature range encountered during a test. Finally, the force required to drive the transducer is relatively large, and it should therefore not be employed in systems where the large driving force will be detrimental.

5.4 MOIRÉ METHOD OF STRAIN ANALYSIS [17, 18]

Discussed in the preceding section were various types of strain gages which permit the determination of an average normal strain along a line segment of length l_0 on the surface of a specimen. The use of these gages, in effect, gives the value of one normal component of strain at one point on the surface of the specimen. A complete determination of the strain field entails the application of a large number of these gages so that ϵ_{xx}, ϵ_{yy}, γ_{xy} can be evaluated over the major portion of the surface. The point-per-point method of evaluation of each of the three cartesian components of strain is a time-consuming and expensive approach to a complete evaluation of a general strain field.

The moiré method of strain analysis, which permits a whole-field examination of the strain in a simple and direct fashion, has been developed to a point where it can be employed in certain problems where the displacements are relatively large. A complete treatment of the moiré method is presented in Chap. 12.

5.5 GRID METHOD OF STRAIN ANALYSIS [19–25]

The grid method of strain analysis is one of the oldest techniques known to experimental stress analysis. The method requires placement of a grid (a series of well-defined parallel lines) on the surface of the specimen. Next, the grid is carefully photographed before and after loading the specimen to obtain negatives which will show the distortion of the grid. Measurements of the distance between the grid lines before and after deformation give lengths l_i and l_f respectively. These lengths may be interpreted in terms of strain in several different ways:

Lagrangian strain:
$$\epsilon = \frac{l_f - l_i}{l_i} \tag{5.17}$$

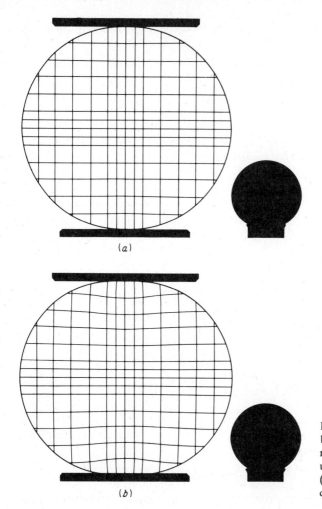

(a)

(b)

Figure 5.16 An embedded rubber-thread grid in a transparent rubber model of a circular disk under diametral compression: (a) before deformation; (b) after deformation.

Eulerian strain:
$$\epsilon = \frac{l_f - l_i}{l_f} \tag{5.18}$$

Natural strain:
$$\epsilon = \ln \frac{l_f}{l_i} \tag{5.19}$$

The exact form used to determine the strain will depend upon the purpose of the analysis and the amount of deformation experienced by the model.

If the grid consists of two series of orthogonal parallel lines over the entire surface of the specimen (see Fig. 5.16), the grid method will give strain components over the entire field. The strain component ϵ_{xx} in the horizontal direction is obtained by comparing the distances between the vertical grid lines before and after deformation. The vertical or ϵ_{yy} component of strain can be obtained from the horizontal array of grid lines, and the $\epsilon_{45°}$ component of strain can be obtained from measurements across the diagonals if the orthogonal grid array is square.

In general, two main difficulties are associated with the utilization of the grid method for measuring strain. First, the strains being measured are usually quite small, and in most instances the displacement readings cannot be made with sufficient precision to keep the accuracy of the strain determinations within reasonable limits. Also, the definition of the grid lines on the negatives is usually poor when magnified, so that appreciable errors are introduced in the displacement readings.

In order to effectively employ the grid method and avoid these two difficulties, the deformations applied to the model must be large enough to impose strain levels of the order of 5 percent. Usually model deformations this large are to be avoided in an elastic stress analysis; hence other experimental methods such as the moiré method or electrical-resistance strain gages are preferred. However, when large deformations are associated with the stress-analysis problem, e.g., plastic deformations or deformations in rubberlike materials, the grid method is one of the most effective methods known. In fact, if the strains exceed 20 to 30 percent, the range of the commonly employed strain gages is exceeded, and only the moiré method or the grid method is adequate for measuring these large strains.

Usually the most critical step in employing the grid method involves the placement of the grid array on the model being investigated. Methods used for this purpose include hand scratching, ink ruling, machine scribing, and photoprinting. More recently, a technique for embedding a rubber-thread grid in a transparent urethane rubber model has been developed by Durelli and Riley which has distinct advantages. With this technique, the grid consists of thin, extruded rubber threads which are stretched to reduce their original diameter of 0.008 in (0.2 mm) to about 0.004 in (0.1 mm). Instead of being cemented to the specimen after it is machined, the grid is threaded in the appropriate array through the sides of a plate mold, and the transparent rubber is cast about it. In this manner, the grid can be located at any desired position in the plane of the model. Upon curing, the sheet of transparent rubber contains the embedded grid and can be machined to the model geometry. This model can then be subjected to load, and both grid and photoelastic data can be recorded simultaneously. These data, together with the elastic constants of the rubber, are sufficient to provide a complete solution of two-dimensional stress problems.

The Replica Technique [25]

A modification of the grid method for determining static strains which eliminates the requirement for accurate placement of a grid array on the specimen has been developed by Hickson. With this method, a family of parallel scratches of uneven spacing and thickness is applied to a polished region of a specimen by drawing fine abrasive paper along the edge of a template. Lines equivalent to those produced by the best engraving techniques are frequently present in such a scratch pattern. A second pattern, applied perpendicular to the first, produces the rectangular array of lines required for grid analyses of deformation and strain. A coarse pattern of broad reference lines is scribed over the scratch pattern to serve as datum lines and to help identify specific scratch lines being used for measurements.

Since it is extremely difficult to make precise measurements of displacement with the specimen under load, a replica technique is used to record the scratch patterns before and after loading. The replicas are made by using a thin film of fusible alloy on a stiffening platen. The alloy is capable of reproducing the finest detail in the scratch pattern. The platen serves to provide dimensional stability and temperature compensation if it is made of the same material as the specimen. Such replicas have been shown to be accurate to at least 40 μin (1 μm).

Displacement measurements are made by comparing a pair of replicas, taken before and after loading, with a microscope fitted with a micrometer eyepiece. The two replicas are mounted in a holder which permits the two scratch patterns to be viewed alternately in the same part of the microscope field. The identification of lines being used for strain determinations is simplified by use of the reference grid and the random nature of the lines in the scratch pattern. Accurate measurement of displacements between the high-quality lines in the patterns permits strain determinations in situations not possible by other means, e.g., near a crack or in two plates after joining by welding.

EXERCISES

5.1 The stress distribution in a thin wide plate with a central circular hole is given by Eq. (3.46) when the plate is subjected to a uniaxial tensile or compressive load. Determine the error made in the determination of the maximum stress σ_{max} on the boundary of the hole if a strain gage having a gage length $l_0 = 6$ mm is used. The diameter of the hole in the plate is 25 mm. $E = 200$ GPa, $v = 0.30$.)

5.2 Determine the error associated with measurements of ϵ_{yy} along the longitudinal axis of the plate of Exercise 5.1 if gages with $l_0/a = \frac{1}{4}$ are located at $y/a = 1.5$, 2.0, and 3.0.

5.3 Determine the error associated with measurements of ϵ_{yy} along the transverse axis of the plate of Exercise 5.1 if gages with a width $w_0/a = \frac{1}{8}$ are located at $x/a = 1.5$, 2.0, and 3.0.

5.4 For the plate of Exercise 5.1, determine the error associated with measurements of ϵ_{yy} along the transverse axis of the plate due to both gage width and gage length. Take $w_0/a = \frac{1}{8}$ and $l_0/a = \frac{1}{4}$ and consider gages located at $x/a = 1.5$, 2.0, and 3.0.

5.5 Consider a strain field given by the expression $\epsilon_{xx} = b \sin(\pi x/a)$. Determine the error made in the measurement of ϵ_{xx} at $x = 0$, $a/4$, and $a/2$ when $l_0/a = \frac{1}{2}, \frac{1}{4}$, and $\frac{1}{10}$.

5.6 Consider a strain field given by the expression $\epsilon_{xx} = b + b \sin(\pi x/a)$. Determine the error made in the measurement of ϵ_{xx} at $x = a/2$ when $l_0/a = \frac{1}{4}$. Compare with the results of Exercise 5.5.

5.7 Consider a strain field given by the expression $\epsilon_{xx} = b + bx + b \sin(\pi x/a)$. Determine the error made in the measurement of ϵ_{xx} at $x = a/2$ when $l_0/a = \frac{1}{4}$. Compare with the results from Exercises 5.5 and 5.6. What is the effect of adding constant and linear contributions to the strain field on the overall error?

5.8 A model of a Huggenberger type of strain gage is to be constructed for a classroom demonstration. The following dimensions have been suggested: $v_1 = 50$ mm, $v_2 = 500$ mm, $w_1 = 100$ mm, and $w_2 = 500$ mm. Determine (a) the magnification factor for the demonstration model and (b) the arc length $\Delta l'$ through which the indicating pointer will move if l_0 is 500 mm and the gage is used to measure a strain of 0.001.

5.9 A scratch gage with a gage length of 500 mm is employed to measure a strain which varies from 0 to 400 μm/m. Determine the total range of motion of the scratch recorded on the target plate. If the target plate is magnified 10 times in a microscope and the range of motion is measured with a filar unit accurate to ± 0.002 mm, determine the expected error. Comment on the influence of the width of the scratch for this type of data reduction.

5.10 A helium-neon laser ($\lambda = 632.8$ nm) is used to illuminate a diffraction-type strain gage with a gage length of 25 mm. The diffraction pattern is displayed on a screen which is located 2 m from the aperture. The initial aperture width w is 0.1 mm.

(a) Determine the density n/y of the diffraction pattern.

(b) If the gage is subjected to a strain of 0.0006, what is the new aperture width and the new density n/y_1?

(c) Suppose the diffraction pattern is sufficiently well defined for the $+8$ and -8 orders of extinction to be clearly observed before and after subjecting the gage to the strain. If distances y_0 and y_1 can be determined from scale measurements on the screen to ± 1 mm, estimate the percent error in the measurement of strain.

5.11 A capacitance strain gage (see Fig. 5.11) has a gage length $l_0 = 20$ mm and $h = 0.20$ mm. Construct a graph of capacitance C versus strain ϵ as the strain is varied from 0 to 100 percent.

(a) Is the output from the gage linear over the entire range?

(b) If not, what is the linear range?

(c) Why was the maximum value in part (b) selected as the limit of the linear range?

5.12 Design a strain extensometer, incorporating a commercially available DCDT, which can be utilized with an xy recorder to plot stress-strain curves automatically during tensile tests of standard ASTM specimens of 1020 carbon steel in a universal testing machine. Select the range of the DCDT, specify the input voltage, decide on the gage length of the extensometer, design the linkage (the specimen may fracture with the extensometer in place), determine the output voltage as a function of stress on the specimen, and specify the characteristics of the xy recorder to be used to plot the stress-strain curves automatically.

5.13 Design an acoustical strain gage which can be installed in a concrete dam and monitored over the life (estimated to exceed two centuries) of the dam. The gage will be monitored periodically to record the change in strain with rising and falling head, to record any change in effective modulus of the concrete due to cracking or other deterioration of the concrete, and to estimate damage to the structure after any natural occurrence such as an earthquake. Items to be considered in the design include selection of materials for the components of the gage, gage length, wire size, wire type, and frequency range during operation. Estimate the accuracy of the strain measurement and comment on the ability of the gage to detect structural damage or deterioration of the concrete.

REFERENCES

1. Cuykendall, T. R., and G. Winter: Characteristics of Huggenberger Strain Gage, *Civ. Eng. (N.Y.)*, vol. 10, no. 7, pp. 448–450, 1940.
2. Huggenberger, A. V.: Mechanical Strain Gage Technique of Separating Strains Due to Normal Forces and Bending Moments, *Proc. SESA*, vol. IV, no. 1, pp. 78–87, 1946.
3. Whittemore, H. L.: The Whittemore Strain Gage, *Instruments*, vol. 1, no. 6, pp. 299–300, 1928.
4. Tatnall, F. G.: Development of the Scratch Gage, *Exp. Mech.*, vol. 9, no. 6, pp. 27N–34N, 1969.
5. Haglage, T. L., and H. A. Wood: Scratch Strain Gage Evaluation, *Tech. Rep.* AFFDL-TR-69-25, Wright-Patterson Air Force Base, Ohio, July 1969.
6. Tuckerman, L. B.: Optical Strain Gages and Extensometers, *Proc. Am. Soc. Test. Mater.*, vol. 23, pt. 2, pp. 602–610, 1923.

7. Weaver, P. R.: An Optical Strain Gage for Use at Elevated Temperatures, *Proc. SESA*, vol. IX, no. 1, pp. 159–162, 1951.
8. Pryor, T. R., and W. P. T. North: The Diffractographic Strain Gage, *Exp. Mech.*, vol. 11, no. 12, pp. 565–568, 1971.
9. Pryor, T. R., O. L. Hageniers, and W. P. T. North: Displacement Measurement along a Line by the Diffractographic Method, *Exp. Mech.*, vol. 12, no. 8, pp. 384–386, 1972.
10. Sharpe, W. N., Jr.: The Interferometric Strain Gage, *Exp. Mech.*, vol. 8, no. 4, pp. 164–170, 1968.
11. Carter, B. C., J. F. Shannon, and J. R. Forshaw: Measurements of Displacement and Strain by Capacity Methods, *Proc. Inst. Mech. Engs. (Lond.)*, vol. 152, pp. 215–221, 1945.
12. Schaevitz, H.: The Linear Variable Differential Transformer, *Proc. SESA*, vol. IV, no. 2, pp. 79–88, 1947.
13. Herceg, E. E.: Handbook of Measurement and Control, Schaevitz Engineering, Pennsauken, N.J., 1972.
14. Shepherd, R.: Strain Measurement Using Vibrating-Wire Gages, *Exp. Mech.*, vol. 4, no. 8, pp. 244–248, 1964.
15. Jerrett, R. S.: The Acoustic Strain Gage, *J. Sci. Instrum.*, vol. 22, no. 2, pp. 29–34, 1945.
16. Potocki, F. P.: Vibrating-Wire Strain Gauge for Long-Term Internal Measurements in Concrete, *Engineer*, vol. 206, pp. 964–967, 1958.
17. Theocaris, P. S.: "Moiré Fringes in Strain Analysis," Pergamon Press, New York, 1969.
18. Durelli, A. J., and V. J. Parks: "Moiré Analysis of Strain," Prentice-Hall, Inc., Englewood Cliffs, N.J., 1970.
19. Durelli, A. J., and I. M. Daniel: A Nondestructive Three-dimensional Strain-Analysis Method, *J. Appl. Mech.*, vol. 28, ser. E, no. 1, pp. 83–86, 1961.
20. Durelli, A. J., J. W. Dally, and W. F. Riley: Developments in the Application of the Grid Method to Dynamic Problems, *J. Appl. Mech.*, vol. 26, no. 4, pp. 629–634, 1959.
21. Durelli, A. J., and W. F. Riley: Developments in the Grid Method of Experimental Stress Analysis, *Proc. SESA*, vol. XIV, no. 2, pp. 91–100, 1957.
22. Miller, J. A.: Improved Photogrid Techniques for Determination of Strain over Short Gage Lengths, *Proc. SESA*, vol. X, no. 1, pp. 29–34, 1952.
23. Parks, V. J., and A. J. Durelli: On the Definitions of Strain and Their Use in Large Strain Analysis, *Exp. Mech.*, vol. 7, no. 6, pp. 279–280, 1967.
24. Parks, V. J., and A. J. Durelli: Various Forms of Strain Displacement Relations Applied to Experimental Strain Analysis, *Exp. Mech.*, vol. 4, no. 2, pp. 37–47, 1964.
25. Hickson, V. M.: A Replica Technique for Measuring Static Strains, *J. Mech. Eng. Sci.*, vol. 1, no. 2, pp. 171–183, 1959.

ELECTRICAL-RESISTANCE STRAIN GAGES

6.1 INTRODUCTION [1–4]

In the preceding ·chapter, several different strain-measuring systems were introduced, and their performance characteristics such as range, sensitivity, gage length, and precision of measurement were discussed. None of these different systems, regardless of the principle upon which the gage was based, exhibits all the properties required for an optimum device; however, the electrical-resistance strain gage approaches the requirements for an optimum system. As such, the electrical-resistance strain gage is the most frequently used device in stress-analysis work throughout the world today. Electrical-resistance strain gages are frequently used also as sensors in transducers designed to measure such quantities as load, torque, pressure, and acceleration.

The discovery of the principle upon which the electrical-resistance strain gage is based was made in 1856 by Lord Kelvin, who loaded copper and iron wires in tension and noted that their resistance increased with the strain applied to the wire. Furthermore, he observed that the iron wire showed a greater increase in resistance than the copper wire when they were both subjected to the same strain. Finally, Lord Kelvin employed a Wheatstone bridge to measure the resistance change. In this classic experiment he established three vital facts which have greatly aided the development of the electrical-resistance strain gage: (1) the resistance of the wire changes as a function of strain; (2) different materials have different sensitivities; and (3) the Wheatstone bridge can be used to measure these resistance changes accurately. It is indeed remarkable that over 80 years passed before strain gages based on Lord Kelvin's experiments became commercially available.

Today, after 40 years of commercial development and extensive utilization by industrial and academic laboratories throughout the world, the bonded-foil gage monitored with a Wheatstone bridge has become a highly perfected measuring system. Precise results for surface strains can be obtained quickly using relatively simple methods and inexpensive gages and instrumentation systems. In spite of the relative ease in employing strain gages, there are many features of the gages which must be thoroughly understood to obtain optimum performance from the measuring system in applied stress analysis. In this chapter, the electrical-resistance strain gage will be examined in detail to illustrate each feature affecting its performance. Strain-gage circuits and recording instruments used in measuring the strain-related resistance changes will be treated in Chaps. 8 and 9.

6.2 STRAIN SENSITIVITY IN METALLIC ALLOYS [5–8]

Lord Kelvin noted that the resistance of a wire increases with increasing strain and decreases with decreasing strain. The question then arises whether this change in resistance is due to the dimensional change in the wire under strain or to the change in resistivity of the wire with strain. It is possible to answer this question by performing a very simple analysis and comparing the results with experimental data which have been compiled on the characteristics of certain metallic alloys. The analysis proceeds in the following manner.

The resistance R of a uniform conductor with a length L, cross-sectional area A, and specific resistance ρ is given by

$$R = \rho \frac{L}{A} \tag{6.1}$$

Differentiating Eq. (6.1) and dividing by the total resistance R leads to

$$\frac{dR}{R} = \frac{d\rho}{\rho} + \frac{dL}{L} - \frac{dA}{A} \tag{a}$$

The term dA represents the change in cross-sectional area of the conductor due to the transverse strain, which is equal to $-v\, dL/L$. If the diameter of the conductor before the application of the axial strain is noted as d_0, then the diameter after the strain is applied is given by

$$d_f = d_0\left(1 - v\frac{dL}{L}\right) \tag{b}$$

and from Eq. (b) it is clear that

$$\frac{dA}{A} = -2v\frac{dL}{L} + v^2\left(\frac{dL}{L}\right)^2 \approx -2v\frac{dL}{L} \tag{c}$$

Substituting Eq. (c) into Eq. (a) gives

$$\frac{dR}{R} = \frac{d\rho}{\rho} + \frac{dL}{L}(1 + 2v)$$

which can be rewritten as

$$S_A = \frac{dR/R}{\epsilon} = 1 + 2v + \frac{d\rho/\rho}{\epsilon} \tag{6.2}$$

where S_A is the sensitivity of the metallic alloy used in the conductor and is defined as the resistance change per unit of initial resistance divided by the applied strain.

Examination of Eq. (6.2) shows that the strain sensitivity of any alloy is due to two factors, namely, the change in the dimensions of the conductor, as expressed by the $1 + 2v$ term, and the change in specific resistance, as represented by $(d\rho/\rho)/\epsilon$. Experimental results show that S_A varies from about 2 to 4 for most metallic alloys. For pure metals, the range is from -12.1 (nickel) to 6.1 (platinum). This fact implies that the change in specific resistance can be quite large for certain metals since $1 + 2v$ usually ranges between 1.4 and 1.7. Apparently, the change in specific resistance has its origin in variations in the number of free electrons and their mobility with applied strain.

A list of some metallic alloys commonly employed in commercial strain gages, together with their sensitivities, is presented in Table 6.1. It should be noted that the sensitivity depends upon the particular alloy being considered. Moreover, the values assigned to S_A in Table 6.1 are not necessarily constants. The value of the strain sensitivity S_A will depend upon the degree of cold working imparted to the conductor in its formation, the impurities in the alloy, and the range of strain over which the measurement of S_A is made.

Most electrical-resistance strain gages produced today are fabricated from the copper-nickel alloy known as Advance or Constantan. A typical curve showing the percent change in resistance $\Delta R/R$ as a function of strain for this alloy is given in Fig. 6.1. This alloy is useful in strain-gage applications for the following reasons:

1. The value of the strain sensitivity S_A is linear over a wide range of strain.

Table 6.1 Strain sensitivity S_A for common strain-gage alloys

Material	Composition, %	S_A
Advance or Constantan	45 Ni, 55 Cu	2.1
Nichrome V	80 Ni, 20 Cr	2.1
Isoelastic	36 Ni, 8 Cr, 0.5 Mo, 55.5 Fe	3.6
Karma	74 Ni, 20 Cr, 3 Al, 3 Fe	2.0
Armour D	70 Fe, 20 Cr, 10 Al	2.0
Platinum tungsten	92 Pt, 8 W	4.0

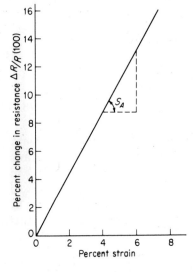

Figure 6.1 Percent change in resistance as a function of percent strain for an Advance alloy.

2. The value of S_A does not change as the material goes plastic.
3. The alloy has a high specific resistance ($\rho = 0.49 \ \mu\Omega \cdot m$).
4. The alloy has excellent thermal stability and is not influenced appreciably by temperature changes when mounted on common structural materials.
5. The small temperature-induced changes in resistance of the alloy can be controlled with trace impurities or by heat treatment.

The first advantage of the Advance-type alloy over other alloys implies that the gage calibration constant will not vary with strain level; therefore, a single calibration constant is adequate for all levels of strain. The wide range of linearity with strain (even into the alloy's plastic region) indicates that it can be employed for measurements of both elastic and plastic strains in most structural materials. The high specific resistance of the alloy is useful when constructing a small gage with a relatively high resistance. Finally, the temperature characteristics of selected melts of the alloy permit the fabrication of temperature-compensating strain gages for each structural material. With temperature-compensated strain gages, the temperature-induced $\Delta R/R$ on a given material can be maintained at less than 10^{-6} per Celsius degree.

The Isoelastic alloy is also employed in commercial gages because of its high sensitivity (3.6 for Isoelastic compared with 2.1 for Advance) and its high fatigue strength. The increased sensitivity is advantageous in dynamic applications where the strain-gage output must be amplified to a considerable degree before recording. The high fatigue strength is useful when the gage is to operate in a cyclic strain field where the alternating strains exceed 1500 μin/in (μm/m). In spite of these two advantages, the Isoelastic alloy has two disadvantages which severely limit its use. First, this alloy is extremely sensitive to temperature changes; and when it is mounted in gage form on a steel specimen, a change in temperature of 1°C will

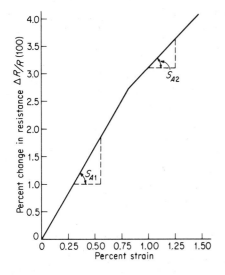

Figure 6.2 Percent change in resistance as a function of percent strain for an Isoelastic alloy.

give an apparent strain indication of 300 to 400 μin/in (μm/m). It can be used in dynamic applications only when the temperature is stable over the time required for the dynamic measurement. A second disadvantage of the Isoelastic alloy is its limited linearity, as illustrated in Fig. 6.2. At a strain level of approximately 0.75 percent, the sensitivity S_A of the alloy changes from approximately 3.6 to 2.5. This fact implies that for strains greater than 7,500 μin/in (μm/m) the calibration factor associated with the gage must be changed to correspond with the reduction in S_A from 3.6 to 2.5.

The Karma alloy is used in temperature-compensated gages in the same manner as the Advance alloy. The range of temperature over which compensation can be achieved, however, is larger for Karma than for Advance. Also, the Karma alloy has a higher resistance to cyclic strain than the Advance alloy.

The other alloys, Nichrome V, Armour D, and the platinum-tungsten alloy, are used for special-purpose gages which permit measurements of strain to be made at temperatures in excess of 450°F (230°C).

6.3 GAGE CONSTRUCTION [9–14]

It is theoretically possible to measure strain with a single length of wire as the sensing element of the strain gage; however, circuit requirements needed to prevent overloading of the power supply and to minimize heat generated by the gage current place a lower limit of approximately 100 Ω on the gage resistance. As a result, a 100-Ω strain gage fabricated from wire having a diameter of 0.001 in (0.025 mm) and a resistance of 25 Ω/in (1000 Ω/m) requires a single length of wire 4 in (100 mm) long.

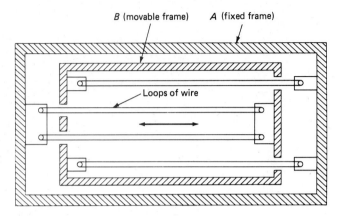

Figure 6.3 An unbounded-wire strain gage.

The very earliest electrical-resistance strain gages were of the unbonded type, where the conductors were straight wires strung between a movable frame and a fixed frame as shown in Fig. 6.3. This gage was large and required knife-edges for mounting, which greatly limited its applicability.

The problem of conductor length and gage mounting was solved in the mid-1930s, when Ruge and Simmons independently developed bonded-wire strain gages. The conductor-length problem was solved by forming the required length of wire into a grid pattern. The attachment problem was solved by bonding the wire grid directly to the specimen with suitable adhesives. Wire gages were produced with both flat-grid and bobbin-type constructions, as illustrated in Fig. 6.4. Bonded-wire strain gages were employed for strain measurements almost

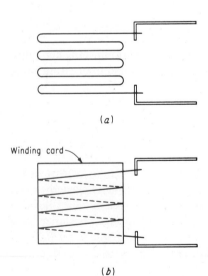

Figure 6.4 (*a*) Flat-grid and (*b*) bobbin-type constructions for bonded-wire-type resistance strain gages.

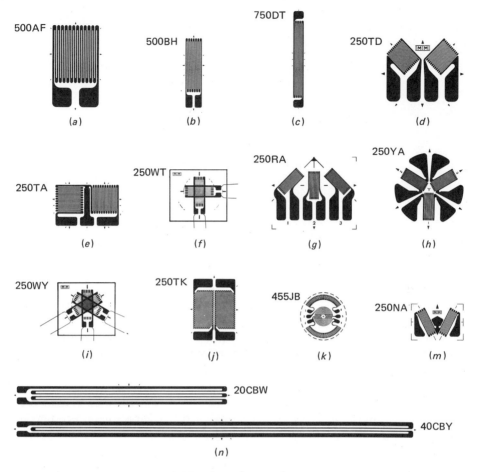

Figure 6.5 Configurations of metal-foil strain gages: (*a*) single-element gage, (*b*) single-element gage, (*c*) single-element gage, (*d*) two-element rosette, (*e*) two-element rosette, (*f*) two-element stacked rosette, (*g*) three-element rosette, (*h*) three-element rosette, (*i*) three-element stacked rosette, (*j*) torque gage, (*k*) diaphragm gage, (*m*) stress gage, (*n*) single-element gage for use on concrete. (*Micro-Measurements.*)

exclusively, from the mid-1930s to the mid-1950s. They are still used occasionally today, but in most instances they have been replaced by the bonded-foil strain gage.

The first metal-foil strain gages were produced in England in 1952 by Saunders and Roe. With this type of gage, the grid configuration is formed from metal foil by a photoetching process. Since the process is quite versatile, a wide variety of gage sizes and grid shapes can be produced. Typical examples of the variety of gages marketed commercially are illustrated in Fig. 6.5. The shortest gage length available in a metal-foil gage is 0.008 in (0.20 mm). The longest gage length is 4.00 in (102 mm). Standard gage resistances are 120 and 350 Ω. Gages having

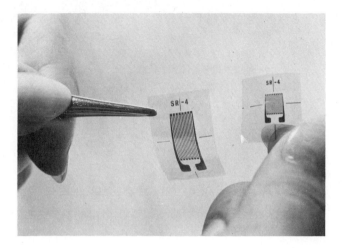

Figure 6.6 Backing sheets are necessary for handling fragile foil strain gages. (*BLH Electronics.*)

lengths greater than 0.060 in (1.52 mm) are also available with a resistance of 1000 Ω.

Multiple-element gages are available with 10 gages arranged along a line. These strip gages are usually installed in fillets where high strain gradients occur and it is difficult to locate the point where the strain is a maximum. Two- and three-element rosettes are available in either the in-plane or stacked configuration in a wide range of sizes for use in biaxial stress fields. Two-element rosettes are used when the directions of the principal stresses (or strains) are known. Three-element rosettes are used when the principal directions are not known. Special gage configurations are also available for use in transducers. A typical example is shown in Fig. 5.6*k* for the diaphragm-type pressure transducer.

The etched metal-film grids are very fragile and easy to distort, wrinkle, or tear. For this reason, the metal film is usually bonded to a thin plastic sheet, which serves as a backing or carrier before photoetching. The carrier material also provides electrical insulation between the gage and the component after the gage is mounted. The use of a backing sheet to serve as a carrier for the grid is illustrated in Fig. 6.6, which shows a gage being handled. Markings for the center-line of the gage length and width are also displayed on the carrier.

Very thin paper was the first carrier material employed in the production of wire-type gages, and its use has been maintained to a limited extent. More recently, a thin (0.001-in, or 0.025-mm) sheet of polyimide, which is a tough and flexible plastic, has been used as a carrier for strain gages intended for general-purpose stress analysis. For transducer applications, where precision and linearity are very important, a very thin high-modulus epoxy is used for the carrier. The epoxy backing is not suitable for general-purpose strain gages since it is brittle and can easily be broken during gage installation. Glass-fiber-reinforced epoxy and/or phenolics are employed as carriers when the strain gage will be exposed to high-

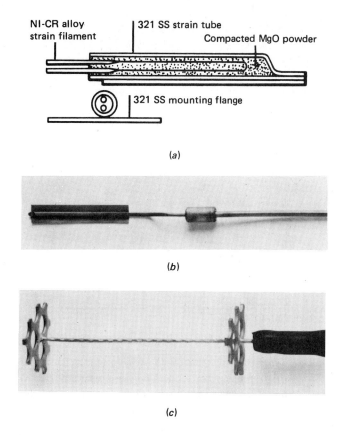

(a)

(b)

(c)

Figure 6.7 Weldable strain gages: (*a*) construction details, (*b*) external appearance of a gage, (*c*) gage for embedment in concrete. (*Ailtech.*)

level cyclic strains and fatigue life of the gage system is important. In this application, the carrier is used to encapsulate the grid. Glass-reinforced epoxy carriers are also used for moderate temperature applications up to 750°F (400°C). For very high temperature applications, a strippable carrier is used. This carrier is removed during application of the gage, and a ceramic adhesive serves to maintain the grid configuration and to insulate the gage.

Another type of gage, originally developed for high-temperature strain measurement, is the weldable strain gage (Fig. 6.7). It consists of a very fine loop of wire which is swaged into a metal case with compacted MgO powder as an insulator. The wire ends are connected to integral lead wires in a metal case. The wire inside the strain tube is reduced in diameter by an etching process; hence, the minimum gage resistance is obtained with a very-small-diameter, high-resistance wire rather than by forming a grid of larger-diameter wire.

Currently the weldable gages are available with resistances which range from 60 to 350 Ω and with lengths which range from 0.375 in (9.5 mm) to 1.09 in

(27.7 mm). The gages are suitable for use from cryogenic to elevated temperatures or within the range from -320 to $1200°F$ (-200 to $650°C$). The gages can be welded onto a number of different metals with a capacitor discharge welder and are ready for use immediately. This simplicity in application is a significant advantage over other high-temperature strain-gage systems, which require rather elaborate installation techniques. Indeed, the simplicity of the welding installation is attractive whenever gages must be mounted in the field under adverse conditions regardless of temperature considerations. The weldable gage is extremely rugged and waterproof. A modification of the gage, shown in Fig. 6.7c, can be embedded in concrete to record strains at interior locations in the structure.

Of the various gages described, the metal-foil strain gage is the most frequently employed for both general-purpose stress analysis and transducer applications. There are occasional special-purpose applications for which unbonded-wire, bonded-wire, weldable, or semiconductor gages are more suitable, but these applications are not common.

6.4 STRAIN-GAGE ADHESIVES AND MOUNTING METHODS [15–21]

The bonded type of resistance strain gage of either wire or foil construction is a high-quality precision resistor which must be attached to the specimen with a suitable adhesive. For precise strain measurements both the correct adhesive and proper mounting procedures must be employed.

The adhesive serves a vital function in the strain-measuring system; it must transmit the strain from the specimen to the gage sensing element without distortion. It may appear that this role can be easily accomplished if the adhesive is suitably strong; however, the characteristics of the polymeric adhesives used to bond strain gages are such that, as will be shown later, the adhesive can influence apparent gage factor, hysteresis characteristics, resistance to stress relaxation, gage resistance, temperature-induced zero drift, and insulation resistance.

The singularly unimpressive feat of bonding a strain gage to a specimen is perhaps one of the most critical steps in the entire process of measuring strain with a bonded resistance strain gage. The improper use of an adhesive costing a few dollars per test can seriously degrade the validity of an experimental stress analysis which may cost thousands of dollars.

When mounting a strain gage, it is important to carefully prepare the surface of the component where the gage is to be located. This preparation consists of sanding away any paint or rust to obtain a smooth but not highly polished surface. Next, solvents are employed to remove all traces of oil or grease. Finally, the clean, sanded, and degreased surface is treated with a basic solution to give the surface the proper chemical affinity for the adhesive.

The gage location is then marked on the specimen and the gage is positioned by using a rigid transparent tape in the manner illustrated in Fig. 6.8. The position

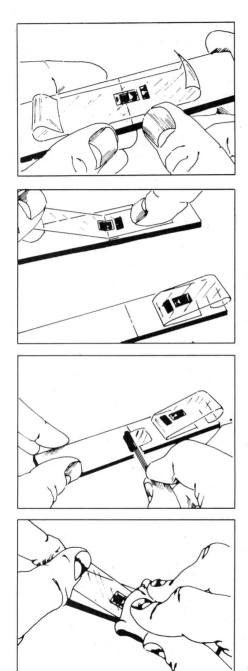

Figure 6.8 The tape method of installation for metal-foil strain gages. (*Micro-Measurements.*)

and orientation of the gage are maintained by the tape as the adhesive is applied and as the gage is pressed into place by squeezing out the excess adhesive.

After the gage is installed, the adhesive must be exposed to a proper combination of pressure and temperature for a suitable length of time to ensure a complete cure. This curing process is quite critical since the adhesive will expand because of heat, experience a volume reduction due to polymerization, exhibit a contraction upon cooling, and on occasion experience postcure shrinkage. Since the adhesive is sufficiently strong to control the deformation of the strain-sensitive element in the gage, any residual stresses set up in the adhesive will influence the output of the strain gage. Of particular importance is the postcure shrinkage, which may influence the gage output long after the adhesive is supposedly completely cured. If a long-term strain measurement is attempted with an incompletely cured adhesive, the stability of the gage will be seriously impaired and the accuracy of the measurements compromised.

A wide variety of adhesives are available for bonding strain gages. Factors influencing the selection of a specific adhesive include the carrier material, the operating temperature, the curing temperature, and the maximum strain to be measured. A discussion of the characteristics of several different adhesive systems in common use follows.

A. Cellulose Nitrate Cement

Cellulose nitrate cements such as Duco or SR-4 are commonly employed to mount paper-backed strain gages. When properly employed, this adhesive system is adequate; however, the adhesive must be totally dried before satisfactory performance is obtained. The primary problem in using cellulose nitrate cement is assisting the solvent to escape from the adhesive layer between the gage and the specimen. Since the cement consists of 85 percent solvent and 15 percent solids, it is evident that an appreciable amount of solvent must be removed by evaporation. Moreover, the gage is often covered with a pad and bounded by the specimen so that the passages for solvent release are limited.

No standard curing procedure can be outlined since the time involved in evaporating the solvent depends upon the gage type and the installation procedure. However, heat applied to the gage by warm, flowing air appreciably shortens the time required for a complete cure. By placing a strain-gaged component in an air-circulating oven at 130°F (54°C) complete cure can be effected in a day or two. If it is not possible to employ heat, the time required for complete cure is greatly extended and depends upon humidity and temperature conditions.

Once the gage is completely dried (particularly if oven-dried), it must be waterproofed immediately. Otherwise the cellulose nitrate cement will begin to expand as a result of water absorption from the atmosphere. These expansions occur slowly since they are diffusion-induced; nevertheless, if the readout period is long, the expansion of the cement can materially influence the gage reading and create instabilities and generally unsatisfactory performance of the measuring system.

Perhaps the best indication of a properly cured adhesive system can be obtained by measuring the resistance between the gage grid and the specimen (resistance through the adhesive layer). A properly dried installation will exhibit a resistance to ground exceeding 10,000 MΩ. Minute traces of either solvent or water in the adhesive will lower the resistance of the adhesive layer and influence gage performance.

B. Epoxy Cements

Epoxies are a class of thermosetting plastics which, in general, exhibit a higher bond strength and a higher level of strain at failure than other types of adhesives used to mount strain gages. Epoxy systems are usually composed of two constituents, a monomer and a hardening agent. The monomer, or base epoxy, is a light amber fluid which is usually quite viscous. A hardening agent mixed with the monomer will induce polymerization. Amine-type curing agents produce an exothermic reaction which releases sufficient heat to accomplish curing at room temperature or at relatively low curing temperatures. Anhydride-type curing agents require the application of heat to promote polymerization. Temperatures in excess of 250°F (120°C) must be applied for several hours to complete polymerization. So many epoxies and curing agents are commercially available today that it is impossible to be specific in the coverage of their properties or their behavior. However, the following remarks will be valid and useful for any system of epoxies used in strain-gage applications.

With both types of curing agent, particularly the amine type, the amount of hardener added to the monomer is extremely important. The adhesive curing temperatures and the residual stresses produced during polymerization can be significantly influenced by as little as 1 or 2 percent variance from the specified values. For this reason, the quantities of both the monomer and the curing agent should be carefully weighed before they are mixed together.

In general, pure epoxies do not liberate volatiles during cure; therefore, postcure heat cycling is not necessary to evaporate chemical by-products released during polymerization. Volatiles should not be added to the epoxies to improve their viscosity for general-purpose applications. A filler material such as Cabosil can be added in moderate quantities (5 to 10 percent by weight) to improve the bond strength and to reduce the temperature coefficient of expansion of the epoxy.

A modest clamping pressure of 5 to 20 lb/in^2 (35 to 140 kPa) is recommended for the epoxies during the cure period to ensure as thin a bond line (adhesive layer) as possible. In transducer applications, solvent-thinned epoxies are frequently specified to reduce viscosity so that extremely thin (less than 200 μin, or 0.005 mm) void-free bond lines can be obtained. Thin bond lines tend to minimize creep, hysteresis, and linearity problems. For these transducer applications, clamping pressures of approximately 50 lb/in^2 (350 kPa) are recommended.

Many different epoxy systems are available from strain-gage manufacturers in kit form with the components preweighed. These systems are recommended since the epoxies are especially formulated for strain-gage applications. They are easy and convenient to use, they have an adequate pot life, and the time-temperature curve for the curing cycle is specified. The use of hardware-store variety two-tube epoxy systems is discouraged since these systems usually incorporate modifiers or plasticizers which although they improve the toughness of the adhesive cause large amounts of creep and hysteresis undesirable for strain-gage applications.

C. Cyanoacrylate Cement (Eastman 910 SL)

A modified form of Eastman 910 adhesive consisting of a methyl-2-cyanoacrylate compound is commonly employed as a strain-gage adhesive. This adhesive system is simple to use, and the strain gage can be employed approximately 10 min after bonding. Chemically, this adhesive is quite unusual in that it requires neither heat nor catalyst to induce polymerization. Apparently when this adhesive is spread in a thin film between two components to be bonded, the minute traces of water or other weak bases on the surfaces of the components are sufficient to trigger the polymerization process. A catalyst can be applied to the bonding surfaces to decrease the reaction times, but it is not essential.

In strain-gage applications, a thin film of the adhesive is placed between the gage and the specimen and a gentle pressure is applied for about 1 or 2 min to induce polymerization. Once initiated, the polymerization will continue at room temperature without the pressure required initially.

The fast room-temperature cure of the cyanoacrylate adhesive makes it ideal for general-purpose strain-gage applications. The performance of this adhesive system, however, will deteriorate with time, moisture absorption, or elevated temperature; therefore, it should not be used where extended life of the gage system is important. Coatings such as microcrystalline wax, silicone rubber, polyurethane, etc., can be used to protect the adhesive from moisture in the air and extend the life of an installation to 1 or 2 years.

D. Ceramic Cements

Two different approaches are used in bonding strain gages with ceramic adhesives. The first utilizes a blend of finely ground ceramic powders such as alumina and silica combined with a phosphoric acid. Usually this blend of powders is mixed with a solvent such as isopropyl alcohol and an organic binder to form a liquid mixture which facilitates handling. A precoat of the ceramic cement is applied and fired to form a thin layer of insulation between the gage grid and the component. A second layer of ceramic cement is then applied to bond the gage. In this application, the carrier for the gage is removed, and the grid is totally encased in ceramic.

Since many of the ceramic cements marketed commercially have proprietary compositions, it is often difficult to select the most suitable product. One cement, developed by the National Bureau of Standards and identified as NBS-x-142, is recommended for high-temperature use since it exhibits a very high resistivity at temperatures up to 1800°F (980°C). Gage resistance to ground with this cement

Table 6.2 Composition of NBS-x-142 ceramic cement

Constituent	Parts by weight
Alumina, Al_2O_3	100
Silica, SiO_2	100
Chrome anhydride, CrO_3	2.5
Colloidal silica solution	200
Orthophosphoric acid, H_3PO_4	30

will normally exceed 6 MΩ. The composition of the NBS-x-142 cement is given in Table 6.2. The ceramic cements are used primarily for high-temperature applications or in radiation environments, where organic adhesives cannot be employed.

A second method for bonding strain gages with a ceramic material utilizes a flame-spraying Rokide process. A special gun (Fig. 6.9) is used to apply the ceramic particles to the gage. For this type of application, gages of wire construction

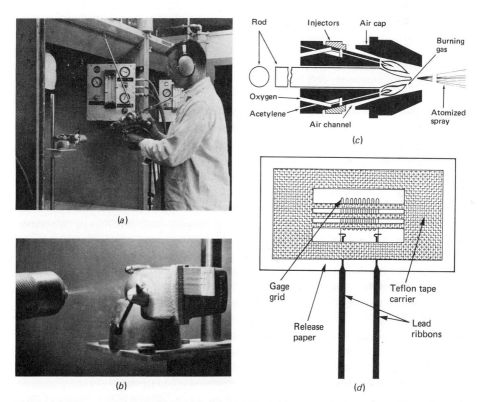

Figure 6.9 Flame spraying ceramic adhesives: (a) flame-spray gun in operation, (b) strain gage with carrier on a specimen, (c) particle formation in the spray gun, (d) free element strain gage with carrier. (*BLH Electronics.*)

mounted on slotted carriers are used. The carrier, which holds the gage and lead wires in position during attachment, is made from a glass-reinforced Teflon tape. The tape is resistant to the molten flame-sprayed ceramic particles; however, after the gage grid is secured to the component, the tape is removed and the grid is completely encased with flame-sprayed ceramic particles.

The flame-spray gun utilizes an oxyacetylene gas mixture in a combustion chamber to produce very high temperatures. Ceramic material in rod form is fed into the combustion chamber; there the rod decomposes into softened semimelted particles which are forced from the chamber by the burning oxyacetylene gas. The particles impinge on the surface of the component and form a continuous coating. Since the particles (even in their softened state) are somewhat abrasive, foil grids are not normally used.

Most gage installations are made with BLH-H rod, which is essentially pure alumina, or with BLH-S rod, which is alumina with about 2 percent silica. For applications involving temperatures less than 800°F (425°C), the S rod is employed since it is easier to melt and apply. The H rod is intended for higher-temperature applications.

E. Testing a Strain-Gage Adhesive System

After a strain gage has been bonded to the surface of a specimen, it must be inspected to determine the adequacy of the bond. The inspection procedure attempts to establish whether any voids exist between the gage and the specimen and whether the adhesive is totally cured. Voids can result from bubbles originally in the cement or from the release of volatiles during the curing process. They can be detected by tapping the gage installation with a soft rubber eraser and observing the effect on a strain indicator. If a strain reading is noted, the gage bond is not satisfactory and voids exist between the sensing element and the specimen.

The stage of the cure of the adhesive is much more difficult to establish. Two procedures are frequently employed which give some indication of the relative completeness of the cure. The first procedure utilizes measurement of the resistance between the gage grid and the specimen as an indication of the stage of the cure. The resistance of the adhesive layer increases as the adhesive cures. Typical gage installations should exhibit a resistance across the adhesive layer of the order of 10,000 MΩ. Resistance values of 100 to 1000 MΩ normally indicate a need to specify further curing treatment for the adhesive.

The second procedure for determining the completeness of the cure involves subjecting the gage installation to a strain cycle while measuring the resistance change. The change in the zero-load strain reading after a strain cycle (zero shift) or the area enclosed by a strain-resistance change curve is a measure of the degree of cure. Strain-gage installations with completely cured adhesives will exhibit zero shifts of less than 2 μin/in (μm/m) of apparent strain. Should larger values of zero shift be observed, the adhesive should be subjected to a postcure temperature cycle.

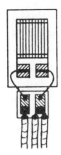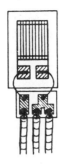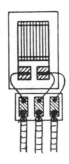

Figure 6.10 Use of anchor terminals to connect lead wires to gages. (*Micro-Measurements.*)

In the event that the gages are mounted on a component or a structure which cannot be strain-cycled before testing, a temperature cycle can be substituted for the strain cycle. If the adhesive in a gage installation has been thoroughly cured, no change in the zero strain will be observed when the gage is returned to its original temperature. If the adhesive cure is incomplete, the temperature cycle will result in additional polymerization with associated shrinkage of the adhesive which causes a shift in the zero reading of the gage.

After the gages have been bonded to the structure or component, it is necessary to attach lead wires so that the change in resistance can be monitored on a suitable instrumentation system. Since the metal-foil strain gages are relatively fragile, care must be exercised in attaching the lead wires to the soldering tabs. Intermediate anchor terminals, which are much more rugged than the metal-foil gage tabs, are usually employed, as illustrated in Fig. 6.10. A small-diameter wire (32- to 36-gage) is used to connect the gage terminal to the anchor terminal. Three lead wires are soldered to the anchor terminals as indicated in Fig. 6.10. The use of three lead wires to ensure temperature compensation in the Wheatstone bridge measuring circuit will be discussed in Sec. 8.6. The size of lead wire employed will depend upon the distance between the gage and the instrumentation system. For normal laboratory installations, where lengths rarely exceed 10 to 15 ft (3 to 5 m), wire sizes between 26 and 30 gage are frequently used. Stranded wire is usually preferred to solid wire since it is more flexible and suffers less breakage due to improper stripping or lead-wire movement during the test.

6.5 Gage Sensitivities and Gage Factor [22–26]

The strain sensitivity of a single, uniform length of a conductor was previously defined as

$$S_A = \frac{dR/R}{\epsilon} \approx \frac{\Delta R/R}{\epsilon} \tag{6.2}$$

where ϵ is a uniform strain along the conductor and in the direction of the axis of the conductor. This sensitivity S_A is a function of the alloy employed to fabricate the conductor and its metallurgical condition. Whenever the conductor is formed into a grid to yield the short gage length required for measuring strain, the gage exhibits a sensitivity to both axial and transverse strain.

With the older flat-grid wire gages, the transverse sensitivity was due primarily to the end loops in the grid pattern, which placed part of the conductor in the transverse direction. In the foil gages, the end loops are enlarged and thus considerably desensitized. The axial segments of the grid pattern, however, have a large width-to-thickness ratio; thus, some amount of transverse strain will be transmitted through the adhesive and carrier material to the axial segments of the grid pattern to produce a response in addition to the axial-strain response. The magnitude of the transverse strain transmitted to the grid segments depends upon the thickness and elastic modulus of the adhesive, the carrier material, the grid material, and the width-to-thickness ratio of the axial segments of the grid.

The response of a bonded strain gage to a biaxial strain field can be expressed as

$$\frac{\Delta R}{R} = S_a \epsilon_a + S_t \epsilon_t + S_s \gamma_{at} \tag{6.3}$$

where ϵ_a = normal strain along axial direction of gage
ϵ_t = normal strain along transverse direction of gage
γ_{at} = shearing strain
S_a = sensitivity of gage to axial strain
S_t = sensitivity of gage to transverse strain
S_s = sensitivity of gage to shearing strain

In general, the gage sensitivity to shearing strain is small and can be neglected. The response of the gage can then be expressed as

NEGLECTING SHEAR STRAIN SENSITIVITY

$$\frac{\Delta R}{R} = S_a(\epsilon_a + K_t \epsilon_t) \tag{6.4}$$

where $K_t = S_t / S_a$ is defined as the transverse sensitivity factor for the gage.

Strain-gage manufacturers provide a calibration constant known as the *gage factor* S_g for each gage. The gage factor S_g relates the resistance change to the axial strain. Thus

S_g →

$$\frac{\Delta R}{R} = S_g \epsilon_a \tag{6.5}$$

The gage factor for each lot of gages produced is determined by mounting sample gages drawn from the lot on a specially designed calibration beam like that illustrated in Fig. 6.11. The beam is then deflected a specified distance to produce a known strain ϵ_a. The resistance change ΔR is then measured and the gage factor S_g is determined by using Eq. (6.5).

With this method of calibration, the strain field experienced by the gage is biaxial, with

$$\epsilon_t = -\nu_0 \epsilon_a \tag{a}$$

Figure 6.11 Beam and Wheatstone bridge circuits used in calibrating strain gages to determine the gage factor S_g. (*Micro-Measurements.*)

where $v_0 = 0.285$ is Poisson's ratio of the beam material. If Eq. (*a*) is substituted into Eq. (6.4), the resistance change in the calibration process is

$$\frac{\Delta R}{R} = S_a \epsilon_a (1 - v_0 K_t) \tag{6.6}$$

Since the resistance changes given by Eqs. (6.5) and (6.6) are identical, the gage factor S_g is related to both S_a and K_t by the expression

$$S_g = S_a (1 - v_0 K_t) \tag{6.7}$$

Typical values of S_g, S_a, and K_t for several different gage configurations are shown in Table 6.3.

Table 6.3 Gage factor S_g, axial sensitivity S_a, transverse sensitivity S_t, and transverse-sensitivity factor K_t for several different foil-type strain gages†‡

Gage type	S_g	S_a	S_t	K_t, %
EA-06-250BG-120	2.11	2.11	0.0084	0.4
WA-06-250BG-120	2.10	2.10	−0.0063	−0.3
WK-06-250BG-350	2.05	2.03	−0.0690	−3.4
EA-50-250BG-120	2.12	2.12	0.0191	0.9
EA-06-250BF-350	2.12	2.12	0.0127	0.6
WK-06-250BF-1000	2.07	2.06	−0.0453	−2.2
EA-06-125AD-120	2.10	2.11	0.0274	1.3
WK-06-125AD-350	2.07	2.06	−0.0288	−1.4
WK-15-125AD-350	2.16	2.15	−0.0409	−1.9
EA-06-125AS-500	2.14	2.14	0.0086	0.4
EA-06-015CK-120	2.13	2.14	0.0385	1.8
EA-06-030TU-120	2.02	2.03	0.0244	1.2
WK-06-030TU-350	1.98	1.98	0.0040	0.2
EA-06-062DY-120	2.03	2.04	0.0286	1.4
WK-06-062DY-350	1.96	1.96	−0.0098	−0.5
EA-06-125RA-120	2.06	2.07	0.0228	1.1
WK-06-125RA-350	1.99	1.98	−0.0297	−1.5
EA-06-500AF-120	2.09	2.09	0.0	0
WA-06-500AF-120	2.09	2.08	−0.0270	−1.3
WK-06-500AF-350	2.04	1.99	−0.1831	−9.2
WK-06-500BH-350	2.05	2.01	−0.1347	−6.7
WK-06-500BL-1000	2.06	2.03	−0.0893	−4.4

† Data from Micro-Measurements.
‡ Values depend on lot of foil.

It is important to recognize that error will occur in a strain-gage measurement when Eq. (6.5) is employed except for the two special cases where either the stress field is uniaxial or where the transverse sensitivity factor K_t for the gage is zero. The magnitude of the error can be determined by considering the response of a gage in a general biaxial field with strains ϵ_a and ϵ_t. Substituting Eq. (6.7) into Eq. (6.4) gives

$$\frac{\Delta R}{R} = \frac{S_g \epsilon_a}{1 - v_0 K_t}\left(1 + K_t \frac{\epsilon_t}{\epsilon_a}\right) \tag{b}$$

From Eq. (b), the true value of the strain ϵ_a can be expressed as

$$\epsilon_a = \frac{\Delta R/R}{S_g} \frac{1 - v_0 K_t}{1 + K_t(\epsilon_t/\epsilon_a)} \tag{c}$$

The apparent strain ϵ_a', which is obtained if only the gage factor is considered, can be determined from Eq. (6.5) as

$$\epsilon_a' = \frac{\Delta R/R}{S_g} \tag{d}$$

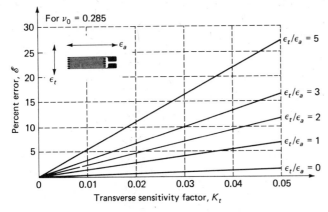

Figure 6.12 Error as a function of transverse sensitivity factor with the biaxial strain ratio as a parameter.

Comparison of Eqs. (c) and (d) shows that

$$\epsilon_a = \epsilon_a' \frac{1 - v_0 K_t}{1 + K_t(\epsilon_t/\epsilon_a)} \qquad (6.8)$$

The percent error $\mathscr{E}$ involved in neglecting the transverse sensitivity of the gage is given by

$$\mathscr{E} = \frac{\epsilon_a' - \epsilon_a}{\epsilon_a}(100) \qquad (6.9)$$

Substituting Eq. (6.8) into Eq. (6.9) yields

$$\mathscr{E} = \frac{K_t(\epsilon_t/\epsilon_a + v_0)}{1 - v_0 K_t}(100) \qquad (6.10)$$

The results of Eq. (6.10), shown graphically in Fig. 6.12, indicate that the error is a function of K_t and the strain biaxiality ratio ϵ_t/ϵ_a. Since the errors can be significant when both K_t and ϵ_t/ϵ_a are large, it is important that corrections be made to account for the transverse sensitivity of the gage. Methods to correct for these errors are presented in Sec. 10.4.

6.6 PERFORMANCE CHARACTERISTICS OF FOIL STRAIN GAGES

Foil strain gages are small precision resistors mounted on a flexible carrier that can be bonded to a component part in a typical application. The gage resistance is accurate to ± 0.4 percent, and the gage factor, based on a lot calibration, is certified to ± 1.5 percent. These specifications indicate that foil-type gages provide

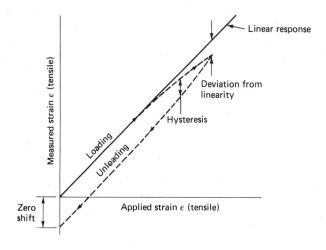

Figure 6.13 A typical strain cycle showing nonlinearity, hysteresis, and zero shift (scale exaggerated).

a means for making precise measurements of strain. The results actually obtained, however, are a function of the installation procedures, the state of strain being measured, and environmental conditions during the test. All these factors affect the performance of a strain-gage system.

A. Strain-Gage Linearity, Hysteresis, and Zero Shift [27–31]

One measure of the performance of a strain-gage system (system here implies gage, adhesive, and instrumentation) involves considerations of linearity, hysteresis, and zero shift. If gage output, in terms of measured strain, is plotted as a function of applied strain as the load on the component is cycled, results similar to those shown in Fig. 6.13 will be obtained. A slight deviation from linearity is typically observed, and the unloading curve normally falls below the loading curve to form a hysteresis loop. Also, when the applied strain is reduced to zero, the gage output indicates a small negative strain, termed *zero shift*. The magnitude of the deviation from linearity, hysteresis, and zero shift depends upon the strain level, the adequacy of the bond, the degree of cold work of the foil material, and the carrier material.

For properly installed gages, deviations from linearity should be approximately 0.1 percent of the maximum strain for polyimide carriers and 0.05 percent for epoxy carriers. First-cycle hysteresis and zero shift are more difficult to predict; however, zero shifts of 1 percent of the maximum strain are frequently observed in typical applications. If possible, strain cycling to 125 percent of the maximum test strain is recommended since the amount of hysteresis and zero shift will decrease to less than 0.2 percent of the maximum strain after 4 or 5 cycles.

B. Temperature Compensation [32–34]

In many test programs, the strain-gage installation is subjected to temperature changes during the test period, and careful consideration must be given to determining whether the change in resistance is due to applied strain or temperature change. When the ambient temperature changes, four effects occur which may alter the performance characteristics of the gage:

1. The strain sensitivity S_A of the metal alloy used for the grid changes.
2. The gage grid either elongates or contracts ($\Delta l/l = \alpha \, \Delta T$).
3. The base material upon which the gage is mounted either elongates or contracts ($\Delta l/l = \beta \, \Delta T$).
4. The resistance of the gage changes because of the influence of the temperature coefficient of resistivity of the gage material ($\Delta R/R = \gamma \, \Delta T$).

The change in the strain sensitivity S_A of Advance and Karma alloys with variations in temperature is shown in Fig. 6.14. These data indicate that $\Delta S_A/\Delta T$ equals 0.00735 and -0.00975 percent per Celsius degree for Advance and Karma alloys, respectively. As a consequence, the variations of S_A with temperature are neglected for room-temperature testing, where the temperature fluctuations rarely exceed $\pm 10°C$. In thermal-stress problems, however, larger temperature variations are possible, and the change in S_A should be taken into account by adjusting the gage factor as the temperature changes during the test period.

The effects of gage-grid elongation, base-material elongation, and increase in gage resistance with increases in temperature combine to produce a temperature-induced change in resistance of the gage $(\Delta R/R)_{\Delta T}$, which can be expressed as

$$\left(\frac{\Delta R}{R}\right)_{\Delta T} = (\beta - \alpha)S_g \, \Delta T + \gamma \, \Delta T \tag{6.11}$$

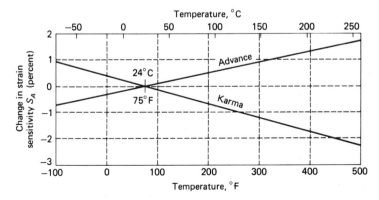

Figure 6.14 Change in alloy sensitivity S_A of Advance and Karma alloys as a function of temperature.

where $\alpha =$ thermal coefficient of expansion of gage material
$\qquad \beta =$ thermal coefficient of expansion of base material
$\qquad \gamma =$ temperature coefficient of resistivity of gage material
$\qquad S_g =$ gage factor

If there is a differential expansion between the gage and the base material due to temperature change (that is, $\alpha \neq \beta$), the gage will be subjected to a mechanical strain $\epsilon = (\beta - \alpha) \Delta T$, which does not occur in the specimen. The gage reacts to this strain by indicating a change in resistance in the same manner that it indicates a change for a strain due to the load applied to the specimen. Unfortunately, it is impossible to separate the apparent strain due to the change in temperature from the strain due to the applied load. If the gage alloy and the base material have identical coefficients of expansion, this component of the thermally induced $(\Delta R/R)_{\Delta T}$ vanishes. The gage may still register a change of resistance with temperature, however, if the coefficient of resistivity γ is not zero. This component of $(\Delta R/R)_{\Delta T}$ indicates an apparent strain which does not exist in the specimen.

Two approaches can be employed to effect temperature compensation in a gage system. The first involves compensation in the gage so that the net effect of the three factors in Eq. (6.11) is canceled out. The second involves compensation for the effects of the temperature change in the electrical system required to measure the output of the gage. The second method will be discussed in Chap. 8, when electrical systems are considered in detail.

Table 6.3 Expansion coefficients available in temperature-compensated gages

Specimen material	Coefficient of expansion, $°F^{-1} \times 10^{-6}$	Self-temperature-compensating number	
		Advance	Karma
Quartz	0.3	00	00
Alumina	3.0	03	03
Zirconium	3.1		
Glass	5.0	05	05
Titanium	5.2		
Cast iron	5.8	06	06
Steel	6.6		
Stainless steel	9.2	09	09
Copper	9.8		
Bronze	10.1		
Brass	11.4	13	13
Aluminum	12.5		
Magnesium	14.4	15	15
Polystyrene	40	41	
Epoxy resin	50	50	
Polymethyl methacrylate	50		
Acrylic resin	100		

In producing temperature-compensated gages it is possible to obtain compensation by perfectly matching the coefficients of expansion of the base material and the gage alloy while holding the temperature coefficient of resistivity at zero. Compensation can also be obtained with a mismatch in the coefficients of expansion if the effect of a finite temperature coefficient of resistivity cancels out the effect of the mismatch in temperature coefficients of expansion.

The values of the factors α and γ influencing the temperature response of the strain gage mounted on a specimen with thermal characteristics specified by the value of β are quite sensitive to the composition of the strain-gage alloy, its impurities, and the degree of cold working used in its manufacture. Recently, it has become common practice for strain-gage manufacturers to determine the thermal response characteristics of sample gages from each lot of alloy material which they employ in their production. Because of variations in α and γ between each melt and each roll of foil, it is possible to select foils of Advance and Karma alloys which are suitable for use with almost any type of base material. The gages produced by using this selection technique are known as *selected-melt* or *temperature-compensated gages* and are commercially available with the self-temperature-compensating numbers listed in Table 6.3. Some widely used materials and approximate values for their temperature coefficients of expansion are also listed in the table.

Unfortunately, these gages are not perfectly compensated over a wide range in temperature because of the nonlinear character of both the expansion coefficients and the resistivity coefficients with temperature. A typical curve showing the apparent strain (temperature-induced) as a function of temperature for a temperature-compensated strain gage fabricated from Advance alloy is shown in Fig. 6.15. These results show that the errors introduced by small changes in temperature in the neighborhood of 75°F (24°C) are quite small with apparent strains

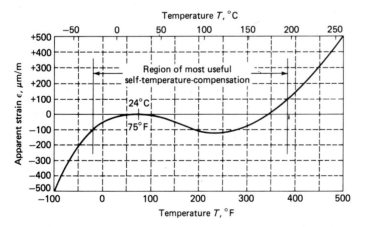

Figure 6.15 Apparent strain as a function of temperature for an Advance alloy temperature-compensated strain gage mounted on a specimen having a matching temperature coefficient of expansion.

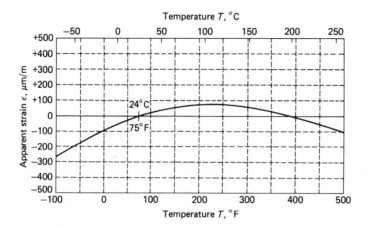

Figure 6.16 Apparent strain as a function of temperature for a Karma alloy temperature-compensated strain gage mounted on a specimen having a matching temperature coefficient of expansion.

of less than 1 μin/(in)(°F) or $\frac{1}{2}$ μm/(m)(°C). However, when the change in temperature is large, the apparent strains can become significant, and corrections to account for the thermally induced apparent strains are necessary. These corrections involve measurement of the test temperature at the gage site with a thermocouple and use of a calibration curve similar to the one shown in Fig. 6.15. A calibration curve is provided by the strain-gage manufacturer for each lot of gages produced.

The range of temperature over which an Advance alloy strain gage can be employed is approximately from -20 to 380°F (-30 to 193°C). For temperatures above and below these respective limits, the gage will function; however, very small changes in temperature will produce large apparent strains which can be difficult to account for properly in the analysis of the data.

An extended range of test temperatures is obtained with gages fabricated with Karma alloy. The thermally induced apparent strains as a function of temperature for the Karma alloy are shown in Fig. 6.16. These data indicate that the strain-temperature curve has a modest slope over the entire range of temperature from -100 to 500°F (-73 to 260°C). Temperature measurements are still required whenever temperature variations are large; however, the modest slope of the calibration makes it possible to account for the thermally induced apparent strains accurately.

C. Elongation Limits [35]

The maximum strain that can be measured with a foil strain gage depends on the gage length, the foil alloy, the carrier material, and the adhesive. The Advance and Karma alloys with polyimide carriers, used for general-purpose strain gages, can be employed to strain limits of ±5 and ±1.5 percent strain, respectively. This

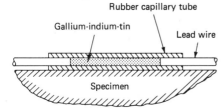

Figure 6.17 A liquid-metal electrical-resistance strain gage.

strain range is adequate for elastic analyses on metallic and ceramic components, where yield or fracture strains rarely exceed 1 percent; however, these limits can easily be exceeded in plastic analyses, where strains in the postyield range can become large. In these instances, a special postyield gage is normally employed; it is fabricated using a double annealed Advance foil grid with a high-elongation polyimide carrier. Urethane-modified epoxy adhesives are generally used to bond postyield gages to the structure. If proper care is exercised in preparing the surface of the specimen, roughening the back of the gage, formulating a high-elongation plasticized adhesive system, and attaching the lead wires without significant stress raisers, it is possible to approach strain levels of 20 percent before cracks begin to occur in the solder tabs or at the ends of the grid loops.

Special-purpose strain-gage alloys are not applicable for measurements of large strains. The Isoelastic alloy will withstand ± 2 percent strain; however, it undergoes a change of sensitivity at strains larger than 0.75 percent (see Fig. 6.2). Armour D and Nichrome V are primarily used for high-temperature measurements and are limited to maximum strain levels of approximately ± 1 percent.

For very large strains, where specimen elongations of 100 percent may be encountered, liquid-metal strain gages can be used. The liquid-metal strain gage is simply a rubber tube filled with mercury or a gallium-indium-tin alloy, as indicated in Fig. 6.17. When the specimen to which the gage is attached is strained, the volume of the tube cavity remains constant since Poisson's ratio of rubber is approximately 0.5. Thus the length of the tube increases ($\Delta l = \epsilon l$) while the diameter of the tube decreases ($\Delta d = -\nu \epsilon d$). The resistance of such a gage increases with strain, and it can be shown that the gage factor is given by

$$S_g = 2 + \epsilon \tag{6.12}$$

The resistance of a liquid-metal gage is very small (less than 1 Ω) since the rubber capillary tubes used in their construction have a relatively large diameter. As a consequence, the gages are usually used in series with a large fixed resistor to form a total resistance of 120 Ω so that the gage can be monitored with a conventional Wheatstone bridge. The response of a liquid-metal gage, as shown in Fig. 6.18, is slightly nonlinear with increasing strain due to the increase in gage factor with strain.

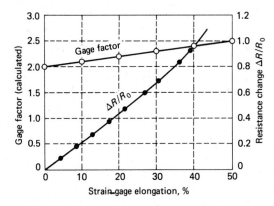

Figure 6.18 Resistance change and gage factor as a function of strain for a liquid-metal strain gage (from data by Harding).

D. Dynamic Response of Strain Gages [36–39]

In dynamic applications of strain gages, the question of their frequency response often arises. This question can be resolved into two parts, namely, the response of the gage in its thickness direction, i.e., how long it takes for an element of the gage to respond to the strain in the specimen beneath it, and the response of the gage due to its length. It is possible to estimate the time required to transmit the strain from the specimen through the adhesive and carrier to the strain-sensing element by considering a gage mounted on a specimen as shown in Fig. 6.19. A strain wave is propagating through the specimen with velocity c_1. This specimen strain wave induces a shear-strain wave in the adhesive and carrier which propagates with a velocity c_2. The transit time, which is given by $t = h/c_2$, equals 5×10^{-8} s for typical carrier and adhesive combinations, where $c_2 = 40{,}000$ in/s (1000 m/s) and $h = 0.002$ in (0.05 mm). The time required for the conductor to respond may exceed this transit time by a factor of 3 to 5; therefore, the response time should be approximately 2×10^{-7} s. Experiments conducted by Oi and an analysis by Bickle indicate that transit times are approximately 0.1 μs.

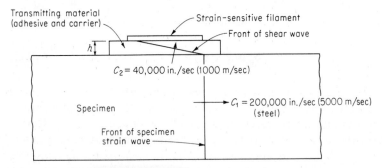

Figure 6.19 Dynamic strain transmission between the specimen and the gage.

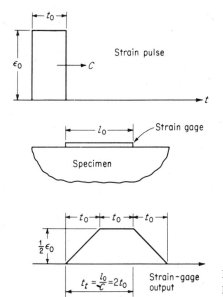

Figure 6.20 Dynamic response of a gage due to a step-pulse input.

The rise time for a strain gage responding to a step pulse is given by

$$t_r = \frac{l_0}{c_1} + 0.1 \times 10^{-6} \tag{6.13}$$

where l_0 is the length of the gage and the 0.1-μs term is added to account for the transmission time through the carrier and adhesive. A typical rise time for a 0.125-in (3.17-mm) gage mounted on a steel bar ($c_1 = 200{,}000$ in/s or 5000 m/s) is $0.6 + 0.1 = 0.7$ μs. Such short rise times are often neglected when long strain pulses are encountered and the rate of change of strain with time is small. When short, steep-fronted strain pulses are encountered, however, the response time of the gage should be considered since the measured strain pulse can be distorted, as shown in Fig. 6.20. In this instance, the gage is mounted on a specimen which is propagating a strain pulse having an amplitude ϵ_0, a time duration t_0, and a velocity c. The front of the pulse will reach and just pass over a gage of length l_0 in a transit time $t_t = l_0/c$. In Fig. 6.20 the transit time t_t equals $2t_0$. If the gage records average strain over its length, its output will rise linearly to a value of $\epsilon_0/2$ over a time period of t_0. The output will then remain constant at $\epsilon_0/2$ for a period of time equal to t_0 and then decrease linearly to zero over a final time period of t_0. The effect of the gage length in this example was to decrease the amplitude of the output by a factor of 2 and to increase the total time duration of the pulse by a factor of 3. The distortion of the pulse as indicated by the gage will depend upon the ratio t_t/t_0; and as the value of this ratio goes to zero, the distortion vanishes.

It is also possible to correct for this distortion. Note that the indicated strain $\bar{\epsilon}$ at time t is given by

$$\bar{\epsilon}(t) = \frac{1}{l_0} \int_{c_1 t - l_0}^{c_1 t} \epsilon(x) \, dx \tag{6.14}$$

Differentiating gives

$$\frac{d\bar{\epsilon}}{dt} = \frac{1}{l_0} \left\{ \int_{c_1 t - l_0}^{c_1 t} \frac{\partial}{\partial t} [\epsilon(x)] \, dx + \epsilon(a)c_1 - \epsilon(b)c_1 \right\} \tag{6.15}$$

where $\epsilon(a)$ and $\epsilon(b)$ are the strains at the two ends of the gage. If the pulse shape remains constant during propagation,

$$\frac{\partial}{\partial t} [\epsilon(x)] = 0$$

and Eq. (6.15) reduces to

$$\frac{d\bar{\epsilon}}{dt} = \frac{c_1}{l_0} [\epsilon(a) - \epsilon(b)] \tag{6.16}$$

Since

$$\epsilon(b, t) = \epsilon \left(a, t - \frac{l_0}{c_1} \right)$$

then

$$\epsilon(a, t) = \frac{l_0}{c_1} \frac{d\bar{\epsilon}}{dt} + \epsilon \left(a, t - \frac{l_0}{c_1} \right) \tag{6.17}$$

An analog circuit can be constructed using operational amplifiers to automatically correct the signal output from the gage as given by Eq. (6.17). The block diagram for such an analog circuit is given in Fig. 6.21.

E. Heat Dissipation [40–43]

It is well recognized that temperature variations can significantly influence the output of strain gages, particularly those which are not temperature-compensated. The temperature of the gage is of course influenced by ambient-temperature variations and by the power dissipated in the gage when it is connected into a Wheatstone bridge or a potentiometer circuit. The power P is dissipated in the form of

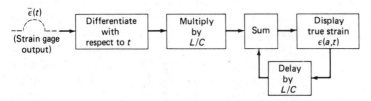

Figure 6.21 Analog circuit which eliminates pulse distortion due to strain-gage length. (*After Bickle.*)

heat; therefore, the temperature of the gage must increase above the ambient temperature to accomplish the heat transfer. The exact temperature increase required is very difficult to specify since many factors influence the heat balance for the gage. The heat to be dissipated depends upon the voltage applied to the gage and the gage resistance. Thus

$$P = \frac{V^2}{R} = I^2 R \tag{6.18}$$

where P = power, W
  I = gage current, A
  R = gage resistance, Ω
  V = voltage across the gage, V
Factors which govern the heat dissipation include

1. Gage size, w_0 and l_0
2. Grid configuration, spacing and size of conducting elements
3. Carrier, type of polymer and thickness
4. Adhesive, type of polymer and thickness
5. Specimen material, thermal diffusivity
6. Specimen volume in the local area of the gage
7. Type and thickness of overcoat used to waterproof the gage
8. Velocity of the air flowing over the gage installation

A parameter often used to characterize the heat-dissipation characteristics of a strain-gage installation is the power density P_D, which is defined as

$$P_D = \frac{P}{A} \tag{6.19}$$

where P is the power that must be dissipated by the gage and A is the area of the grid of the gage. Power densities that can be tolerated by a gage are strongly related to the specimen which serves as the heat sink. Recommended values of P_D for different materials and conditions are listed in Table 6.4.

Table 6.4 Allowable power densities

Power density P_D		Specimen conditions
W/in²	W/mm²	
5–10	0.008–0.016	Heavy aluminum or copper sections
2–5	0.003–0.008	Heavy steel sections
1–2	0.0015–0.003	Thin steel sections
0.2–0.5	0.0003–0.0008	Fiber glass, glass, ceramics
0.02–0.05	0.00003–0.00008	Unfilled plastics

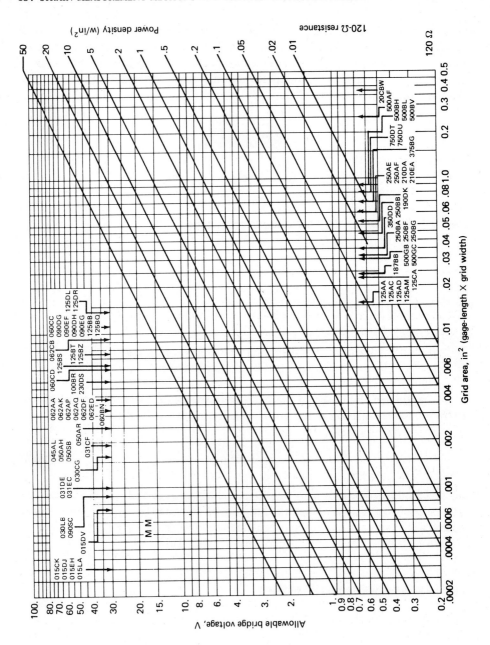

Figure 6.22 Allowable bridge voltage as a function of grid area with power density as a parameter for 120-Ω electrical-resistance strain gages. (*Micro-Measurements.*)

When a Wheatstone bridge with four equal arms is employed, the bridge excitation voltage V_B is related to the power density on the strain gage by

$$V_B^2 = 4AP_DR \tag{6.20}$$

Allowable bridge voltages for specified grid areas and power densities are shown for 120-Ω gages in Fig. 6.22. Typical grid configurations are identified along the abscissa in the figure to illustrate the effect of gage size on allowable bridge excitation. It should be noted that small gages mounted on a poor heat sink ($P_D < 1$ W/in² or 0.0015 W/mm²) result in lower allowable bridge voltages than those employed in most commercial strain indicators (3 to 5 V). In these instances it is necessary to use a higher-resistance gage (350 Ω in place of 120 Ω), a gage with a larger grid area, or a fixed resistor in series with the gage to reduce the voltage on the gage in both the active and compensating arms of the bridge.

F. Stability [44–47]

In certain strain-gage applications it is necessary to record strains over a period of months or years without having the opportunity to unload the specimen and recheck the zero reading. The duration of the readout period is important and makes this application of strain gages one of the most difficult. All the factors which can influence the behavior of the gage have an opportunity to do so; moreover, there is enough time for the individual contribution to the error from each of the factors to become quite significant. For this reason it is imperative that every precaution be taken in employing the resistance-type gage if meaningful data are to be obtained.

Drift in the zero reading from an electrical-resistance strain-gage installation is due to the effects of moisture or humidity variations on the carrier and the adhesive, the effects of long-term stress relaxation of the adhesive, the carrier, and the strain-gage alloy, and the instabilities in the resistors in the inactive arms of the Wheatstone bridge.

Results of an interesting series of stability tests by Freynik are presented in Fig. 6.23. In evaluating a typical general-purpose strain gage with a grid fabricated from Advance alloy and a polyimide carrier, zero shifts of 270 μin/in (μm/m) were observed after 30 days. Since this installation was carefully waterproofed, the large drifts were attributed to stress relaxation in the polyimide carrier over the period of observation.

Results from a second strain-gage installation with a grid fabricated from Advance alloy and a carrier from glass-fiber-reinforced phenolic were more satisfactory with zero drift of approximately 100 μin/in (μm/m) after 50 days. The presence of the glass fibers essentially eliminated drift due to stress relaxation in the carrier material; the drift measured was attributed to instabilities in the Advance alloy grid at the test temperature of 167°F (75°C). The final and most satisfactory results were obtained with a Karma grid and an encapsulating glass-reinforced epoxy-phenolic carrier. In this case, the zero shift averaged only 30 μin/in (μm/m) after an observation period of 900 days. In similar tests with this

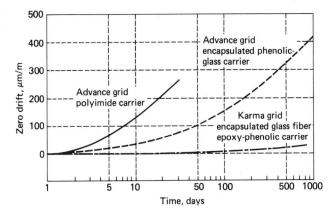

Figure 6.23 Zero drift as a function of time for several gage types at 167°F (75°C) (from data by Freynik).

gage installation at room temperature, the zero drift was only −25 μin/in (μm/m). These results show that electrical-resistance strain gages can be used for long-term measurements provided Karma grids with glass-fiber-reinforced epoxy-phenolic carriers are employed with a well-cured epoxy adhesive system. The gage installation should be waterproofed to minimize the effects of humidity. It is also important to specify hermatically sealed bridge-completion resistors to ensure stability of the bridge over the long observation periods.

6.7 ENVIRONMENTAL EFFECTS

The performance of resistance strain gages is markedly affected by the environment. Moisture, temperature extremes, hydrostatic pressure, nuclear radiation, and cyclic loading produce changes in gage behavior which must be accounted for in the installation of the gage and in the analysis of the data to obtain meaningful results. Each of these parameters is discussed in the following subsections.

A. Effects of Moisture and Humidity [48–50]

A strain-gage installation can be detrimentally affected by direct contact with water or by the water vapor normally present in the air. The water is absorbed by both the adhesive and the carrier, and the gage performance is affected in several ways. First, the moisture decreases the gage-to-ground resistance. If this value of resistance is reduced sufficiently, the effect is the same as that of placing a shunt resistor across the active gage. The water also degrades the strength and rigidity of the bond and reduces the effectiveness of the adhesive in transmitting the strain from the specimen to the gage. If this loss in adhesive strength or rigidity is

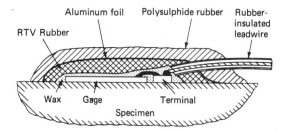

Figure 6.24 Waterproofing a gage for severe seawater exposure.

sufficient, the gage will not develop its stated calibration factors and measuring errors are introduced.

Plastics also expand when they absorb water and contract when they release it; thus, any change in the moisture concentration in the adhesive will produce strains in the adhesive, which will in turn be transmitted to the strain gage. These moisture-induced adhesive strains will produce a strain-gage response that cannot be separated from the response due to the applied mechanical strain. Finally, the presence of water in the adhesive will cause electrolysis when current passes through the gage. During the electrolysis process, the gage filament will erode and a significant increase in resistance will occur. Again, the strain gage will indicate a tensile strain due to this electrolysis which cannot be differentiated from the applied mechanical strain.

Many methods for waterproofing strain gages have been developed; however, the extent of the measures taken to protect the gage from moisture depends to a large degree on the application and the extent of the gage exposure to water. For normal laboratory work, where the readout time is relatively short, a thin layer of microcrystalline wax or an air-drying polyurethane coating is usually sufficient to protect the gage installation from moisture in the air. For much more severe applications, e.g., prolonged exposure to seawater, it is necessary to build up a seal out of soft wax, synthetic rubber, metal foil, and a final coat of rubber. A cross section of a well-protected gage installation is shown in Fig. 6.24. Care should be exercised in forming the seal at the lead-wire terminal since the seal usually fails at this location. Also, the insulation for the lead wires should be rubber, and splices in the cable should be avoided.

B. Effects of Hydrostatic Pressure [51–55]

In the stress analysis of pressure vessels and piping systems, strain gages are frequently employed on interior surfaces where they are exposed to a gas or fluid pressure which acts directly on the sensing element of the gage. Under such conditions, pressure-induced resistance changes occur which must be accounted for in the analysis of the strain-gage data.

Milligan and Brace independently studied this effect of pressure by mounting a gage on a small specimen, placing the specimen in a special high-pressure vessel,

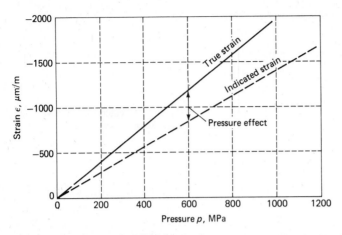

Figure 6.25 True and indicated strain as a function of pressure for an Advance foil gage with epoxy backing on a hydrostatically loaded specimen of Armco iron (from data by Brace).

and monitoring the strain as the pressure was increased to 140,000 lb/in^2 (965 MPa). In this type of experiment, the hydrostatic pressure p produces a strain in the specimen which is given by Eq. (2.16) as

$$\epsilon = \frac{-p}{E} - \frac{v}{E}[(-p) + (-p)] = -\frac{1 - 2v}{E}p = K_T p \qquad (6.21)$$

where $K_T = -(1 - 2v)/E$ is often referred to as the *compressibility constant* for a material. The strain gages were monitored during the pressure cycle, and it was observed that the indicated strains were less than the true strains predicted by Eq. (6.21). Some typical results are shown in Fig. 6.25. The difference between the true strains and the indicated strains was attributed to the pressure effect.

The pressure effect can be characterized by defining the slope of the indicated pressure-strain curve as an apparent compressibility constant for the material. Thus

$$K_I = -\frac{\Delta \epsilon}{\Delta p} \qquad (6.22)$$

The difference due to pressure D_p can then be computed as

$$D_p = \frac{K_T - K_I}{K_T} \qquad (6.23)$$

Experimental results by Milligan, shown in Fig. 6.26, indicate that D_p depends upon the compressibility constant for the specimen material, the curvature of the specimen where the gage is mounted, and the type of strain-sensing alloy used in fabricating the gage. In spite of relatively large values for D_p, however, it was observed that the pressure-strain response of the gage remained linear.

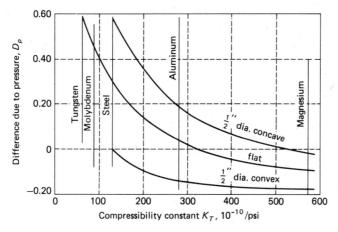

Figure 6.26 Difference due to pressure D_p as a function of compressibility constant K_T for Advance foil gages on specimens with different curvatures (from data by Milligan).

Thus, true strain can be expressed in terms of indicated strain and a correction term

$$\epsilon_t = \epsilon_i - \epsilon_{cp} \tag{6.24}$$

The magnitude of the correction term ϵ_{cp} can be expressed in terms of the experimentally determined values of D_p as

$$\epsilon_{cp} = D_p K_T p \tag{6.25}$$

For a flat steel specimen, $K_T = 133 \times 10^{-10}$ in²/lb (1.93×10^{-12} m²/N) and $D_p = 0.3$; hence, the correction term for the strain is approximately 4 μin/in (μm/m) for a pressure of 1000 lb/in² (7 MPa). Since this is a very small correction, it is possible to neglect pressure effects for pressures less than approximately 3000 lb/in² (20 MPa).

For hydrostatic pressure applications, foil strain gages with the thinnest possible carriers should be employed. The gage should be mounted on a smooth surface with a thinned adhesive to obtain the thinnest possible bond line. Bubbles in the adhesive layer cannot be tolerated since the pressure normal to the surface of the gage will force the sensing element into any void beneath the gage and erroneous resistance changes will result.

C. Effects of Nuclear Radiation [56–58]

Several difficult problems are encountered when electrical-resistance strain gages are employed in nuclear-radiation fields. The most serious difficulty involves the change in electrical resistivity of the strain gage and lead wires as a result of the fast-neutron dose. This effect is significant; changes of 2 to 3 percent in $\Delta R/R$ have been observed with a neutron dose of $10^{18} nvt$. These changes in resistivity produce

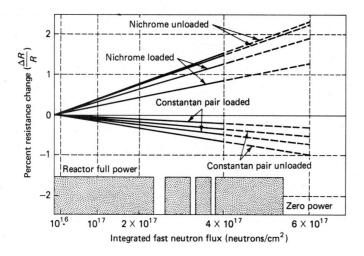

Figure 6.27 Percent resistance change as a function of exposure to neutron flux (from data by Tallman).

zero drift with time which can be as large as an apparent strain of 10,000 to 15,000 $\mu in/in$ ($\mu m/m$). The exact rate of change of resistivity is a function of the strain-gage and lead-wire materials, the state of strain in the gage, and the temperature. Typical changes in resistance with integrated fast-neutron flux are shown in Fig. 6.27. Since the changes in resistivity are a function of the magnitude and sign of the strain, the use of dummy gages to cancel the effects of radiation exposure is not effective. Since the change in electrical resistivity appears to be a linear function of the logarithm of the dose, the most satisfactory solution to neutron-induced zero drift is to employ preexposed strain-gage installations and to reduce test times to a minimum. Unloading and reestablishing the zero resistance of the gage is essential.

The neutrons also produce a change in the sensitivity of the strain-gage alloys. Typical variations in S_A for Advance alloy, shown in Fig. 6.28, range from 15 to -10 percent as the integrated neutron exposure increases from 10^{16} to $6 \times 10^{17} nvt$.

The fast neutrons also produce mechanical effects which are detrimental to strain-gage installations. With exposure to the fast neutrons, the strain-gage alloy exhibits an increase in its yield strength and modulus of elasticity and a decrease in elongation capability. Radiation-induced cross-linking in polymers also destroys the original organic structure of the bond. For this reason, ceramic adhesives are normally employed in any long-term tests where exposure will accumulate.

In nuclear-radiation fields with high gamma flux, considerable energy is transferred to the gage and specimen; therefore, temperature changes can be significant. For precise measurements, the temperature change must be predicted or determined so that the strain-gage results can be corrected for the effects of temperature.

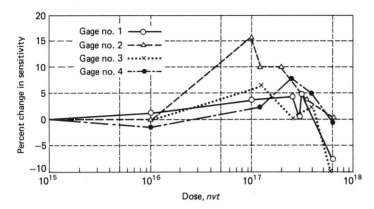

Figure 6.28 Percent change in sensitivity of an Advance alloy strain gage with exposure to fast neutrons (from data by Tallman).

Electromagnetic effects are important when the strain-gage circuit is subjected to transient flux fields for short periods of time. If gamma particles are present, potential differences of several thousand volts can be generated. The specimen must be adequately grounded to prevent the development of these high voltages.

Transient radiation pulses also produce Compton replacement currents, which are injected into the strain-gage circuit by the lead wires and the strain gages. The Compton effect results from an unbalance between the number of electrons being captured and the number being emitted by the cable conductor. These currents, when injected into the Wheatstone bridge circuit, produce spurious output signals which can be misinterpreted as strain pulses.

There are two precautions which can be taken to minimize the effects of Compton replacement currents. (1) Coaxial cables can be used for the lead wires. The shields on these conductors are effective in limiting the injected currents on the center conductor. (2) The compensating Wheatstone bridge circuit, shown in Fig. 6.29, can be used to reduce the output signal generated by the injected currents I_1, I_2, and I_3. If the active strain gage is placed in arm R_1 and the dummy

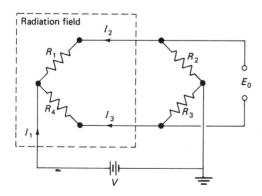

Figure 6.29 Compensating Wheatstone bridge to reduce the effects of Compton replacement currents.

gage in arm R_4, both gages are exposed to the same radiation transient and the effects of the injection currents I_1, I_2, and I_3 are minimized. The current I_1 finds a balanced path through $R_1 + R_2$ and $R_4 + R_3$ to ground and does not generate a differential voltage E_0. Currents I_2 and I_3 are equal since gages R_1 and R_4 are identical and are subjected to the same field. As a result, these currents do not produce a differential voltage E_0.

It is possible to measure strains in strong radiation fields which are either transient or steady-state. In either instance, special precautions must be taken or serious errors will occur.

D. Effects of High Temperature [59–67]

Resistance-type strain gages can be employed at elevated temperatures for both static and dynamic stress analyses; however, the measurements require many special precautions which depend primarily on the temperature and the time of observation. At elevated temperatures, the resistance R of a strain gage must be considered to be a function of temperature T and time t in addition to strain ϵ. Thus

$$R = f(\epsilon, T, t) \tag{6.26}$$

The resistance change $\Delta R/R$ is then given by

$$\frac{\Delta R}{R} = \frac{1}{R}\frac{\partial f}{\partial \epsilon}\Delta\epsilon + \frac{1}{R}\frac{\partial f}{\partial T}\Delta T + \frac{1}{R}\frac{\partial f}{\partial t}\Delta t \tag{6.27}$$

where $\dfrac{1}{R}\dfrac{\partial f}{\partial \epsilon} = S_g$ = gage sensitivity to strain (gage factor)

$\dfrac{1}{R}\dfrac{\partial f}{\partial T} = S_T$ = gage sensitivity to temperature

$\dfrac{1}{R}\dfrac{\partial f}{\partial t} = S_t$ = gage sensitivity to time

Equation (6.27) can then be expressed in terms of the three sensitivity factors as

$$\frac{\Delta R}{R} = S_g\,\Delta\epsilon + S_T\,\Delta T + S_t\,\Delta t \tag{6.28}$$

In the discussion of performance characteristics of foil strain gages in Secs. 6.5B and 6.5F, it was shown that sensitivity of the gages to temperature and time was minimized at normal operating temperatures of 0 to 150°F (-18 to 65°C) by proper selection of the strain-gage alloy and carrier materials. As the test temperature increases above this level, however, the performance of the gage changes, and S_T and S_t are not usually negligible.

As the temperature increases, temperature compensation is less effective, and corrections must be made to account for the apparent strain, as shown in

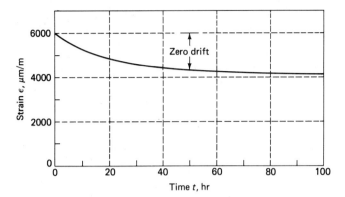

Figure 6.30 Typical zero drift as a function of time for a Karma alloy strain gage with a glass-fiber–epoxy–phenolic carrier at 560°F (293°C) (from data by Hayes).

Figs. 6.15 and 6.16. Comparisons of these results indicate that the Karma strain-gage alloy is more suitable for the higher-temperature applications than Advance. The Karma gages can be employed to temperatures up to about 500°F (260°C) without encountering excessively large temperature-induced apparent strains.

The stability of a strain-gage installation is also affected by temperature; and strain-gage drift becomes a more serious problem as the temperature and the time of observation are increased. Stability is affected by stress relaxation in the adhesive bond and in the carrier material and by metallurgical changes (phase transformations and annealing) in the strain-gage alloy. The upper temperature limit on commercially available Karma gages is controlled by the carrier material. Glass-reinforced epoxy-phenolic carriers are rated at 550°F (288°C); however, Karma gages with this type of carrier drift with time as shown in Fig. 6.30. If the time of loading and observation is long, corrections must be made for the zero drift. Drift rates will depend upon both the strain level and the temperature. For high-temperature strain analyses, it is recommended that a series of strain-time calibration curves, similar to the one shown in Fig. 6.30, be developed to cover the range of strains and temperatures to be encountered. Zero-drift corrections can then be taken from the appropriate curve.

The problem of gage stability and apparent strain due to temperature changes is greatly reduced if the period of observation is short. For dynamic analyses, where relatively short strain-time records are used (times usually less than 1 s), the temperature does not have sufficient time to change and temperature effects are therefore small. For this reason, dynamic strain-gage analyses can be made at very high temperatures with good precision and with relative ease.

Resistance-strain-gage measurements at temperatures higher than 550°F (288°C) require special gages and special techniques for mounting and monitoring the strain-gage signal. At these higher temperatures, polymeric materials can no longer be used for the carrier or the adhesive. The gages must be mounted to the specimen with the ceramic cements described in Sec. 6.4D. The carrier is either

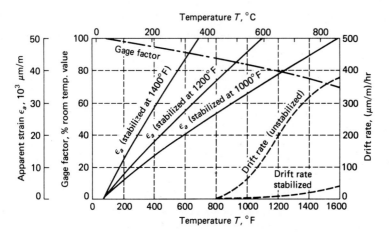

Figure 6.31 Performance characteristics of a platinum-tungsten alloy strain gage as a function of temperature (from data by Wnuk).

removed entirely or replaced with a thin stainless-steel shim. Strain gages for the very high temperature applications are fabricated using materials such as Armour D (70 Fe, 20 Cr, and 10 Al) or alloy 479 (92 Pt and 8 W).

Most commercially available strain gages rated for high-temperature use are fabricated from the platinum-tungsten alloy because of its inherent oxidation resistance and metallurgical stability. Performance characteristics of this alloy are shown as a function of temperature in Fig. 6.31. These results indicate that the gage factor drops about 30 percent as the temperature increases from 70 to 1600°F

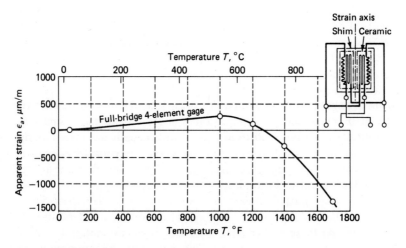

Figure 6.32 Apparent strain as a function of temperature for a four-element full-bridge gage (from data by Wnuk).

(21 to 871°C). Also, the temperature coefficient of resistance of this alloy is very high; therefore, large apparent strains [40 to 80 μin/(in)(°F) or 70 to 140 μm/(m)(°C)] are indicated by the gage when the temperature changes. The material is stable, and zero drift is negligible to temperatures of 800°F (427°C). At temperatures above 800°F (427°C), drift will occur, the rate of drift increasing with temperature.

The effects of drift rate can be minimized by stabilizing the alloy. The stabilization process consists of annealing the alloy at the test temperature for 12 to 16 h. The stabilized drift is relatively low, with rates of 20 μin/(in)(h) or μm/(m)(h) reported at 1400°F (760°C). The effect of apparent strain is much more difficult to treat and usually requires the use of dual-element or four-element gages, which compensate for the effects of temperature by signal subtraction in the Wheatstone bridge.

An example of apparent strain as a function of temperature for a four-element (complete-bridge) gage is shown in Fig. 6.32. Temperature compensation has been achieved for temperatures less than approximately 1400°F (760°C). The gage can be used at higher temperatures by correcting for apparent strain if the ceramic adhesive continues to perform adequately.

E. Effects of Cryogenic Temperatures [68–70]

Strain can be measured with electrical-resistance strain gages to liquid-nitrogen temperatures (−320°F or −196°C) and below. Two factors must be carefully accounted for in using strain gages at this temperature extreme. The first effect is the change in gage factor with temperature, illustrated in Fig. 6.33. These results

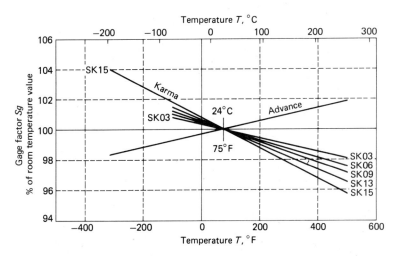

Figure 6.33 Changes in gage factor with temperature (from data by Telinde).

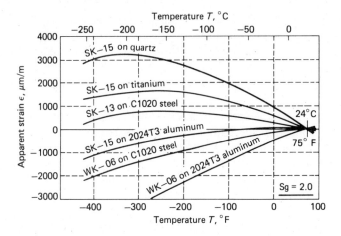

Figure 6.34 Apparent strain as a function of temperature for several selected-melt Karma alloys on different materials (from data by Telinde).

indicate that this correction is small, being only -1.8 and $+3.8$ percent, respectively, for Advance and the selected-melt Karma alloy SK-15 at $-320°F$ $(-196°C)$.

The second effect, which is extremely important to account for when measuring strains at cryogenic temperatures, is the very large apparent strains introduced by small changes in temperature. Karma alloys are preferred to Advance alloys because of their better stability at the temperature extremes, as shown in Figs. 6.15 and 6.16. Temperature-induced apparent strains for Karma strain gages mounted on a variety of materials are shown in Fig. 6.34.

Although both Advance and Karma strain gages are temperature-compensated, the compensation is limited to the temperature range from -100 to $500°F$ $(-73$ to $260°C)$. As the test temperature is reduced below $-100°F$ $(-73°C)$, the strain gages become more sensitive to temperature changes and large apparent strains are produced. It is possible to minimize the effects of temperature changes at cryogenic temperatures by selecting compensating gages which are mismatched relative to the component material. For instance, the Karma gage compensated for a material with a temperature coefficient of expansion equal to 13×10^{-6} per Fahrenheit degree (23.4×10^{-6} per Celsius degree) exhibits a relatively stable response to temperature variations from -100 to $-400°F$ $(-73$ to $-240°C)$ when mounted on a C-1020 steel specimen having a temperature coefficient of expansion of 6×10^{-6} per Fahrenheit degree (10.8×10^{-6} per Celsius degree) (see Fig. 6.34).

If temperature sensors are mounted with the strain gages, the measured temperatures can be used with curves similar to those shown in Fig. 6.34 to correct for the apparent strains induced by temperature changes during the test. It is also possible to use a four-element gage which is wired so that signal cancellation in

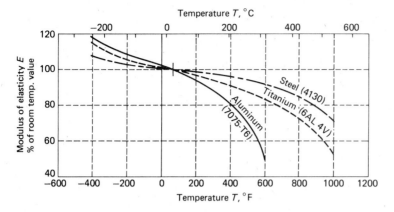

Figure 6.35 Change in elastic modulus as a function of temperature (from data by Telinde).

the Wheatstone bridge compensates for the temperature-induced response of the strain gages.

Cryogenic temperatures are usually obtained in tests by using liquid nitrogen, liquid hydrogen, or liquid helium. All three are excellent insulating materials, and it is not necessary or advisable to use electrical insulating compounds between the gage grid and the cryogenic fluid. The gage should be coated with a silicone grease to provide a heat shield and to eliminate the possibility of boiling of the liquid over the gage grid during the test.

Special consideration must also be given to changes in the mechanical properties of component materials at cryogenic temperatures. The effect of cryogenic temperatures is to increase the elastic modulus from 5 to 20 percent, as indicated in Fig. 6.35. The higher values of the elastic modulus must be employed in the stress-strain equations when converting the strain data to stress.

F. Effects of Strain Cycling [72–73]

Strain gages are frequently mounted on components subjected to fatigue or cyclic loading, and the life of the component during which the gage must be monitored can exceed several million cycles. Three factors which must be considered for this particular type of application are zero shift, change in gage factor, and failure of the gage in fatigue.

As the strain gage is subjected to repeated cyclic strain, the gage gridwork hardens and its specific resistance changes. The specific resistance change produces a zero shift. The amount of the zero shift depends on the magnitude of the strain, the number of cycles, the grid alloy, the original state of cold work of the grid alloy, and the type of carrier employed in the gage construction. Typical examples of zero shift as a function of number of strain cycles are shown in Fig. 6.36 for four different gages. The poorest gage for this type of application is the open-faced Advance grid (EA) gage, which begins to exhibit noticeable zero

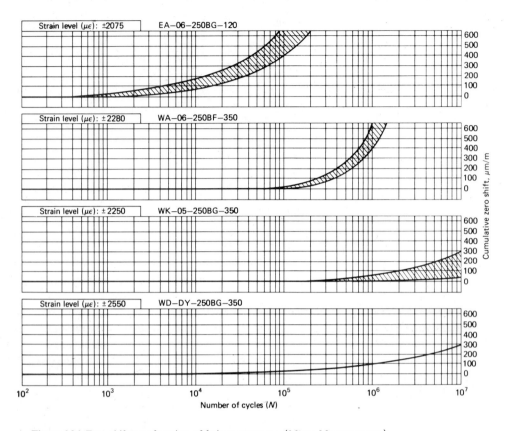

Figure 6.36 Zero shift as a function of fatigue exposure. (*Micro-Measurements.*)

shift at 10^3 cycles. Encapsulating the grid in a glass-reinforced epoxy-phenolic resin (WA type) improves the life, and exposures of 10^5 cycles at ± 2100 μin/in (μm/m) occur before zero shift becomes apparent. Further improvement can be achieved by using Karma (K) or isoelastic (D) alloys fully encapsulated in the glass-reinforced epoxy-phenolic carrier material with factory-installed lead wires.

The changes in gage factor are quite small due to strain cycling and in general can be neglected if the zero shift is less than a few hundred microinches per inch or micrometers per meter.

A very important factor to consider when using strain gages under fatigue conditions is the increase in sensitivity which occurs once a fatigue crack develops in the grid of the gage. The crack will usually develop in the grid near the lead wire on the tab. The presence of the crack produces a small increase in the resistance of the gage, which is monitored as an apparent strain. Thus the apparent gage factor has increased with the development and subsequent propagation of the fatigue crack.

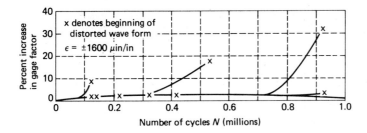

Figure 6.37 Increase in gage factor as a function of fatigue exposure for an annealed Advance foil gage. (*Micro-Measurements.*)

Typical results showing this increase in gage factor are shown in Fig. 6.37. It is important to note that the gage factor increased from 10 to 32 percent for the three gages which failed in fatigue. Also, the error in "calibration" existed over an appreciable fraction of the life of the gage.

EXERCISES

6.1 The strain sensitivity of most metallic alloys is about 2.0. What portion of this sensitivity is due to dimension changes in the conductor? What portion is due to changes in the number of free electrons and their mobility?

6.2 Advance or Constantan alloy exhibits a linear response of $\Delta R/R$ to strain ϵ for strains as large as 8 percent. Discuss why this is a remarkably large range of linearity with respect to strain when the elastic limit of the material may be as low as 21,000 lb/in^2 (145 MPa) and the modulus of elasticity is 22×10^6 lb/in^2 (152 GPa). Base the discussion on Eq. (6.2) and recall that Poisson's ratio for the material will increase from 0.3 to 0.5 as the material transforms from the elastic to the plastic state.

6.3 Describe the advantages and disadvantages of a strain gage fabricated using an isoelastic alloy.

6.4 Give several reasons why the foil strain gage is preferred over the bonded-wire strain gage.

6.5 List four different carrier materials used in strain-gage construction and give the reasons for their use.

6.6 Briefly discuss the conditions which would dictate using the following cements in a strain-gage installation:

    (*a*) Cellulose nitrate cement   (*b*) Epoxy cement
    (*c*) Cyanoacrylate cement   (*d*) NBS-x-142 ceramic cement

6.7 Describe two procedures used to evaluate the completeness of the cure of an adhesive being used to bond a strain gage to a specimen.

6.8 Determine the transverse sensitivity factor for a strain gage fabricated using Constantan alloy with $S_A = 2.11$ if it has been calibrated and found to exhibit a gage factor $S_g = 2.04$.

6.9 The transverse sensitivity factor K_t for a new grid configuration must be determined. It has been suggested that a simple tension specimen can be used with gages oriented in the axial and transverse directions to make the determination. Develop an expression for K_t in terms of Poisson's ratio of the specimen material and the ratio of the outputs from the two strain gages.

6.10 A value of $K_t = 0.03$ was obtained for the gage in Exercise 6.9 by assuming Poisson's ratio $= 0.30$ for the specimen material. The true value of Poisson's ratio was 0.29. Determine the percent error introduced in the determination of K_t.

6.11 Determine the percent error introduced by neglecting transverse sensitivity if an WK-06-250BG-350 type strain gage is used to measure the longitudinal strain in a steel beam having Poisson's ratio $= 0.30$.

6.12 Determine the percent error introduced by neglecting transverse sensitivity if a WK-06-500AF-350 type strain gage is used to measure the maximum tensile strain on the surface of a circular steel shaft (Poisson's ratio $= 0.30$) subjected to a pure torque.

6.13 Determine the percent error introduced by neglecting transverse sensitivity if an EA-06-125RA-120 type strain gage is used to measure the maximum tensile strain on the outside surface of a thin-walled cylindrical pressure vessel which is being subjected to an internal gas pressure. The vessel is fabricated from steel with a Poisson's ratio of 0.29.

6.14 Specify a test procedure which will determine the zero shift associated with a strain-gage installation. Specify a second test procedure which will minimize this zero shift.

6.15 Determine the change in alloy sensitivity S_A of Constantan as the temperature is increased from 32 to 392°F (0 to 200°C).

6.16 A strain-gage analysis is to be conducted on a component which will be maintained at 150 ± 36°F (65 ± 20°C) throughout the test period. Which alloy should be specified for the strain-gage grid? If temperature-compensated gages were specified, what errors would occur due to the temperature fluctuations?

6.17 Specify the special procedures required for a strain-gage analysis where plastic strains of the order of 15 percent are anticipated.

6.18 For the liquid-metal strain gage illustrated in Fig. 6.17, show that the gage factor S_g is given by $S_g = 2 + \epsilon$.

6.19 A strain wave is propagating in the x direction through a specimen with a characteristic velocity $c_1 = 5000$ m/s. The strain wave may be represented by the equation $\epsilon_1 = A \sin (2\pi/\lambda)(x - c_1 t)$, where A is the amplitude and λ is the wavelength of the wave. If $\lambda = nl_0$, determine the amplitude of strain recorded by a strain gage having a length l_0.

6.20 Determine the allowable bridge voltage for the grid configurations 015LA, 060BN, 125DL, 250BA, 375BG, and 500BH if the gages are mounted on:

(a) An aluminum engine head   (b) A large steel beam
(c) A fiberglass tank

6.21 Write a test specification outlining special procedures to be used in measuring strain over a 2-year period on a large steel smokestack subjected to wind loads.

6.22 Write a test specification to be followed by a laboratory technician in waterproofing a strain gage for (a) a short-duration test in the laboratory and (b) a long-term test under seawater exposure.

6.23 A strain gage is mounted on the inside surface of a large-diameter steel pressure vessel which is subjected to an internal pressure of 75 MPa. Determine the true strain in the hoop direction if the inside diameter is 450 mm and the outside diameter is 800 mm. Compute the error involved in neglecting pressure effects on the output of the strain gages.

6.24 Describe a procedure to minimize the effect of neutron-induced zero drift in a strain-gage installation subjected to intense radiation.

6.25 Specify a Wheatstone bridge arrangement which will minimize the effects of Compton replacement currents produced by transient radiation pulses. Include lead-wire treatment in your specification.

6.26 Describe a procedure for measuring the strain on a mechanical component at 500°F (260°C) if the loading cycle requires:

(a) 1 s   (b) 1 min   (c) 1 h   (d) 1 day

6.27 Describe a procedure for measuring the strain on a component at a temperature of 1000°F (538°C) if the strain cycle occurs in 1 h.

6.28 Describe a procedure for measuring the strain on a mechanical component at a temperature of 1600°F (872°C) if the strain is transient and a cycle requires only 0.01 s.

6.29 A strain-gage test is to be conducted with a Karma strain gage at −300°F (−184°C). The gage factor is specified as 2.05 at room temperature. What gage factor should be used in reducing the value of $\Delta R/R$ to strain?

6.30 Specify the self-temperature-compensating number for a Karma strain gage to be employed on a stainless-steel tank at liquid-nitrogen temperatures.

6.31 Select a gage to operate under cyclic strain conditions where unloading of the structure is not practical. The strain is:

    (a) ± 1000 μm/m for 10^5 cycles   (b) ± 2000 μm/m for 10^6 cycles

6.32 Will a continuity check on the gage resistance indicate that a fatigue crack has initiated in the grid and that the gage will indicate strain with a supersensitive gage factor? Why?

REFERENCES

1. Thomson, W. (Lord Kelvin): On the Electrodynamic Qualities of Metals, *Proc. R. Soc.*, 1856.
2. Tomlinson, H.: The Influence of Stress and Strain on the Action of Physical Forces, *Phil. Trans. R. Soc. Lond.*, vol. 174, pp. 1–172, 1883.
3. Simmons, E. E., Jr.: Material Testing Apparatus, U.S. Patent 2,292,549, Feb. 23, 1940.
4. Mason, W. P., and R. N. Thurston: Use of Piezoresistive Materials in the Measurement of Displacement, Force, and Torque, *J. Acoust. Soc. Am.*, vol. 29, no. 10, pp. 1096–1101, 1957.
5. Kammer, E. W., and T. E. Pardue: Electric Resistance Changes of Fine Wires during Elastic and Plastic Strains, *Proc. SESA*, vol. VII, no. 1, pp. 7–20, 1949.
6. Maslen, K. R.: Resistance Strain Characteristics of Fine Wires, *R. Aircr. Establ. Tech. Note Instr.* 127, 1952.
7. Sette, W. J., L. D. Anderson, and J. G. McGinley: Resistance Strain Characteristics of Stretched Fine Wires, *David Taylor Model Basin Rep.* R-212, 1945.
8. Stein, P. K.: Constantan as a Strain Sensitive Alloy, *Strain Gage Readings*, vol. 1, no. 1, p. 43, 1958.
9. Meyer, R. D.: Application of Unbonded-Type Resistance Gages, *Instruments*, vol. 19, no. 3, pp. 136–139, 1946.
10. Campbell, W. R.: Performance of Wire Strain Gages: I, Calibration Factors in Tension, *NACA Tech. Note* 954, 1944; II, Calibration Factors in Compression, *NACA Tech. Note* 978, 1945.
11. DeForest, A. V.: Characteristics and Aircraft Applications of Wire Wound Resistance Strain Gages, *Instruments*, vol. 15, no. 4, pp. 112, 1942.
12. Jones, E., and K. R. Maslen: The Physical Characteristics of Wire Resistance Strain Gages, *R. Aircrt. Establ. R and M* 2661, November 1948.
13. Jackson, P.: The Foil Strain Gauges, *Instrum. Pract.*, August 1953.
14. Maslen, K. R., and I. G. Scott: Some Characteristics of Foil Strain Gauges, *R. Aircr. Establ. Tech. Note Instr.* 134, 1953.
15. Dorsey, J.: New Developments in Strain Gages, *Proc. West. Reg. Strain Gage Comm. 1968*, pp. 1–8.
16. Wnuk, S. P.: Recent Developments in Flame Sprayed Strain Gages, *Proc. West. Reg. Strain Gage Comm. 1965*, pp. 1–6.
17. Stein, P. K.: Adhesives: How They Determine and Limit Strain Gage Performance, "Semiconductor and Conventional Strain Gages," pp. 45–72 in M. Dean and R. D. Douglas (eds.), Academic Press, Inc., New York, 1962.
18. Pitts, J. W., and D. G. Moore: Development of High Temperature Strain Gages, *Nat. Bur. Std. Monogr.* 26, 1966.
19. Micro-measurements, Strain Gage Installations, *Instr. Bull.* B-130-2, July 1972.

20. Lee, H., and K. Neville: "Epoxy Resins: Their Applications and Technology," McGraw-Hill Book Company, New York, 1957.
21. Bodnar, M. J., and W. H. Schroder: Bonding Properties of a Solventless Cyanoacrylate Adhesive, *Mod. Plastics*, September 1958.
22. Campbell, W. R.: Performance Tests of Wire Strain Gages: IV, Axial and Transverse Sensitivities, *NACA Tech. Note* 1042, June 1946.
23. Baumberger, R., and F. Hines: Practical Reduction Formulas for Use on Bonded Wire Strain Gages, *Proc. SESA*, vol. II, no. 1, pp. 113–127, 1944.
24. Meier, J. H.: On the Transverse Strain Sensitivity of Foil Gages, *Expt. Mech.*, vol. 1, no. 7, pp. 39–40, 1961.
25. Wu, C. T.: Transverse Sensitivity of Bonded Strain Gages, *Expt. Mech.*, vol. 2, no. 11, pp. 338–344, 1962.
26. Transverse Sensitivity Errors, *Micro-Measurements Tech. Note* 137, pp. 1–16.
27. Boodberg, A., E. D. Howe, and B. York: Stability of SR-4 Electric Strain Gages and Methods for Their Waterproofing and Protection in Field Service, *Trans. ASME*, vol. 70, no. 8, pp. 915–920, 1948.
28. Matlock, H., and S. A. Thompson: Creep in Bonded Electrical Strain Gages, *Proc. SESA*, vol. XII, no. 2, pp. 181–188, 1955.
29. McWhirter, M., and B. W. Duggin: Minimizing Creep of Paper-Base SR-4 Strain-Gages, *Proc. SESA*, vol. XIV, no. 2, pp. 149–154, 1957.
30. Miller, B. L., L. D. Anderson, and H. Shaub: The Performance of Wire Resistance Strain Gages as Influenced by the Drying Time of Three Mounting Cements, *David Taylor Model Basin Rep.* R-213, January 1946.
31. Bloss, R. L.: Characteristics of Resistance Strain Gages, pp. 123–142 in M. Dean and R. D. Douglas (eds.), "Semiconductor and Conventional Strain Gages," Academic Press, Inc., New York, 1962.
32. Barker, R. S.: Self-Temperature Compensating SR-4 Strain Gages, *Proc. SESA*, vol. XI, no. 1, pp. 119–128, 1953.
33. Campbell, W. R.: Errors in Indicating Strains for a Typical Wire Strain Gage Caused by Temperature and Weathering, *NACA Tech. Bull.* 1011, 1946.
34. Hines, F. F., and L. J. Weymouth: Practical Aspects of Temperature Effects on Resistance Strain Gages, *Strain Gage Readings*, vol. IV, no. 1, 1961.
35. Harding, D.: High Elongation Measurements with Foil and Liquid Metal Strain Gages, *Proc. West. Reg. Strain Gage Comm.* 1965, pp. 23–31.
36. Mason, W. P., J. J. Forst, and L. M. Tornillo: Recent Developments in Semiconductor Strain Transducers, *Instrum. Soc. Am. Meet. September 1960.*
37. Stein, P. K.: "Advanced Strain Gage Techniques," chap. 2, Stein Engineering Services, Phoenix, Ariz., 1962.
38. Bickle, L. W.: The Response of Strain Gages to Longitudinally Sweeping Strain Pulses, *Expt. Mech.*, vol. 10, no. 8, pp. 333–337, 1970.
39. Oi, K.: Transient Response of Bonded Strain Gages, *Expt. Mech.*, vol. 6, no. 9, pp. 463–469, 1966.
40. Freynik, H. S., Jr.: "Investigation of Current Carrying Capacity of Bonded Resistance Strain Gages," M.S. Thesis, Massachusetts Institute of Technology, Cambridge, Mass., 1961.
41. Frocht, M. M., D. Landsberg, and B. G. Wang: On the Effect of Current on Gage Factor, *Proc. SESA*, vol. XVI, no. 2, pp. 69–72, 1959.
42. Owens, J. B. B.: The Effect of Current Variation on the out of Balance of Strain Gage Bridges, *R. Aircr. Establ. R and M* 2497, 1947.
43. Optimizing Strain Gage Excitation Levels, *Micro-Measurements Tech. Note* 127, August 1968.
44. Beyer, F. R., and M. J. Lebow: Long-Time Strain Measurements in Reinforced Concrete, *Proc. SESA*, vol. XI, no. 2, pp. 141–152, 1954.
45. Rohrback, C., and N. Czarka: The Creep of Strain Gages as a Rheological Problem, *Materialpruefung*, vol. 2, no. 3, March 1960.
46. Tschebotarioff, G. P.: Use of Electric Resistivity Strain Gages over Long Periods of Time, *Proc. SESA*, vol. III, no. 2, pp. 47–52, 1946.

47. Freynik, H. S., and G. R. Dittbenner: Strain Gage Stability Measurements for a Year at 75°C in Air, *Univ. Calif. Radiat. Lab. Rep.* 76039, 1975.

48. Dean, M.: Strain Gage Waterproofing Methods and Installation of Gages in Propeller Strut of *U.S.S. Saratoga, Proc. SESA*, vol. XVI, no. 1, pp. 137–150, 1958.

49. Palermo, P. M.: Methods of Waterproofing SR-4 Strain Gages, *Proc. SESA*, vol. XIII, no. 2, pp. 79–88, 1956.

50. Wells, F. E.: A Rapid Method of Waterproofing Bonded Wire Strain Gages, *Proc. SESA*, vol. XV, no. 2, pp. 107–110, 1958.

51. Gerdeen, J. C.: Effects of Pressure on Small Foil Strain Gages, *Expt. Mech.* vol. 3, no. 3, pp. 73–80, 1963.

52. Guerard, J. P., and G. P. Weissmann: Effect of Hydrostatic Pressure on SR-4 Strain Gages, *Proc. SESA*, vol. XVI, no. 1, pp. 151–156, 1958.

53. Milligan, R. V.: The Effects of High Pressure on Foil Strain Gages, *Expt. Mech.*, vol. 4, no. 2, pp. 25–36, 1964.

54. Brace, W. F.: Effect of Pressure on Electrical-Resistance Strain Gages, *Expt. Mech.*, vol. 4, no. 7, pp. 212–216, 1964.

55. Milligan, R. V.: Effects of High Pressure on Foil Strain Gages on Convex and Concave Surfaces, *Expt. Mech.*, vol. 5, no. 2, pp. 59–64, 1965.

56. Anderson, S. D., and R. C. Strahm: Nuclear Radiation Effects on Strain Gages, *Proc. West. Reg. Strain Gage Comm. 1968*, pp. 9–16.

57. Tallman, C. R.: Nuclear Radiation Effects on Strain Gages, *Proc. West. Reg. Strain Gage Comm. 1968*, pp. 17–25.

58. Wnuk, S. P.: Progress in High Temperature and Radiation Resistant Strain Gage Development, *Proc. West. Reg. Strain Gage Comm. 1964*, pp. 41–47.

59. Carpenter, J. E., and L. D. Morris: A Wire Resistance Strain Gage for the Measurement of Static Strains at Temperatures up to 1600°F, *Proc. SESA*, vol. IX, no. 1, pp. 191–200, 1951.

60. Day, E. E.: Characteristics of Electric Strain Gages at Elevated Temperatures, *Proc. SESA*, vol. IX, no. 1, pp. 141–150, 1951.

61. Day, E. E., and A. H. Sevand: Characteristics of Electric Strain Gages at Low Temperatures, *Proc. SESA*, vol. VIII, no. 1, pp. 133–142, 1950.

62. Gorton, R. E.: Development and Use of High-Temperature Strain Gages, *Proc. SESA*, vol. IX, no. 1, pp. 163–176, 1951.

63. Wnuk, S. P.: New Strain Gage Developments, *Proc. West. Reg. Strain Gage Comm. 1964*, pp. 1–7.

64. Hayes, J. K., and B. Roberts: Measurements of Stresses under Elastic-Plastic Strain Conditions at Elevated Temperatures, *Proc. Tech. Comm. Strain Gages 1969*, pp. 20–39.

65. Denyssen, I. P.: Platinum-Tungsten Foil Strain Gages for High-Temperature Applications, *Proc. West. Reg. Strain Gage Comm. 1964*, pp. 19–23.

66. Weymouth, L. J.: Strain Measurement in Hostile Environment, *Appl. Mech. Rev.*, vol. 18, no. 1, pp. 1–4, 1965.

67. Sharpe, W. N.: Strain Gages for Long-Term High-Temperature Strain Measurement, *Rep. Pressure Vessel Res. Comm.*, December 1973.

68. Kaufman, A.: Investigation of Strain Gages for Use at Cryogenic Temperatures, *Expt. Mech.*, vol. 3, no. 8, pp. 177–183, 1963.

69. Telinde, J. C.: Strain Gages in Cryogenic Environment, *Expt. Mech.*, vol. 10, no. 9, pp. 394–400, 1970.

70. Nicholls, D.: Strain Gages in Cryogenic Applications, *Proc. West. Reg. Strain Gage Comm. 1968*, pp. 37–43.

71. Telinde, J. C.: Strain Gages in Cryogenics and Hard Vacuum, *Proc. West. Reg. Strain Gage Comm. 1968*, pp. 45–54.

72. Dorsey, J.: New Developments and Strain Gage Progress, *Proc. West. Reg. Strain Gage Comm. 1965*, pp. 1–10.

73. Fatigue Characteristics of Micro-measurement Strain Gages, *Micro-Measurements Tech. Note* 130-2, 1974.

SEVEN

SEMICONDUCTOR STRAIN GAGES

7.1 INTRODUCTION [1–5]

The development of semiconductor strain gages was an outgrowth of the research at the Bell Telephone Laboratories which led to the introduction of the transistor in the early 1950s. The piezoresistive properties of semiconducting silicon and germanium were determined by Smith in 1954. Further development of semiconductor transducers by Mason and Thurston in 1957 eventually led to the commercial marketing of piezoresistive strain gages in 1960.

Basically the semiconductor strain gage consists of a small, ultrathin rectangular filament of a single crystal of silicon, which is usually mounted on a carrier to simplify handling. The semiconducting materials exhibit a very high strain sensitivity, with values of S_A ranging from 50 to 175 depending upon the type and amount of impurity diffused into the pure silicon crystal. The resistivity ρ of a single-crystal semiconductor with impurity concentrations of the order of 10^{16} to 10^{20} atoms/cm^3 is given by

$$\rho = \frac{1}{eN\mu} \tag{7.1}$$

where e = electron charge, which depends on type of impurity
$\quad\quad N$ = number of charge carriers, which depends on concentration of impurity
$\quad\quad \mu$ = mobility of charge carriers, which depends on strain and its direction relative to crystal axes

The importance of Eq. (7.1) can be better understood if it is considered in terms of Eq. (6.2), where the sensitivity of a conductor was defined as

$$S_A = 1 + 2v + \frac{d\rho/\rho}{\epsilon} \tag{6.2}$$

For metallic conductors, the term $1 + 2v$ is approximately equal to 1.6; the term $(d\rho/\rho)/\epsilon$ ranges from 0.4 to 2.0 for the common strain-gage alloys. For semiconductor materials $(d\rho/\rho)/\epsilon$ can be varied between -125 and 175 by selecting the type and concentration of the impurity. Thus, very high conductor sensitivities are possible where the resistance change is about 100 times larger than that obtained for the same strain with metallic-alloy gages. Also, negative gage factors are possible which permit large electrical outputs from Wheatstone bridge circuits where multiple strain gages are employed.

In producing semiconductor strain gages, ultrapure single-crystal silicon is usually employed. Boron is used as the trace impurity in producing the P-type (positive gage factor) piezoresistive material, and arsenic is used to produce the N-type (negative gage factor) material.

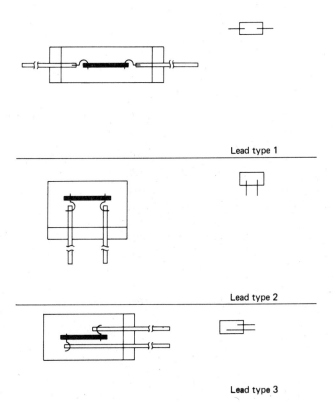

Figure 7.1 Semiconductor elements with different lead-wire configurations. Gage lengths vary from 0.06 to 0.17 in (1.5 to 4.3 mm). (*BLH Electronics.*)

The semiconductor materials have another advantage over metallic alloys for strain-gage applications. The resistivity of P-type silicon is of the order of 500 $\mu\Omega \cdot$ m; this is 1000 times greater than the resistivity of Constantan, which is 0.49 $\mu\Omega \cdot$ m. Because of this very high resistivity, semiconductor strain gages do not utilize grids. They are usually very short single elements with leads oriented in one of the ways illustrated in Fig. 7.1.

The very high sensitivity of semiconductor gages to strain and their high resistivity have led to their application in measuring extremely small strains, in miniaturized transducers, and in very high signal output transducers.

7.2 PIEZORESISTIVE PROPERTIES OF SEMICONDUCTORS [6, 7]

A crystal of a semiconducting material is electrically anisotropic; consequently, the relation between the potential gradient $\mathbf{E}$ and the current density $\mathbf{i}$ is formulated relative to the directions of the crystal axes to give components E_1, E_2, and E_3 of the vector $\mathbf{E}$. Thus

$$E_1 = \rho_{11} i_1 + \rho_{12} i_2 + \rho_{13} i_3$$
$$E_2 = \rho_{21} i_1 + \rho_{22} i_2 + \rho_{23} i_3$$
$$E_3 = \rho_{31} i_1 + \rho_{32} i_2 + \rho_{33} i_3 \tag{7.2}$$

where the first subscript of each resistivity coefficient indicates the component of the voltage field to which it contributes and the second identifies the component of current. The single crystal will permit isotropic conduction only if

$$\rho_{11} = \rho_{22} = \rho_{33} = \rho$$

and
$$\rho_{12} = \rho_{13} = \rho_{21} = \rho_{23} = \rho_{31} = \rho_{32} = 0$$

Since this situation exists for unstressed cubic crystals, Eqs. (7.2) reduce to

$$E_1 = \rho i_1 \qquad E_2 = \rho i_2 \qquad E_3 = \rho i_3 \tag{7.3}$$

When a state of stress is imposed on the crystal, it responds by exhibiting a piezoresistive effect which can be described by the expression

$$\rho_{ij} = \delta_{ij}\rho + \pi_{ijkl}\tau_{kl} \tag{7.4}$$

where subscripts i, j, k, and l range from 1 to 3 and π_{ijkl} is a fourth-rank piezoresistivity tensor, which is a function of the crystal and the level and type of the impurities. A complete discussion of the results obtained from Eq. (7.4) is beyond the scope of this text. Fortunately, several reductions in the complexity of Eq. (7.4) are possible for the type of crystals used in the fabrication of semiconductor strain gages. Recall from Chap. 1 [Eqs. (1.4)] that the stress tensor τ_{kl} is symmetric. If the analysis is limited to cubic crystals, the resistivity tensor ρ_{ij} is also symmetric.

Finally due to the properties of a silicon crystal, the piezoresistive coefficients can be shown to reduce to

$$\pi_{1111} = \pi_{2222} = \pi_{3333} = \rho\pi_{11}$$

$$\pi_{1122} = \pi_{2211} = \pi_{1133} = \pi_{3311} = \pi_{2233} = \pi_{3322} = \rho\pi_{12}$$

$$\pi_{1212} = \pi_{2121} = \pi_{2323} = \pi_{3232} = \pi_{1313} = \pi_{3131} = \frac{\rho\pi_{44}}{2}$$

The simplified matrix representing the piezoresistive coefficients for silicon, since all other coefficients are zero, is then

$$\begin{bmatrix} \pi_{11} & \pi_{12} & \pi_{12} & 0 & 0 & 0 \\ \pi_{12} & \pi_{11} & \pi_{12} & 0 & 0 & 0 \\ \pi_{12} & \pi_{12} & \pi_{11} & 0 & 0 & 0 \\ 0 & 0 & 0 & \pi_{44} & 0 & 0 \\ 0 & 0 & 0 & 0 & \pi_{44} & 0 \\ 0 & 0 & 0 & 0 & 0 & \pi_{44} \end{bmatrix}$$

Equation (7.4) can then be expressed in terms of the simplified notation as

$$\rho_{11} = \rho[1 + \pi_{11}\sigma_{11} + \pi_{12}(\sigma_{22} + \sigma_{33})]$$

$$\rho_{22} = \rho[1 + \pi_{11}\sigma_{22} + \pi_{12}(\sigma_{33} + \sigma_{11})]$$

$$\rho_{33} = \rho[1 + \pi_{11}\sigma_{33} + \pi_{12}(\sigma_{11} + \sigma_{22})]$$

$$\rho_{12} = \rho\pi_{44}\tau_{12} \qquad \rho_{23} = \rho\pi_{44}\tau_{23} \qquad \rho_{31} = \rho\pi_{44}\tau_{31} \qquad (7.5)$$

Substitution of Eqs. (7.5) into Eqs. (7.2) yields

$$\frac{E_1}{\rho} = i_1[1 + \pi_{11}\sigma_{11} + \pi_{12}(\sigma_{22} + \sigma_{33})] + \pi_{44}(i_2\tau_{12} + i_3\tau_{31})$$

$$\frac{E_2}{\rho} = i_2[1 + \pi_{11}\sigma_{22} + \pi_{12}(\sigma_{33} + \sigma_{11})] + \pi_{44}(i_3\tau_{23} + i_1\tau_{12})$$

$$\frac{E_3}{\rho} = i_3[1 + \pi_{11}\sigma_{33} + \pi_{12}(\sigma_{11} + \sigma_{22})] + \pi_{44}(i_1\tau_{31} + i_2\tau_{23}) \qquad (7.6)$$

These results indicate that the voltage drop across a gage fabricated from a piezoelectric element will depend upon the current density, the stress field, and the three piezoresistivity coefficients.

Consider now a strain gage consisting of a single conductor, which is long relative to its lateral dimensions and cut in an arbitrary direction from a cubic crystal, as illustrated in Fig. 7.2. The direction of the gage conductor relative to the crystal axes is given by the unit vector **g**. Consider that lead wires are attached to the ends of this element in the manner illustrated in Fig. 7.1 and that a current density i_g is applied along the gage axis. The strain transmitted to the gage after it is bonded to a component will produce a stress σ_g along the axis of the element.

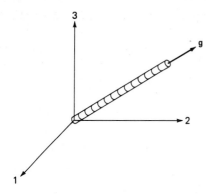

Figure 7.2 Orientation of strain-gage element relative to the crystal axes 1, 2, and 3.

Since **i** is a vector quantity, the gage current exhibits components along the crystal axes. Thus

$$i_1 = i_g \cos (g, 1) = l i_g$$
$$i_2 = i_g \cos (g, 2) = m i_g$$
$$i_3 = i_g \cos (g, 3) = n i_g \qquad (7.7)$$

where l, m, and n are the direction cosines associated with the unit vector **g**.

The stresses along the crystal axes in terms of the stress along the axis of the gage are given by Eqs. (1.6) as

$$\sigma_{11} = l^2 \sigma_g \qquad \sigma_{22} = m^2 \sigma_g \qquad \sigma_{33} = n^2 \sigma_g$$
$$\tau_{12} = lm \sigma_g \qquad \tau_{23} = mn \sigma_g \qquad \tau_{31} = nl \sigma_g \qquad (7.8)$$

From the vector properties of the voltage gradient it is also obvious that

$$E_g = l E_1 + m E_2 + n E_3 \qquad (7.9)$$

Substituting Eqs. (7.6) to (7.8) into Eq. (7.9) and simplifying gives

$$\frac{E_g}{\rho} = i_g \{ 1 + \sigma_g [\pi_{11} + 2(\pi_{12} + \pi_{44} - \pi_{11})(l^2 m^2 + m^2 n^2 + n^2 l^2)] \} \qquad (7.10)$$

which can be rewritten as

$$\frac{E_g}{\rho} = i_g (1 + \pi_g \sigma_g) \qquad (7.11)$$

where π_g represents the sensitivity of the gage to stress and is related to the piezoresistive coefficients by

$$\pi_g = A + B(l^2 m^2 + m^2 n^2 + n^2 l^2) \qquad (7.12)$$

where $\qquad A = \pi_{11} \qquad B = 2(\pi_{12} + \pi_{44} - \pi_{11})$

Examination of Eq. (7.12) indicates that the sensitivity π_g of the gage to stress can be varied by changing the orientation g or by changing the type (N or P) and the level of doping of the silicon. Conditions needed to maximize π_g for a given semiconductor material can be obtained by differentiating π_g relative to the two independent direction cosines. Thus

$$\frac{\partial \pi_g}{\partial l} = l(2 - 4l^2 - 2m^2) = 0 \qquad \frac{\partial \pi_g}{\partial m} = m(2 - 4m^2 - 2l^2) = 0 \qquad (a)$$

Examination of Eqs. (a) indicates that four sets of direction cosines must be considered to maximize π_g:

Case 1, $l = 0$, $m = 0$: $\qquad\qquad\qquad\qquad\qquad\qquad \pi_g = A$

Case 2, $l = 0$, $m = \pm\sqrt{2}/2$: $\qquad\qquad\qquad\qquad\quad \pi_g = A + \dfrac{B}{4}$

Case 3, $l = \pm\sqrt{2}/2$, $m = 0$: $\qquad\qquad\qquad\qquad\quad \pi_g = A + \dfrac{B}{4}$

Case 4, $l = \pm\sqrt{3}/3$, $m = \pm\sqrt{3}/3$: $\qquad\qquad\qquad \pi_g = A + \dfrac{B}{3}$

The final selection of the direction cosines depends upon the characteristics of the semiconducting silicon. With P-type silicon, A and B have the same sign; therefore, case 4 gives the maximum value, with $\pi_g = A + B/3$ and $l = m = n = \pm\sqrt{3}/3$. With N-type silicon, A and B have opposite signs with $|B| > 6|A|$; therefore, case 1 provides the maximum sensitivity, with $\pi_g = A$ and $l = m = n = 0$.

From Eq. (7.11), the difference in voltage gradient ΔE_g across the semiconductor element before and after stressing is

$$\frac{\Delta E_g}{\rho} = i_g(1 + \pi_g \sigma_g) - i_g = i_g \pi_g \sigma_g$$

Normalizing this relationship with respect to the voltage gradient in the unstressed state gives

$$\frac{\Delta E_g}{E_g} = \frac{\Delta R_g}{R_g} = \pi_g \sigma_g \qquad (7.13)$$

Since a uniaxial state of stress exists in the gage element,

$$\sigma_g = E\epsilon \qquad (b)$$

where E is the modulus of elasticity of the semiconducting silicon and ϵ is the strain transmitted to the element from the specimen. Substituting Eq. (b) into Eq. (7.13) gives

$$\frac{\Delta R_g}{R_g} = \pi_g E\epsilon = S_{sc}\epsilon \qquad (7.14)$$

where $S_{sc} = \pi_g E$ is the strain sensitivity of the piezoresistive material due to strain-induced changes in resistivity.

The form of the resistance-strain relationship for a metallic-alloy conductor was previously defined in Chap. 6 as

$$\frac{\Delta R_g}{R_g} = \frac{d\rho}{\rho} + (1 + 2v)\epsilon = S_A \epsilon \tag{6.2}$$

In a similar manner, the alloy sensitivity S_A for a semiconductor element can be defined as

$$S_A = 1 + 2v + S_{sc} \tag{7.15}$$

The term $1 + 2v$ is due to dimensional changes of the semiconducting element and must be added to S_{sc} to obtain the alloy sensitivity S_A. Since the value of S_{sc} is in the range from 40 to 200, the contribution of dimensional change to alloy sensitivity S_A in semiconducting elements is small (approximately 1.6) in comparison with its effect in metal alloys.

7.3 PERFORMANCE CHARACTERISTICS OF SEMICONDUCTOR GAGES

A. Temperature Effects [8, 9]

The results of the previous subsection indicate that the response of a semiconductor strain gage is linear with respect to strain. Unfortunately, this is a simplification which is not true in the general case. For lightly doped semiconducting materials (10^{19} atoms/cm^3 or less), the sensitivity S_A is markedly dependent upon both strain and temperature with

$$S_A = \frac{T_0}{T}(S_A)_0 + C_1\left(\frac{T_0}{T}\right)^2 \epsilon + C_2\left(\frac{T_0}{T}\right)^3 \epsilon^2 + \cdots \tag{7.16}$$

where $(S_A)_0$ = room-temperature zero-strain sensitivity, as defined in Eq. (7.15)
  T = temperature, with $T_0 = 294°$K
  C_1, C_2 = constants, which depend upon type of impurity, level of doping, and orientation of element with respect to crystal axes
  The variation of alloy sensitivity S_A as a function of doping level for P-type silicon is shown in Fig. 7.3. As the impurity concentration is increased from 10^{16} to 10^{20} atoms/cm^3, the sensitivity S_A decreases from 155 to 50. Two significant advantages related to the temperature effect compensate for this loss of sensitivity: (1) the effect of temperature on the sensitivity S_A is greatly diminished (since the temperature coefficient of sensitivity approaches zero) when the impurity concentration approaches 10^{20} atoms/cm^3, as shown in Fig. 7.3; (2) the temperature coefficient of resistance decreases from 0.005 per Fahrenheit degree (0.009 per Celsius degree) for impurity concentrations of 10^{16} atoms/cm^3 to 0.0002 per Fahrenheit degree (0.00036 per Celsius degree) for impurity concentrations of

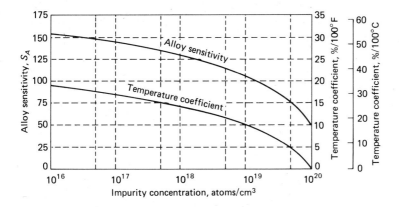

Figure 7.3 Alloy sensitivity and temperature coefficient of sensitivity as a function of impurity concentration of *P*-type silicon.

10^{19} atoms/cm³. The effect of impurity concentration on resistance change as a function of temperature for *P*-type silicon is indicated in Fig. 7.4.

From the results of Fig. 7.4, it is evident that the effects of temperature are minimized with impurity concentrations of 10^{19} atoms/cm³. This concentration yields an alloy sensitivity of 105.

Temperature compensation of single-element *P*-type semiconductor strain gages is not possible; therefore, gage response resulting in apparent strain (see Sec. 6.5B) occurs when the temperature changes during the test period. Two methods exist for achieving temperature compensation with semiconductor gages.

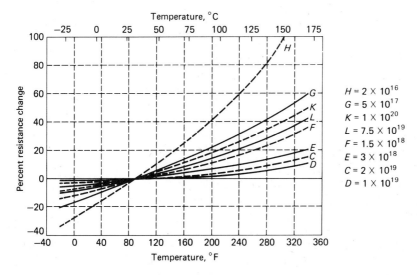

Figure 7.4 Resistance change as a function of temperature for several *P*-type silicon semiconductors with different impurity concentrations, relative to 81°F. (*Kulite Semiconductor Products, Inc.*)

Both make use of the N-type piezoresistive material, which exhibits negative alloy sensitivity.

The first approach for temperature compensation makes use of two gages (one of P-type material and the other of N-type material) which exhibit the same temperature coefficient of resistance. The two gages are connected into adjacent arms of a Wheatstone bridge, where the output resulting from resistance change due to temperature is canceled while the output resulting from strain is added due to the positive and negative sensitivities of the two gages. Typical dual-element gages, which are compensated for apparent strain to within ± 0.5 μm/(m)($^\circ$C) over a temperature range from 10 to 65°C (50 to 150°F), exhibit strain sensitivities in the range from 220 to 265.

The second approach to temperature compensation with semiconductor gages involves the use of N-type material with a positive temperature coefficient of resistance. This positive coefficient is selected by controlling the impurity concentration to cancel the negative change in resistance produced by differential expansion between the gage and specimen materials [see Eq. (6.11)]. Single-element N-type gages are available which are compensated for specimen materials with coefficients of expansion of 4, 6, 9, and 12 $\times 10^{-6}$ per Fahrenheit degree (7.2, 10.8, 16.2, and 21.6 $\times 10^{-6}$ per Celsius degree).

B. Linearity [8, 10]

It is clear from Eq. (7.16) that the response of semiconductor strain gages may be nonlinear with respect to strain. For piezoresistive materials with a low impurity concentration, the nonlinearities are significant, as illustrated in Fig. 7.5. Increas-

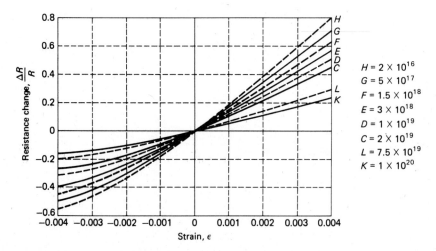

Figure 7.5 Linearity of P-type silicon semiconductor gages with different impurity concentrations. (*Kulite Semiconductor Products, Inc.*)

ing the impurity level to 10^{19} to 10^{20} atoms/cm^3 markedly improves the linearity, and performance is usually satisfactory when *P*-type gages are employed in tensile fields and *N*-type gages are employed in compressive fields. The linearity characteristics of dual element gages (one *P*-type and one *N*-type) are usually satisfactory since the nonlinear components are canceled in the Wheatstone bridge circuit.

C. Single- and Multiple-Cycle Strain Limits [10, 11]

The small elements used in fabricating semiconductor strain gages are removed from a single crystal of silicon by a slicing procedure similar to the one illustrated in Fig. 7.6. The single crystal is sectioned and then sliced and subsliced with a diamond saw to produce small rectangular bars of material. The small sensing elements are then usually etched to reduce their cross section further and improve the quality of their surfaces.

The silicon material is glasslike and must be treated with care during mounting. On flat surfaces, installation presents no problems, and semiconductor gages can be subjected to approximately 3000 μin/in(μm/m) of strain before the gages begin to fail by brittle rupture. When semiconductor gages are installed in a fillet and bending stresses are imposed during installation, extreme care must be exercised to avoid rupture of the element. Fillet radii from 0.1 to 0.25 in (2.5 to 6 mm) can be gaged; however, the strains induced in the element during installation reduce its strain range for subsequent tests.

The fatigue life of semiconductor gages is rated at 10^7 cycles for cyclic strains

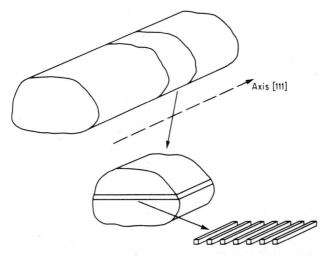

Axis [111]

Figure 7.6 Sectioning the crystal to remove the strain-gage sensing elements.

of ± 500 μin/in(μm/m). This is considerably less than the capability of metallic-foil-type gages; however, semiconductor strain gages are usually employed in low-magnitude strain fields, so that the relatively low strain limits placed on single-cycle and multiple-cycle measurements do not cause problems.

D. Power Dissipation [10, 11]

The maximum excitation voltage that can be placed across a semiconductor gage is limited by the amount of heat that can be dissipated from the gage to the specimen and to the surrounding air. The heat dissipation is a function of gage size, thickness of the adhesive, thermal mass of the specimen, thermal diffusivity of the specimen, and the coupling to the air. For semiconductor strain gages, the power that can be dissipated per unit length P/l_0 ranges from 0.1 to 0.2 W/in (4 to 8 W/m). When the test specimen has poor thermal characteristics, the lower ratio should be employed. For specimens with good heat-sink conditions, the higher ratio is frequently satisfactory. Thermal characteristics of different types of specimens are listed in Table 6.4.

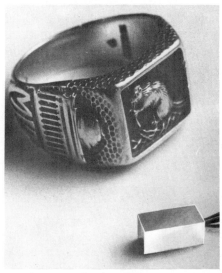

Accelerometer

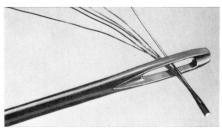

Pressure Transducer

Figure 7.7 Subminiature piezoresistive transducers. (*Kulite Semiconductor Products.*)

7.4 GAGE AVAILABILITY [11, 12]

Semiconductor strain gages of both the P and N type are commercially available in gage lengths from 0.02 to 0.25 in (0.50 to 6.25 mm). The gage factors for the P-type gages are varied from 45 to 130 by controlling the concentration of impurities from 1×10^{19} to 2×10^{16} atoms/cm^3. For the N-type gages, negative gage factors ranging from -100 to -135 are available. Dual-element gages, which have one P and one N element connected together to form one-half of a Wheatstone bridge, provide a strain sensor with a sensitivity of 250 which is temperature-compensated for apparent strain over the range from 50 to 150°F (10 to 65°C) to within 1 μin/(in)(°F) or 1.8 μm/(m)(°C).

Specialized sensors containing four semiconducting elements are formed by diffusing the impurities into the surface layer of chips of single-crystal silicon. The chips represent small beams and/or diaphragms which are used in the fabrication of acceleration, displacement, or pressure transducers. The semiconductor strain sensor permits the design of a wide range of subminiature transducers which have a high natural frequency and a high output signal. Examples of a subminiature accelerometer (the rectangular transducer) and a subminiature pressure transducer (the cylindrical transducer) are shown in Fig. 7.7.

EXERCISES

7.1 Show the matrix which represents the 81 terms in the fourth-rank piezoresistivity tensor π_{ijkl} given in Eq. (7.4).

7.2 Show that Eq. (7.4) reduces to Eq. (7.5) when the tensors ρ_{ij} and τ_{ij} are symmetrical and that the piezoresistive coefficients are reduced to the three terms defined in the simplified π matrix.

7.3 Verify Eq. (7.10) by substituting Eqs. (7.6) to (7.8) into Eq. (7.9) and simplifying the resulting relationship.

7.4 One of the advantages of semiconductor strain gages is related to the fact that both positive and negative gage factors can be achieved with P- and N-type silicon. Show the gain in sensitivity, based on gage factor, achieved in the design of a beam type deflection transducer when four gages (two of the P type and two of the N type) are employed.

7.5 Outline the two methods which can be used to compensate for temperature-induced apparent strains in semiconductor strain gages.

7.6 Describe procedures for improving the linearity of the resistance ratio $\Delta R/R$ of semiconductor strain gages relative to strain.

7.7 Describe the precautions which must be taken in installing and cycling semiconductor strain gages due to their inherently brittle characteristics.

7.8 Determine the power which can be dissipated by a semiconductor strain gage 1.5 mm long. If the gage resistance is 500 Ω, determine the current passing through the gage and the voltage drop across the gage when the maximum power is being dissipated.

REFERENCES

1. Mason, W. P., and R. N. Thurston: Piezoresistive Materials in Measuring Displacement, Force, and Torque, *J. Acoust. Soc. Am.*, vol. 29, no. 10, pp. 1096–1101, 1957.
2. Geyling, F. T., and J. J. Forst: Semiconductor Strain Transducers, *Bell Syst. Tech. J.*, vol. 39, 1960.
3. Mason, W. P.: Semiconductors in Strain Gages, *Bell Lab. Rec.*, vol. 37, no. 1, pp. 7–9, 1959.
4. Sanchez, J. C.: The Semi-conductor Strain Gage, *Proc. 1st Int. Congr. Exp. Mech. 1963*, pp. 255–274.
5. Smith, C. S.: Piezoresistive Effect in Germanium and Silicon, *Phys. Rev.*, vol. 94, pp. 42–49, 1954.
6. Padgett, E. D., and W. V. Wright: Silicon Piezoresistive Devices, pp. 1–20, in M. Dean and R. D. Douglas (eds.), "Semiconductor and Conventional Strain Gages," Academic Press, Inc., New York, 1962.
7. O'Regan, R.: Development of the Semiconductor Strain Gage and Some of Its Applications, pp. 245–257, in M. Dean and R. D. Douglas (eds.), "Semiconductor and Conventional Strain Gages," Academic Press, Inc., New York, 1962.
8. Mason, W. P., J. J. Forst, and L. M. Tornillo: Recent Developments in Semiconductor Strain Transducers, pp. 109–120, in M. Dean and R. D. Douglas (eds.), "Semiconductor and Conventional Strain Gages," Academic Press, Inc., New York, 1962.
9. Kurtz, A. D.: Adjusting Crystal Characteristics to Minimize Temperature Dependence, pp. 259–272, in M. Dean and R. D. Douglas (eds.), "Semiconductor and Conventional Strain Gages," Academic Press, Inc., New York, 1962.
10. Sanchez, J. C., and W. V. Wright: Recent Developments in Flexible Silicon Strain Gages, pp. 307–346, in M. Dean and R. D. Douglas (eds.), "Semiconductor and Conventional Strain Gages," Academic Press, Inc., New York, 1962.
11. Semiconductor Strain Gages, *Kulite Bull.* KSG-5A.
12. Dorsey, J.: Semiconductor Strain Gage Handbook, BLH Electronics, Waltham, Mass., 1963.

EIGHT

STRAIN-GAGE CIRCUITS

8.1 INTRODUCTION

An electrical-resistance strain gage will change in resistance due to applied strain according to Eq. (6.5), which indicates

$$\frac{\Delta R}{R} = S_g \epsilon_{xx} \qquad (6.5)$$

where the gage axis coincides with the x axis and $\epsilon_{yy} = -0.285\epsilon_{xx}$. In order to apply the electrical-resistance strain gage in any experimental stress analysis, the quantity $\Delta R/R$ must be measured and converted to the strain which produced the resistance change. Two electrical circuits, the potentiometer and the Wheatstone bridge, are commonly employed to convert the value of $\Delta R/R$ to a voltage signal (denoted here as ΔE) which can be measured with a recording instrument.

In this chapter the basic theory for the two circuits is presented, and the circuit sensitivities and ranges are derived. In addition, temperature compensation, signal addition, and loading effects are discussed in detail. Also covered are the constant-current circuits recommended for semiconductor strain gages. Methods of calibrating both the potentiometer and the Wheatstone bridge circuits are covered, and the effects of lead wires on noise, calibration, and temperature compensation are discussed.

Insofar as possible, this discussion has been kept fundamental without appreciable reference to commercially available circuits. The material presented here is applicable to all commercial circuits since their design is based on the fundamental principles covered in this chapter.

The methods of measuring the circuit output voltages ΔE are discussed in Chap. 9.

8.2 THE POTENTIOMETER AND ITS APPLICATION TO STRAIN MEASUREMENT [1]

The potentiometer circuit, which is often employed in dynamic strain-gage applications to convert the gage output $\Delta R/R$ to a voltage signal ΔE, is shown in Fig. 8.1. For fixed-value resistors R_1 and R_2 in the circuit, the open-circuit output voltage E can be expressed as

$$E = \frac{R_1}{R_1 + R_2} V = \frac{1}{1 + r} V \tag{8.1}$$

where V is the input voltage and $r = R_2/R_1$ is the resistance ratio for the circuit. If incremental changes ΔR_1 and ΔR_2 occur in the value of the resistors R_1 and R_2, the change ΔE of the output voltage E can be computed by using Eq. (8.1) as follows:

$$E + \Delta E = \frac{R_1 + \Delta R_1}{R_1 + \Delta R_1 + R_2 + \Delta R_2} V \tag{a}$$

Solving Eq. (a) for ΔE gives

$$\Delta E = \left(\frac{R_1 + \Delta R_1}{R_1 + \Delta R_1 + R_2 + \Delta R_2} - \frac{R_1}{R_1 + R_2} \right) V \tag{b}$$

which can be expressed in the following form by introducing $r = R_2/R_1$:

$$\Delta E = \frac{[r/(1 + r)^2](\Delta R_1/R_1 - \Delta R_2/R_2)}{1 + [1/(1 + r)][\Delta R_1/R_1 + r(\Delta R_2/R_2)]} V \tag{8.2}$$

Examination of Eq. (8.2) shows that the voltage signal ΔE from the potentiometer

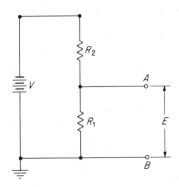

Figure 8.1 The potentiometer circuit.

circuit is a nonlinear function of $\Delta R_1/R_1$ and $\Delta R_2/R_2$. To inspect the nonlinear aspects of this circuit further, it is possible to rewrite Eq. (8.2) in the form

$$\Delta E = \frac{r}{(1 + r)^2}\left(\frac{\Delta R_1}{R_1} - \frac{\Delta R_2}{R_2}\right)(1 - \eta)\,V \qquad (8.3a)$$

where the nonlinear term η is expressed as

$$\eta = 1 - \frac{1}{1 + [1/(1 + r)][\Delta R_1/R_1 + r(\Delta R_2/R_2)]} \qquad (8.3b)$$

Equations (8.3) are the basic relationships which govern the behavior of the potentiometer circuit, and as such they can be used to establish the applicability of this circuit for strain-gage measurements.

A. Range and Sensitivity of the Potentiometer Circuit

The range of the potentiometer circuit is effectively established by the nonlinear term η, as expressed in Eq. (8.3b). If a strain gage of resistance R_g is used in the position of R_1, and if a fixed-ballast resistor R_b is employed in the position of R_2, then

$$R_1 = R_g \qquad \Delta R_1 = \Delta R_g \qquad R_2 = R_b \qquad \Delta R_2 = 0$$

and Eq. (8.3b) becomes

$$\eta = 1 - \frac{1}{1 + [1/(1 + r)](\Delta R_g/R_g)} \qquad (a)$$

By substituting Eq. (6.5) into Eq. (a), the nonlinear term is given in terms of strain ϵ, the gage factor S_g, and the resistance ratio $r = R_b/R_g$ as

$$\eta = 1 - \frac{1}{1 + [1/(1 + r)]S_g\epsilon} \qquad (b)$$

This expression can be evaluated by means of a series expansion where

$$\eta = 1 - \frac{1}{1 + x} = x - x^2 + x^3 - x^4 + \cdots$$

$$x = \frac{1}{1 + r}S_g\epsilon \qquad (8.4)$$

The range of the potentiometer circuit depends upon the error which can be tolerated due to the nonlinearities introduced by the circuit. In Fig. 8.2, the nonlinear term η is given as a function of strain for various values of r measured with a metallic strain gage with a gage factor S_g equal to 2. Inspection of this figure shows that the error introduced by the nonlinearity of the circuit is sufficiently small to be negligible for normal strain-gage measurements of the order of a few thousand microinches per inch (micrometers per meter). The error does, however, approach

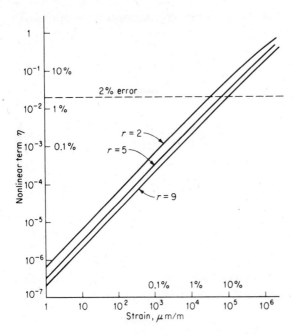

Figure 8.2 Dependence of the nonlinear term η on the strain level.

2 percent for strains between 1 and 10 percent, depending upon the value of the resistance ratio r. It is clear that increasing the value of r decreases the error due to circuit nonlinearities. The range of the potentiometer circuit, based on an allowable error of 2 percent, varies from 10 to 2 percent strain as r varies from a value of 9 to 2. Obviously the range of the potentiometer circuit is sufficient for determining elastic strains in metallic components. However, elastic-strain determinations in plastic materials and plastic-strain determinations in metallic materials may require corrections of the output signal.

The sensitivity of the potentiometer circuit is defined as the ratio of the output voltage divided by the strain:

$$S_c = \frac{\Delta E}{\epsilon} \tag{8.5}$$

It is apparent that the relative merit of a circuit can be judged by the magnitude of the circuit sensitivity S_c. In particular for dynamic applications, where relatively small strains are being measured, it is quite important that S_c be made as large as possible to reduce the degree of amplification necessary to read the output signal.

The sensitivity of the potentiometer circuit can be established by substituting Eq. (8.3a) into Eq. (8.5) to obtain

$$S_c = \frac{r}{(1+r)^2}\left(\frac{\Delta R_1}{R_1} - \frac{\Delta R_2}{R_2}\right)\frac{V}{\epsilon} \tag{8.6}$$

where $\eta \ll 1$.

If a strain gage is placed in the circuit in place of R_1 and a ballast resistor is employed for R_2, then

$$R_g = R_1 \qquad \Delta R_g = \Delta R_1 \qquad R_b = R_2 \qquad \Delta R_2 = 0$$

and by Eqs. (6.5) and (8.6) the circuit sensitivity becomes

$$S_c = \frac{r}{(1+r)^2} S_g V \tag{c}$$

This equation indicates that the circuit sensitivity is controlled by the resistance ratio r and the circuit voltage V. However, if V can be varied, the limiting factor becomes the power which can be dissipated by the gage. This power must be held within the limits set forth in Sec. 6.6E. To account for the influence of the power P_g dissipated by the gage, the relationship between P_g and V given by Eq. (6.18) is applied to the potentiometer circuit to give

$$V = I_g R_g (1+r) = \sqrt{P_g R_g}(1+r) \tag{8.7}$$

Substituting Eq. (8.7) into Eq. (c) yields

$$S_c = \frac{r}{1+r} S_g \sqrt{R_g P_g} \tag{8.8}$$

This equation shows that the circuit sensitivity is controlled by two independent parameters, $r/(1+r)$ and $S_g\sqrt{R_g P_g}$. The first parameter $r/(1+r)$ is related to the circuit and is dictated by the selection of R_b, which fixes the value of r. It is clear that as r becomes very large, $r/(1+r)$ approaches 1 and maximum circuit efficiency is obtained. However, for very high values of r, the voltage required to drive the potentiometer becomes excessive. Values of r of about 9 are commonly used, which gives a circuit efficiency of approximately 90 percent. The circuit sensitivity is strongly influenced by the second parameter $S_g\sqrt{R_g P_g}$, which is determined entirely by the selection of the strain gage and is totally independent of the elements used in the design of the potentiometer circuit. In fact, if very high circuit sensitivities are required, much can be gained by careful gage selection, since $S_g\sqrt{R_g P_g}$ can be varied over wide limits, i.e., from 3 to about 700, while circuit efficiency can be varied over a much more limited range ($\frac{1}{2}$ to about 1).

The output voltage from any circuit containing a strain gage is inherently low, and as a consequence S_c is quite often as low as 5 or 10 μV per (μin/in) (μm/m) of strain. Exercise 8.3 shows typical values of S_c as well as the voltage required to drive a potentiometer circuit with a circuit efficiency of 90 percent.

B. Temperature Compensation and Signal Addition in the Potentiometer Circuit

It is possible to effect a certain degree of temperature compensation in the strain-measuring system by employing two strain gages in the potentiometer circuit. The gage used in place of R_1 is known as the *active strain gage* and is used to measure the strain at a given point and orientation on a specimen. The gage used to replace R_2 is known as the *dummy gage* and is mounted on a small block of material identical to that of the specimen and is exposed to the same thermal environment as the active gage. In the active gage the total change in resistance will be due to strain ΔR_ϵ and to changes in temperature ΔR_T. Hence,

$$\frac{\Delta R_1}{R_1} = \frac{\Delta R_\epsilon}{R_1} + \frac{\Delta R_T}{R_1} \tag{d}$$

However, in the dummy gage the change in resistance is due to a change in temperature alone; hence

$$\frac{\Delta R_2}{R_2} = \frac{\Delta R_T}{R_2} \tag{e}$$

Substituting Eqs. (d) and (e) into Eq. (8.3a) give

$$\Delta E = \frac{r}{(1+r)^2} \left(\frac{\Delta R_\epsilon}{R_1} + \frac{\Delta R_T}{R_1} - \frac{\Delta R_T}{R_2} \right) (1 - \eta) V \tag{8.9}$$

Since $R_1 = R_2 = R_g$, it is evident that the ΔR_T terms cancel out of Eq. (8.9) and the output signal ΔE is due to the ΔR_ϵ term alone.

This type of temperature compensation is effective if the dummy and the active gages are from the same lot of gages, if the materials upon which each gage is mounted are identical, and if the temperature changes due to ambient conditions and gage currents on both gages are equal. If any one of these three conditions is violated, complete temperature compensation cannot be achieved.

A loss in circuit efficiency, as determined by $r/(1 + r)$, is the price which must be paid for temperature compensation. Since $R_1 = R_2 = R_g$ is a requirement for temperature compensation, $r = 1$, and the circuit efficiency is fixed at 50 percent. If circuit sensitivity is extremely important, temperature compensation by this means should not be specified, since temperature-compensated gages (discussed in Sec. 6.6B) can be employed to accomplish the same purpose. Moreover, the potentiometer circuit can be used only to record dynamic strains, and very little time is available during the readout period for the ambient temperature to change.

Resistance changes from two active strain gages placed in positions R_1 and R_2 can be used to increase the output signal in the potentiometer circuit in certain cases. As an example, consider the application shown in Fig. 8.3, where two gages

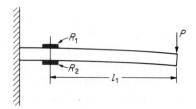

Figure 8.3 Strain gages mounted on the top and bottom surfaces of a cantilever beam.

are mounted on a beam in bending. The gage mounted on the top surface of the beam will exhibit a resistance change which can be indicated by

$$\frac{\Delta R_1}{R_1} = S_g f(P) \tag{f}$$

and the gage mounted on the bottom surface of the beam will exhibit a negative resistance change of

$$\frac{\Delta R_2}{R_2} = -S_g f(P) \qquad \text{where} \qquad f(P) = \frac{6Pl_1}{bh^2 E} = \epsilon \tag{g}$$

The signal output from the potentiometer circuit containing these two active gages can be computed from Eqs. (8.3) as

$$\Delta E = \frac{r}{(1 + r)^2} (2S_g \epsilon) V \tag{h}$$

which can be reduced to

$$\Delta E = \frac{r}{1 + r} 2S_g \sqrt{R_g P_g} \, \epsilon = S_g \sqrt{R_g P_g} \, \epsilon \tag{8.10}$$

where $r = 1$ since $R_1 = R_2 = R_g$. If one active gage had been employed, the relation for ΔE would have been

$$\Delta E = \frac{r}{1 + r} S_g \sqrt{R_g P_g} \, \epsilon \tag{i}$$

Comparison of Eq. (8.10) and Eq. (i) indicates that the output signal has been doubled, based on a value of $r = 1$. However, if r is increased to, say, 9 in Eq. (i), the advantage of using two gages to increase circuit sensitivity is almost completely nullified.

Multiple-gage circuits are more important in transducer applications of strain gages, where the signals due to unwanted components of load are canceled in the circuit. This application is illustrated in Sec. 8.8.

C. Potentiometer Output

The output of a potentiometer circuit is measured across terminals A and B, as indicated in Fig. 8.1, to give a total voltage $E + \Delta E$. The output of a strain gage mounted to a specimen with a sinusoidally varying strain field is shown in Fig. 8.4.

Normally the voltage E is of the order of a few volts, say 2 to 10, and ΔE is measured in terms of microvolts. The voltage which is related to the strain is ΔE, and this quantity must be measured independently of the large superimposed voltage E. Unfortunately, voltage-measuring instruments are not usually available with sufficient range and sensitivity to measure ΔE when it is superimposed on E. Thus, the potentiometer circuit is inadequate for static strain-gage applications where both E and ΔE are constant with respect to time. The potentiometer circuit

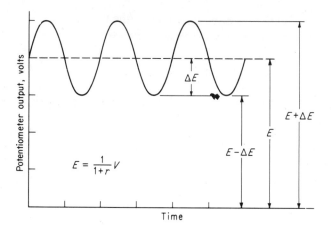

$$E = \frac{1}{1+r} V$$

Figure 8.4 Potentiometer output as a function of time for a sinusoidally varying strain.

is quite useful, however, in dynamic applications where ΔE varies with time and E is a constant. In these applications it is possible to block the dc voltage E by the use of a suitable filter while passing the voltage ΔE without significant distortion. The filter serves to reduce the required range of the measuring instrument from $E + \Delta E$ to ΔE and permits the use of highly sensitive but limited-range measuring instruments. A typical filter used with a potentiometer circuit is illustrated in Fig. 8.5.

If R_M is large in comparison with $R_1 R_2/(R_1 + R_2)$, the voltage E' measured across the resistance of the measuring instrument R_M is

$$\frac{E'}{E + \Delta E} = \frac{R_M}{\sqrt{R_M^2 + 1/\omega^2 C^2}} \tag{8.11}$$

where C is the capacitance used in the filter and ω is the angular frequency of the voltage $E + \Delta E$. If one considers that the voltage $E + \Delta E$ is made up of many frequencies varying from zero (direct current) to several thousand hertz, it is

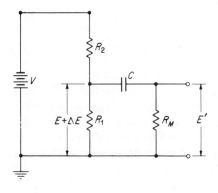

Figure 8.5 Filter for blocking out the dc component of the potentiometer-circuit output.

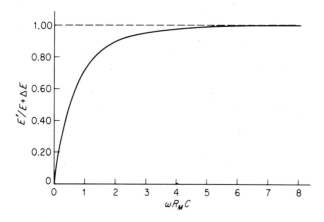

Figure 8.6 Typical response curve for a high-pass filter.

possible to plot $E'/(E + \Delta E)$ as a function of ω, as shown in Fig. 8.6. The student is referred to Exercise 8.6, where the validity of Fig. 8.6 is established.

Since the E component of the potentiometer output voltage $E + \Delta E$ is constant with time (that is, $\omega = 0$), this dc component of the signal is entirely blocked by the filter. If $R_M = 1$ MΩ and $C = 0.1$ μF, the RC constant for the filter is 10^{-1}, which indicates that frequency components associated with the ΔE voltage will pass through the filter with negligible distortion provided they exceed 30 rad/s or about 5 Hz. This filter arrangement can then be employed to successfully adapt the potentiometer circuit to dynamic strain-gage applications whenever the dynamic strains are composed of frequency components which exceed about 5 Hz.

D. Load Effects on the Potentiometer Circuit

In the derivation of the output voltage obtained from the potentiometer circuit given in Sec. 8.2, the effect of the resistance of the voltage-measuring instrument was neglected. That is to say, the measuring instrument was considered to have an infinite resistance. In practice, however, the voltage-measuring instrument does have a finite resistance, and this resistance can influence the output voltage of the potentiometer circuit since some of the current which normally passes through the gage will be diverted through the measuring instrument.

To show the effect of the resistance or, more correctly, the impedance of the measuring instrument, consider the circuit shown in Fig. 8.7, where a single active gage is used. This circuit can be reduced to an equivalent circuit with an equivalent resistor R_e replacing resistors R_M and R_g, as is also indicated in Fig. 8.7. If R_b is considered as a fixed-value ballast resistor and the nonlinear term η is much less than 1, then Eqs. (8.3) may be rewritten as

$$\Delta E \bigg|_{R_M} = \frac{r}{(1 + r)^2} \frac{\Delta R_e}{R_e} V = \frac{R_b R_e}{(R_e + R_b)^2} \frac{\Delta R_e}{R_e} V \tag{8.12}$$

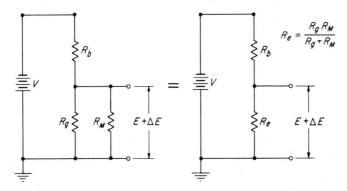

Figure 8.7 Potentiometer circuit with a resistive load applied across its output terminals and an equivalent circuit.

This value must be compared with the open-circuit (i.e., infinite resistance for the measuring instrument) output voltage of

$$\Delta E \bigg|_{R_M = \infty} = \frac{R_b R_g}{(R_g + R_b)^2} \frac{\Delta R_g}{R_g} V \tag{8.13}$$

To show the difference between these two equations, consider the expansion of the terms $R_b R_e/(R_e + R_b)^2$ and $\Delta R_e/R_e$ in terms of R_b, R_g, and R_M. By expanding $R_b R_e/(R_e + R_b)^2$, it is possible to show by using

$$R_e = \frac{R_g R_M}{R_g + R_M} \tag{j}$$

that

$$\frac{R_b R_e}{(R_e + R_b)^2} = \frac{R_b R_g (R_g/R_M + 1)}{[R_g + R_b(R_g/R_M) + R_b]^2} \tag{k}$$

Next, if $\Delta R_e/R_e$ is expanded, one obtains

$$\frac{\Delta R_e}{R_e} = \frac{(R_g + \Delta R_g)/(R_g + \Delta R_g + R_M) - R_g/(R_g + R_M)}{R_g/(R_g + R_M)}$$

$$= \frac{1}{R_g/R_M + 1} \frac{\Delta R_g}{R_g} \tag{l}$$

Substituting Eqs. (k) and (l) into Eq. (8.12) gives

$$\Delta E \bigg|_{R_M} = \frac{R_b R_g}{[R_g + R_b(R_g/R_M) + R_b]^2} \frac{\Delta R_g}{R_g} V \tag{8.14}$$

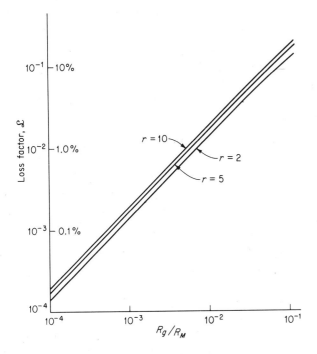

Figure 8.8 Loss factor $\mathscr{L}$ as a function of R_g/R_M.

If $\Delta E|_{R_M}$ is compared with $\Delta E|_{R_M = \infty}$ in the following manner:

$$\Delta E\bigg|_{R_M} = \Delta E\bigg|_{R_M = \infty} (1 - \mathscr{L}) \tag{8.15}$$

it is possible to determine an expression for $\mathscr{L}$ which represents the loss in voltage output ΔE due to the resistive load R_M.

Substituting Eqs. (8.13) and (8.14) into Eq. (8.15) and solving for $\mathscr{L}$, while noting that $r = R_b/R_g$, gives

$$\mathscr{L} = \frac{2r(R_g/R_M)\{1 + r[1 + \frac{1}{2}(R_g/R_M)]\}}{[1 + r(1 + R_g/R_M)]^2} \tag{8.16}$$

The loss in the output voltage, as expressed by the loss factor $\mathscr{L}$, is dependent on the resistance ratios R_b/R_g and R_g/R_M. The loss factor $\mathscr{L}$ is shown as a function of R_g/R_M for various values of R_b/R_g in Fig. 8.8. It is clear from the results shown in this figure that the loss in sensitivity is less than 2 percent if the resistance of the measuring instrument is about 100 times that of the gage resistance. Thus, if a 120-Ω strain gage is employed in the potentiometer circuit, the measuring instrument should have an input resistance of 12 kΩ. This value of resistance is not high for most measuring instruments; in fact, most cathode-ray oscilloscopes, which are frequently used with a potentiometer circuit, have input resistances of the order of 1 MΩ.

E. Summary of the Potentiometer Circuit

The equations which govern the behavior of the potentiometer circuit as it is applied to dynamic strain measurement are summarized below:

$$E = \frac{1}{1+r} V \tag{8.1}$$

$$\Delta E = \frac{r}{(1+r)^2}\left(\frac{\Delta R_1}{R_1} - \frac{\Delta R_2}{R_2}\right)(1-\eta)V \tag{8.3a}$$

$$\eta = 1 - \frac{1}{1 + [1/(1+r)][\Delta R_1/R_1 + r(\Delta R_2/R_2)]} \tag{8.3b}$$

$$V = \sqrt{P_g R_g}(1+r) \tag{8.7}$$

$$S_c = \frac{r}{1+r} S_g \sqrt{P_g R_g} \tag{8.8}$$

$$E' = \frac{R_M}{\sqrt{R_M^2 + 1/\omega^2 C^2}}(E + \Delta E) \tag{8.11}$$

$$\mathscr{L} = \frac{2r(R_g/R_M)\{1 + r[1 + \frac{1}{2}(R_g/R_M)]\}}{[1 + r(1 + R_g/R_M)]^2} \tag{8.16}$$

The potentiometer circuit cannot be employed to measure the resistance change of a gage due to static strains because of the voltage E, which appears across its output terminals and swamps out the small voltage ΔE which is related to the resistance change. The circuit can, however, be employed to measure the resistance change in the gage due to dynamic strains if a suitable filter is used to block out the dc voltage E while passing without distortion the voltage pulse ΔE.

For most strain-gage applications the range over which the potentiometer circuit can respond is larger than the strains which are to be measured. However, for more specialized problems where strains of about 10 percent are to be measured, corrections must be made to account for nonlinearities introduced by the circuit.

The sensitivity of a potentiometer measuring system is controlled by the circuit efficiency given by $r/(1+r)$ and the gage selection which dictates the value of $S_g\sqrt{P_g R_g}$. Of the two factors, the gage selection is more important because of its greater variability (that is, 3 to about 700). The circuit efficiency can usually be maintained at about 90 percent by selection of $R_b = 9R_g$ and the provision of a sufficiently high voltage V to drive the allowable current through the gage.

The potentiometer circuit can be used to add or subtract signals from a multiple-strain-gage installation. However, the gain in circuit sensitivity by the addition of two strain-gage signals is usually not sufficient to warrant the use of two gages in place of one. For transducer applications, multiple-gage circuits should be considered to eliminate gage response from components (of, say, the load) which are not sought in the measurement.

Temperature compensation can be introduced into the circuit by employing both a dummy and an active gage. However, the circuit efficiency is reduced to 50 percent in this circuit arrangement. If low-magnitude strains are to be measured, it is usually more advisable to employ temperature-compensated gages and seek higher circuit efficiencies by using a fixed-value ballast resistor in place of the dummy gage required for temperature compensation within the circuit.

The output of the potentiometer circuit ΔE is directly proportional to the supply voltage V. Thus, it is imperative that the voltage supply provide a stable voltage V over the period of readout. Batteries serve as excellent power supplies for the potentiometer circuit, provided their rate of decay is small over the period of readout (this period of readout is usually quite short in dynamic applications).

Finally, the potentiometer circuit can be grounded together with the amplifier. This feature represents a real advantage when the signal-to-noise ratio is low and it is important to reduce the electronic noise level.

8.3 THE WHEATSTONE BRIDGE [2]

The Wheatstone bridge is a second circuit which can be employed to determine the change in resistance which a gage undergoes when it is subjected to a strain. Unlike the potentiometer, the Wheatstone bridge can be used to determine both dynamic and static strain-gage readings. The bridge may be used as a direct-readout device, where the output voltage ΔE is measured and related to strain. Also, the bridge may be used as a null-balance system, where the output voltage ΔE is adjusted to a zero value by adjusting the resistive balance of the bridge. In either of the modes of operation the bridge can be effectively employed in a wide variety of strain-gage applications.

To show the principle of operation of the Wheatstone bridge as a direct-readout device (where ΔE is measured to determine the strain), consider the circuit shown in Fig. 8.9. The voltage drop across R_1 is denoted as V_{AB} and is given as

$$V_{AB} = \frac{R_1}{R_1 + R_2} V \qquad (a)$$

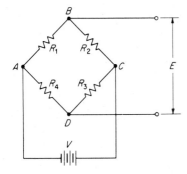

Figure 8.9 The Wheatstone bridge circuit.

and, similarly, the voltage drop across R_4 is denoted as V_{AD} and is given by

$$V_{AD} = \frac{R_4}{R_3 + R_4} V \tag{b}$$

The output voltage from the bridge E is equivalent to V_{BD}, which is

$$E = V_{BD} = V_{AB} - V_{AD} \tag{c}$$

Substituting Eqs. (a) and (b) into Eq. (c) and simplifying give

$$E = \frac{R_1 R_3 - R_2 R_4}{(R_1 + R_2)(R_3 + R_4)} V \tag{8.17}$$

The voltage E will go to zero and the bridge will be considered in balance when

$$R_1 R_3 = R_2 R_4 \tag{8.18}$$

It is this feature of balancing which permits the Wheatstone bridge to be employed for static strain measurements. The bridge is initially balanced before strains are applied to the gages in the bridge; thus the voltage E is initially zero, and the strain-induced voltage ΔE can be measured directly for both static and dynamic applications.

Consider an initially balanced bridge with $R_1 R_3 = R_2 R_4$ so that $E = 0$ and then change each value of resistance R_1, R_2, R_3, and R_4 by an incremental amount ΔR_1, ΔR_2, ΔR_3, and ΔR_4. The voltage output ΔE of the bridge can be obtained from Eq. (8.17), which becomes

$$\Delta E = V \frac{\begin{vmatrix} R_1 + \Delta R_1 & R_2 + \Delta R_2 \\ R_4 + \Delta R_4 & R_3 + \Delta R_3 \end{vmatrix}}{\begin{vmatrix} R_1 + \Delta R_1 + R_2 + \Delta R_2 & 0 \\ 0 & R_3 + \Delta R_3 + R_4 + \Delta R_4 \end{vmatrix}} = V \frac{A}{B} \tag{d}$$

where A is the determinant in the numerator and B is the determinant in the denominator. By expanding each of these determinants, neglecting second-order terms, and noting that $R_1 R_3 = R_2 R_4$, it is possible to show that

$$A = R_1 R_3 \left(\frac{\Delta R_1}{R_1} - \frac{\Delta R_2}{R_2} + \frac{\Delta R_3}{R_3} - \frac{\Delta R_4}{R_4} \right) \tag{e}$$

$$B = \frac{R_1 R_3 (R_1 + R_2)^2}{R_1 R_2} \tag{f}$$

Combining Eqs. (d) to (f) yields

$$\Delta E = V \frac{R_1 R_2}{(R_1 + R_2)^2} \left(\frac{\Delta R_1}{R_1} - \frac{\Delta R_2}{R_2} + \frac{\Delta R_3}{R_3} - \frac{\Delta R_4}{R_4} \right) \tag{8.19}†$$

---

† See Exercise 8.9.

By letting $R_2/R_1 = r$ it is possible to rewrite Eq. (8.19) as

$$\Delta E = V \frac{r}{(1+r)^2}\left(\frac{\Delta R_1}{R_1} - \frac{\Delta R_2}{R_2} + \frac{\Delta R_3}{R_3} - \frac{\Delta R_4}{R_4}\right) \tag{8.20}$$

In reality, Eqs. (8.19) and (8.20) both carry a nonlinear term $1 - \eta$, as defined in Exercise 8.8. However, the influence of the nonlinear term is quite small and can be neglected, provided the strains being measured are less than 5 percent. Equation (8.20) thus represents the basic equation which governs the behavior of the Wheatstone bridge in strain measurement.

A. Wheatstone-Bridge Sensitivity

The sensitivity of the Wheatstone bridge must be considered from two points of view: (1) with a fixed voltage applied to the bridge regardless of gage current (a condition which exists in most commercially available instrumentation) and (2) with a variable voltage whose upper limit is determined by the power dissipated by the particular arm of the bridge which contains the strain gage. By recalling the definition for the circuit sensitivity given in Eq. (8.15) and using the basic bridge relationship given in Eq. (8.20), it is clear that the circuit sensitivity is

$$S_c = \frac{\Delta E}{\epsilon} = \frac{V}{\epsilon}\frac{r}{(1+r)^2}\left(\frac{\Delta R_1}{R_1} - \frac{\Delta R_2}{R_2} + \frac{\Delta R_3}{R_3} - \frac{\Delta R_4}{R_4}\right) \tag{g}$$

If a multiple-gage circuit is considered with n gages (where $n = 1, 2, 3$, or 4) whose outputs sum when placed in the bridge circuit, it is possible to write

$$\sum_{m=1}^{m=n}\frac{\Delta R_m}{R_m} = n\frac{\Delta R}{R} \tag{h}$$

which by Eq. (6.5) becomes

$$\sum_{m=1}^{m=n}\frac{\Delta R_m}{R_m} = nS_g\epsilon \tag{i}$$

Substituting Eq. (i) into Eq. (g) gives the circuit sensitivity as

$$S_c = V\frac{r}{(1+r)^2}nS_g \tag{8.21}$$

This sensitivity equation is applicable in those cases where the bridge voltage V is fixed and independent of gage current. The equation shows that the sensitivity of the bridge depends upon the number n of active arms employed, the gage factor S_g, the input voltage, and the ratio of the resistances R_1/R_2. A plot of r versus $r/(1+r)^2$ (the circuit efficiency) in Fig. 8.10 shows that maximum efficiency and hence maximum circuit sensitivity occur when $r = 1$. With four active arms in this bridge, a circuit sensitivity of $S_g V$ can be achieved, whereas with one active gage a circuit sensitivity of only $S_g V/4$ can be obtained.

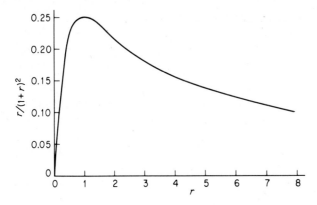

Figure 8.10 Circuit efficiency as a function of r for a fixed-voltage bridge.

When the bridge supply voltage V is selected to drive the gages in the bridge so that they dissipate the maximum allowable power, a different sensitivity equation must be employed. Since the gage current is a limiting factor in this approach, the number of gages used in the bridge and their relative position are important. To show this fact, consider the following four cases, illustrated in Fig. 8.11, which represent the most common bridge arrangements.

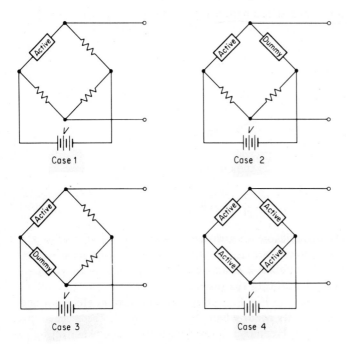

Figure 8.11 Four common bridge arrangements.

Case 1 This bridge arrangement consists of a single active gage in position R_1 and is often employed for many dynamic and some static strain measurements where temperature compensation in the circuit is not critical. The value of R_1 is, of course, equal to R_g, but the value of the other three resistors may be selected to maximize circuit sensitivity, provided the initial balance condition $R_1 R_3 = R_2 R_4$ is maintained. The power dissipated by the gage can be determined from

$$V = I_g(R_1 + R_2) = I_g R_g(1 + r) = (1 + r)\sqrt{P_g R_g} \tag{j}$$

By combining Eqs. (8.21) and (j) and recalling that $r = R_2/R_1$, it is possible to obtain the circuit sensitivity in the following form:

$$S_c = \frac{r}{1 + r} S_g \sqrt{P_g R_g} \tag{8.22}$$

Here it is evident that the circuit sensitivity of the bridge is due to two factors, namely, the circuit efficiency, which is given by $r/(1 + r)$, and the gage selection, represented by the term $S_g\sqrt{P_g R_g}$. For this bridge arrangement, r should be selected as high as possible to increase circuit efficiency but not high enough to increase the supply voltage beyond reasonable limits. A value of $r = 9$ will give a circuit efficiency of 90 percent and, with a 120-Ω gage dissipating 0.15 W requires a supply voltage $V = 42.4$ V.

Case 2 This bridge arrangement employs one active gage in arm R_1 and one dummy gage in arm R_2 which is utilized for temperature compensation (see Exercise 8.13). The value of the gage current which passes through both gages (note $R_1 = R_2 = R_g$) is given by

$$V = 2I_g R_g \tag{k}$$

Substituting Eq. (k) into Eq. (8.21) and noting for this case that $n = 1$ and $r = 1$ gives the circuit sensitivity as

$$S_c = \tfrac{1}{2} S_g \sqrt{P_g R_g} \tag{8.23}$$

In this instance the circuit efficiency is fixed at a value of $\tfrac{1}{2}$ since the condition $R_1 = R_2 = R_g$ requires that $r = 1$. Thus, it is clear that the placement of a dummy gage in position R_2 to effect temperature compensation reduces circuit efficiency to 50 percent. The gage selection, of course, maintains its importance in this arrangement since the term $S_g\sqrt{P_g R_g}$ again appears in the circuit-sensitivity equation. In fact, this term will appear in the same form in all four bridge arrangements covered in this section. It also appeared in the circuit sensitivity of the potentiometer circuit in Eq. (8.8).

Case 3 The bridge arrangement in this case incorporates an active gage in the R_1 position and a dummy gage in the R_4 position. Fixed resistors of any value are placed in positions R_2 and R_3. With these gage positions, the bridge is temperature-compensated since the temperature-introduced resistance changes

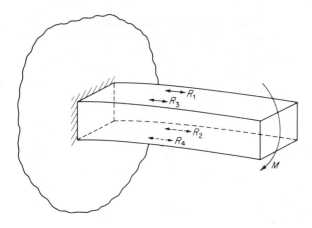

Figure 8.12 Positions of four gages employed on a beam in bending to give a bridge factor of 4.

are canceled out in the Wheatstone bridge circuit. The current through both the active gage and the dummy gage is given by

$$V = I_g(R_1 + R_2) = (1 + r)\sqrt{P_g R_g} \tag{l}$$

Substituting Eq. (*l*) into Eq. (8.21) and recalling for this case that $n = 1$ gives the circuit sensitivity as

$$S_c = \frac{r}{1 + r} S_g \sqrt{P_g R_g} \tag{8.24}$$

The circuit sensitivity for this bridge arrangement (case 3) is identical to that which can be achieved with the bridge circuit shown in case 1. Thus, temperature compensation can be obtained without any loss in circuit sensitivity if the dummy gage is placed in position R_4 rather than position R_2.

Case 4 In this bridge arrangement, four active gages are placed in the bridge, with one gage in each of the four arms. If the gages are placed on, say, a beam in bending, as shown in Fig. 8.12, the signals from each of the four gages will add and the value of n given in Eq. (8.21) will be equal to 4. The power dissipated by each of the four gages is given by

$$V = 2I_g R_g = 2\sqrt{P_g R_g} \tag{m}$$

Also, since the resistance is the same for all four gages,

$$R_1 = R_2 = R_3 = R_4 = R_g$$

and $r = 1$. The circuit sensitivity for this bridge arrangement is obtained by substituting Eq. (*m*) into Eq. (8.21):

$$S_c = 2I_g R_g S_g = 2S_g \sqrt{P_g R_g} \tag{8.25}$$

A four-active-arm bridge is slightly more than twice as sensitive as a single-active-arm bridge (case 1 or case 3). Also, this bridge arrangement is temperature-compensated. The four gages employed are a relatively high price to pay for this increased sensitivity. The two-active-arm bridge presented in Exercise 8.14 exhibits a sensitivity which approaches that given in Eq. (8.25).

Examination of these four bridge arrangements plus the fifth arrangement presented in Exercise 8.14 shows that the circuit sensitivity can be varied between $\frac{1}{2}$ and 2 times $S_g \sqrt{P_g R_g}$. Temperature compensation for a single active gage in position R_1 can be effected without loss in sensitivity by placing the dummy gage in position R_4. If the dummy gage is placed in position R_2, the circuit sensitivity is reduced by a factor of 2. Circuit sensitivities can be improved by the use of multiple-active-arm bridges, as illustrated in cases 4 and 5 (see Exercise 8.14). However, the cost of an additional gage or gages for a given strain measurement is rarely warranted except for transducer applications. For experimental stress analyses, single-active-arm bridges are normally employed, and the signal from the bridge is amplified from 10 to 1000 times before recording.

C. Null-Balance Bridges

The Wheatstone bridge arrangements described previously are normally employed in dynamic strain-gage applications where the bridge voltage ΔE is measured directly and related to the strain level. In static applications it is possible to employ a null-balance bridge where the resistance of one or more arms in the bridge is changed to match the effect of the change in resistance of the active gage. This null-balance system is usually more accurate than the direct-readout bridge and requires less expensive equipment for its operation.

A relatively simple null-balance type of Wheatstone bridge is illustrated in Fig. 8.13. A slide-wire resistance or helical potentiometer is placed across the bridge from points B to D. The center tap of the potentiometer is connected to point C. Active strain gages may be placed in any or all arms of the bridge. In the following discussion a single active gage will be considered in the R_1 position. A voltage-measuring instrument with a high sensitivity near the null point is also placed across the bridge between points B and D. Assume that the bridge is initially balanced with $R_1 R_3 = R_2 R_4$ and $R_5 = R_6$. The meter G is at null or zero

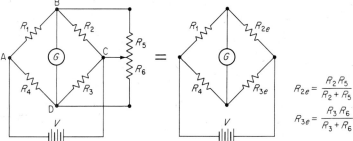

$$R_{2e} = \frac{R_2 R_5}{R_2 + R_5}$$

$$R_{3e} = \frac{R_3 R_6}{R_3 + R_6}$$

Figure 8.13 Parallel-balance circuit and an equivalent circuit for the null-balance Wheatstone bridge.

voltage. Now consider a resistance change in the arm R_1 which upsets this balance, causing a voltage indication on meter G. The slide wire or potentiometer is adjusted, making $R_5 \neq R_6$, until the bridge is again in balance. The potentiometer adjustment, which is calibrated, is proportional to the resistance change in the active gage. Thus the mechanical movement of the potentiometer serves as the readout means, and the voltage is measured only to establish a zero or null point. In this null-balance condition the adjustment of the potentiometer is independent of the input voltage V.

This circuit can be analyzed by considering the equivalent circuit, also shown in Fig. 8.13. In the equivalent circuit the equivalent resistances R_{2e} and R_{3e} are related to R_2, R_5 and R_3, R_6, respectively, by

$$R_{2e} = \frac{R_2 R_5}{R_2 + R_5} \qquad R_{3e} = \frac{R_3 R_6}{R_3 + R_6} \tag{n}$$

An adjustment of the potentiometer will produce a change in the resistance of R_5 equal to ΔR_5 and of R_6 equal to ΔR_6. Moreover, it is clear that

$$\Delta R_5 = -\Delta R_6 \tag{o}$$

since the total resistance of the potentiometer $R_p = R_5 + R_6$ is constant regardless of the adjustments.

The change in the effective resistance $\Delta R_{2e}/R_{2e}$ can be computed by employing finite differences, as shown below,

$$\frac{\Delta R_{2e}}{R_{2e}} = \frac{[R_2(R_5 + \Delta R_5)]/(R_2 + R_5 + \Delta R_5) - R_2 R_5/(R_2 + R_5)}{R_2 R_5/(R_2 + R_5)} \tag{p}$$

where the value of R_2 is considered fixed, which is consistent with the placement of a single active gage in position R_1 and a dummy gage in position R_4.

Simplifying this equation yields

$$\frac{\Delta R_{2e}}{R_{2e}} = \frac{\Delta R_5}{R_5} \frac{1}{1 + (R_5/R_2)(1 + \Delta R_5/R_5)} \tag{q}$$

Since the bridge is symmetric about its horizontal axis, it is possible to let subscripts $2 \rightarrow 3$ and $5 \rightarrow 6$ to obtain

$$\frac{\Delta R_{3e}}{R_{3e}} = \frac{\Delta R_6}{R_6} \frac{1}{1 + (R_6/R_3)(1 + \Delta R_6/R_6)} \tag{r}$$

Substituting Eqs. (q) and (r) into Eq. (8.19) gives

$$\Delta E = V \frac{R_1 R_{2e}}{(R_1 + R_{2e})^2} \left(\frac{\Delta R_1}{R_1} - \frac{\Delta R_{2e}}{R_{2e}} + \frac{\Delta R_{3e}}{R_{3e}} \right) = 0 \tag{s}$$

In Eq. (s), $\Delta R_{2e}/R_{2e}$ and $\Delta R_{3e}/R_{3e}$ are adjusted until $\Delta E \to 0$; hence, it is evident from Eq. (6.5) that

$$\frac{\Delta R_1}{R_1} = S_g \epsilon = \frac{\Delta R_{2e}}{R_{2e}} - \frac{\Delta R_{3e}}{R_{3e}}$$

$$= \frac{\Delta R_5}{R_5} \frac{1}{1 + (R_5/R_2)(1 + \Delta R_5/R_5)} - \frac{\Delta R_6}{R_6} \frac{1}{1 + (R_6/R_3)(1 + \Delta R_6/R_6)} \quad (t)$$

However, for an initially balanced bridge, where $R_5 = R_6$, $R_2 = R_3$, and $\Delta R_5 = -\Delta R_6$, Eq. (t) reduces appreciably, so that

$$\frac{\Delta R_1}{R_1} = S_g \epsilon = 2 \frac{1 + R_5/R_2}{1 + 2(R_5/R_2) + (R_5/R_2)^2[1 - (\Delta R_5/R_5)^2]} \frac{\Delta R_5}{R_5} \tag{8.26}$$

Examination of Eq. (8.26) shows that the strain reading ϵ obtained by using a parallel-balance circuit is nonlinear in terms of the adjustment of R_5 and R_6, as indicated by the presence of the $(\Delta R_5/R_5)^2$ term. To minimize nonlinearities, it is necessary to limit the range of $\Delta R_5/R_5$, which in turn limits the overall range of the measurement of strain which can be achieved with this circuit.

If the value of $\Delta R_5/R_5$ is limited in range so that

$$0 < \frac{\Delta R_5}{R_5} < 0.1 \quad \text{and} \quad 0 < \left(\frac{\Delta R_5}{R_5}\right)^2 < 0.01 \tag{u}$$

the nonlinear term $(\Delta R_5/R_5)^2$ is negligible in comparison with 1 and Eq. (8.26) reduces to

$$\epsilon = \frac{2}{S_g} \frac{1}{1 + R_5/R_2} \frac{\Delta R_5}{R_5} \tag{8.27}$$

Equation (8.27) can be used to determine the strain directly from the adjustment of the potentiometer or slide wire if $\Delta R_5/R_5 < 0.1$. The range of measurement of strain which is possible with this parallel-balance circuit can be determined from Eq. (8.27) by setting $\Delta R_5/R_5 = 0.1$ and solving for R_5/R_2 in terms of strain to obtain

$$\frac{R_5}{R_2} = \frac{0.2}{S_g \epsilon} - 1 \tag{8.28}$$

The range in the strain measurement is clearly a function of R_5/R_2 and the gage factor S_g. For the particular case where $S_g = 2$ and $\Delta R_5/R_5 = 0.1$, the range of strain is illustrated as a function of R_5/R_2 in Fig. 8.14. The selection of R_5/R_2 to extend the range of measurement cannot be made without considering sensitivity, since R_5/R_2 also influences the sensitivity of the measurement. The sensitivity factor S_{pb} is defined as

$$S_{pb} = \frac{\Delta R_5/R_5}{\epsilon} \tag{8.29}$$

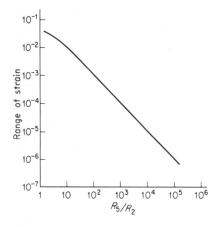

Figure 8.14 Range of strain which can be measured with a parallel-balance circuit.

and by Eq. (8.27), S_{pb} may be expressed as

$$S_{pb} = \frac{S_g}{2}\left(1 + \frac{R_5}{R_2}\right) \tag{8.30}$$

From Eq. (8.30) it is clear that the sensitivity of the parallel-balance circuit is increased by using higher values of R_5/R_2; however, as shown in Fig. 8.14, increasing R_5/R_2 results in a decrease in the allowable range of measurement. Thus a trade-off is necessary to establish a value of R_5/R_2 which gives a suitable range with an adequate sensitivity.

In practice, the parallel-balance arm is fabricated from two fixed resistors R_a and R_b and a potentiometer R_p, as shown in Fig. 8.15. A 20-turn potentiometer divided into 100 readable divisions per turn will permit $\Delta R_5/R_5$ to be established

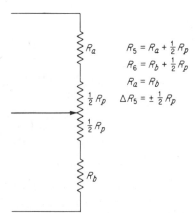

Figure 8.15 The parallel arm containing two fixed resistors and a potentiometer.

within ± 0.001 over the entire range of the potentiometer. It is reasonable then to require that

$$S_{pb}\epsilon = 0.001 \tag{8.31}$$

to fully utilize the sensitivity of the potentiometer readout.

By properly selecting R_a, R_b, R_p, and R_5/R_2, a number of different parallel bridge circuits can be designed. Those circuits with a high sensitivity are limited in range; however, this range can be increased if sensitivity is decreased or if the nonlinear effects are taken into account. Exercises 8.16 and 8.17 illustrate the design aspects of this parallel-balance bridge arrangement.

D. Commercial Strain Indicators

One of the most commonly employed bridge arrangements for static strain measurements is the reference bridge, which is used in several commercial instruments. A schematic illustration of the reference-bridge arrangement is presented in Fig. 8.16, where two bridges are used together to provide a null-balance system.

In this reference-bridge arrangement the bridge on the left-hand side is employed to contain the strain gage or gages, and the bridge on the right-hand side contains either fixed or variable resistors. When gages are placed in the left-hand bridge, the variable resistance in the right-hand bridge is adjusted to obtain initial balance. Strains producing resistance changes in the left-hand bridge cause an unbalance between the two bridges and an associated meter reading.

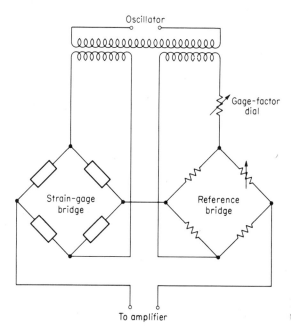

Figure 8.16 Schematic illustration of the reference-bridge arrangement.

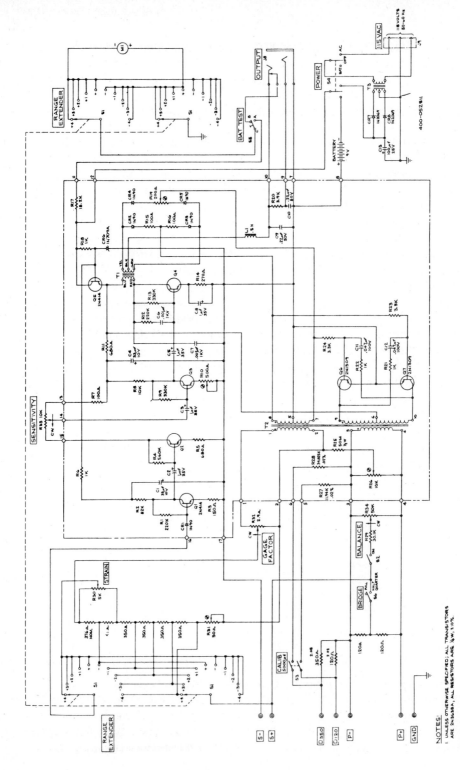

Figure 8.17 Circuit diagram for the Vishay Instrument Model P-350A strain indicator. (*Vishay Instruments, Inc.*)

This meter reading is nulled out by further adjustments of the variable resistor in the right-hand bridge. The advantage of the dual-bridge arrangement is that the left-hand bridge is left completely free for the strain gages and all adjustments for both initial and null balance are performed on the right-hand bridge.

In practice, the reference bridge is somewhat more complex than the schematic arrangement shown in Fig. 8.16. A more detailed circuit, used in Vishay Instrument's P-350A strain indicator, is shown in Fig. 8.17. While a detailed analysis of this circuit is beyond the scope of this text, its operating characteristics can be discussed in some detail. The strain-gage bridge is powered by an oscillator with a 1000-Hz square-wave output of 1.5 V (rms). There is no control over the magnitude of this input voltage; however, for null-balance systems, a fixed-value input voltage is not considered a serious limitation since in the null position the readout is independent of V. The indicator will function over a range of gage resistances from 50 to 2000 Ω. When the gage resistance is less than 50 Ω, the oscillator becomes overloaded. When the gage resistance is greater than 2000 Ω, the load on the bridge circuit due to the amplifier becomes excessive. The gage factor can be set to give readings which are directly calibrated in terms of strain provided $1.5 \leq S_g \leq 4.5$. The indicator can be read to ± 2 μin/in (μm/m) and is accurate to ± 0.1 percent of the reading or 5 μin/in (μm/m) whichever is greater. The range is $\pm 50,000$ μin/in (μm/m). The unit, shown in Fig. 8.18, is small, lightweight, and portable. It is simple to operate and is adequate for almost all static strain-gage applications.

E. Summary of the Wheatstone Bridge Circuit

The equations which govern the behavior of the Wheatstone bridge circuit, under initial balance conditions, for both static and dynamic strain measurements are summarized below:

$$\Delta E = V \frac{R_1 R_2}{(R_1 + R_2)^2}\left(\frac{\Delta R_1}{R_1} - \frac{\Delta R_2}{R_2} + \frac{\Delta R_3}{R_3} - \frac{\Delta R_4}{R_4}\right) \tag{8.19}$$

$$\Delta E = V \frac{r}{(1 + r)^2}\left(\frac{\Delta R_1}{R_1} - \frac{\Delta R_2}{R_2} + \frac{\Delta R_3}{R_3} - \frac{\Delta R_4}{R_4}\right) \tag{8.20}$$

$$S_c = V \frac{r}{(1 + r)^2} n S_g \tag{8.21}$$

$$S_c = \frac{r}{1 + r} S_g \sqrt{P_g R_g} \tag{8.24}$$

The Wheatstone bridge circuit can be employed for both static and dynamic strain measurements since it can be initially balanced to yield a zero output voltage. The potentiometer circuit, on the other hand, is suitable only for dynamic measurements. Comparison of Eqs. (8.8) and (8.24) shows that the potentiometer circuit and the Wheatstone bridge circuit have the same circuit sensitivity. For this

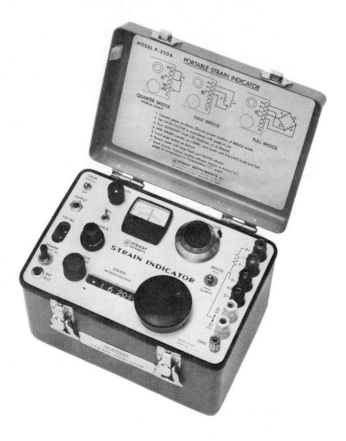

Figure 8.18 Model P-350A strain indicator. (*Vishay Instruments, Inc.*)

reason, the potentiometer circuit, because of its simplicity, is usually employed for dynamic measurements.

The output voltage ΔE from the Wheatstone bridge is nonlinear with respect to resistance change ΔR, and the measuring instrument used to detect the output voltage can produce loading effects which are similar to those produced with the potentiometer circuit. No emphasis was placed on nonlinearity or on loading effects since they are not usually troublesome in the typical experimental stress analysis. Nonlinear effects can become significant when semiconductor gages are used to measure relatively large strains because of the large ΔR's involved. In these applications, the constant-current circuits described in Sec. 8.4 should be used. Loading effects should never be a problem since high-quality, low-cost measuring instruments with high input impedances are readily available.

The Wheatstone bridge can be used to add or subtract signals from multiple-strain-gage installations. A four-active-arm bridge is slightly more than twice as sensitive as an optimum single-active-arm bridge. For transducer applications, the gain in sensitivity as well as cancellation of gage response from components of load which are not being measured are important factors; therefore, the Wheatstone bridge is used almost exclusively for this purpose.

Temperature compensation can be employed without loss of sensitivity provided a separate dummy gage or another active gage is employed in arm R_4 of the bridge. Grounding of the Wheatstone bridge can be accomplished only at point C (see Fig. 8.9); consequently, noise can become more of a problem with the Wheatstone bridge than with the potentiometer circuit. For further details on the treatment of noise in strain-gage circuits, see Sec. 8.7.

8.4 CONSTANT-CURRENT CIRCUITS [3–4]

The potentiometer and Wheatstone bridge circuits described in Secs. 8.2 and 8.3 were driven with a voltage source which ideally remains constant with changes in the resistance of the circuit. These voltage-driven circuits exhibit nonlinear outputs whenever $\Delta R/R$ is large [see Eq. (8.3b)]. This nonlinear behavior limits their applicability to semiconductor strain gages. It is possible to replace the constant-voltage source with a constant-current source, and it can be shown that improvements in both linearity and sensitivity result.

Constant-current power supplies with sufficient regulation for strain-gage applications are relatively new and have been made possible by advances in solid-state electronics. Basically, the constant-current power supply is a high-impedance (1 to 10 MΩ) device which changes output voltage with changing resistive load to maintain a constant current.

A. Constant-Current Potentiometer Circuit

Consider the constant-current potentiometer circuit shown in Fig. 8.19a. When a very high impedance meter is placed across resistance R_1, the measured output voltage E is

$$E = IR_1 \qquad (8.32)$$

When resistances R_1 and R_2 change by ΔR_1 and ΔR_2, the output voltage becomes

$$E + \Delta E = I(R_1 + \Delta R_1) \qquad (a)$$

Thus from Eqs. (8.32) and (a)

$$\Delta E = (E + \Delta E) - E = I\,\Delta R_1 = IR_1\,\frac{\Delta R_1}{R_1} \qquad (8.33)$$

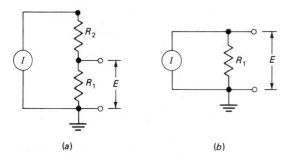

(a) (b)

Figure 8.19 Constant-current potentiometer circuits.

It should be noted that ΔR_2 does not affect the signal output. Indeed, even R_2 is not involved in the output voltage, and hence it can be eliminated to give the very simple potentiometer circuit shown in Fig. 8.19b.

Substituting Eq. (6.5) into Eq. (8.33) yields

$$\Delta E = I R_g S_g \epsilon \tag{8.34}$$

Equations (8.33) and (8.34) show that the output signal ΔE is linear with respect to resistance change ΔR and strain ϵ. It is this feature of the constant-current potentiometer circuit which makes it more suitable for use with semiconductor strain gages than the constant-voltage potentiometer circuit.

The circuit sensitivity $S_c = \Delta E/\epsilon$ reduces to

$$S_c = I R_g S_g \tag{8.35}$$

If the constant-current source is adjustable, so that the current I can be increased to the power-dissipation limit of the strain gage, then $I = I_g$ and Eq. (8.35) can be rewritten as

$$S_c = \sqrt{P_g R_g} \, S_g \tag{8.36}$$

Thus, the circuit sensitivity is totally dependent on the strain-gage parameters P_g and R_g, and S_g and is totally independent of circuit parameters except for the capability to adjust the current source. Comparison of Eqs. (8.8) and (8.36) shows that the sensitivities differ by the $r/(1 + r)$ multiplier for the constant-voltage potentiometer; thus, S_c will always be higher for the constant-current potentiometer.

It was noted in deriving Eq. (8.33) that R_2 and ΔR_2 did not affect the signal output of the constant-current potentiometer. This indicates that temperature compensation by signal cancellation in the strain-gage circuit or signal addition cannot be performed. It is possible to maintain the advantages of high sensitivity and perfect linearity of this circuit and to obtain the capability of signal addition or subtraction by using a double constant-current potentiometer circuit.

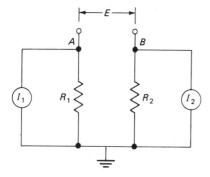

Figure 8.20 Double constant-current potentiometer circuit.

B. Double Constant-Current Potentiometer Circuit

Consider the double constant-current potentiometer circuit shown in Fig. 8.20. In this circuit, two constant-current generators I_1 and I_2 are employed, and the output voltage is measured with a very high impedance meter between points A and B. The voltage at points A and B will be

$$V_A = I_1 R_1 \qquad V_B = I_2 R_2$$

and the output voltage E is

$$E = V_A - V_B = I_1 R_1 - I_2 R_2 \tag{8.37}$$

The circuit can be balanced to give a null output $(E = 0)$ initially if the current can be adjusted so that

$$I_1 R_1 = I_2 R_2 \tag{8.38}$$

If resistances R_1 and R_2 change by ΔR_1 and ΔR_2, Eq. (8.37) yields

$$E + \Delta E = I_1 (R_1 + \Delta R_1) - I_2 (R_2 + \Delta R_2) \tag{b}$$

For a circuit which is initially balanced, $E = 0$ and Eq. (b) reduces to

$$\Delta E = I_1 \, \Delta R_1 - I_2 \, \Delta R_2 = I_1 R_1 \frac{\Delta R_1}{R_1} - I_2 R_2 \frac{\Delta R_2}{R_2} \tag{8.39}$$

Examination of Eq. (8.39) shows that the output voltage ΔE is linear with respect to resistance change ΔR; thus, the circuit can be used with semiconductor gages where large values of $\Delta R/R$ are experienced. Also, since the circuit can be initially balanced, the voltage change ΔE can easily be measured and the circuit can be used for both static and dynamic measurements.

Application of the double potentiometer circuit to strain-gage measurements usually involves one of the four cases illustrated in Fig. 8.21.

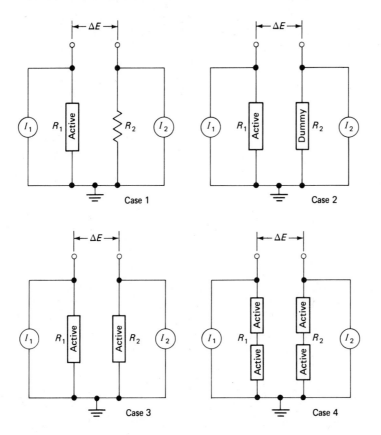

Figure 8.21 Four common double potentiometer circuits.

Case 1 This simple circuit incorporates an active gage in position R_1 and a fixed resistor in position R_2. The currents are adjusted so that $R_1 I_1 = R_2 I_2$. Since R_2 is fixed, $\Delta R_2 = 0$ and Eq. (8.39) reduces to

$$\Delta E = I_1 R_1 \frac{\Delta R_1}{R_1} = I_g R_g S_g \epsilon \qquad (c)$$

and the circuit sensitivity is

$$S_c = \frac{\Delta E}{\epsilon} = I_g R_g S_g = \sqrt{P_g R_g} \, S_g \qquad (8.40)$$

which is the same sensitivity as that achieved with the ordinary constant-current potentiometer circuit. The advantage of using the double constant-current potentiometer circuit in this case is to eliminate the output voltage E so that the voltage ΔE can be determined directly and related to the strain produced by static loadings.

Case 2 Figure 8.21 shows that this circuit incorporates an active gage in the R_1 position and a dummy gage in the R_2 position. The currents are adjusted to achieve initial balance of the circuit. In this case, $R_1 = R_2 = R_g$, and if both gages are exposed to the same thermal environment,

$$\Delta R_1 = \Delta R_\epsilon + \Delta R_T \qquad \Delta R_2 = \Delta R_T \qquad (d)$$

Substitution of Eq. (*d*) into Eq. (8.39) gives

$$\Delta E = I_g R_g \left(\frac{\Delta R_\epsilon}{R_g} + \frac{\Delta R_T}{R_g} - \frac{\Delta R_T}{R_g} \right) \qquad (e)$$

which reduces to

$$\Delta E = I_g R_g S_g \epsilon \qquad (f)$$

From Eq. (*f*) it is clear that the circuit sensitivity is

$$S_c = I_g R_g S_g = \sqrt{P_g R_g}\, S_g \qquad (8.41)$$

With active and dummy gages in the R_1 and R_2 positions, respectively, it is clear from Eq. (*e*) that temperature compensation is achieved. It is also evident by comparing Eqs. (8.40) and (8.41) that temperature compensation has been achieved with the double potentiometer circuit without loss in circuit sensitivity.

Case 3 Case 3, illustrated in Fig. 8.21, shows active gages in both positions R_1 and R_2 so that $R_1 = R_2 = R_g$. The currents are adjusted to achieve initial balance of the circuit. If the gages are placed on a beam in bending with the R_1 gage on the tension side and the R_2 gage on the compression side, then $\Delta R_1 = -\Delta R_2 = \Delta R_g$. For this case, $I_1 = I_2 = I_g$, and Eq. (8.39) becomes

$$\Delta E = 2R_g I_g S_g \epsilon \qquad (g)$$

The circuit sensitivity becomes

$$S_c = 2R_g I_g S_g = 2\sqrt{P_g R_g}\, S_g \qquad (8.42)$$

Examination of Eqs. (*g*) and (8.42) indicates that two active gages in the double potentiometer circuit provide a circuit sensitivity which is twice the value obtained with a single active gage. Temperature compensation is also achieved in this case.

Case 4 In certain applications, signal addition and the associated increase in output voltage are extremely important. Signal addition can be accomplished with a double constant-current potentiometer circuit by using multiple gages as shown in case 4 of Fig. 8.21. In this instance, $R_1 = R_2 = 2R_g$, $\Delta R_1 = 2\Delta R_g$, and $\Delta R_2 = -2\Delta R_g$. Substitution into Eq. (8.39) gives

$$\Delta E = 2I_g R_g \left(\frac{2\Delta R_g}{2R_g} + \frac{2\Delta R_g}{2R_g} \right) = 4I_g R_g \frac{\Delta R_g}{R_g} \qquad (h)$$

and the circuit sensitivity becomes

$$S_c = 4I_g R_g S_g = 4\sqrt{P_g R_g}\, S_g \tag{8.43}$$

Clearly, multiple-gage installations can be used with the double constant-current potentiometer circuit to increase the output voltage to relatively high values in transducer applications. Signal addition or subtraction can be achieved and temperature compensation is automatic if identical gages are used and exposed to the same thermal environment.

The advantages of the double constant-current potentiometer circuit can be summarized as follows:

1. The output voltage ΔE is perfectly linear with respect to $\Delta R/R$, and, as such, the circuit is ideal for use with semiconductor strain gages where large values of $\Delta R/R$ may be encountered.
2. The sensitivity of the double constant-current potentiometer circuit is equal to or better than that of the Wheatstone bridge circuit (constant voltage supply) when one or two active gages are used. The sensitivity is twice that of the normal Wheatstone bridge when four active gages are employed.
3. Grounding is possible for noise elimination.
4. The circuit is the ultimate in simplicity. It can easily be used for either static or dynamic measurements if the current can be adjusted to achieve initial balance and to drive the gages at their maximum power capabilities.

The disadvantage of the circuit pertains to the ripple requirement for the constant-current power supply; however, with battery-driven constant-current supplies, ripple can be limited to 20 parts per million (ppm) for carefully designed circuits. Also, two constant-current sources are required, which increases the costs for an otherwise extremely simple circuit.

C. Constant-Current Wheatstone Bridge Circuits

Since Wheatstone bridges have been used to measure resistance changes in strain gages since the discovery of the basic phenomenon by Lord Kelvin in 1856, it is logical to consider a bridge driven by a constant-current supply, as shown in Fig. 8.22. The current I delivered by the supply divides at point A of the bridge into currents I_1 and I_2, where $I = I_1 + I_2$. The voltage drop between points A and B of the bridge is

$$V_{AB} = I_1 R_1 \tag{i}$$

and the voltage drop between points A and D is

$$V_{AD} = I_2 R_4 \tag{j}$$

Thus the output voltage E from the bridge can be expressed as

$$E = V_{BD} = V_{AB} - V_{AD} = I_1 R_1 - I_2 R_4 \tag{8.44}$$

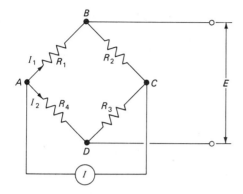

Figure 8.22 Wheatstone bridge with a constant current supply.

For the bridge to be in balance ($E = 0$) under no-load conditions,

$$I_1 R_1 = I_2 R_4 \tag{8.45}$$

Consider next the voltage V_{AC} and note that

$$V_{AC} = I_1(R_1 + R_2) = I_2(R_3 + R_4) \tag{k}$$

from which

$$I_1 = \frac{R_3 + R_4}{R_1 + R_2} I_2 \tag{l}$$

Since

$$I = I_1 + I_2 \tag{m}$$

Eq. (l) can be substituted into Eq. (m) to obtain

$$I_1 = \frac{R_3 + R_4}{R_1 + R_2 + R_3 + R_4} I \qquad I_2 = \frac{R_1 + R_2}{R_1 + R_2 + R_3 + R_4} I \tag{n}$$

Substituting Eqs. (n) into Eq. (8.44) yields

$$E = \frac{I}{R_1 + R_2 + R_3 + R_4}(R_1 R_3 - R_2 R_4) \tag{8.46}$$

From Eq. (8.46) it is evident that the balance condition ($E = 0$) for the constant-current Wheatstone bridge is the same as that for the constant-voltage Wheatstone bridge, namely,

$$R_1 R_3 = R_2 R_4 \tag{8.47}$$

If resistances R_1, R_2, R_3, and R_4 change by the amounts $\Delta R_1, \Delta R_2, \Delta R_3$, and ΔR_4, the voltage $E + \Delta E$ measured with a very high impedance meter is

$$E + \Delta E = \frac{I}{\sum R + \sum \Delta R}[(R_1 + \Delta R_1)(R_3 + \Delta R_3) - (R_2 + \Delta R_2)(R_4 + \Delta R_4)] \tag{o}$$

where $\sum R = R_1 + R_2 + R_3 + R_4$ $\qquad$ $\sum \Delta R = \Delta R_1 + \Delta R_2 + \Delta R_3 + \Delta R_4$

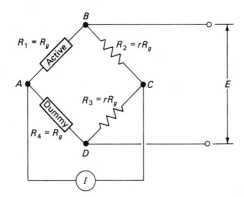

Figure 8.23 Constant-current Wheatstone bridge designed to minimize nonlinear effects.

Expanding Eq. (*o*) and simplifying after assuming the initial balance condition gives

$$\Delta E = \frac{IR_1 R_3}{\sum R + \sum \Delta R}\left(\frac{\Delta R_1}{R_1} - \frac{\Delta R_2}{R_2} + \frac{\Delta R_3}{R_3} - \frac{\Delta R_4}{R_4} + \frac{\Delta R_1}{R_1}\frac{\Delta R_3}{R_3} - \frac{\Delta R_2}{R_2}\frac{\Delta R_4}{R_4}\right)$$

$$(8.48)$$

Inspection of Eq. (8.48) shows that the output signal ΔE is nonlinear with respect to ΔR because of the term $\sum \Delta R$ in the denominator and because of the second-order terms $(\Delta R_1/R_1)(\Delta R_3/R_3)$ and $(\Delta R_2/R_2)(\Delta R_4/R_4)$ within the bracketed quantity. The nonlinearity of the constant-current Wheatstone bridge, however, is less than that with the constant-voltage bridge. Indeed, if the constant-current Wheatstone bridge is properly designed, the nonlinear terms can be made insignificant even for the large $\Delta R/R$'s encountered with semiconductor strain gages.

The nonlinear effects in a typical situation can be evaluated by considering the constant-current Wheatstone bridge shown in Fig. 8.23. A single active gage is employed in arm R_1, and a temperature-compensating dummy gage is employed in arm R_4. Fixed resistors are employed in arms R_2 and R_3. Thus

$$R_1 = R_4 = R_g \qquad R_2 = R_3 = rR_g \qquad \Delta R_2 = \Delta R_3 = 0$$

Under stable thermal environments, $\Delta R_1 = \Delta R_g$ and $\Delta R_4 = 0$. Equation (8.48) then reduces to

$$\Delta E = \frac{IR_g r}{2(1 + r) + \Delta R_g/R_g}\frac{\Delta R_g}{R_g}$$

$$(8.49)$$

Again it is evident that Eq. (8.49) is nonlinear due to the presence of the term $\Delta R_g/R_g$ in the denominator. To determine the degree of the nonlinearity let

$$\frac{IR_g r}{2(1 + r) + \Delta R_g/R_g}\frac{\Delta R_g}{R_g} = \frac{IR_g r}{2(1 + r)}\frac{\Delta R_g}{R_g}(1 - \eta)$$

where η, the nonlinear term, is

$$\eta = \frac{\Delta R_g/R_g}{2(1+r) + \Delta R_g/R_g} = \frac{S_g\epsilon}{2(1+r) + S_g\epsilon} \tag{8.50}$$

Inspection of Eq. (8.50) shows that the nonlinear term η can be minimized by increasing r (making the fixed resistors R_2 and R_3, say, nine times the value of R_g). In this case, the nonlinear term η will depend on the gage factor and on the magnitude of the strain. Consider, for example, a semiconductor strain gage with $S_g = 100$; then η will be less than 1 percent for strains less than 2000 μin/in (μm/m). Since strains from 1000 to 2000 μin/in (μm/m) represent the upper limit for semiconductor strain gages, they can be used with the constant-current Wheatstone bridge if it is properly designed ($r \geq 9$).

8.5 CALIBRATING STRAIN-GAGE CIRCUITS [5]

A strain-measuring system usually consists of a strain gage, a power supply (either constant voltage or constant current), circuit-completion resistors, an amplifier, and a recording instrument of some type. A schematic illustration of a typical system is shown in Fig. 8.24. It is possible to calibrate each component of the system and determine the voltage-strain relationship from the equations developed in Sec. 8.3. However, this procedure is time-consuming and subject to calibration errors associated with each of the components involved in the system. A more precise and direct procedure is to obtain a single calibration for the complete system so that readings from the recording instrument can be directly related to the strains which produced them.

   Direct system calibration can be achieved by shunting a fixed resistor R_c across one arm (say R_2) of the Wheatstone bridge, as shown in Fig. 8.24. If the

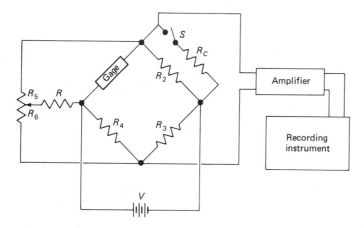

Figure 8.24 Typical strain-recording system.

bridge is initially balanced and the switch is closed to place R_c in parallel with R_2, the effective resistance of this arm of the bridge is

$$R_{2e} = \frac{R_2 R_c}{R_2 + R_c} \qquad (a)$$

Because of the shunt resistance R_c, the ratio of the change in resistance to the original resistance in arm R_2 of the bridge is

$$\frac{\Delta R_2}{R_2} = \frac{R_{2e} - R_2}{R_2} \qquad (b)$$

Combining Eqs. (a) and (b) gives

$$\frac{\Delta R_2}{R_2} = \frac{-R_2}{R_2 + R_c} \qquad (c)$$

Substituting Eq. (c) into Eq. (8.19) gives the output voltage of the bridge produced by R_c. Thus

$$\Delta E = \frac{R_1 R_2}{(R_1 + R_2)^2} \frac{R_2}{R_2 + R_c} V \qquad (d)$$

Note also that a single active gage in position R_1 of the bridge would produce an output due to a strain ϵ of

$$\Delta E = \frac{R_1 R_2}{(R_1 + R_2)^2} (S_g \epsilon) V \qquad (e)$$

Equating Eqs. (d) and (e) gives

$$\epsilon_c = \frac{R_2}{S_g (R_2 + R_c)} \qquad (8.51)$$

where ϵ_c is the calibration strain which would produce the same voltage output from the bridge as the calibration resistor R_c.

For example, consider a bridge with $R_2 = R_g = 350 \ \Omega$ and $S_g = 2.05$. If $R_c = 100 \ \text{k}\Omega$, the calibration strain $\epsilon_c = 1700 \ \mu\text{in/in} \ (\mu\text{m/m})$. If the recording instrument is operated during the period of time when the switch S is closed, an instrument deflection $d_c = 1700 \ \mu\text{in/in} \ (\mu\text{m/m})$ will be recorded as shown in Fig. 8.25. The switch can then be opened and the load-induced strain can be recorded in the normal manner to give a strain-time pulse similar to the one illustrated

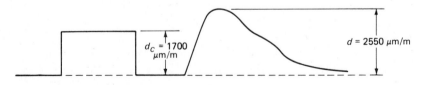

Figure 8.25 Calibrated strain-time trace.

in Fig. 8.25. The peak strain associated with this strain-time pulse produces an instrument deflection d which can be numerically evaluated as

$$\epsilon = \frac{d}{d_c} \epsilon_c \tag{8.52}$$

This method of shunt calibration is accurate and easy to employ. It provides a means of calibrating the complete system regardless of the number of components in the system. The calibration strain produces a reading on the recording instrument; all other readings are linearly related to this calibration value.

8.6 EFFECTS OF LEAD WIRES, SWITCHES, AND SLIP RINGS

The resistance change for a metallic foil strain gage is quite small [0.7 mΩ per μin/in (μm/m) of strain for a 350-Ω gage]. As a consequence, anything that produces a resistance change within the Wheatstone bridge is extremely important. The components within a Wheatstone bridge almost always include lead wires, soldered joints, and binding posts. Frequently, switches and slip rings are also included. The effects of each of these components on the output of the Wheatstone bridge circuit are discussed in the following sections.

A. Effect of Lead Wires [6, 7]

Consider first a two-lead wire system, illustrated in Fig. 8.26, where a single active gage is positioned on a test structure at a location remote from the bridge and recording system. If the length of the two-lead wire system is long, three detrimental effects occur: signal attenuation, loss of balancing capability, and loss of temperature compensation.

To show that signal attenuation may occur note that

$$R_1 = R_g + 2R_L \tag{a}$$

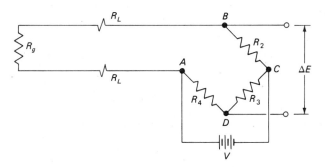

Figure 8.26 Two-lead wire system.

where R_L is the resistance of a single lead wire. Note also that

$$\frac{\Delta R_1}{R_1} = \frac{\Delta R_g}{R_g + 2R_L} = \frac{\Delta R_g/R_g}{1 + 2R_L/R_g} \tag{b}$$

Equation (b) may be expressed in terms of a signal loss factor $\mathscr{L}$. Thus

$$\frac{\Delta R_1}{R_1} = \frac{\Delta R_g}{R_g}(1 - \mathscr{L}) \tag{c}$$

From Eqs. (b) and (c), the signal loss factor $\mathscr{L}$ for the two-lead wire system can be expressed as

$$\mathscr{L} \begin{cases} = \dfrac{2R_L/R_g}{1 + 2R_L/R_g} \\[4mm] \approx \dfrac{2R_L}{R_g} \quad \text{if} \quad \dfrac{2R_L}{R_g} \ll 1 \end{cases} \tag{8.53}$$

The signal loss factor $\mathscr{L}$ is shown as a function of the ratio of lead resistance to gage resistance in Fig. 8.27. It is clear from this relationship that $\mathscr{L}$ increases rapidly as R_L becomes a significant fraction of R_g. In order to limit lead-wire losses to less than 2 percent, $R_L/R_g \leq 0.01$. If test conditions dictate long leads, then large-gage wire must be employed to limit R_L and it is advantageous to use 350-Ω gages in place of 120-Ω gages. The resistance of a 100-ft (30.5-m) length of lead wire as a function of wire gage size is listed in Table 8.1 for solid copper wire.

The second detrimental effect of the two-lead wire system is loss of the ability to initially balance the bridge. For the bridge shown in Fig. 8.26,

$$R_1 = R_g + 2R_L \qquad R_2 = R_3 = rR_g \qquad \text{fixed resistors}$$
$$R_4 = R_g \qquad \Delta R_2 = \Delta R_3 = 0$$

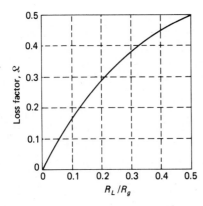

Figure 8.27 Loss factor as a function of the ratio of lead wire to gage resistance, two-lead wire system.

Table 8.1 Resistance of solid-conductor copper wire, Ω per 100 ft (30.5 m)

Gage size	Resistance	Gage size	Resistance	Gage size	Resistance
12	0.159	22	1.614	32	16.41
14	0.253	24	2.567	34	26.09
16	0.402	26	4.081	36	41.48
18	0.639	28	6.490	38	65.96
20	1.015	30	10.310	40	104.90

With the above resistances present in the bridge, it is obvious that the initial balance condition $R_1 R_3 = R_2 R_4$ is not satisfied. Of course, a parallel-balance resistor similar to the one shown in Fig. 8.24 is available in most commercial bridges to obtain initial balance; however, if $R_L/R_g > 0.02$, the range of the balance potentiometer is exceeded and initial balance of the bridge cannot be achieved.

The third detrimental effect of the two-lead wire system is that temperature compensation of the measuring circuit is lost. First, consider a temperature-compensating dummy gage with relatively short leads in arm R_4 of the bridge shown in Fig. 8.26. The output voltage due to resistance changes in arms R_1 and R_4 as obtained from Eq. (8.20) is

$$\Delta E = V \frac{r}{(1+r)^2} \left(\frac{\Delta R_1}{R_1} - \frac{\Delta R_4}{R_4} \right) \tag{d}$$

If the gages are subjected to a temperature difference ΔT at the same time that the active gage is subjected to a strain ϵ, then Eq. (d) becomes

$$\Delta E = V \frac{r}{(1+r)^2} \left[\left(\frac{\Delta R_g}{R_g + 2R_L} \right)_\epsilon + \left(\frac{\Delta R_g}{R_g + 2R_L} \right)_{\Delta T} + \left(\frac{2\Delta R_L}{R_g + 2R_L} \right)_{\Delta T} - \left(\frac{\Delta R_g}{R_g} \right)_{\Delta T} \right] \tag{8.54}$$

Examination of Eq. (8.54) shows that temperature compensation is not achieved in the Wheatstone bridge since the second and fourth terms in the bracketed quantity are not equal. Also, the lead wires can suffer significant resistance changes due to temperature; therefore, the third term in the bracketed quantity can produce significant errors in the measurement of strain with the two-lead-wire system.

The detrimental effects of long lead wires can be minimized by employing the three-lead-wire system shown in Fig. 8.28. In this circuit, both the active and dummy gages are placed at the remote location. One of the three wires is used to transfer terminal A of the bridge to the remote location. It is not considered a lead wire since it is not within either arm R_1 or arm R_4 of the bridge. The active and dummy gages each have one long lead wire with resistance R_L and one short lead

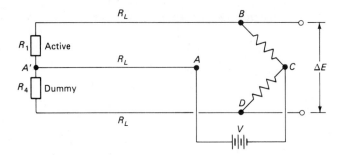

Figure 8.28 Three-lead wire system.

wire with negligible resistance (see Fig. 8.28). With the three-lead-wire system, the signal loss factor $\mathscr{L}$ is reduced to

$$\mathscr{L}\begin{cases} = \dfrac{R_L/R_g}{1 + R_L/R_g} \\[2ex] \approx \dfrac{R_L}{R_g} \quad \text{if} \quad \dfrac{R_L}{R_g} \ll 1 \end{cases} \tag{8.55}$$

The bridge retains its initial balance capability since the resistance of both arms R_1 and R_4 is increased by R_L. With the three-lead-wire system, Eq. (8.54) becomes

$$\Delta E = V \frac{r}{(1+r)^2} \left[\left(\frac{\Delta R_g}{R_g + R_L} \right)_\epsilon + \left(\frac{\Delta R_g}{R_g + R_L} \right)_{\Delta T} + \left(\frac{\Delta R_L}{R_g + R_L} \right)_{\Delta T} \right.$$
$$\left. - \left(\frac{\Delta R_g}{R_g + R_L} \right)_{\Delta T} - \left(\frac{\Delta R_L}{R_g + R_L} \right)_{\Delta T} \right]$$

Temperature compensation is achieved since all the temperature-related terms in the bracketed quantity cancel.

B. Effect of Switches [8]

In most strain-gage applications, many strain gages are installed and are monitored several times during the test. When the number of gages is large, it is not economically feasible to employ a separate recording instrument for each gage. Instead, a single recording system is used, and the gages are switched in and out of the instrument system. Two different methods of switching are commonly used in multiple-gage installations.

The first method, illustrated in Fig. 8.29, involves switching one side of each active gage, in turn, into arm R_1 of the bridge. The other side of each of the active gages is connected to terminal A of the bridge by a common lead wire. A single dummy gage or fixed resistor is used in arm R_4 of the bridge. With this arrangement, the switch is located within arm R_1 of the bridge; therefore, an extremely

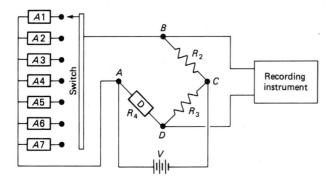

Figure 8.29 Switching active gages in arm R_1 of the Wheatstone bridge.

high quality switch with negligible resistance (less than 500 $\mu\Omega$) must be used. If the switching resistance is not negligible, the switch resistance adds to ΔR_g to produce an error in the strain measurement. The quality of a switch can be checked quite easily, since excessive switch resistance makes it impossible to reproduce zero strain readings.

The second method, illustrated in Fig. 8.30, involves switching the complete bridge. In this method, a three-pole switch is employed in the leads between the bridge and the power supply and the recording instrument. Since the switch is not located in the arms of the bridge, switching resistance is not particularly important. Switching the complete bridge is more expensive, however, since a separate dummy gage and two bridge-completion resistors are required for each active gage.

C. Effect of Slip Rings [9]

Strain gages are frequently used on rotating machinery, where it is impossible to use ordinary lead wires to connect the active gages to the recording instrument. Slip rings are employed in these applications to provide lead-wire connections. The rings are mounted on a shaft which is attached to the rotating member in such a way that the axis of the shaft coincides with the axis of rotation of the member. The shell of the slip-ring assembly is stationary and usually carries several brushes for each slip ring. Lead wires from the strain-gage bridge, which rotates with the member, are connected to the slip rings. Lead wires from the power supply and recording instrument, which are stationary, are connected through the brushes to the appropriate slip ring. Depending on the design of the slip-ring assembly, satisfactory operation at rotary speeds to 24,000 r/min can be achieved.

Dirt collecting on the slip rings and brush jump tends to generate electrical noise in this type of strain-gage system. The use of multiple brushes in parallel helps to minimize the noise; however, the resistance changes between the rings and the brushes are usually so large that slip rings are not recommended for use in an arm of the bridge. Instead, a complete bridge should be assembled on the

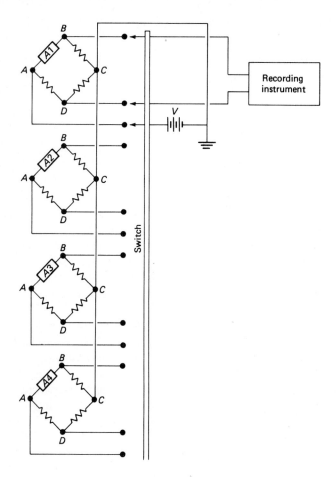

Figure 8.30 Switching the complete bridge.

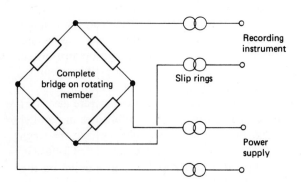

Figure 8.31 Slip rings connecting a complete bridge to the recording instrument and power supply.

rotating member, as shown in Fig. 8.31, and the slip rings should be used to connect the bridge to the power supply and recording instrument. In this way, the effects of slip-ring resistance changes on the strain measurements are minimized.

8.7 ELECTRICAL NOISE [10]

The voltages ΔE from strain gage circuits are quite small [usually less than 10 μV per μin/in (μm/m) of strain]. As a consequence, electrical noise is an important consideration in strain-gage circuit design. Electrical noise in strain-gage circuits is produced by the magnetic fields generated when currents flow through disturbing wires in close proximity to the strain-gage lead wires, as shown in Fig. 8.32. When an alternating current flows in the disturbing wire, a time-varying magnetic field is produced which cuts both wires of the signal circuit and induces a voltage in the signal loop. The induced voltage is proportional to the current I and the area enclosed by the signal loop but inversely proportional to the distance from the disturbing wire to the signal circuit.

Since the distances d_1 and d_2 in Fig. 8.32 are not equal, the difference in magnetic fields at the two signal leads induces a noise voltage E_N which is superimposed on the strain-gage signal voltage ΔE. In certain instances, where the disturbing fields are large, the noise voltage becomes significant and makes separation of the true strain-gage signal from the noise signal quite difficult.

There are three procedures which should normally be employed to reduce noise to a minimum.

Procedure 1 All lead wires should be tightly twisted together to minimize the area in the signal loop and make the distances d_1 and d_2 equal. In this way, the noise voltage is minimized.

Procedure 2 Only shielded cables should be used, and the shields should be connected only to the signal ground as indicated in Fig. 8.33. If the shield is connected

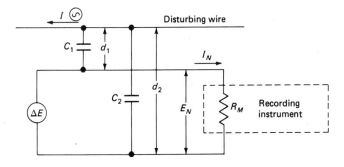

Figure 8.32 Generation of electrical noise.

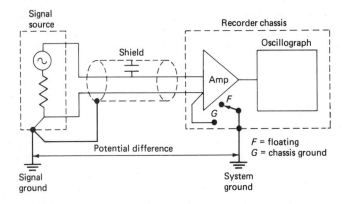

Figure 8.33 Proper method of grounding a shielded cable.

to both the signal ground and the system ground, as shown in Fig. 8.34, a ground loop is formed. Since two different grounds are seldom at the same absolute voltage, a noise signal can be generated by the potential difference which exists between the two to combine with the strain-gage signal in the lower lead of Fig. 8.34. A second ground loop, from the signal source through the cable shield to the amplifier, also occurs with the grounding method shown in Fig. 8.34. Alternating currents in the shield, due to this second ground loop, are coupled to the signal pair through the distributed capacitance in the signal cable (see Fig. 8.34). Either of these ground loops is capable of generating a noise signal 100 times larger than the strain gage signal.

In most recording instruments, the third conductor in the power cord is used to provide the system ground. Since this ground is connected to the electronic enclosure for the instrument, care should be exercised to insulate the enclosure

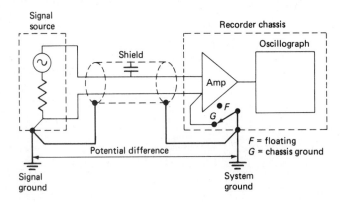

Figure 8.34 Incorrect method of grounding a shielded cable, resulting in a ground loop.

from any other building ground. In most modern recording instruments, the amplifier can be operated in either a floating mode or a grounded mode, as shown in Figs. 8.33 and 8.34, respectively. In the floating mode, the amplifier is insulated from the system ground; and, with the signal source also isolated from the system ground, the arrangement is correct for minimizing the noise signal. The proper point of attachment for the signal ground on both the potentiometer circuit and the Wheatstone bridge circuit is at the negative terminal of the power supply. The power supply itself should be floated relative to the system ground to avoid a ground loop at the supply.

Procedure 3 The third way to eliminate noise is by common-mode rejection. Here the lead wires are arranged so that any noise signals will appear equally and simultaneously in both the lead wires. If a differential amplifier is employed in the recording system, the noise signals are rejected and the strain signals are amplified. Unfortunately, the common-mode rejection of the very best differential amplifiers is not perfect, and a very small portion of the common-mode voltage is transmitted by the amplifier. The common-mode rejection for good low-level data amplifiers is about 10^6 to 1 at 60 Hz; thus, it is evident that significant noise suppression can be achieved in this manner.

8.8 TRANSDUCER APPLICATIONS

Since the electrical-resistance strain gage is such a remarkable measuring device—small, lightweight, linear, precise, and inexpensive—it is used as the sensor in a wide variety of transducers. In a transducer such as a load cell, the load is measured by subjecting a mechanical member to the load and measuring the strain developed in the mechanical member. Since the load is linearly related to the strain, as long as the mechanical member remains elastic, the load cell can be calibrated so that the output signal is interpreted as a load reading.

Transducers of many different types and models are commercially available. Included are load cells, torque meters, pressure gages, displacement gages, and accelerometers. In addition, many different special-purpose transducers are custom-designed, with strain gages as the sensing device, to measure other quantities.

A. Load Cells

The design of strain-gage transducers for general-purpose measurement is relatively easy. Consider, for instance, a tensile load cell fabricated from a tension specimen, as shown in Fig. 8.35. To convert the simple tension bar into a load cell, four strain gages are mounted on the central region of the bar with two opposite gages in the axial direction and two opposite gages in the transverse direction, as

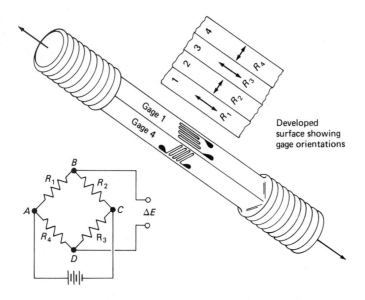

Figure 8.35 Strain gages mounted on a simple tension specimen to produce a load cell.

shown in Fig. 8.35. When a load P is applied to the tension member, the axial and transverse strains produced are

$$\epsilon_a = \frac{P}{AE} \qquad \epsilon_t = -\frac{vP}{AE} \tag{8.56}$$

where A = cross-sectional area of tension member
E = modulus of elasticity of the material used to fabricate member
v = Poisson's ratio of material used to fabricate member

If the four gages are positioned in the Wheatstone bridge as shown in Fig. 8.35, the ratio of output voltage to supply voltage $\Delta E/V$ is given by Eq. (8.20) as

$$\frac{\Delta E}{V} = \frac{1}{4}\left(\frac{\Delta R_1}{R_1} - \frac{\Delta R_2}{R_2} + \frac{\Delta R_3}{R_3} - \frac{\Delta R_4}{R_4}\right) \tag{8.57}$$

The changes in resistance of the four gages on the tension member are obtained from Eqs. (6.5) and (8.56) as

$$\frac{\Delta R_1}{R_1} = \frac{\Delta R_3}{R_3} = S_g\epsilon_a = \frac{S_gP}{AE} \qquad \frac{\Delta R_2}{R_2} = \frac{\Delta R_4}{R_4} = S_g\epsilon_t = -\frac{vS_gP}{AE}$$

Substituting into Eq. (8.57) gives

$$\frac{\Delta E}{V}\begin{cases} = \dfrac{S_gP}{2AE}(1+v) \\[2mm] \approx \dfrac{P}{AE}(1+v) \qquad \text{when } S_g \approx 2.00 \end{cases} \tag{8.58}$$

From Eq. (8.58), it is evident that the output signal $\Delta E/V$ is linearly related to the load P. The magnitude of $\Delta E/V$ will depend upon the design of the tension member, i.e., its cross-sectional area A and the material constants E and v. In most commercial load cells $\Delta E/V$ varies between 0.001 and 0.003. Steel with $E = 30 \times 10^6$ lb/in² (207 GPa) and $v = 0.30$ is usually used to fabricate the tension member. The range of the load cell P_R is then

$$P_R = A \frac{\Delta E}{V} \frac{E}{1 + v} \tag{8.59}$$

The upper limit on the output signal $\Delta E/V$ is determined by the strength of the tensile member and the fatigue limit of the strain gages. The maximum stress in the tension member is obtained from Eq. (8.59) as

$$\sigma = \frac{P}{A} = \frac{\Delta E}{V} \frac{E}{1 + v} \tag{8.60}$$

With steel tension members and $\Delta E/V = 0.003$, the stress $\sigma = 69,000$ lb/in² (476 MPa) is well within the fatigue limit of heat-treated alloy steel such as 4340. However, the axial strain $\epsilon_a = 2300$ μin/in (μm/m) is near the fatigue limit of most strain gages.

Placement of the strain gages on the four sides of the tension member, as shown in Fig. 8.35, provides a load cell which is essentially independent of either bending or torsional loads. Consider a bending moment M applied to the tension member either by a transverse load or by an eccentrically applied axial load. The moment M may have any direction relative to the axes of symmetry of the cross section as shown in Fig. 8.36. The components M_1 and M_2 of the moment M will produce resistance changes in the gages as follows:

$$\left.\frac{\Delta R_2}{R_2}\right|_{M_1} = -\left.\frac{\Delta R_4}{R_4}\right|_{M_1} \quad \text{and} \quad \left.\frac{\Delta R_1}{R_1}\right|_{M_1} = \left.\frac{\Delta R_3}{R_3}\right|_{M_1} = 0$$

$$\left.\frac{\Delta R_3}{R_3}\right|_{M_2} = -\left.\frac{\Delta R_1}{R_1}\right|_{M_2} \quad \text{and} \quad \left.\frac{\Delta R_2}{R_2}\right|_{M_2} = \left.\frac{\Delta R_4}{R_4}\right|_{M_2} = 0$$

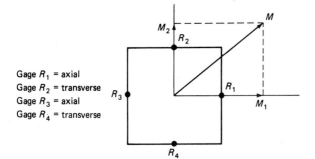

Gage R_1 = axial
Gage R_2 = transverse
Gage R_3 = axial
Gage R_4 = transverse

Figure 8.36 Resolution of the moment M into components M_1 and M_2.

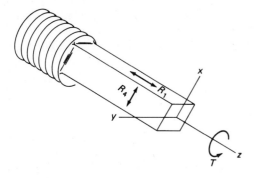

Figure 8.37 Torque T applied to the tension member of an axial-load cell.

Substitution of these relations into Eq. (8.57) shows that the effects of bending moments applied to the load cell are canceled in the Wheatstone bridge since $\Delta E/V$ vanishes for both M_1 and M_2.

Consider next the tension member subjected to a torque T, as shown in Fig. 8.37. The state of stress in the tension member for this form of loading has been shown to be

$$\tau_{max} = \frac{4.81 T}{a^3} \qquad \sigma_{xx} = \sigma_{yy} = \sigma_{zz} = 0$$

Thus

$$\epsilon_{xx} = \epsilon_{yy} = \epsilon_{zz} = 0 \qquad (a)$$

When Eqs. (a) are substituted into Eq. (6.5),

$$\frac{\Delta R_1}{R_1} = \frac{\Delta R_3}{R_3} = S_g \epsilon_{zz} = 0 \qquad \frac{\Delta R_2}{R_2} = \frac{\Delta R_4}{R_4} = S_g \epsilon_{xx} = 0 \qquad (b)$$

Substitution of Eqs. (b) into Eq. (8.57) indicates that the output of the tensile load cell is independent of the applied torque since $\Delta E/V$ is again equal to zero. Temperature compensation is also achieved with the four active strain gages in the bridge.

B. Diaphragm Pressure Transducers [11–13]

A second type of transducer which utilizes a strain gage as the sensing element is the diaphragm type of pressure transducer. Here a special-purpose strain gage is mounted on one side of the diaphragm while the other side is exposed to the pressure, as shown in Fig. 8.38. The diaphragm pressure transducer is small, easy to fabricate, and inexpensive and has a relatively high natural frequency.

The special-purpose diaphragm strain gage, shown in Fig. 8.39, has been designed to maximize the output voltage from the transducer. The strain distribution in the diaphragm is given by

$$\epsilon_{rr} = \frac{3p(1 - v^2)}{8Et^2}(R_o^2 - 3r^2) \qquad \epsilon_{\theta\theta} = \frac{3p(1 - v^2)}{8Et^2}(R_o^2 - r^2) \qquad (8.61)$$

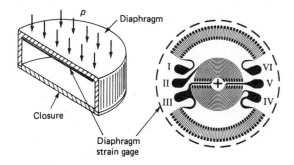

Figure 8.38 Diaphragm-type pressure transducer.

where p = pressure

t = thickness of diaphragm

R_o = outside radius of diaphragm

r = position parameter

Examination of this strain distribution indicates that the circumferential strain $\epsilon_{\theta\theta}$ is always positive and assumes its maximum value at $r = 0$. The radial strain ϵ_{rr} is positive in some regions but negative in others and assumes its maximum negative value at $r = R_o$. Both these distributions are shown in Fig. 8.39. The special-purpose diaphragm strain gage has been designed to take advantage of this distribution. Circumferential grids are employed in the central region of the diaphragm, where $\epsilon_{\theta\theta}$ is a maximum. Similarly, radial grids are employed near the edge of the diaphragm, where ϵ_{rr} is a maximum. It should also be noted that the circumferential and radial grids are each divided into two parts so that the special-purpose gage actually consists of four separate gages. Terminals are provided which permit the individual gages to be connected into a bridge with the circumferential elements in arms R_1 and R_3 and the radial elements in arms R_2 and R_4.

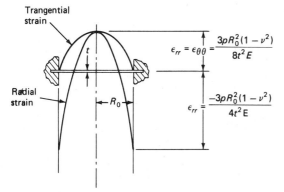

Trangential strain

t

Radial strain

R_0

$$\epsilon_{rr} = \epsilon_{\theta\theta} = \frac{3pR_0^2(1 - \nu^2)}{8t^2 E}$$

$$\epsilon_{rr} = \frac{-3pR_0^2(1 - \nu^2)}{4t^2 E}$$

Figure 8.39 Distribution of strain over a diaphragm.

If the strains are averaged over the areas of the circumferential and radial grids, and if the average values of $\Delta R/R$ obtained are substituted into Eq. (8.57), the signal output can be approximated by

$$\frac{\Delta E}{V} = 0.82 \frac{pR_o^2(1 - v^2)}{t^2 E} \tag{8.62}$$

Special-purpose diaphragm strain gages are commercially available in seven sizes ranging from 0.187 to 1.25 in (4.75 to 31.8 mm) in diameter.

Under the action of the pressure, the diaphragm deflects and changes from a flat circular plate to a segment of a large-radius shell. As a consequence, the strain in the diaphragm is nonlinear with respect to the applied pressure. Acceptable linearity can be maintained by limiting the deflection of the diaphragm. The center deflection w_c of the diaphragm can be expressed as

$$w_c = \frac{3pR_o^4(1 - v^2)}{16t^3 E} \tag{8.63}$$

If $w_c \leq t/4$, $\Delta E/V$ will be linear to within 0.3 percent over the pressure range of the transducer.

Frequently, diaphragm pressure transducers are employed to measure pressure transients. In these dynamic applications, the natural frequency of the diaphragm should be considerably higher than the highest frequency present in the pressure pulse. Depending upon the degree of damping incorporated in the design of the transducer, an undamped natural frequency 5 to 10 times greater than the highest applied frequency should be sufficient to avoid resonance effects. The natural frequency f_n of the diaphragm can be expressed as

$$\omega_n = 2\pi f_n = \frac{10.21t}{R_o^2} \sqrt{\frac{gE}{12(1 - v^2)\gamma}} \tag{8.64}$$

where γ is the density of the diaphragm material and g is the gravitational constant. If the thickness of the diaphragm cannot be determined accurately, the natural frequency can be determined experimentally by tapping the transducer on the center of the diaphragm and recording the oscillatory signal on an oscilloscope. The peak-to-peak period is the reciprocal of the natural frequency.

EXERCISES

8.1 Verify Eqs. (8.3a) and (8.3b) from Eq. (8.2).

8.2 Determine the magnitude of the nonlinear term as a function of strain for a potentiometer circuit ($r = 9$) with a single active gage ($S_g = 2.00$). Discuss the results obtained.

8.3 Select the following gages from a strain-gage catalog:
(a) A gage having the smallest gage length listed
(b) A gage having the largest gage length listed
(c) A gage having an area of approximately 0.06 in² (38 mm²), a resistance of 120 Ω, and a gage factor of 2.00

(*d*) A gage having an area of approximately 0.03 in² (19 mm²), a resistance of 1000 Ω, and a gage factor of 2.00

(*e*) A gage having an area of approximately 0.03 in² (19 mm²), a resistance of 350 Ω, and a gage factor of 3.5

If the allowable power density is 0.5 W/in² (0.78 mW/mm²) for each of these gages, determine S_c, V, and R_b for a potentiometer circuit with $r = 9$. Discuss the results.

8.4 For the best and poorest cases in Exercise 8.3, construct a plot of output voltage ΔE as a function of strain. For strain levels associated with the yield strength of low-carbon steels, determine the magnitude of ΔE. Is this a high- or low-level signal?

8.5 The sawtooth voltage pulse shown in Fig. E8.5 is fed into the filter as indicated. Resolve this voltage pulse into its Fourier components and consider the pulse distortion as the first five components pass through the filter. Discuss the results obtained.

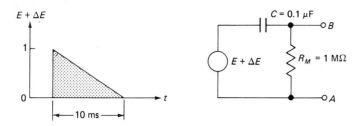

Figure E8.5

8.6 How could the simple filter shown in Fig. E8.5 be changed to improve the fidelity of the signal?

8.7 The triangular voltage pulse shown in Fig. E8.7 is fed into the filter as indicated. Resolve this voltage pulse into its Fourier components and consider the pulse distortion as the first five components pass through the filter. Discuss the results obtained and indicate how this filter design can be improved.

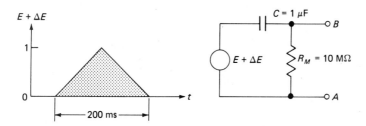

Figure E8.7

8.8 Beginning with Eqs. (8.12) and (8.13), verify Eqs. (8.14) and (8.16).

8.9 Verify Eq. (8.19) and the preceding Eqs. (*e*) and (*f*). Note that if the second-order terms are neglected, Eq. (8.19) is linear in terms of $\Delta R/R$. However, if the second-order terms are retained,

$$\Delta E = V \frac{R_1 R_2}{(R_1 + R_2)^2} \left(\frac{\Delta R_1}{R_1} - \frac{\Delta R_2}{R_2} + \frac{\Delta R_3}{R_3} - \frac{\Delta R_4}{R_4} \right)(1 - \eta)$$

where
$$\eta = \frac{1}{1 + \dfrac{r+1}{\Delta R_1/R_1 + \Delta R_4/R_4 + r(\Delta R_2/R_2 + \Delta R_3/R_3)}} \qquad \text{and} \qquad r = \frac{R_2}{R_1}$$

8.10 Three strain gages are placed in series in arm R_1 of a fixed-voltage Wheatstone bridge, as shown in Fig. E8.10. If the strain experienced by all gages is the same, determine the increase in circuit sensitivity over that obtained with a single gage.

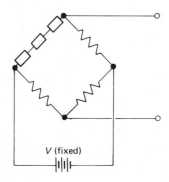

V (fixed)

Figure E8.10

8.11 If the voltage can be varied in the Wheatstone bridge of Fig. E8.10, does the insertion of the additional gages improve the sensitivity? If $r = 9$, determine the percent change in sensitivity for the three gages versus a single gage. Compute the required voltage V for the three gages and for a single gage if $R_g = 350\ \Omega$ and $P_g = 0.02$ W.

8.12 Compare the circuit sensitivities of the two bridges shown in Fig. E8.12. Compute the required voltage for bridge (b).

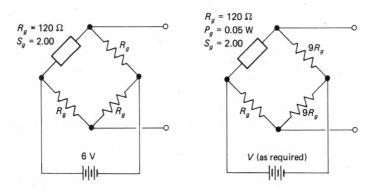

$R_g = 120\ \Omega$
$S_g = 2.00$

R_g

R_g R_g

6 V

$R_g = 120\ \Omega$
$P_g = 0.05$ W
$S_g = 2.00$

$9R_g$

R_g $9R_g$

V (as required)

Figure E8.12

8.13 Show that the bridge arrangements represented in cases 2 to 4 of Fig. 8.11 are all temperature-compensating. Follow the procedure outlined in Sec. 8.2B.

8.14 Determine the circuit sensitivity for the bridge arrangement shown in Fig. E8.14. Compare this sensitivity with that given by Eq. (8.25) for case 4 of Fig. 8.11. Discuss the results obtained.

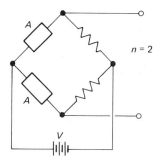

n = 2

Case 5   **Figure E8.14**

8.15 Use Eq. (8.26) to determine $\Delta R_1/R_1$ as a function of $\Delta R_5/R_5$ while varying R_5/R_2 as a parameter equal to 10, 100, 1000, and 10,000. Discuss the nonlinearities and place limits on $\Delta R_5/R_5$ which will hold these nonlinearities to 1, 2, or 5 percent.

8.16 Design a parallel-balance Wheatstone bridge with the capability of measuring strain in increments of 1 μm/m with a 120-Ω gage having $S_g = 2.00$. Determine the range of this instrument.

8.17 Design a parallel-balance Wheatstone bridge with the capability of measuring strains over a range of ± 10 percent with a 120-Ω gage having $S_g = 2.00$. Determine the sensitivity of the instrument.

8.18 For the constant-current potentiometer circuit shown in Fig. 8.19b, show that

$$\Delta E = R_1 I \frac{\Delta I}{I} + R_1 I \frac{\Delta R_1}{R_1}$$

where ΔI is the ripple current from an imperfect constant-current power supply.

8.19 For a semiconductor strain gage employed as R_1 in Fig. 8.19b, specify the ripple requirement $\Delta I/I$ to limit the ripple signal to an equivalent strain of 1 μm/m. Assume $S_g = 100$.

8.20 Rederive Eq. (8.39) after permitting I_1 and I_2 to change by ΔI_1 and ΔI_2 to obtain

$$\Delta E = I_1 R_1 \left(\frac{\Delta I_1}{I_1} + \frac{\Delta R_1}{R_1}\right) - I_2 R_2 \left(\frac{\Delta I_2}{I_2} + \frac{\Delta R_2}{R_2}\right)$$

Will the terms $I_1 R_1 \Delta I_1/I_1$ and $I_2 R_2 \Delta I_2/I_2$ tend to cancel? Why?

8.21 Determine ΔE if two semiconductor gages are employed in the double potentiometer circuit shown under case 3 of Fig. 8.21. The gage in R_1 is a P-type gage with $R_g = 500$ Ω and $S_g = 100$. The gage in R_2 is an N-type gage with $R_g = 500$ Ω and $S_g = -100$. Assume $I_g = 10$ mA and $\epsilon = 100$ μm/m for both gages.

8.22 Beginning with Eq. (8.49), verify Eq. (8.50).

8.23 Prepare a graph showing ϵ_c as a function of R_c for $R_2 = 120, 350, 500$, and 1000 Ω. Assume $S_g = 2.00$.

8.24 Could the calibration resistance be placed across arm R_1 of the bridge which contains the active gage? When would this procedure be recommended?

8.25 Determine the signal loss factor for a two-lead wire system if 18-gage wire is employed and the recording instrument and strain gage are separated by 600 ft (183 m). Determine the signal loss factor if a three-lead-wire system is employed.

8.26 For the two-lead-wire system of Exercise 8.25, determine the apparent strain introduced by an average temperature change of 18°F (10°C) over the length of the lead wires. Assume $S_g = 2.00$.

8.27 Specify a switch which would be adequate to use in the switching arrangement shown in Fig. 8.29.

8.28 Specify a switch which would be adequate to use in the switching arrangement shown in Fig. 8.30.

8.29 Briefly describe electrical noise and indicate why it occurs.

8.30 Outline three procedures which should be followed to minimize electrical noise in a strain-gage recording system.

8.31 Describe signal ground, system ground, and building ground. Describe a ground loop. How are ground loops avoided? How many grounds are usually employed in a recording system? How is a ground loop avoided if more than one ground is employed?

8.32 Design a torque cell to measure torques which range from 0 to 10 kN·m. Specify the size and shape of the mechanical member, the gage layout, the circuit design, and the voltage required. The torque cell must be insensitive to axial loads and bending moments.

8.33 Prove that the torque cell designed in Exercise 8.32 is insensitive to axial loads and bending moments. Refer to Sec. 8.8A for the procedure.

8.34 Design a load cell to measure a 50-MN axial tensile load. Specify the size and shape of the mechanical member, the gage layout, the circuit design, and the voltage required.

8.35 A very compact compression-load cell can be designed by using the mechanical member shown in Fig. E8.35. If the maximum load is 225 kN and the ratio $\Delta E/V = 0.002$, specify the size of the mechanical member, the gage layout, the circuit design, and the voltage required.

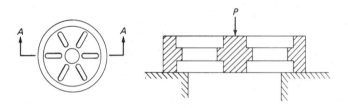

Figure E8.35

8.36 Design a displacement transducer having a cantilever beam as the mechanical member. The transducer must be capable of measuring the displacement to an accuracy of ± 0.002 mm over a range of 2.5 mm. Specify the size of the mechanical member, the gage layout, the circuit design, and the voltage required.

8.37 Design a strain extensiometer with a strain range of 10 percent, a gage length of 50 mm, and a ratio $\Delta E/V = 0.003$. Use the thin semicircular ring segment shown in Fig. E8.37 as the mechanical member.

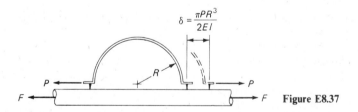

$$\delta = \frac{\pi PR^3}{2EI}$$

Figure E8.37

8.38 Design a diaphragm type of pressure transducer to measure pressures to 1500 kPa. If the diameter of the transducer is 15 mm and the diaphragm is fabricated from stainless steel, specify the thickness t for $\Delta E/V = 0.003$. Determine the displacement at the center of the diaphragm at the maximum pressure. Determine the natural frequency of the diaphragm.

8.39 Four electrical-resistance strain gages have been mounted on a torque wrench and wired into a fixed-voltage Wheatstone bridge, as shown in Fig. E8.39. The designer of the system claims that the output signal from the bridge is proportional to the torque applied by the wrench irrespective of the point of application of the load P so long as it remains to the right of the gages. Analyze the system and prove or disprove his claim.

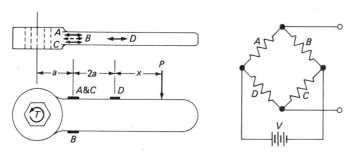

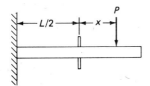

Figure E8.39

8.40 Design a load-measuring transducer having a cantilever beam as the mechanical member and four electrical-resistance strain gages as the sensing elements. Position the gages on the beam and wire them into a fixed-voltage Wheatstone bridge to achieve maximum sensitivity. The output from the system must be proportional to the load P, which can be positioned anywhere on the right half of the beam, as shown in Fig. E8.40.

Figure E8.40

REFERENCES

1. Geldmacher, R. C.: Ballast Circuit Design, *Proc. SESA*, vol. XII, no. 1, pp. 27–32, 1954.
2. Wheatstone, C.: An Account of Several New Instruments and Processes for Determining the Constants of a Voltaic Circuit, *Phil. Trans. R. Soc. (Lond.)*, vol. 133, pp. 303ff., 1843.
3. Stein, P. K.: Strain Gage Circuits for Semiconductor Gages, pp. 273–282, in M. Dean and R. D. Douglas (eds.), "Semiconductor and Conventional Strain Gages," Academic Press, Inc., New York, 1962.
4. Frank, E.: Strain Indicator for Semiconductor Gages, pp. 283–306, in M. Dean and R. D. Douglas (eds.), "Semiconductor and Conventional Strain Gages, Academic Press, Inc., New York, 1962.
5. Haakana, C. H.: Shunt Bridge Balancing in Strain Gage Indicators, *Electronics*, vol. 32, no. 30, pp. 50–51, 1959.
6. Murray, W. M., and P. K. Stein: "Strain Gage Techniques," 1960. Copyright by authors; lectures and laboratory exercises by authors at M.I.T. Cambridge, Mass.
7. Ruge, A. C.: Elimination of Lead Wire Errors in the SR-4 Temperature Compensated Strain Gage, *Test. Top.*, vol. 6, 1951.

8. Warshawsky, I.: A Multiple Bridge for Elimination of Contact-Resistance Errors in Resistance Strain Gage Measurements, *NACA Tech. Note* 1031, 1946.

9. Perry, C. C., and H. R. Lissner: "The Strain Gage Primer," 2d ed., pp. 200–217, McGraw-Hill Book Company, New York, 1962.

10. Elimination of Noise in Low-Level Circuits, report of Brush Instrument Division of Clevite Corporation, Cleveland, Ohio.

11. Design Considerations for Diaphragm Pressure Transducers, *Micro-Measurements Tech. Note* 129, 1968.

12. Timoshenko, S.: "Strength of Materials," pt II, "Advanced Theory and Problems," 3d ed., pp. 96–97, D. Van Nostrand Company, Inc., Princeton, N.J., 1956.

13. Timoshenko, S.: "Vibration Problems in Engineering, 3d ed., pp. 449–451, D. Van Nostrand Company, Inc., Princeton, N.J., 1955.

NINE

RECORDING INSTRUMENTS

9.1 INTRODUCTION [1–4]

In strain-gage applications, the signal generated by the strain gage through the use of a potentiometer, Wheatstone bridge, or constant-current circuit must be measured accurately to determine the strain. This chapter deals with instruments commonly employed by the experimental stress analyst to measure the small signal voltages produced by the gages. In selecting the type of voltage-measuring instrument, the first criterion to be met is adequate frequency response. If the strain signal is static, the time available for recording is relatively long and null-balance methods or potentiometer recorders can be employed to record the signal output. In recent years, high-quality, low-cost, integrating digital voltmeters have become available and have led to the development of static data-logging systems. With such systems, the output signals from several gages (systems accommodating 100 or more gages are common) are measured with the integrating digital volt-meter and automatically recorded in digital form on either punched or magnetic tape for processing with relatively simple computer programs.

If the strain-gage signal is dynamic, the time available to measure the output voltage is limited and recording is much more difficult and expensive. Also, automatic processing of dynamic data requires highly specialized equipment not usually available in the normal stress-analysis laboratory. In treating dynamic signals, the frequency of the strain signal is an important consideration in selecting the most appropriate recording system. For very low frequencies (0 to 3 Hz), quasi-static recording instruments such as the integrating digital voltmeter, the potentiometer recorder, and the xy-recorder can be used very effectively. For intermediate frequencies (0 to 10,000 Hz), the oscillograph with either a pen (0 to 100 Hz) or a light-writing (0 to 10,000 Hz) galvanometer is the most common recording method. For the high-frequency range (0 to 20,000 Hz), an FM instrument tape recorder can be used to monitor the signal. While this method of recording is expensive, it has an advantage in that the analog signal on the tape can be digitized and processed automatically. Finally, for very high frequencies (greater than 20,000 Hz), the cathode-ray oscilloscope is recommended. If the number of gages is small, this is a very effective recording system.

Each of these recording methods is described in detail in the following sections. In most instances, the description is qualitative with available equipment shown and capabilities, as they pertain to a given recording system, listed. However, coverage of the galvanometer and the oscillograph is detailed since the galvanometer is a low-input-impedance recording device which must be properly matched to the bridge to give the correct frequency response and the maximum sensitivity.

Telemetry is covered in the final section. While telemetry is not a recording method, it is closely aligned with recording since frequency modulation, multiplexing, and demodulation are techniques common to recording on magnetic tape and to automatic data processing. The description presented covers a relatively simple, single-bridge, short-range transmission of data. A brief discussion of multichannel, long-range transmission is also provided.

9.2 STATIC RECORDING AND DATA LOGGING

A. Manual Null-Balance Strain Indicators

The null-balance strain indicator previously described in Sec. 8.3D and shown in Fig. 8.18 is the most common recording method used for static strain signals. This type of strain indicator, which is commercially available from several firms at modest cost, is accurate and simple to use in quarter-, half-, or full-bridge arrangements. The indicator can be used for multiple-strain-gage installations by adding a switch or a switch and balance circuit. Since the gage factor can be dialed directly into the indicator, the balance resistor is calibrated to read out directly in terms of strain. Balancing adjustments require approximately 30 s per gage: therefore, strain readings can be obtained from an installation involving 100 gages in approximately 1 h. Normally, the output from each gage is recorded manually on a data sheet. The data are usually processed by hand or with a simple programmable calculator.

B. Manual Direct-Reading Strain Indicator [5]

A more sophisticated manual strain indicator, which employs an integrating digital voltmeter, is shown in Fig. 9.1. With this system, the Wheatstone bridge is initially balanced, and then the voltage output due to strain is amplified and read out on a digital voltmeter. The system incorporates a constant-current power supply which can be adjusted to match the power-dissipative characteristics of the strain gage. The Wheatstone bridge is calibrated with shunt resistors and the output from the amplifier is attenuated so that the digital voltmeter reads directly in terms of strain.

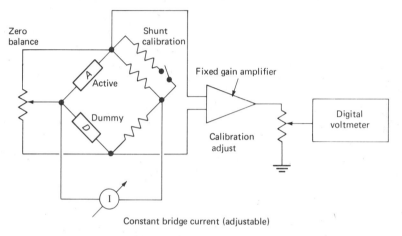

Figure 9.1 Direct-reading digital strain indicator. (*Vishay Instruments, Inc.*)

Multiple-strain-gage installations can be monitored with a direct-reading indicator; however, a separate switch and balance unit is needed for each gage since the bridge should be initially balanced at zero load. Since digital voltmeters can operate at rates exceeding 40 readings per second, their response time is never a factor in manual operations. Instead, the time required to scan a multiple-gage installation is controlled by the operator, who must manually switch the gages into the circuit and manually record the strain-gage readings. Normally 10 to 20 gages per minute can be monitored manually.

The direct-reading indicators can also be used to log data automatically since the output from the digital voltmeter can be printed on paper or recorded in digital form on either punched tape or magnetic tape. The primary advantage of utilizing the more sophisticated (and more costly) direct-reading indicators is that data are obtained in a form which can be automatically processed. Data recorded in digital form on punched tape or magnetic tape can be processed by a computer to yield stresses, graphs of stress distributions, etc., or to provide failure or yield predictions.

C. Automatic Data-Acquisition Systems [6–8]

Automatic data-acquisition systems should be considered by laboratories where relatively large strain-gage installations are employed on a regular basis. Automatic data-acquisition systems suitable for use with large numbers of strain gages are expensive, are much more complex to operate, and require considerably more maintenance than the simpler manual systems. However, if the volume of strain-gage readings is sufficiently large, the larger capital expenditure can be justified by the reduction in time required to test and analyze the data and by the immediate availability of test results.

Automatic data-acquisition systems usually involve so many optional components that each can essentially be considered custom designed. In each system, however, there are always four basic subsystems which include the controller, the signal conditioner-scanner, the analog-to-digital (A/D) converter (which is usually an integrating digital voltmeter), and the readout devices. An example of a 500-channel automatic data-acquisition system is illustrated in Fig. 9.2.

The signal conditioner-scanner consists of the power supply, the Wheatstone bridges, and the switches used to connect a large number of gages in turn to the single voltage-recording instrument. A typical schematic of the signal-conditioning circuit, shown in Fig. 9.3, indicates that each channel contains a separate Wheatstone bridge. Usually, the bridges are contained on a plug-in circuit board, which can be modified by adding or deleting fixed resistors to provide for quarter-, half-, or full-bridge arrangements. A single power source is often used to power a number of individual bridges (from 10 to 100, depending upon the design of the system). Usually, the power supply is a highly regulated constant-voltage supply which can be adjusted to provide from 5 to 15 V.

Each bridge has a small control panel with a balance adjustment, a span adjustment, and a calibration switch. Initial balance adjustment and shunt cali-

Model IC1613-13H Strain Gage Signal Conditioner/Scanner (40 ch/sec)

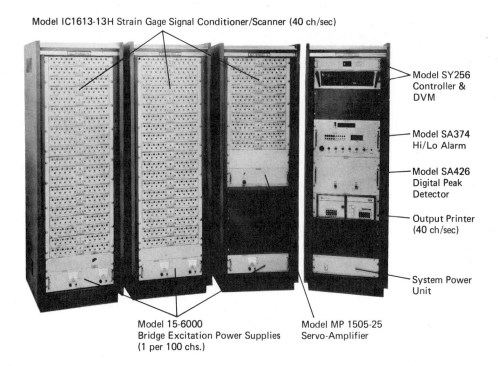

Model SY256
Controller &
DVM

Model SA374
Hi/Lo Alarm

Model SA426
Digital Peak
Detector

Output Printer
(40 ch/sec)

System Power
Unit

Model 15-6000
Bridge Excitation Power Supplies
(1 per 100 chs.)

Model MP 1505-25
Servo-Amplifier

Figure 9.2 A 500-channel digital data-acquisition system. (*B & F Instruments, Inc.*)

bration have already been discussed in Chap. 8. The same methods are used in these complex systems. The span adjustment is simply a potentiometer which reduces the voltage across the bridge to adjust for the specific gage factor associated with the strain gage employed in the bridge.

The scanner portion of the signal conditioner-scanner subassembly consists of two parts: (1) A bank of switches (usually three-pole) serves to switch the two output leads and the cable shield from the bridge to the integrating digital voltmeter. In modern systems, solid-state switching devices (*J* field-effect transistors) are employed. (2) The scanner contains the logic circuits which control the switching sequence. The modes of system operation which are possible include single point, single scan, continuous scan, and periodic scan. In the single-point mode, a preselected channel is monitored at the reading rate of the system (2 to 40 times per second). In the single-scan mode, the scanner makes a single sequential sweep through a preselected group of channels and measures and displays the voltage on a given channel before advancing to the next channel. The channels included in the sequential sweep are selected by dialing in the first and last channel numbers on the scanner control unit, as illustrated in Fig. 9.4. The continuous-scan mode is identical to the single-scan mode except that the system automatically resets and recycles on completion of the previous scan. The periodic scan is simply a single

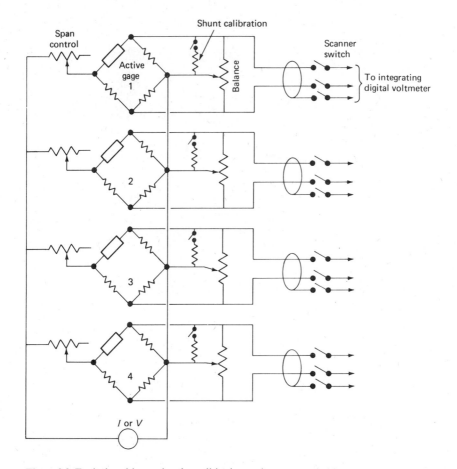

Figure 9.3 Typical multigage signal conditioning and scanner switching.

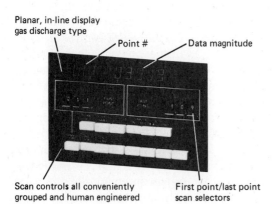

Figure 9.4 Scanner control panel on a Doric data-acquisition system. (*Doric Scientific Corporation.*)

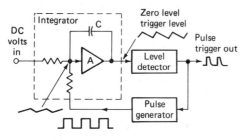

(a) Voltage-frequency conversion

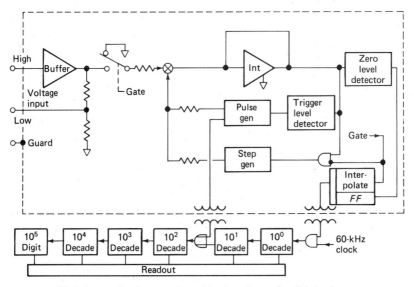

(b) Pulse counting for measuring voltage in an integrating digital voltmeter

Figure 9.5 Schematic diagrams for integrating digital voltmeters.

scan which is initiated at preselected time intervals such as 1, 5, 15, 30, and 60 min. The scanner also provides a visual display of the channel number and a binary-code signal to the controller to identify the strain gage being monitored.

The output from the Wheatstone bridge is switched into an A/D converter, which is usually a high-quality integrating digital voltmeter. An integrating digital voltmeter measures the average of the input voltage over a fixed measuring period. A widely used procedure for integrating utilizes the voltage-to-frequency converter shown schematically in Fig. 9.5. The integrator in the circuit converts the dc input into a ramp which activates a level detector. The level detector triggers a pulse generator in the feedback loop which provides a rectangular pulse train to the input of the integrator. The rate of pulse generation required to make the average voltage of the rectangular pulse train equal to the dc input voltage over the integrating period gives an accurate measurement of the input voltage.

The major advantage of this type of A/D converter is its ability to measure accurately with large superimposed noise. This occurs because the effect of the noise is integrated out over the sampling period. An accuracy of 0.01 percent can easily be achieved; however, accuracy, speed of measurement, and range are related. With modern systems, a resolution of 1 μV within a range of ± 30 V can be achieved. The speed of measurement depends on the length of the integration period and the time required to scan and store the reading. Since the length of the integration period is often 100 ms, reading rates of approximately 10 per second are common. Higher reading rates (20 to 40 per second) are possible with high-speed A/D converters. In these units, the integration period is reduced to 16.67 ms. The output from an integrating digital voltmeter is in binary code, and 1, 2, 4, 8 logic is common. This output is usually stored momentarily in the controller.

The controller is essentially the brains of the automatic data-acquisition system and usually contains a microprocessor with several memory devices (RAMs, ROMs, and PROMs). The controller has two main functions: (1) it activates the scanner and controls the time sequence of the switching from one channel to the next, and (2) it stores the output from the integrating digital voltmeter, the channel number, and the time when the reading is made in its operating memory. The data can be either conditioned or reduced, depending upon the computing capabilities of the microprocessor in the controller. The final form of the data is then transmitted to the readout devices, which may include planar displays, strip printers, magnetic-tape recorders, paper-tape punches, or minicomputers.

The planar displays are usually seven-segment light-emitting devices, as illustrated in Fig. 9.4. The planar display usually shows the output from the microprocessor (voltage, strain, or stress) and the identification of the channel. The planar displays can be monitored during the test and provide the operator with direct visual readout.

If the data are to be processed after the test is completed, the readout can be recorded with a digital printer, a paper-tape punch, or an incremental magnetic-tape recorder. The digital printers operate at rates between 2 and 40 readings per second and give the channel number and sign and magnitude of the quantity being measured. Usually the data are processed manually when a printer is used.

Data recorded on a punched tape or on an incremental magnetic tape are usually processed with a computer at some convenient time after the test is completed. Magnetic-tape units are compatible with IBM-type digital computers. A typical incremental magnetic-tape recording subsystem (the Kennedy 1600/5) has 8.5-in (216-mm) reels capable of handling 1200 ft (336 m) of $\frac{1}{2}$-in (12.7-mm) 1.5-mil tape. The tape format is seven-track IBM-compatible NRZI recording. The packing density is 556 bits per inch (21.9 per millimeter). The tape recorder operates in synchronization with the controller and writes with a sufficiently high rate (300 characters per second) to permit scanning at 20 channels per second.

In some instances, tests are conducted with the computer operating on line in real time. For a large computer facility, the connection to the data-acquisition system is often made with a teletype and data phone links. Recently, small but capable minicomputers have become available which can be directly coupled to the data-acquisition system. With further developments in digital circuits, the storage and computing capabilities will obviously become an integral part of the data-acquisition system.

9.3 DYNAMIC RECORDING AT VERY LOW FREQUENCIES [8]

The methods used to record dynamic signals from strain gage circuits depend upon the frequencies of the components which make up the transient strain pulse. This coverage of dynamic recording methods has been divided into four subsections which deal with frequencies classified as very low (0 to 2 Hz), intermediate (0 to 5000 Hz), high (0 to 20,000 Hz), and very high (0 to 50,000 Hz). The division into four frequency ranges was made to correspond with typical frequency limits on four different types of recording instruments.

For very low frequencies, instruments such as potentiometer (or strip-chart) recorders and xy recorders, which employ servomotors together with feedback control and null-balance positioning, can be used to measure the output voltage from the strain-gage bridge. The operating principle used in this type of instrument is illustrated in Fig. 9.6.

Potentiometer recorders are multirange instruments which can be used to measure voltages from 1 μV to 100 V. The sensitivity is achieved by the stable and high-gain amplification provided by the servo drive system. Since no current flows in the potentiometer circuit in the balanced condition, the voltage measurements are independent of the length of the lead wires. The response of the potentiometer recorder is inherently slow because of the inertia effects of the balancing motor. Most commercial instruments require between $\frac{1}{4}$ and 5 s for the pointer to traverse

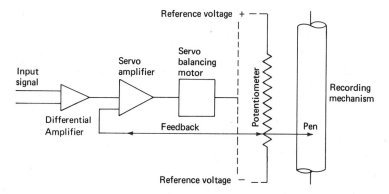

Figure 9.6 Schematic diagram of a servomotor-driven null-balance (potentiometer) circuit for voltage measurement.

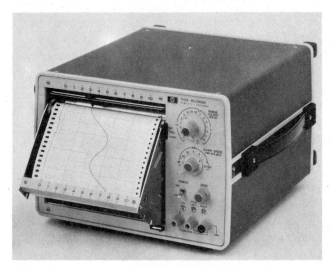

Figure 9.7 A potentiometer recorder. (*Hewlett-Packard.*)

the full scale of the indicator. Because of this low-frequency response, the potentiometer cannot be employed in strain-gage applications where the strain signal has frequency components greater than about 1 Hz. However, since the stability of the amplifier and servo system is excellent, accurate measurements of bridge voltages can be made over extended periods of time. The chart speeds can be varied over a wide range [1 in/h to 2 in/s (25 mm/h to 50 mm/s)] to provide an extremely versatile recording system for monitoring strain versus time for slowly changing loads on structures. The cost of approximately $1000 per channel is relatively low in comparison with other dynamic recording instruments. A typical strip-chart recorder is shown in Fig. 9.7.

Another instrument which utilizes servo-driven pen motors is the xy recorder illustrated in Fig. 9.8. The xy recorder is very similar to the strip-chart recorder except that it simultaneously records one voltage along the x axis and another along the y axis. The two potentiometers (slide wires) are mounted along the traveling paths of the arm and pen carriage. The xy recorder is also a multirange device which can record voltages from about 5 μV to 100 V, depending upon the range selected. The time response of an xy recorder is determined by the slewing speed of the pen and the distance through which the pen must travel to record the voltage inputs. Slewing speeds of 20 in/s (500 mm/s) are common, and high-speed units are available with slewing speeds of 30 in/s (750 mm/s). Prices for high-sensitivity, high-speed xy recorders are between $2000 and $2500, depending upon the optional features specified. The xy recorder is quite useful in monitoring load versus strain and is often employed in material-testing applications to record stress-strain curves. In this application, the output from a load cell is used to drive the y (stress) axis, and the output from an extensometer is used to drive the x (strain) axis.

Figure 9.8 An *xy* recorder. (*Hewlett-Packard.*)

9.4 DYNAMIC RECORDING AT INTERMEDIATE FREQUENCIES

A. The Oscillograph and The Galvanometer

Oscillographs employing galvanometers as the recording device are the most widely used method for recording dynamic strain signals. There are two different types of oscillographs, the pen-writing type, where the galvanometer drives a pen or hot stylus, and the light-writing type, where the galvanometer drives a mirror and a beam of light is used to write on a photosensitive paper. The pen-writing galvanometer can be used for frequencies up to about 60 Hz, and the hot-stylus type can record frequencies up to about 100 Hz. The inertia of the pen arm limits the frequency response of this type of instrument. For these lower frequencies, the pen-type oscillograph is preferred over the light-writing oscillograph since the recordings on chart paper by either a pen or a hot stylus are of higher quality, more permanent, and less expensive than comparable recordings made on photosensitive paper with a light-writing oscillograph.

When the dynamic strain signal contains frequency components from dc to 10 kHz, the light-writing oscillograph is generally used as the recording method. A schematic diagram illustrating the operating principle of a light-writing oscillograph is presented in Fig. 9.9. A late-model oscillograph currently used in many laboratories is shown in Fig. 9.10. The oscillograph houses a series of moving-coil light-beam galvanometers mounted in magnetic blocks. The galvanometer drives

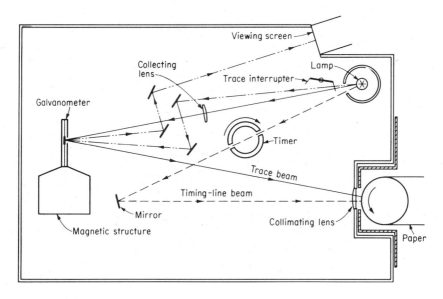

Figure 9.9 Schematic illustration of the operating principle of a light-writing oscillograph.

a mirror, which reflects light through a lens system onto a light-sensitive paper. The deflection of the galvanometer is amplified optically and provides a y displacement on the record proportional to the strain magnitude. The paper speed is controlled by the motor and gear train driving the paper-feed mechanism and can be adjusted to give various time scales on the abscissa of the record. The light-writing oscillograph may be employed for both static and dynamic strain-gage applications, although it is most commonly utilized to record a dynamic strain signal. The frequency response of this type of oscillograph will vary, depending upon the galvanometer employed; however, frequencies from 0 to approximately 10 kHz can be recorded if the proper galvanometer and paper speeds are employed.

Since the galvanometer represents the heart of the oscillograph and can be changed to vary the instrument's sensitivity and frequency response, it is important that the experimentalist thoroughly understand its principle of operation. The essential elements of an electromagnetic galvanometer, which consists of a filament suspension system, a rotating coil, a mirror, and a pair of pole pieces, are shown in Fig. 9.11.

A current I, which passes through the filament suspension and the coil, produces a pair of equal and opposite forces F, which act to rotate the coil through an angle θ, as illustrated in Fig. 9.11. The magnitude of these current-induced forces is given by

$$F = nBIl \qquad\qquad (a)$$

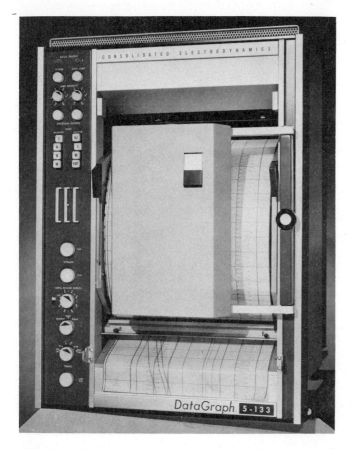

Figure 9.10 Model 5-133 oscillograph. (*Consolidated Electrodynamics Corporation.*)

where n = number of turns on coil

   B = strength of magnetic field

   l = length of coil

These coil forces produce a torque T_1, which is dependent on the angle of rotation θ and the coil width w as follows:

$$T_1 = Fw \cos \theta \qquad (b)$$

By Eqs. (*a*) and (*b*) the coil torque T_1 can be expressed in terms of the electrical and mechanical parameters of the galvanometer as

$$T_1 = nBIA \cos \theta \qquad (9.1)$$

where A is the area of the coil, equal to wl.

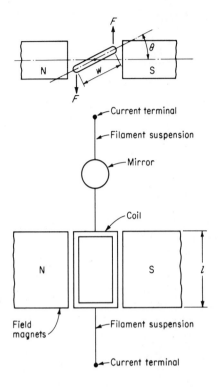

Figure 9.11 Schematic illustration showing the essential elements of an electromagnetic galvanometer.

The rotation of the coil is resisted by the torsional constraint of the filament suspension system. The opposing torque T_2 can be expressed as

$$T_2 = G\theta \tag{9.2}$$

Under static conditions, that is, I independent of time, the two torques T_1 and T_2 must be equal to ensure rotational equilibrium. Hence

$$nBIA \cos \theta = G\theta \tag{c}$$

Solving Eq. (c) for the rotational deflection θ of the galvanometer gives

$$\frac{\theta}{\cos \theta} = \frac{nBA}{G} I = CI \tag{9.3}$$

where C is the galvanometer constant, equal to nBA/G.

The rotational deflection θ as expressed by Eq. (9.3) is a nonlinear function of the galvanometer current I due to the cos θ term. This nonlinearity is illustrated in Fig. 9.12, where θ is shown plotted as a function of CI. The errors introduced by the nonlinearities of Eq. (9.3) can be avoided if the angular deflection θ is kept sufficiently small to ensure that

$$\frac{\theta}{\cos \theta} \approx \theta \tag{d}$$

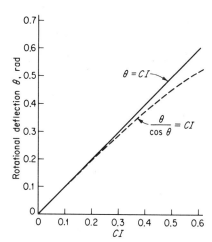

Figure 9.12 Nonlinear character of θ with respect to CI.

In order to increase the sensitivity of the galvanometer to measure very small currents accurately, it is necessary to increase the galvanometer constant C to as large a value as possible. This increase in sensitivity can be accomplished by increasing the number of turns in the coil and the coil area; however, this approach increases the inertia of the coil and reduces the frequency response of the galvanometer in dynamic applications. The galvanometer constant C can also be increased by increasing the field strength B. Unfortunately, the materials from which the permanent magnets are constructed limit the value of B which can be achieved. Finally, the sensitivity of the galvanometer can be enhanced by reducing the torsional spring constant G of the filament suspension system. This is accomplished by reducing the diameter of the wire in the suspension system, but if carried too far will result in a delicate and fragile galvanometer.

In spite of these difficulties, sensitive and reasonably rugged galvanometers are commercially available for use in oscillographs. A sectioned view showing construction details of a commercially available unit is presented in Fig. 9.13. In dynamic strain-gage applications, sensitivity of the galvanometer is usually sacrificed to some degree to provide adequate frequency response. The dynamic response of a galvanometer will be covered in more detail in the next two subsections.

B. Transient Response of Galvanometers

Since galvanometers are largely employed in dynamic strain-gage applications, it is appropriate to consider their response to a step pulse of current, which can be applied by closing the switch in the circuit shown in Fig. 9.14.

The equation of motion which describes the angular deflection of the coil as a function of time after closing the switch is given by

$$T_1 - T_2 - T_3 = J\frac{d^2\theta}{dt^2} \tag{a}$$

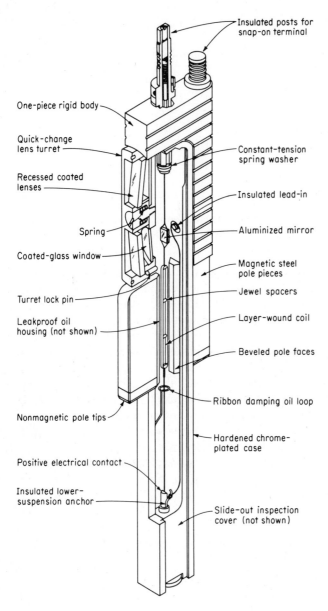

Insulated posts for
snap-on terminal

One-piece rigid body

Quick-change
lens turret

Constant-tension
spring washer

Recessed coated
lenses

Insulated lead-in

Spring

Aluminized mirror

Coated-glass window

Magnetic steel
pole pieces

Turret lock pin

Jewel spacers

Leakproof oil
housing (not shown)

Layer-wound coil

Beveled pole faces

Nonmagnetic pole tips

Ribbon damping oil loop

Hardened chrome-
plated case

Positive electrical contact

Insulated lower-
suspension anchor

Slide-out inspection
cover (not shown)

Figure 9.13 Construction details of a galvanometer. (*Consolidated Electrodynamics Corporation.*)

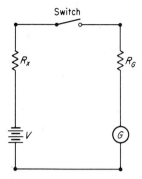

Switch

R_x

R_G

V

G

R_G = Galvanometer resistance
R_x = External resistance

Figure 9.14 Circuit for applying a step pulse of current to a galvanometer.

where T_1 = applied torque [Eq. (9.1)]
$\quad T_2$ = restraining torque due to the suspension system [Eq. (9.2)]
$\quad T_3$ = restraining torque due to fluid damping = $D_0 \, d\theta/dt$
$\quad D_0$ = fluid-damping constant

Substituting the three torques T_1, T_2, and T_3 in Eq. (a) gives

$$J\frac{d^2\theta}{dt^2} + D_0\frac{d\theta}{dt} + G\theta = nBIA \cos\theta \qquad (b)$$

By considering only small values of θ, where $\cos\theta \approx 1$, and using Eq. (9.3) it is possible to rewrite Eq. (b) as

$$J\frac{d^2\theta}{dt^2} + D_0\frac{d\theta}{dt} + G\theta = GCI \qquad (9.4)$$

where I is the instantaneous current in the galvanometer coil.

The instantaneous current in the coil can be computed from the circuit shown in Fig. 9.14 by employing Kirchoff's law

$$I(R_x + R_G) = V - E_m \qquad (c)$$

where E_m, the back electromotive force, is equal to

$$E_m = nBA\frac{d\theta}{dt} = GC\frac{d\theta}{dt}$$

By solving Eq. (c) for I and noting the equivalence of E_m to $GC(d\theta/dt)$, it is clear that

$$I = \frac{V}{R_x + R_G} - \frac{GC}{R_x + R_G}\frac{d\theta}{dt} = I_s - \frac{GC}{R_x + R_G}\frac{d\theta}{dt} \qquad (d)$$

where I_s is the steady-state current through the galvanometer. Substituting Eq. (d) into Eq. (9.4) and simplifying gives

$$J\frac{d^2\theta}{dt^2} + D\frac{d\theta}{dt} + G\theta = GCI_s \qquad (9.5)$$

where the damping coefficient D is a combination of fluid and electromechanical damping given by

$$D = D_0 + \frac{G^2C^2}{R_x + R_G} \qquad (9.6)$$

As will be shown later, the ability to vary the damping coefficient D by controlling the external resistance R_x is quite important in adjusting the dynamic response of a given galvanometer.

The complementary solution of the differential equation given in Eq. (9.5) depends upon the constants J, D, and G. Three cases which arise must be treated individually, as summarized below:

Case 1, overdamped: $\qquad\qquad\qquad\qquad\qquad\left(\dfrac{D}{2J}\right)^2 > \dfrac{G}{J}$

Case 2, critically damped: $\qquad\qquad\qquad\quad\left(\dfrac{D}{2J}\right)^2 = \dfrac{G}{J}$

Case 3, underdamped: $\qquad\qquad\qquad\qquad\left(\dfrac{D}{2J}\right)^2 < \dfrac{G}{J}$

The solutions of Eq. (9.5) for these three cases are:

Case 1, overdamped: $\quad \dfrac{\theta}{\theta_s} = 1 - e^{-\alpha t}\left(\dfrac{\alpha}{\beta}\sinh \beta t + \cosh \beta t\right) \qquad (9.7)$

where $\alpha = D/2J$
$\qquad \beta^2 = (D/2J)^2 - G/J$
$\qquad \theta_s$ = steady-state deflection of galvanometer after sufficient period of time

In this, the overdamped case, the response of the galvanometer is sluggish, as indicated in Fig. 9.15, where the time required for the galvanometer to assume its steady-state position is quite long. In most applications the overdamped galvanometer is to be avoided since the time required for a stable galvanometer reading is unduly long.

Case 2, critically damped: $\quad \dfrac{\theta}{\theta_s} = 1 - e^{-\alpha t}(1 + \alpha t) \qquad (9.8)$

When the galvanometer is critically damped, $\alpha^2 = G/J$ and the response of the galvanometer is improved in comparison with the overdamped case. The galvanometer deflection θ approaches the steady-state deflection θ_s but never exceeds this value. This characteristic response of the galvanometer is also shown in Fig. 9.15.

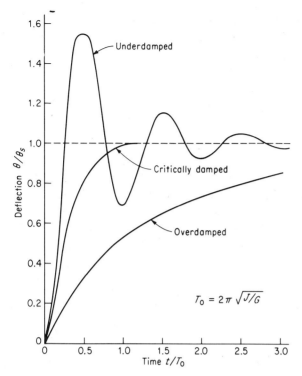

Figure 9.15 Response of overdamped, critically damped, and underdamped galvanometers to an applied step pulse.

Case 3, underdamped: $\dfrac{\theta}{\theta_s} = 1 - e^{-\alpha t}\left(\dfrac{\alpha}{\beta}\sin \beta t + \cos \beta t\right)$ (9.9)

Underdamped galvanometers respond vigorously to a transient signal and initially overshoot the steady-state deflection then oscillate about θ_s for some time. Figure 9.15 also shows a plot of θ/θ_s for the underdamped case, which illustrates the character of the oscillation of the galvanometer about θ_s.

The degree of damping d

$$d = \dfrac{\alpha}{\sqrt{G/J}}$$ (9.10)

which is desirable to reduce the time of response of the galvanometer, depends upon the accuracy required in the measurement. This fact is illustrated in Fig. 9.16, where accuracy limits of ± 10 percent are placed on the value of θ/θ_s. The dimensionless time required for a critically damped galvanometer ($d = 1$) to record θ totally within this accuracy band is defined at point A. For an underdamped galvanometer with $d = 0.2$ and $d = 0.5$, the time required for the galvanometer to reach and remain in the ± 10 percent limit bands is given by points B and C, respectively. The relative positions of points A, B, C will, of course, depend upon the accuracy requirements which establish the bandwidth. The locations of these points are important because they establish the response time of the galvanometer.

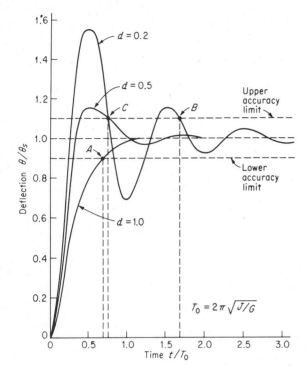

Figure 9.16 Accuracy limits superimposed on the response curves of a galvanometer.

In most applications galvanometers are employed in a slightly underdamped condition, where the first overshoot and subsequent oscillations about θ_s do not exceed the bandwidth imposed by the accuracy limits. The degree of damping is normally selected so that the overshoot makes a tangent at its peak with the upper accuracy limit. The response time is therefore minimized by the selection of this particular degree of damping. In Fig. 9.17 the minimum response time is shown as a function of the percent accuracy required. Superimposed on this curve at selected points is the degree of damping required to minimize the response time. Since galvanometers are electromechanically damped, the damping constant D and thus the degree of damping can be varied by the proper selection of R_x, shown in Fig. 9.14 and Eq. (9.6). The value of R_x can often be varied to provide the minimum response time for the galvanometer.

C. Response of a Galvanometer to a Sinusoidal Signal

When a galvanometer is used to measure a sinusoidal signal which is periodic and of constant magnitude, the initial transient response of the galvanometer is of little importance. Instead, it is more important to determine how well the galvanometer deflection θ is following the input signal. The response of the galvanometer to this

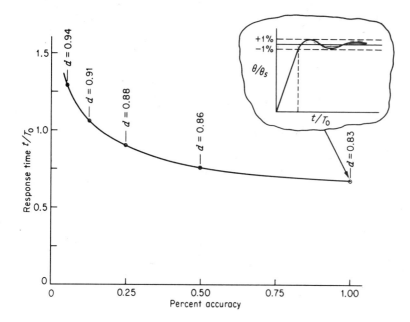

Figure 9.17 Response time as a function of percent accuracy for the optimum degree of damping.

type of signal can be determined by considering its equation of motion,

$$J\frac{d^2\theta}{dt^2} + D\frac{d\theta}{dt} + G\theta = GCI_s \sin \omega t \qquad (9.11)$$

where the instantaneous current in the galvanometer coil is given by

$$I = \frac{V}{R_x + R_G}\sin \omega t - \frac{GC}{R_x + R_G}\frac{d\theta}{dt} = I_s \sin \omega t - \frac{GC}{R_x + R_G}\frac{d\theta}{dt}$$

This differential equation is identical to that given by Eq. (9.5) except for the term on the right-hand side. As such, the complementary solution is also the same; however, in this measurement the solution sought is the steady-state response of the galvanometer, which is obtained from the particular solution of Eq. (9.11). Solving Eq. (9.11) for the particular solution leads to

$$\theta_s = \frac{GCI_s}{(\omega D)^2 + (G - \omega^2 J)^2}[(G - \omega^2 J) \sin \omega t - \omega D \cos \omega t] \qquad (9.12)$$

To reduce Eq. (9.12) to a more concise form, let

$$\sin (\omega t - \phi) = \frac{(G - \omega^2 J) \sin \omega t - \omega D \cos \omega t}{\sqrt{(\omega D)^2 + (G - \omega^2 J)^2}} \qquad (9.13)$$

which implies that

$$\phi = \arctan \frac{\omega D}{G - \omega^2 J} \tag{a}$$

Then, by Eqs. (9.12) and (9.13), the steady-state oscillations of the galvanometer can be written as

$$\theta_s = \frac{GCI_s}{\sqrt{(\omega D)^2 + (G - \omega^2 J)^2}} \sin (\omega t - \phi) \tag{9.14}$$

Consider now the free, undamped vibration of the galvanometer, which can be defined by its free angular frequency ω_0 and free period T_0, so that

$$\omega_0 = \sqrt{\frac{G}{J}} \qquad T_0 = \frac{2\pi}{\omega_0} \tag{9.15}$$

Consider also its degree of damping

$$d = \frac{D}{2\sqrt{JG}} \tag{9.10a}$$

Substitution of Eqs. (9.15) and (9.10) into Eqs. (a) and Eq. (9.14) gives

$$\phi = \arctan \frac{1}{(1/2d)(\omega_0/\omega - \omega/\omega_0)} \tag{9.16}$$

and

$$\frac{\theta_s}{CI_s} = \frac{1/2d}{(\omega/\omega_0)\sqrt{1 + (1/2d)^2(\omega_0/\omega - \omega/\omega_0)^2}} \sin (\omega t - \phi) \tag{9.17}$$

The angle ϕ represented in Eq. (9.16) is a phase angle, which indicates that the galvanometer deflection is following the signal by some phase angle ϕ. Provided this angle ϕ remains constant, it is not of serious concern since it represents a shift on the time scale of the oscillograph record. The phase angle ϕ is dependent upon the galvanometer characteristics (as defined by ω_0) and the degree of damping d. The results of Eq. (9.16) are shown in graphical form in Fig. 9.18, where ϕ is plotted as a function of the frequency ratio ω/ω_0 for different values of d.

In recording any complex waveform the pulse will not be distorted if the phase angle increases linearly with frequency. This lack of distortion is due to the fact that with a linear phase angle the second harmonic will be shifted twice as many degrees as the fundamental harmonic, etc. The implication here is that all the harmonics will experience the same time lag in seconds as the fundamental, and no distortion of the strain-time record will occur.

The frequency response of the galvanometer is described in Eq. (9.17), where the amplitude ratio θ_s/CI_s is shown to be a function of the frequency ratio ω/ω_0 and the degree of damping. The results of this equation are shown in Fig. 9.19, where the amplitude ratio θ_s/CI_s is given as a function of ω/ω_0.

The curves shown in Fig. 9.19 represent the theoretical deflection of a galvanometer to a sinusoidal current which is constant in magnitude (I_s) but of variable

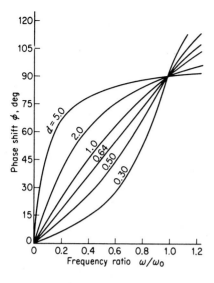

Figure 9.18 Phase shift as a function of frequency ratio.

frequency (ω). The frequency response of the galvanometer is dependent upon the accuracy limit and the degree of damping associated with the circuit. In Fig. 9.19 accuracy limits of ± 5 percent have been imposed upon the response curves over the entire frequency range considered. For this accuracy requirement it is clear that the degree of damping $d = 0.59$ maximizes the frequency response of the galvanometer; hence the response of the galvanometer will be flat, i.e., will be within the ± 5 percent band, for frequencies ω from 0 to 85 percent of ω_0. For other degrees of damping the frequency response is reduced appreciably. For instance, with $d = 1.0$ the response of the galvanometer will lie within the ± 5 percent accuracy limits for frequencies from 0 to 22 percent of ω_0. As the accuracy limits are decreased, the range of the frequency response of the galvanometer also decreases. Accuracy requirements of ± 2 percent achieved with $d = 0.64$ can be realized over a frequency range from 0 to 67 percent of ω_0. This frequency response of $0.67\omega_0$ compares with $0.85\omega_0$ for ± 5 percent accuracy requirements.

D. The Wheatstone Bridge and the Galvanometer

When a galvanometer is used to measure the output signal generated in a Wheatstone bridge, due consideration must be given to the fact that the galvanometer is a low-impedance measuring instrument. The coverage of the Wheatstone bridge presented in Chap. 8 was based upon the use of a high-impedance instrument to measure the bridge voltage; hence the bridge analysis presented there must be modified to account for the characteristics of the galvanometer. Also, it has been shown that the frequency response of the galvanometer is quite dependent upon the degree of damping d. This degree of damping is in turn influenced by the external resistance R_x in the galvanometer circuit. Since the Wheatstone bridge forms a portion of the external circuit, the value of the resistors placed in the

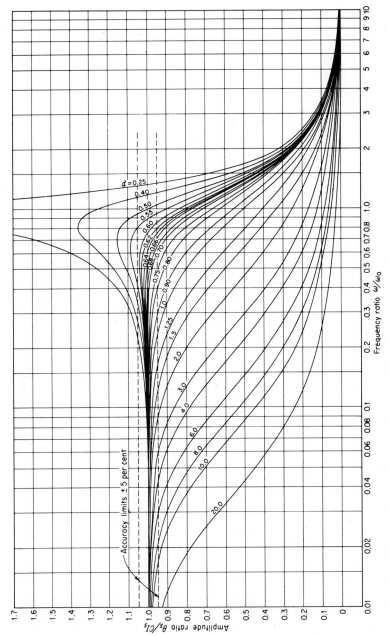

Figure 9.19 Amplitude ratio as a function of frequency ratio.

296

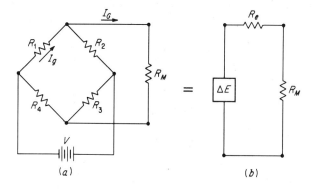

Figure 9.20 (*a*) Wheatstone bridge with a low-impedance measuring system; (*b*) equivalent circuit.

bridge cannot be selected on an arbitrary basis. Finally, the deflection θ of the galvanometer is dependent upon the input current I, which can be maximized by proper design of the bridge and proper selection of the strain gage.

To adequately specify a strain-measuring system with maximum sensitivity and maximum frequency response, the galvanometer and Wheatstone bridge must be considered together since the characteristics of the galvanometer will influence the behavior of the bridge and vice versa. A typical electric circuit showing the Wheatstone bridge with an attached resistance R_M due to the measuring system is illustrated in Fig. 9.20. By employing Thévenin's theorem this circuit can be reduced to a simpler equivalent circuit containing only a voltage source ΔE, an equivalent resistance R_e, and the resistance of the measuring system R_M. This equivalent circuit is also illustrated in Fig. 9.20, and the use of Thévenin's theorem to obtain the equivalent resistance R_e of the Wheatstone bridge is shown in Fig. 9.21. For this latter figure it is clear that

$$R_e = \frac{R_1 R_2}{R_1 + R_2} + \frac{R_3 R_4}{R_3 + R_4} \tag{a}$$

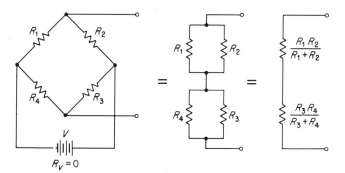

Figure 9.21 Reduction of the Wheatstone bridge to an equivalent resistance R_e.

provided the resistance of the voltage supply $R_v = 0$.

If one defines

$$r = \frac{R_2}{R_1} \quad \text{and} \quad q = \frac{R_2}{R_3} \tag{b}$$

and considers an initially balanced bridge where $R_1 R_3 = R_2 R_4$, then Eq. (a) can be written as

$$R_e = R_1 \frac{r}{1+r} \frac{1+q}{q} \tag{9.18}$$

In a four-active-arm bridge, the voltage output ΔE is given by Eq. (8.20) as

$$\Delta E = \frac{r}{(1+r)^2} \left(\frac{\Delta R_1}{R_1} - \frac{\Delta R_2}{R_2} + \frac{\Delta R_3}{R_3} - \frac{\Delta R_4}{R_4} \right) V$$

which, by Eq. (6.5), can be reduced to

$$\Delta E = \frac{r}{(1+r)^2} n S_g \epsilon V \tag{c}$$

where n is the number of active arms of the bridge which contribute positively to the voltage output. The voltage V which can be applied to the bridge depends upon the power which can be dissipated by the gage. Thus

$$V = (1+r)\sqrt{P_g R_g} \tag{d}$$

Combining Eqs. (c) and (d) yields

$$\Delta E = \frac{r}{1+r} n S_g \sqrt{P_g R_g} \epsilon \tag{e}$$

The current I_G which passes through the galvanometer coil can now be computed from the equivalent circuit shown in Fig. 9.20 as

$$I_G = \frac{\Delta E}{R_e + R_M} = \frac{\Delta E}{R_e} \frac{R_e}{R_e + R_M}$$

which reduces, upon substitution of Eqs. (9.18) and (e), to

$$I_G = \frac{q}{1+q} n S_g \sqrt{\frac{P_g}{R_g}} \epsilon \frac{R_e}{R_e + R_M} \tag{9.19}$$

Examination of Eq. (9.19) shows that the galvanometer current I_G is controlled by the selection of the strain gage, which fixes the value of $S_g \sqrt{P_g/R_g}$, by the strain level ϵ, by the number of active arms n, by the design of the Wheatstone bridge circuit as indicated by $q/(1+q)$ and R_e, and finally by the resistance of the measuring circuit R_M. The specification of the Wheatstone bridge circuit cannot be made on the basis of sensitivity alone since its equivalent resistance R_e influences the degree of damping of the galvanometer. With this restriction in mind it is possible to define the combined Wheatstone bridge and galvanometer circuit sensitivity S_{CG} as

$$S_{CG} = \frac{I_G}{\epsilon} = \frac{q}{1+q} n S_g \sqrt{\frac{P_g}{R_g} \frac{R_e}{R_e + R_M}} \tag{9.20}$$

If the sensitivity of the galvanometer is taken into account by application of Eq. (9.3) and if the sensitivity S_θ is defined in terms of the deflection angle θ so that

$$S_\theta = \frac{\theta}{\epsilon} \tag{9.21}$$

it is possible to show that

$$S_\theta = \frac{I_G C}{\epsilon} = \frac{q}{1+q} n S_g \sqrt{\frac{P_g}{R_g}} C \frac{R_e}{R_e + R_M} \tag{9.22}$$

This equation, unlike Eq. (9.20), includes the characteristic sensitivity of the galvanometer, expressed in terms of the instrument constant C, and gives the angular deflection of the galvanometer per unit strain. The experimentalist normally attempts to maximize S_θ without sacrificing the frequency response of the measuring system.

Methods which can be employed to match the galvanometer to the Wheatstone bridge with the proper degree of damping are described in the next section.

E. Frequency Response of the Wheatstone Bridge and Galvanometer System

The frequency response of a galvanometer subjected to a sinusoidal input was discussed in Sec. 9.4C, and it was shown that the range of frequency over which the galvanometer would function within a given accuracy band depended upon the degree of damping. The degree of damping depends upon the inherent fluid damping in the galvanometer and an electromechanical damping which is a function of the resistance of the galvanometer and the external circuit resistance, according to Eq. (9.6). In practice, the galvanometers are normally operated at about 64 percent of critical damping to provide a flat frequency response (that is, ± 2 percent) for frequencies from 0 to 67 percent of the natural frequency of the galvanometer. For commercially available galvanometers, manufacturers usually provide the following information:

1. Nominal undamped natural frequency ω_0
2. Nominal coil resistance R_G
3. Nominal external resistance R_x required to achieve 64 percent of critical damping

In matching the galvanometer to the Wheatstone bridge it is important that the value of R_x be maintained at the specified value to obtain an adequate range of frequency response. In this matching process three possible cases can arise.

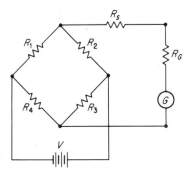

Figure 9.22 Series resistor added to the Wheatstone bridge–galvanometer circuit to achieve the required external resistance R_x.

Case 1 $R_e < R_x$ In this instance the equivalent resistance of the bridge [see Eq. (9.18)] is less than the required external resistance. To provide the necessary additional resistance, a series resistor R_s is added to the circuit, as shown in Fig. 9.22. The required value of the resistance R_s is given by

$$R_s = R_x - R_e = R_x - R_1 \frac{r}{1+r} \frac{1+q}{q} \tag{9.23}$$

With the addition of the series resistance R_s the resistance of the measuring circuit R_M, as defined in Fig. 9.20a, becomes

$$R_M = R_s + R_G$$

and the system sensitivity S_θ is, by Eq. (9.22),

$$S_\theta = \frac{q}{1+q} nS_g \sqrt{\frac{P_g}{R_g}} C \frac{R_e}{R_e + R_s + R_G} \tag{9.24}$$

Finally, the frequency response of the system will be flat within a ±2 percent accuracy band to $\frac{2}{3}\omega_0$.

Case 2 $R_e = R_x$ In this instance the equivalent resistance of the Wheatstone bridge is precisely equal to the required external resistance. From Eq. (9.18) it is apparent that

$$R_e = R_1 \frac{r}{1+r} \frac{1+q}{q} = R_x \tag{9.25}$$

and, from Eq. (9.22), the sensitivity of the system is given by

$$S_\theta = \frac{q}{1+q} nS_g \sqrt{\frac{P_g}{R_g}} C \frac{R_x}{R_x + R_G} \tag{9.26}$$

Again the frequency response of the system will be flat within a ±2 percent accuracy band to $\frac{2}{3}\omega_0$.

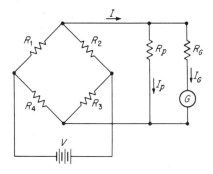

Figure 9.23 Parallel resistor added to the Wheatstone bridge–galvanometer circuit to achieve the required external resistance.

Case 3 $R_e > R_x$ In this instance the equivalent resistance of the Wheatstone bridge is greater than the required external resistance. To decrease the value of the external circuit resistance which the galvanometer sees, it is necessary to place a resistance R_p in parallel with the bridge, as shown in Fig. 9.23.

The value of the parallel resistance R_p which must be placed in the circuit is given by

$$R_x = \frac{R_p R_e}{R_e + R_p} \quad \text{or} \quad R_p = \frac{R_e R_x}{R_e - R_x} \tag{9.27}$$

With the addition of the parallel resistance R_p, the resistance of the measuring circuit R_M, as defined in Fig. 9.20a, becomes

$$R_M = \frac{R_p R_G}{R_p + R_G} \tag{a}$$

The current I divides after leaving the Wheatstone bridge, with I_p passing through the parallel resistor and I_G passing through the galvanometer. That portion of the current passing through the galvanometer is

$$I_G = I \frac{R_p}{R_p + R_G} \tag{9.28}$$

The system sensitivity S_θ is given by substituting Eq. (*a*) into Eq. (9.22) and accounting for the current passing through the galvanometer by Eq. (9.28)

$$S_\theta = \frac{q}{1+q} n S_g \sqrt{\frac{P_g}{R_g}} C \frac{R_e}{R_e + [R_p R_G/(R_p + R_G)]} \frac{R_p}{R_p + R_G} \tag{9.29}$$

which can be rewritten as

$$S_\theta = \frac{q}{1+q} n S_g \sqrt{\frac{P_g}{R_g}} C \frac{R_e R_p}{R_e R_p + R_e R_G + R_p R_G} \tag{9.30}$$

If it is assumed for all three cases that q, n, S_g, $\sqrt{P_g/R_g}$, and C are identical, then it can be stated that case 2 gives the highest sensitivity and case 3 the lowest.

However, to better illustrate the concept of matching a Wheatstone bridge to a given galvanometer for optimum performance, a number of specific examples are given as exercises.

9.5 DYNAMIC RECORDING AT HIGH FREQUENCIES [8, 9]

Magnetic-tape analog-data recording systems of the type illustrated in Fig. 9.24 offer an extremely powerful method of recording and storing dynamic strain-gage signals. Multichannel recording of signals, ranging in frequency from dc to 80 kHz, can be accomplished with this type of equipment. Data recorded and stored on magnetic tape are usually played back and displayed on an oscillograph. By varying the tape speed during playback, the time base can be extended or compressed. Detailed data analysis can be accomplished by utilizing an A/D converter and processing the digital data directly on a computer. The storage feature of magnetic tape is also important since it permits repeated examination of the data at any convenient time after completion of the test. Information stored on

Figure 9.24 Magnetic-tape analog recording system (*Honeywell, Inc.*)

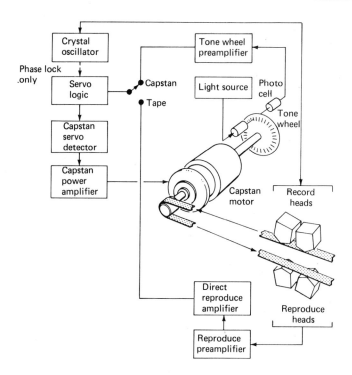

Figure 9.25 DC capstan magnetic-tape drive. (*Honeywell, Inc.*)

magnetic tape can be reliably retrieved any number of times; therefore, it can be subjected to many different types of analyses as the need arises.

In a magnetic-tape recorder, the tape [$\frac{1}{2}$ to 1 in (12.7 to 25.4 mm) wide] is driven at a constant speed by a servoed dc capstan motor over either the record or reproduce heads, as shown in Fig. 9.25. The speed of the capstan motor is monitored with a photocell (see Fig. 9.25) and compared with the frequency from a crystal oscillator to provide the feedback signal in a closed-loop servo system designed to maintain constant tape speeds. Tape speeds have been standardized at $1\frac{7}{8}$, $3\frac{3}{4}$, $7\frac{1}{2}$, 15, 30, 60, and 120 in/s (47.6, 191, 381, 762, 1524, and 3048 mm/s).

The signal written on the tape by the record magnetic head assemblies is in the form of variations in the level of magnetism imposed on the magnetic coating of the tape. The reproduce heads convert these variations in magnetism back into electrical signals. Multichannel recorders employ four head assemblies which are precisely positioned on a single baseplate to ensure alignment with the tape. Two of the heads are record heads and the other two are reproduce heads, as shown in Fig. 9.25. As the tape passes the first head assembly, odd-channel data are recorded; it then passes the second head assembly, where even-channel data are recorded. This procedure minimizes interchannel cross talk by maximizing the spacing between individual heads in the stacked-head assembly. Use of two

Table 9.1 Bandwidth and signal-to-noise ratios for direct and FM recording

IRIG band	Direct recording		FM recording	
	BW, kHz	S/N, dB	BW, kHz	S/N, dB
Low	0.1–100	40	dc–10	45
Intermediate	0.3–300	40	dc–20	48
Wide	0.4–1500	30	dc–400	30

recording-head stacks permits seven channels of recordings on $\frac{1}{2}$-in (12.7-mm) tape and 14 channels of recordings on 1-in (25.4-mm) tape.

Most instrument tape systems can be used in either direct-recording or frequency-modulation (FM) modes. With direct recording, the intensity of magnetization on the tape is proportional to the instantaneous amplitude of the input signal. Direct recording is easy to accomplish with simple, moderately priced electronics. Extremely high frequencies (1.5 MHz) can be recorded. In playback, however, a signal is generated in the reproduce head only in response to changes in the magnetic flux on the tape; consequently, the direct-record process cannot reproduce the dc components of the input signal. The direct-recording method is also subject to significant signal variations caused by randomly distributed surface inhomogeneities in the tape. Because of these irregularities, direct recording is usually limited to audio recording where the human ear, on playback, can average the amplitude errors, or to recordings where the signal frequency and not the signal amplitude is of primary importance.

Practically all strain-gage signals are written on magnetic tape with the FM method of recording. With FM recording, a carrier oscillator is frequency-modulated by the input signal. The oscillator has a center frequency which corresponds to a zero input signal. Deviations from the center frequency are proportional to the input signal. The polarity of the input signal determines the direction of the deviation. FM recording preserves the dc component in the signal and is much more accurate than direct recording. The price paid for these improvements is the sharp reduction in frequency response, illustrated in Table 9.1, which shows bandwidth for direct and FM recording with three different IRIG (inter range instrumentation group) bands.

The detailed discussion of an instrument-quality analog tape recording system is beyond the scope of this text. The systems are complex and require considerable care in specifying the tape system and the associated electronics. A typical system adapted for strain-gage-data recording includes bridge-conditioning units, an input-signal coupler, a control panel, direct-record data amplifiers, FM-record data amplifiers, an FM frequency plug-in for each tape speed, direct-reproduce data amplifiers, FM-reproduce data amplifiers, the tape system, and related visual recorders. Complete system prices range from $20,000 to $40,000 depending on the IRIG band specified and the associated accessory equipment included. Even for the most affluent laboratory, the analog instrument tape recorder represents a large investment which should be carefully specified to meet the needs of the laboratory while minimizing the cost of the system.

9.6 DYNAMIC RECORDING AT VERY HIGH FREQUENCIES [10]

Very high frequency recording can be accomplished with cathode-ray oscilloscopes which have bandwidths up to 500 MHz. The cathode-ray oscilloscope is, in effect, a voltmeter which can be employed to measure transient voltage signals. The essential element in the oscilloscope is the cathode-ray tube, illustrated in Fig. 9.26. The cathode-ray tube is a highly evacuated tube in which electrons are produced by heating a cathode. These electrons are then collected, accelerated, and focused on a fluorescent screen on the face of the tube. The impinging stream of electrons forms a bright point of light on the fluorescent screen. Voltages applied to horizontal and vertical deflection plates located in the cathode-ray tube will deflect the stream of electrons and move the bright point of light on the face of the tube. It is this ability to deflect the stream of electrons which permits the cathode-ray tube to be employed as a dynamic voltmeter with an inertialess indicating system.

In recent years, advances in cathode-ray tube technology have provided storage-type tubes, which retain and display the image of an electrical waveform on the tube face after the waveform ceases to exist. The image retention may be from a few seconds to several hours. The stored display can be instantaneously erased to ready the tube for display of the next waveform. Since the storage tubes utilize a bistable phosphor, they can be operated in either the storage or conventional (nonstorage) mode.

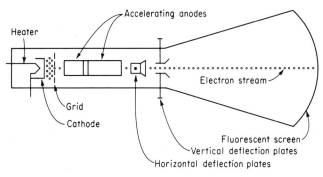

Figure 9.26 Basic elements of a cathode-ray tube.

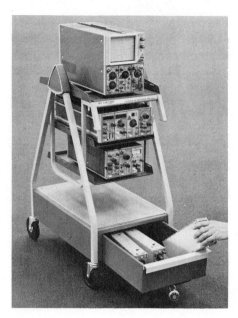

Figure 9.27 Cathode-ray oscilloscope with plug-in amplifiers, auxiliary measuring instruments, and cart. (*Tektronix, Inc.*)

Storage oscilloscopes provide clear visual displays of slowly changing phenomena that would appear as a slow-moving dot on a conventional tube. The storage tube also gives visual displays of rapidly changing nonrepetitive waveforms and allows for on-scope evaluation of the recorded data. This feature reduces the number of photographs which must be taken to give strain-time traces. The unwanted displays are erased and only the important displays are photographed.

For dynamic strain-gage applications, the cathode-ray oscilloscope is an ideal voltage-measuring instrument. The input impedance of the instrument is quite high (usually of the order of 1 MΩ); thus, there is no appreciable interaction between the Wheatstone bridge and the measuring instrument. The frequency response of an oscilloscope is usually quite high, and even a relatively low-frequency model (800-kHz bandpass) greatly exceeds the requirements for mechanical strain measurements, which are rarely more than 50 kHz.

The sensitivity of the cathode-ray tube is normally quite low, and it often requires about 100 V to produce 1 in (25 mm) of deflection on the face of the tube. Thus it is clear that the strain signal voltage, which is rarely larger than a few millivolts, will require appreciable amplification before it is applied to the deflection plates of the cathode-ray tube. Fortunately, most good-quality oscilloscopes are equipped with high-gain ac and dc amplifiers which are adequate for strain-gage applications. In fact, many oscilloscopes incorporate a plug-in amplifier (see Fig. 9.27), which can be changed quickly to vary the operating characteristics of the oscilloscope over an extremely wide range of amplifier gain and frequency response.

In most strain-gage work, the voltage from the Wheatstone bridge is displayed on the y axis, or ordinate, of the tube face. The x-axis display, which is controlled by the horizontal deflection plates, is generally used to represent a time base. This is accomplished internally in the oscilloscope by applying the voltage from a sawtooth oscillator to the horizontal deflection plates. As the voltage increases to its maximum, the light spot sweeps across the face of the tube with a uniform velocity. When the oscillator voltage reaches its maximum and drops abruptly to zero, the spot of light rapidly returns to its starting point and the cycle is repeated. By varying the frequency of the sawtooth oscillator, the sweep rate of the oscilloscope can be changed over the range from 0.1 μs to 5 s per division.

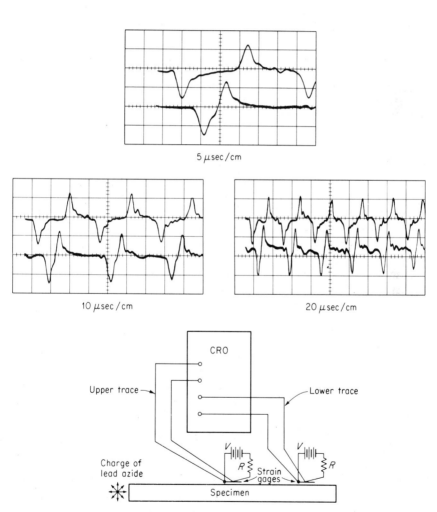

Figure 9.28 Dynamic strain-time records obtained with an oscilloscope. (*From a study by B. W. Abbott, IIT Research Institute.*)

The strain-gage record showing strain as a function of time is obtained by photographing the face of the cathode-ray tube as the spot of light traverses the fluorescent screen. If a still camera is employed, the time of the event which can be recorded in this manner varies between about 1.0 μs and 50 s, depending upon the sweep rate. A typical strain-time record obtained from an oscilloscope is shown in Fig. 9.28.

9.7 TELEMETRY SYSTEMS [11, 12]

Test conditions sometimes preclude the use of lead wires, e.g., on rotating machinery, measurements in space, or measurements at great distances. In these instances, it is often necessary to transmit the signal from the measuring point to the recording site by telemetry. In relatively simple telemetry systems, the output from a single bridge is used to modulate a radio signal. The strain gage, bridge, power supply, and radio transmitter are located at the measuring point, and the receiver and recorder are at the recording site. The distance of transmission is a function of the power of the transmitter. For low-power transmission, where the range of operation is limited to a few feet, licensing is not required and the carrier frequency used for transmission is not limited.

An example of a telemetry system used to convert an operating shaft into a torsional load cell is illustrated in Fig. 9.29. This system incorporates a split collar which fits over the shaft and houses the dc power supply for the strain-gage bridge, the modulator, and the voltage-controlled oscillator (VCO). The signal from the bridge is used to pulse width modulate a constant-amplitude 5-kHz square wave (where the width of the output pulse is directly proportional to the voltage output from the bridge). This square wave serves to vary the frequency of the VCO, which is centered at 10.7 MHz. The VCO signal is transmitted by a rotating antenna mounted on the split collar and is received by a stationary loop antenna which encircles the shaft. The signal is demodulated, filtered, and amplified before being displayed. The transmitting unit is completely self-contained. The power to drive the bridge, modulator, and VCO is obtained by inductively coupling a 160-kHz signal to the power supply through the stationary loop antenna.

For longer-range multiple-transducer applications, the systems become more complex. First, licensing is required since the frequencies available for transmission are limited and crowded. Telemetry transmissions have been restricted to the bands from 1435 to 1535 and 2200 to 2300 MHz in the United States. Next, as the number of transducers increases at the measuring site, it becomes less practical to provide a separate transmitter for each; therefore, most of the long-range multiple-transducer systems combine the signals from a number of transducers into a single signal for transmission by one radio transmitter. The process of combining the different signals is called *multiplexing*. Each individual signal in the group retains its original information by using either frequency-division or time-division multiplexing. At the radio receiver, the individual signals are separated

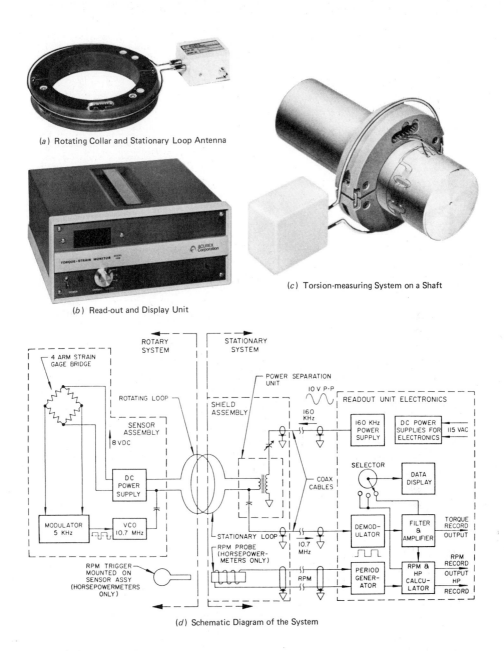

(a) Rotating Collar and Stationary Loop Antenna

(b) Read-out and Display Unit

(c) Torsion-measuring System on a Shaft

(d) Schematic Diagram of the System

Figure 9.29 Torsion-measuring system utilizing telemetry for data transmission. (*Acurex Corporation.*)

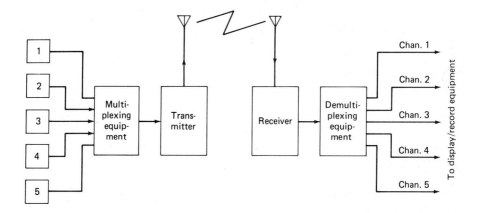

Figure 9.30 Block diagram showing a telemetry system which utilizes multiplexing and de-multiplexing equipment to transmit data simultaneously from several transducers.

from the composite signal and recorded on individual recording instruments, as shown in Fig. 9.30.

When the output voltages from two transducers are used to modulate two separate carrier frequencies, two distinct frequency bands are formed. If the frequencies of the carriers are far enough apart, the two bands can be transmitted together with no interaction between the bands. Frequency-sensitive filters are used after the signals have been received at the receiving station to filter out the individual carrier frequencies, which are then demodulated to obtain the transducer data. This process in telemetry is known as *frequency-division multiplexing*.

As an example of frequency-division multiplexing, consider a system where three transducer outputs modulate three subcarrier frequencies. As shown in Fig. 9.31, the oscillator for transducer 1 is centered at 400 Hz, and the applied data signal produces a maximum deviation of ± 30 Hz (± 7.5 percent). Similarly, the oscillator for transducer 2 is centered at 560 Hz with a maximum deviation of ± 42 Hz, and the oscillator for transducer 3 is centered at 730 Hz with a maximum deviation of ± 55 Hz. These three channels can be mixed to form a frequency band ranging from 370 to 785 Hz and then transmitted over a radio link at a transmitting frequency of, say, 2200 MHz. It should be noted in Fig. 9.31 that there is no overlapping between the three channels and that *guard bands* lie between them to ensure separation. At the receiving station, bandpass filters separate the channels and send the recovered frequency bands to individual discriminators for demodulation and recovery of the transducer signal for display on recording instruments.

With *time-division multiplexing*, all the channels use the same portion of the frequency spectrum but not at the same time. As illustrated in Fig. 9.32, each transducer or channel is sampled in a repeated sequence by a commutator to give a composite signal containing time-spaced signals from each transducer. Since the individual channels are not monitored continuously, the sampling rate must be

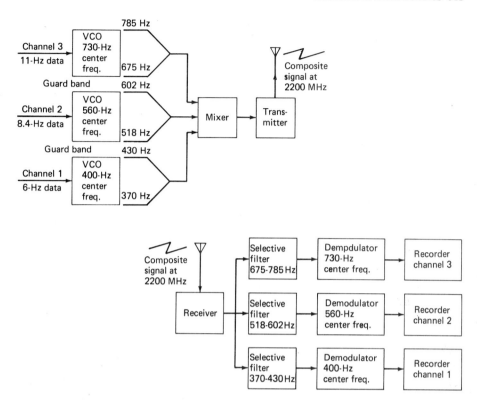

Figure 9.31 Schematic illustration of frequency-division multiplexing.

sufficient to ensure that the signal amplitude does not change appreciably during the time between samples. Sampling rates about five times greater than the highest frequency component in the transducer signal are used in most telemetry systems. For instance, if the highest frequency component in a set of six transducers is 50 Hz, the individual channels should be sampled 250 times per second. The commutator would then have to process samples at a rate of 1500 samples per second.

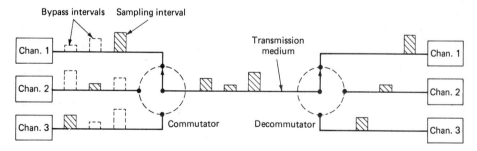

Figure 9.32 Schematic diagram showing the essential features of time-division multiplexing. The commutator and decommutator are electronic switches.

At the receiving station, a decommutator, operating at precisely the same frequency as the commutator, serves to distribute the time-sequenced signals to the correct output channels. Time-division multiplexing is based on precise timing and it is essential that the commutator and decommutator be precisely synchronized to avoid mixing the signal samples from the various transducers.

EXERCISES

9.1 Four strain gages are mounted on a steel tension specimen having a cross-sectional area of 6.0 mm² to produce a load cell. If two of the gages are mounted in the axial direction and the other two in the transverse direction, and if the gage factor for all gages is $S_g = 2.05$, what gage factor should be dialed into a manual null-balance strain indicator so that it will give a direct reading of the load P applied to the tension load cell?

9.2 Two 350-Ω strain gages are mounted in the bridge shown in Fig. 9.1, one to serve as an active gage and the other to serve as a dummy gage. The gage factor for both of the gages is $S_g = 2.04$, and a strain of 1500 μm/m is anticipated in the test. Referring to Fig. 9.1, describe the balance and calibration procedure to be followed so that the digital readout will be direct in terms of strain in units of micrometers per meter.

9.3 A pressure vessel is being subjected to a series of hydrostatic tests, where the pressure changes from zero to maximum in 30 min on each run, and 200 strain gages have been installed on the vessel. Describe in complete detail how an automatic data-processing system would be employed to monitor, process, and analyze the data from each test run.

9.4 Assume that a 100-kN load cell with an output of 3 mV/V and a 2.5-mm strain extensiometer with a 50-mm gage length and an output of 3 mV/V are used to monitor the stress and strain in a standard mild-steel tension specimen. Select appropriate input voltages for each transducer and select the amplifier setting on an xy recorder so that the stress-strain curve can be recorded directly on 180 by 250 mm sheet of graph paper.

9.5 Verify the equation of motion of the galvanometer as given by Eq. (9.4). Then show the intermediate steps required to obtain Eq. (9.5).

9.6 Solve Eq. (9.5) for the overdamped, critically damped, and underdamped conditions and verify Eqs. (9.7) to (9.9).

9.7 For a ± 4 percent accuracy band on a step-pulse input, determine the minimum response time of a galvanometer characterized by constants α, J, and G. Also find the degree of damping d associated with this minimum response time.

9.8 A galvanometer with an undamped frequency ω_0 is to be used to measure a sinusoidal current to an accuracy of:

   (a) ± 1 percent   (b) ± 2 percent
   (c) ± 5 percent   (d) ± 10 percent

In each case, determine the frequency response and the degree of damping required to maximize this frequency response.

9.9 Verify Eq. (9.12) by solving the differential equation (9.11).

9.10 Verify Eq. (9.18).

9.11 For a static or quasi-static strain-gage application, where the external resistance of the galvanometer circuit is not critical, verify the sensitivity factors $S_\theta/(S_g C \sqrt{P_g/R_g})$ for the bridge arrangements shown in Fig. E9.11.

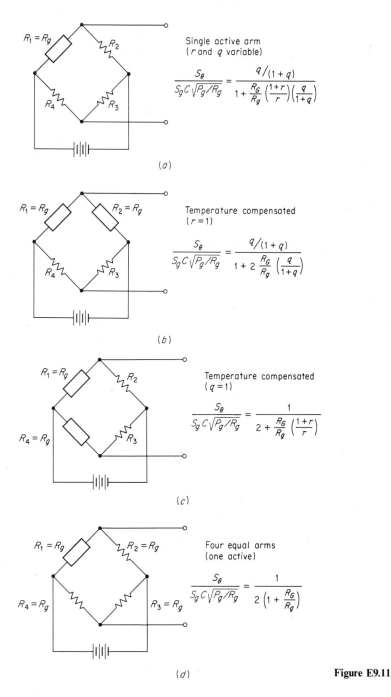

Single active arm
(r and q variable)

$$\frac{S_\theta}{S_g C \sqrt{P_g/R_g}} = \frac{q/(1+q)}{1 + \frac{R_G}{R_g}\left(\frac{1+r}{r}\right)\left(\frac{q}{1+q}\right)}$$

(a)

Temperature compensated
($r = 1$)

$$\frac{S_\theta}{S_g C \sqrt{P_g/R_g}} = \frac{q/(1+q)}{1 + 2\frac{R_G}{R_g}\left(\frac{q}{1+q}\right)}$$

(b)

Temperature compensated
($q = 1$)

$$\frac{S_\theta}{S_g C \sqrt{P_g/R_g}} = \frac{1}{2 + \frac{R_G}{R_g}\left(\frac{1+r}{r}\right)}$$

(c)

Four equal arms
(one active)

$$\frac{S_\theta}{S_g C \sqrt{P_g/R_g}} = \frac{1}{2\left(1 + \frac{R_G}{R_g}\right)}$$

(d)

Figure E9.11

9.12 Prepare a graph showing sensitivity factor $S_\theta/(S_g C \sqrt{P_g/R_g})$ as a function of R_G/R_g for the four bridge arrangements shown in Fig. E9.11. From these results determine (a) the resistance ratio R_G/R_g for maximum sensitivity, (b) the best bridge arrangement for temperature compensation, and (c) the sensitivity of the four-equal-arm bridge.

Table 9.2 Galvanometer Characteristics (Metric)

Electromagnetic-damped types

Type no.	Nominal undamped natural frequency (Hz)	Flat (±5%) frequency response (Hz)	Required external damping resistance (Ω)	Nominal coil resistance (Ω)	Current (±5%) I_g (μA/cm)	Current (±5%) $\frac{1}{I_g}$ (cm/μA)	Voltage Galvanometer V_g (mV/cm)	Voltage Galvanometer $\frac{1}{V_g}$ (cm/mV)	Voltage Circuit V_c (mV/cm)	Voltage Circuit $\frac{1}{V_c}$ (cm/mV)	Maximum safe current (mA)	Maximum P-P deflection, $w/\pm2\%$ linear (cm)	Balance standard (cm/g)	Precision (cm/g)
M18-5000	18	0–12	4200	220.0	0.169	5.920	0.037	37.2	0.749	1.33	5	20	NA	NA
M24-350A	24	0–15	350	135.0	0.59	1.700	0.080	12.4	0.288	3.48	5	20	0.124	NA
M40-350A	40	0–24	350	66.0	1.62	0.617	0.107	9.4	0.673	1.49	15	20	0.089	0.045
M100-350	100	0–60	350	75.5	2.48	0.401	0.187	5.33	1.06	0.95	10	20	0.056	0.028
M200-350	200	0–180	350	62.0	10.1	0.100	0.626	1.61	4.13	0.24	10	20	0.041	0.020
M400-350	400	0–360	350	125.0	30.4	0.033	3.80	0.264	14.4	0.068	10	20	0.041	0.020
M600-350	600	0–540	350	320.0	51.2	0.020	16.4	0.061	34.2	0.028	15	20	0.041	0.020
M40-120A	40	0–24	120	30.0	3.15	0.317	0.095	10.6	0.472	2.12	15	20	0.089	0.045
M100-120A	100	0–60	120	52.2	3.94	0.254	0.206	4.8	0.671	1.48	15	20	0.040	0.020
M200-120	200	0–120	120	62.0	10.1	0.100	0.626	1.61	1.83	0.548	15	20	0.040	0.020
M400-120	400	0–240	120	125.0	30.4	0.033	3.80	2.64	7.54	0.134	10	20	0.040	0.020

Sensitivity in 30-cm optical arm of all Honeywell oscillographs

Fluid-damped types

	Hz	Hz	†	Ω	mA/cm	cm/mA	V/cm	cm/V		mA	cm	cm/g
M1000	100	0–600		39.0	1.04	0.965	0.040	2.50		100	20	0.020
M1650	1,650	0–1,000		26.8	3.66	0.272	0.098	1.02		100	20	0.020
M3300	3,300	0–2,000		32.0	7.88	0.127	0.252	0.396		100	15	0.020
M3300T	‡	‡		32.0	7.88	0.127	0.252	0.396		100	15	0.020
M5000	5,000	0–3,000		39.5	12.3	0.080	0.485	0.206		100	9	0.020
M8000	8,000	0–4,800		35.0	15.7	0.063	0.552	0.182		100	5	0.020
M10000	10,000	0–6,000 ± 8%		35.0	15.7	0.069	0.552	0.182		100	5	0.020
M13000	13,000	0–13,000 ± 8%		81.6	32.3	0.030	2.61	0.038		55	2.5	0.020

† No requirement with fluid-damped galvanometers.
‡ No specification.

9.13 For a static or quasi-static strain-gage application, where the external resistance of the galvanometer circuit is not critical, determine the sensitivity factor $S_\theta/(S_g C\sqrt{P_g/R_g})$ for the bridge arrangements shown in Fig. E9.13. Compare these results with the results shown in Fig. E9.11 and draw conclusions regarding the use of multiple-gage installations to improve sensitivity.

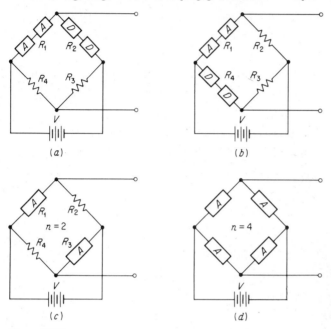

Figure E9.13

9.14 A single active strain gage ($R_1 = R_g = 120\ \Omega$) and a dummy gage are employed in a four-equal-arm bridge; that is, $r = q = 1$. An M40-120A galvanometer is used to measure the bridge output. For this combination of bridge and galvanometer, determine the system sensitivity $S_\theta/(S_g\sqrt{P_g/R_g})$. If the strain-gage characteristics are $S_g = 2.06$ and $P_g = 0.05$ W, determine the galvanometer deflection for a strain of 600 μm/m. The characteristics of a "line" of galvanometers are shown in Table 9.2.

9.15 A single active gage ($R_1 = R_g = 350\ \Omega$) and a dummy gage are employed in a four-equal-arm bridge; that is, $r = q = 1$. An M40-350A galvanometer is used to measure the bridge output. For this combination of bridge and galvanometer, determine the system sensitivity $S_\theta/(S_g\sqrt{P_g/R_g})$ and compare the results with those obtained in Exercise 9.14.

9.16 Determine the system sensitivity $S_\theta/(S_g\sqrt{P_g/R_g})$ if an M400-120 galvanometer is used in place of the M40-120A galvanometer in Exercise 9.14. Compare the two systems on the basis of sensitivity and frequency response.

9.17 Determine the system sensitivity $S_\theta/(S_g\sqrt{P_g/R_g})$ if an M400-350 galvanometer is used in place of the M40-350A galvanometer in Exercise 9.15. Compare the two systems on the basis of sensitivity and frequency response.

9.18 Determine the system sensitivity $S_\theta/(S_g\sqrt{P_g/R_g})$ if an M40-120A galvanometer must be used in place of the M40-350A galvanometer in Exercise 9.15. Compare with the results of Exercise 9.15 and discuss.

9.19 Determine the system sensitivity $S_\theta/(S_g\sqrt{P_g/R_g})$ if an M40-350A galvanometer must be used in place of the M40-120A galvanometer in Exercise 9.14. Compare with the results of Exercise 9.14 and discuss.

9.20 For the combination of bridge and galvanometer given in Exercise 9.14, determine the maximum deflection of the trace associated with a sinusoidal strain of 300 μm/m for frequencies of 10, 26, 40, and 60 Hz. Describe the behavior of the system when it is used to record strains with frequencies greater than 26 Hz.

9.21 For an equal-arm bridge with one active gage ($S_g = 2.00$ and $R_g = 350$ Ω) recording a dynamic strain of 1000 μm/m, determine the amplitude of the trace recorded with an M13000 galvanometer. Assume a bridge voltage of 10 V. Discuss the results.

9.22 Briefly describe the difference between direct (AM) and frequency modulation (FM) recording and give the reasons for not using AM recording with instrument-quality magnetic tape.

9.23 Briefly discuss the advantages and disadvantages of magnetic-tape recording systems relative to the oscillograph recording method.

9.24 Consider a dynamic application where only two or three strain gages are to be monitored as a function of time. Discuss the relative merits of oscillograph, oscilloscope, and magnetic-tape recording.

9.25 Completely specify the test conditions for the strain gages, potentiometer circuits, and oscilloscope in Fig. 9.28 if the amplitude of the strain pulse is 1200 μm/m and its bar velocity is 5000 m/s.

REFERENCES

1. Beckwith, T. G., and N. L. Buck: "Mechanical Measurements," Addison-Wesley Publishing Company, Inc., Reading, Mass., 1961.
2. Cerni, R. H., and L. E. Foster: "Instrumentation for Engineering Measurement," John Wiley & Sons, Inc., New York, 1962.
3. Christian, D.: BL-310 Strain Analyzer, *Proc. SESA*, vol. VII, no. 1, pp. 21–29, 1949.
4. Considine, D. M. (ed.): "Process Instruments and Controls Handbook," McGraw-Hill Book Company, New York, 1957.
5. Specifications for Model V/E-20 Direct Reading Digital Strain Indicator, Vishay Instruments, Malvern, Pa.
6. Specifications for B & F-SY 256 Digital Data Acquisition System, *B & F* Instrum. Bull. 450A, Cornwell Heights, Pa.
7. Specifications for Digitrend—220 μP—Data Logger, Doric Scientific Corp., San Diego, Calif.
8. Electronics for Measurement, Analysis, and Computation, Hewlett Packard, Palo Alto, Calif.
9. Honeywell Instrumentation Handbook, Honeywell, Inc., Denver, Colorado.
10. Tektronix Products, Tektronix, Inc., Beaverton, Oreg.
11. Specifications for Model 1200A Torsion Measurement System, Acurex Corp., Mountain View, Calif.
12. Telemeter, *EMR Telemetry*, no. 1, Sarasota, Fla., 1965.

TEN

ANALYSIS OF STRAIN-GAGE DATA

10.1 INTRODUCTION

Electrical-resistance strain gages are normally employed on the free surface of a specimen to establish the stress at a particular point on this surface. In general it is necessary to measure three strains at a point to completely define either the stress or the strain field. In terms of principal strains it is necessary to measure ϵ_1, ϵ_2, and the direction of ϵ_1 relative to the x axis as given by the principal angle ϕ. Conversion of the strains into stresses requires, in addition, a knowledge of the elastic constants E and v of the specimen material.

In certain special cases the state of stress can be established with a single strain gage. Consider first a uniaxial state of stress where $\sigma_{yy} = \tau_{xy} = 0$ and the direction of σ_{xx} is known. In this case a single-element strain gage is mounted with its axis coincident with the x axis. The stress σ_{xx} is given by

$$\sigma_{xx} = E\epsilon_{xx} \tag{10.1}$$

Next, consider an isotropic state of stress where $\sigma_{xx} = \sigma_{yy} = \sigma_1 = \sigma_2$ and $\tau_{xy} = 0$. In this case a strain gage may be mounted in any direction, and the magnitude of the stresses can be established from

$$\sigma_{xx} = \sigma_{yy} = \sigma_1 = \sigma_2 = \frac{E}{1-v}\epsilon_\theta \tag{10.2}$$

where ϵ_θ is the strain measured in any direction in the isotropic stress field.

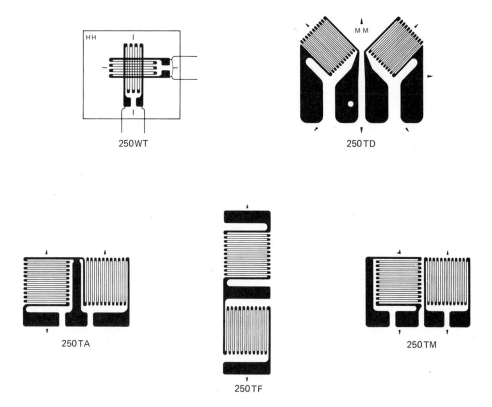

Figure 10.1 Two-element rectangular rosettes for use when the principal directions are known.

When less is known beforehand regarding the state of stress in the specimen, it is necessary to employ multiple-element strain gages to establish the magnitude of the stress field. If the specimen being investigated has an axis of symmetry, or if a brittle-coating analysis has been conducted to establish the principal-stress directions, this knowledge can be used to reduce the number of gage elements required from three to two. A two-element rectangular rosette similar to those illustrated in Fig. 10.1 is mounted on the specimen with its axes coincident with the principal directions. The two principal strains ϵ_1 and ϵ_2 obtained from the gages can be employed to give the principal stresses σ_1 and σ_2:

$$\sigma_1 = \frac{E}{1 - v^2}(\epsilon_1 + v\epsilon_2) \qquad \sigma_2 = \frac{E}{1 - v^2}(\epsilon_2 + v\epsilon_1) \qquad (10.3)$$

These relations give the complete state of stress since the principal directions are known a priori. The stresses on any plane can be established by employing Eqs. (1.16) with the results obtained from Eq. (10.3).

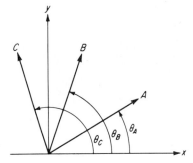

Figure 10.2 Three gage elements placed at arbitrary angles relative to the x and y axes.

In the most general case, no knowledge of the stress field or its directions is available before the experimental analysis is conducted. Three-element rosettes are required in these instances to completely establish the stress field. To show that three strain measurements are sufficient, consider three strain gages aligned along axes A, B, and C, as shown in Fig. 10.2.

From Eqs. (2.18) it is evident that

$$\epsilon_A = \epsilon_{xx} \cos^2 \theta_A + \epsilon_{yy} \sin^2 \theta_A + \gamma_{xy} \sin \theta_A \cos \theta_A$$
$$\epsilon_B = \epsilon_{xx} \cos^2 \theta_B + \epsilon_{yy} \sin^2 \theta_B + \gamma_{xy} \sin \theta_B \cos \theta_B$$
$$\epsilon_C = \epsilon_{xx} \cos^2 \theta_C + \epsilon_{yy} \sin^2 \theta_C + \gamma_{xy} \sin \theta_C \cos \theta_C \quad (10.4)$$

The cartesian components of strain ϵ_{xx}, ϵ_{yy}, and γ_{xy} can be determined from a simultaneous solution of Eqs. (10.4). The principal strains and the principal directions can then be established by employing Eqs. (2.7), (1.12), and (1.14). The results are

$$\epsilon_1 = \tfrac{1}{2}(\epsilon_{xx} + \epsilon_{yy}) + \tfrac{1}{2}\sqrt{(\epsilon_{xx} - \epsilon_{yy})^2 + \gamma_{xy}^2}$$
$$\epsilon_2 = \tfrac{1}{2}(\epsilon_{xx} + \epsilon_{yy}) - \tfrac{1}{2}\sqrt{(\epsilon_{xx} - \epsilon_{yy})^2 + \gamma_{xy}^2} \quad (10.5)$$

$$\tan 2\phi = \frac{\gamma_{xy}}{\epsilon_{xx} - \epsilon_{yy}}$$

where ϕ is the angle between the principal axis (σ_1) and the x axis. The principal stresses can then be computed from the principal strains by utilizing Eqs. (10.3).

In actual practice, three-element rosettes with fixed angles (that is, θ_A, θ_B, and θ_C fixed at specified values) are employed to provide sufficient data to completely define the stress field. These rosettes are defined by the fixed angles as the rectangular rosette, the delta rosette, and the tee-delta rosette. Examples of commercially available three-element rosettes are presented in Fig. 10.3. Also shown in this figure is a stress gage which can be employed to give the stress in an arbitrary direction. Each of these rosette gages will be discussed in detail in subsequent sections of this chapter.

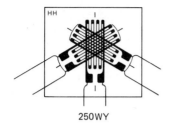
250WY

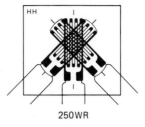
250WR

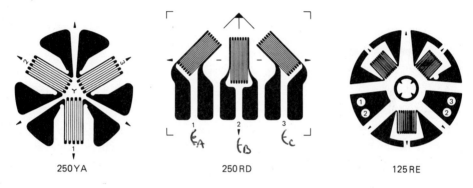

250YA   250RD   125RE

Figure 10.3 Three-element rectangular rosettes for use when the principal directions are unknown.

10.2 THE THREE-ELEMENT RECTANGULAR ROSETTE

The three-element rectangular rosette employs gages placed at the 0, 45, and 90° positions, as indicated in Fig. 10.4.

For this particular rosette it is clear from Eqs. (10.4) that

$$\epsilon_A = \epsilon_{xx} \qquad \epsilon_B = \tfrac{1}{2}(\epsilon_{xx} + \epsilon_{yy} + \gamma_{xy}) \qquad \epsilon_C = \epsilon_{yy} \qquad (10.6)$$

and that

$$\gamma_{xy} = 2\epsilon_B - \epsilon_A - \epsilon_C$$

Thus by measuring the strains ϵ_A, ϵ_B, and ϵ_C the cartesian components of strain ϵ_{xx}, ϵ_{yy}, and γ_{xy} can be quickly and simply established through the use of Eqs. (10.6). Next, by utilizing Eqs. (10.5), the principal strains ϵ_1 and ϵ_2 can be established as

$$\epsilon_1 = \tfrac{1}{2}(\epsilon_A + \epsilon_C) + \tfrac{1}{2}\sqrt{(\epsilon_A - \epsilon_C)^2 + (2\epsilon_B - \epsilon_A - \epsilon_C)^2}$$

$$\epsilon_2 = \tfrac{1}{2}(\epsilon_A + \epsilon_C) - \tfrac{1}{2}\sqrt{(\epsilon_A - \epsilon_C)^2 + (2\epsilon_B - \epsilon_A - \epsilon_C)^2} \qquad (10.7a)$$

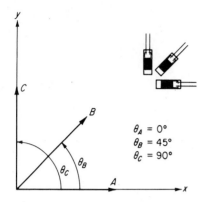

$\theta_A = 0°$
$\theta_B = 45°$
$\theta_C = 90°$

Figure 10.4 Gage positions in a three-element rectangular rosette.

and the principal angle ϕ is given by

$$\tan 2\phi = \frac{2\epsilon_B - \epsilon_A - \epsilon_C}{\epsilon_A - \epsilon_C} \tag{10.7b}$$

The solution of Eq. (10.7b) gives two values for the angle ϕ, namely, ϕ_1, which refers to the angle between the x axis and the axis of the maximum principal strain ϵ_1, and ϕ_2, which is the angle between the x axis and the axis of the minimum principal strain ϵ_2. These angles are illustrated in the Mohr's strain circle shown in Fig. 10.5. It is possible to show (see Exercise 10.4) that the principal axes can be identified by applying the following rules:

$$0 < \phi_1 < 90° \qquad \text{when } \epsilon_B > \tfrac{1}{2}(\epsilon_A + \epsilon_C)$$
$$-90° < \phi_1 < 0 \qquad \text{when } \epsilon_B < \tfrac{1}{2}(\epsilon_A + \epsilon_C)$$
$$\phi_1 = 0 \qquad \text{when } \epsilon_A > \epsilon_C \text{ and } \epsilon_A = \epsilon_1$$
$$\phi_1 = \pm 90° \qquad \text{when } \epsilon_A < \epsilon_C \text{ and } \epsilon_A = \epsilon_2 \tag{10.8}$$

Finally, the principal stresses occurring in the component can be established by employing Eqs. (10.7) together with Eqs. (10.3) to obtain

$$\sigma_1 = E\left[\frac{\epsilon_A + \epsilon_C}{2(1-v)} + \frac{1}{2(1+v)}\sqrt{(\epsilon_A - \epsilon_C)^2 + (2\epsilon_B - \epsilon_A - \epsilon_C)^2}\right]$$

$$\sigma_2 = E\left[\frac{\epsilon_A + \epsilon_C}{2(1-v)} - \frac{1}{2(1+v)}\sqrt{(\epsilon_A - \epsilon_C)^2 + (2\epsilon_B - \epsilon_A - \epsilon_C)^2}\right] \tag{10.9}$$

The use of Eqs. (10.6) to (10.9) permits a determination of the cartesian components of strain, the principal strains and their directions, and the principal stresses by a totally analytical approach. However, it is also possible to determine these

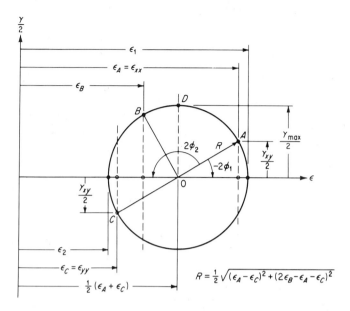

Figure 10.5 Graphical solution for the principal strains and their directions from a rectangular rosette.

quantities with a graphical approach, as illustrated in Fig. 10.5. A Mohr's strain circle is initiated by laying out the ϵ (abscissa) and the $\frac{1}{2}\gamma$ (ordinate) axes. The three strains ϵ_A, ϵ_B, and ϵ_C are then plotted as points on the abscissa. Vertical lines are then drawn through these three points. The shearing strain γ_{xy} is computed from Eqs. (10.6), and $\frac{1}{2}\gamma_{xy}$ is plotted positive downward or negative upward along the vertical line drawn through ϵ_A to establish point A. This shearing strain may also be plotted as positive upward or negative downward along the vertical line through ϵ_C to establish point C. The diameter of the circle is then determined by drawing a line between points A and C which intersects the abscissa and defines the center of the circle at a distance $\frac{1}{2}(\epsilon_A + \epsilon_C)$ from the origin. A circle is then drawn from this center passing through points A and C. The circle will intersect the vertical line drawn through ϵ_B, and this point of intersection is labeled as B. A straight line through the center of the circle and point B should be a perpendicular bisector of the diameter AC. The values of the principal strains ϵ_1 and ϵ_2 are given by the intersections of the circle with the abscissa. The principal angle $2\phi_1$ is given by the angle $AO\epsilon_1$ and is negative if point A lies above the ϵ axis. The principal angle $2\phi_2$ is given by the angle $AO\epsilon_2$ and is positive if point A lies above the ϵ axis. The maximum shearing strain is established by a vertical line drawn through the center of the circle to give point D at the intersection. The projection of point D onto the $\frac{1}{2}\gamma$ axis determines the value of $\frac{1}{2}\gamma_{max}$. The principal stresses can be determined directly from the principal strains by employing Eqs. (10.3).

This graphical approach for reducing the data obtained from a rectangular rosette is quite applicable when the data to be reduced are limited to a few gages. However, if large amounts of data must be reduced, the analytical approach is normally preferred since it requires less time.

Special strain-gage computers and nomographs are available for use when data from a very large number of rosettes must be reduced. Also, the analytical approach for obtaining the principal strains and stresses and their directions can be programmed for a digital computer. With a digital instrumentation system where the strain-gage output is punched on a tape, the digital computer can be employed quite effectively. The punched tape is used as the input for the digital computer, and results from several hundred rosettes can be reduced in a matter of minutes.

10.3 THE DELTA ROSETTE

The delta rosette employs three gages placed at the 0, 120, and 240° positions, as indicated in Fig. 10.6.

For the angular layout of the delta rosette it is clear from Eqs. (10.4) that

$$\epsilon_A = \epsilon_{xx}$$

$$\epsilon_B = \tfrac{1}{4}(\epsilon_{xx} + 3\epsilon_{yy} - \sqrt{3}\,\gamma_{xy})$$

$$\epsilon_C = \tfrac{1}{4}(\epsilon_{xx} + 3\epsilon_{yy} + \sqrt{3}\,\gamma_{xy}) \tag{10.10}$$

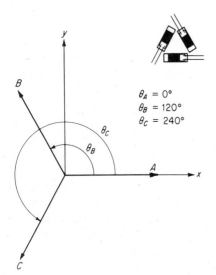

$\theta_A = 0°$
$\theta_B = 120°$
$\theta_C = 240°$

Figure 10.6 Gage positions in a three-element delta rosette.

Solving Eqs. (10.10) for ϵ_{xx}, ϵ_{yy}, and γ_{xy} in terms of ϵ_A, ϵ_B, and ϵ_C gives

$$\epsilon_{xx} = \epsilon_A \qquad \epsilon_{yy} = \tfrac{1}{3}[2(\epsilon_B + \epsilon_C) - \epsilon_A] \qquad \gamma_{xy} = \frac{2\sqrt{3}}{3}(\epsilon_C - \epsilon_B) \qquad (10.11)$$

Also from Eqs. (10.5) the principal strains ϵ_1 and ϵ_2 can be written in terms of ϵ_A, ϵ_B, and ϵ_C as

$$\epsilon_1 = \frac{1}{3}(\epsilon_A + \epsilon_B + \epsilon_C) + \frac{\sqrt{2}}{3}\sqrt{(\epsilon_A - \epsilon_B)^2 + (\epsilon_B - \epsilon_C)^2 + (\epsilon_C - \epsilon_A)^2}$$

$$\epsilon_2 = \frac{1}{3}(\epsilon_A + \epsilon_B + \epsilon_C) - \frac{\sqrt{2}}{3}\sqrt{(\epsilon_A - \epsilon_B)^2 + (\epsilon_B - \epsilon_C)^2 + (\epsilon_C - \epsilon_A)^2}$$

$$(10.12)$$

The principal angle ϕ can be determined from Eqs. (10.5) as

$$\tan 2\phi = \frac{(2/\sqrt{3})(\epsilon_C - \epsilon_B)}{2[\epsilon_A - \tfrac{1}{3}(\epsilon_A + \epsilon_B + \epsilon_C)]} = \frac{\sqrt{3}(\epsilon_C - \epsilon_B)}{2\epsilon_A - (\epsilon_B + \epsilon_C)} \qquad (10.13)$$

The solution of Eq. (10.13) gives two values for the principal angle ϕ, as was the case for the rectangular rosette. The angles ϕ_1 and ϕ_2 are illustrated in the Mohr's strain circle shown in Fig. 10.7. It is possible to show (see Exercise 10.7) that the

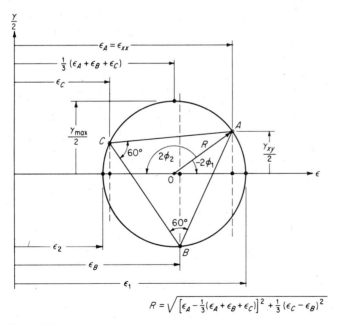

$$R = \sqrt{\left[\epsilon_A - \tfrac{1}{3}(\epsilon_A + \epsilon_B + \epsilon_C)\right]^2 + \tfrac{1}{3}(\epsilon_C - \epsilon_B)^2}$$

Figure 10.7 Graphical solution for the principal strains and their directions from a delta rosette.

principal angles can be identified by applying the following rules:

$$0° < \phi_1 < 90° \qquad \text{when } \epsilon_C > \epsilon_B$$
$$-90° < \phi_1 < 0° \qquad \text{when } \epsilon_C < \epsilon_B$$
$$\phi_1 = 0 \qquad \text{when } \epsilon_B = \epsilon_C \text{ and } \epsilon_A > \epsilon_B = \epsilon_C$$
$$\phi_1 = \pm 90° \qquad \text{when } \epsilon_B = \epsilon_C \text{ and } \epsilon_A < \epsilon_B = \epsilon_C \qquad (10.14)$$

Finally, the principal stresses can be determined from the principal strains by employing Eqs. (10.3) to obtain

$$\sigma_1 = E\left\{ \left| \frac{\epsilon_A + \epsilon_B + \epsilon_C}{3(1 - v)} \right. \right.$$

$$\left. + \frac{\sqrt{2}}{3(1 + v)}\sqrt{(\epsilon_A - \epsilon_B)^2 + (\epsilon_B - \epsilon_C)^2 + (\epsilon_C - \epsilon_A)^2} \right\}$$

$$\sigma_2 = E\left\{ \left| \frac{\epsilon_A + \epsilon_B + \epsilon_C}{3(1 - v)} \right. \right.$$

$$\left. - \frac{\sqrt{2}}{3(1 + v)}\sqrt{(\epsilon_A - \epsilon_B)^2 + (\epsilon_B - \epsilon_C)^2 + (\epsilon_C - \epsilon_A)^2} \right\} \qquad (10.15)$$

By employing Eqs. (10.11) to (10.15), it is possible to determine the cartesian components of strain, the principal strains and their directions, and the principal stresses from the three observations of strain made with a delta rosette. This approach is purely analytical and does not require the construction of a Mohr's strain diagram. However, it is possible to achieve the same results by graphical construction of the Mohr's strain diagram, as illustrated in Fig. 10.7.

The normal strain ϵ is represented along the abscissa, and one-half of the shearing strain $\frac{1}{2}\gamma$ is represented along the ordinate. The center of the circle (point O) is established by plotting the point corresponding to $\frac{1}{3}(\epsilon_A + \epsilon_B + \epsilon_C)$ along the abscissa. Also, the three strains ϵ_A, ϵ_B, and ϵ_C are plotted on the abscissa, and a vertical line is drawn through each of these three points. Next γ_{xy} is computed from Eqs. (10.11) and $\frac{1}{2}\gamma_{xy}$ is plotted along the vertical line through ϵ_A. A positive value of γ_{xy} is plotted downward, and a negative value is plotted upward to establish point A, as indicated in Fig. 10.7. A line drawn between points O and A establishes the radius of the circle. The circle is drawn, and its intercepts on the abscissa establish the principal strains ϵ_1 and ϵ_2. The accuracy of the construction of the circle can be checked by identifying the circle intercepts with the vertical lines through ϵ_B and ϵ_C as points B and C, respectively. If straight lines AB, AC, and BC are drawn, an equilateral triangle will be formed inside the circle if the construction is correct.

The principal angles are established by measuring the angle $AO\epsilon_1$, which gives $2\phi_1$, and angle $AO\epsilon_2$, which gives $2\phi_2$. If point A is located on the lower half of the circle, ϕ_1 will be positive and ϕ_2 negative. If point A is located on the upper half of the circle, ϕ_1 will be negative and ϕ_2 positive, as indicated in Fig. 10.7. The

Table 10.1 A summary of the equations used to determine principal strains, principal stresses, and their directions from four types of rosettes

Type of rosette	Gage arrangement	Principal strain and principal stress	Principal angle	Identification $0 < \phi_1 < 90°$
Three-element, rectangular		$\epsilon_{1,} = \dfrac{\epsilon_A + \epsilon_C}{2} \pm \dfrac{1}{2}\sqrt{(\epsilon_A - \epsilon_C)^2 + (2\epsilon_B - \epsilon_A - \epsilon_C)^2}$ $\sigma_{1,2} = \dfrac{E}{2}\left[\dfrac{\epsilon_A + \epsilon_C}{1-\nu} \pm \dfrac{1}{1+\nu}\sqrt{(\epsilon_A - \epsilon_C)^2 + (2\epsilon_B - \epsilon_A - \epsilon_C)^2}\right]$	$\tan 2\phi_1$ $= \dfrac{2\epsilon_B - \epsilon_A - \epsilon_C}{\epsilon_A - \epsilon_C}$	$\epsilon_B > \dfrac{\epsilon_A + \epsilon_C}{2}$
Delta		$\epsilon_{1,2} = \dfrac{\epsilon_A + \epsilon_B + \epsilon_C}{3} \pm \dfrac{\sqrt{2}}{3}\sqrt{(\epsilon_A - \epsilon_B)^2 + (\epsilon_B - \epsilon_C)^2 + (\epsilon_C - \epsilon_A)^2}$ $\sigma_{1,2} = \dfrac{E}{3}\left[\dfrac{\epsilon_A + \epsilon_B + \epsilon_C}{(1-\nu)} \pm \dfrac{\sqrt{2}}{(1+\nu)}\sqrt{(\epsilon_A - \epsilon_B)^2 + (\epsilon_B - \epsilon_C)^2 + (\epsilon_C - \epsilon_A)^2}\right]$	$\tan 2\phi_1$ $= \dfrac{\sqrt{3}(\epsilon_C - \epsilon_B)}{2\epsilon_A - (\epsilon_B + \epsilon_C)}$	$\epsilon_C > \epsilon_B$
Four-element, rectangular		$\epsilon_{1,2} = \dfrac{\epsilon_A + \epsilon_B + \epsilon_C + \epsilon_D}{4} \pm \dfrac{1}{2}\sqrt{(\epsilon_A - \epsilon_C)^2 + (\epsilon_B - \epsilon_D)^2}$ $\sigma_{1,2} = \dfrac{E}{2}\left[\dfrac{\epsilon_A + \epsilon_B + \epsilon_C + \epsilon_D}{2(1-\nu)} \pm \dfrac{1}{1+\nu}\sqrt{(\epsilon_A - \epsilon_C)^2 + (\epsilon_B - \epsilon_D)^2}\right]$	$\tan 2\phi_1 = \dfrac{\epsilon_B - \epsilon_D}{\epsilon_A - \epsilon_C}$	$\epsilon_B > \epsilon_D$
Tee-delta		$\epsilon_{1,2} = \dfrac{\epsilon_A + \epsilon_D}{2} \pm \dfrac{1}{2}\sqrt{(\epsilon_A - \epsilon_D)^2 + \dfrac{4}{3}(\epsilon_C - \epsilon_B)^2}$ $\sigma_{1,2} = \dfrac{E}{2}\left[\dfrac{\epsilon_A + \epsilon_D}{1-\nu} \pm \dfrac{1}{1+\nu}\sqrt{(\epsilon_A - \epsilon_D)^2 + \dfrac{4}{3}(\epsilon_C - \epsilon_B)^2}\right]$	$\tan 2\phi_1$ $= \dfrac{2(\epsilon_C - \epsilon_B)}{\sqrt{3}(\epsilon_A - \epsilon_D)}$	$\epsilon_C > \epsilon_B$

maximum shearing strain is given by the projection of the circle radius onto the ordinate. The principal stresses can be established from the principal strains by employing Eqs. (10.3).

Both the delta and the rectangular rosette can be employed with about equal facility for determining the principal stresses and their directions on the surface of the specimen. The selection of one type of rosette over the other will depend primarily on the nature of the stress field at the point of interest and on the type of rosette commercially available in the size (gage length) required.

A complete summary of the analytical expressions used to reduce the data obtained from four different rosettes is presented in Table 10.1.

10.4 CORRECTIONS FOR TRANSVERSE STRAIN EFFECTS [11–14]

In Sec. 6.5 it was noted that foil-type resistance strain gages exhibit a sensitivity S_t to transverse strains. Reference to Fig. 6.12 shows that in certain instances this transverse sensitivity can lead to large errors, and it is important to correct the data to eliminate this effect. Two different procedures for correcting data have been developed.

The first procedure requires a priori knowledge of the ratio ϵ_t/ϵ_a of the strain field. The correction factor is evident in Eq. (6.8), where

$$\epsilon_a = \epsilon_a' \frac{1 - v_0 K_t}{1 + K_t \epsilon_t/\epsilon_a} \tag{6.8}$$

The term ϵ_a' is the apparent strain, and the correction factor CF is given by

$$\text{CF} = \frac{1 - v_0 K_t}{1 + K_t \epsilon_t/\epsilon_a} \tag{10.16}$$

It is possible to correct the strain gage for this transverse sensitivity by adjusting its gage factor. The corrected gage factor S_g^* which should be dialed into the measuring instrument is

$$S_g^* = S_g \frac{1 + K_t \epsilon_t/\epsilon_a}{1 - v_0 K_t} \tag{10.17}$$

Correction for the cross-sensitivity effect when the strain field is unknown is more involved and requires the experimental determination of strain in both the x and y directions. If ϵ_{xx}' and ϵ_{yy}' are the apparent strains recorded in the x and y directions, respectively, then from Eq. (6.8) it is evident that

$$\epsilon_{xx}' = \frac{1}{1 - v_0 K_t}(\epsilon_{xx} + K_t \epsilon_{yy}) \qquad \epsilon_{yy}' = \frac{1}{1 - v_0 K_t}(\epsilon_{yy} + K_t \epsilon_{xx}) \tag{10.18}$$

where the unprimed quantities ϵ_{xx} and ϵ_{yy} are the true strains. Solving Eqs. (10.18) for ϵ_{xx} and ϵ_y gives

$$\epsilon_{xx} = \frac{1 - v_0 K_t}{1 - K_t^2}(\epsilon'_{xx} - K_t \epsilon'_{yy}) \qquad \epsilon_{yy} = \frac{1 - v_0 K_t}{1 - K_t^2}(\epsilon'_{yy} - K_t \epsilon'_{xx}) \qquad (10.19)$$

Equations (10.19) give the true strains ϵ_{xx} and ϵ_{yy} in terms of the apparent strains ϵ'_{xx} and ϵ'_{yy}. Correction equations for transverse strains in two- and three-element rosettes are given in Ref. 14.

10.5 THE STRESS GAGE [15–18]

The transverse sensitivity which was shown in the previous section to result in errors in strain measurements can be employed to produce a special-purpose transducer known as a *stress gage*. The stress gage looks very much like a strain gage (see Fig. 10.3c) except that its grid is designed to give a select value of K_t so that the output $\Delta R/R$ is proportional to the stress along the axis of the gage. The stress gage serves a very useful purpose when a stress determination in a particular direction is the ultimate objective of the analysis, for it can be obtained with a single gage rather than a three-element rosette.

The principle upon which a stress gage is based is exhibited in the following derivation. The output of a gage $\Delta R/R$ as expressed by Eq. (6.4) is

$$\frac{\Delta R}{R} = S_a(\epsilon_a + K_t \epsilon_t)$$

The relationship between stress and strain for a plane state of stress is given by Eqs. (2.19) as

$$\epsilon_a = \frac{1}{E}(\sigma_a - v\sigma_t) \qquad \epsilon_t = \frac{1}{E}(\sigma_t - v\sigma_a)$$

Substituting Eqs. (2.19) into Eq. (6.4) yields

$$\frac{\Delta R}{R} = \frac{S_a}{E}(\sigma_a - v\sigma_t) + \frac{K_t S_a}{E}(\sigma_t - v\sigma_a)$$

$$= \frac{\sigma_a S_a}{E}(1 - vK_t) + \frac{\sigma_t S_a}{E}(K_t - v) \qquad (a)$$

Examination of Eq. (a) indicates that the output of the gage $\Delta R/R$ will be independent of σ_t if $K_t = v$. It can also be shown that the axial sensitivity S_a of a gage is related to the alloy sensitivity S_A by the expression

$$S_a = \frac{S_A}{1 + K_t} \qquad (b)$$

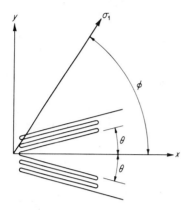

Figure 10.8 The stress gage relative to the x axis and the principal axis corresponding to σ_1.

Substituting Eq. (*b*) into Eq. (*a*) and letting $K_t = v$ leads to

$$\sigma_a = \frac{E}{S_A(1 - v)}\frac{\Delta R}{R} \tag{10.20}$$

Since the factor $E/S_A(1 - v)$ is a constant for a given gage alloy and specimen material, the gage output in terms of $\Delta R/R$ is linearly proportional to stress.

In practice the stress gage is made with a V-type grid configuration, as shown in Fig. 6.5*m*. Further analysis of the stress gage is necessary to understand its operation in a strain field which is unknown and in which the strain gage is placed in an arbitrary direction. Consider the placement of the gage, as shown in Fig. 10.8, along an arbitrary x axis which is at some unknown angle ϕ with the principal axis corresponding to σ_1. The grid elements are at a known angle θ relative to the x axis.

The strain along the top grid element is given by a modified form of Eq. (10.4) as

$$\epsilon_{\phi - \theta} = \tfrac{1}{2}(\epsilon_1 + \epsilon_2) + \tfrac{1}{2}(\epsilon_1 - \epsilon_2)\cos 2(\phi - \theta) \tag{c}$$

The strain along the lower grid element is

$$\epsilon_{\phi + \theta} = \tfrac{1}{2}(\epsilon_1 + \epsilon_2) + \tfrac{1}{2}(\epsilon_1 - \epsilon_2)\cos 2(\phi + \theta) \tag{d}$$

Summing Eqs. (*c*) and (*d*) and expanding the cosine terms yield

$$\epsilon_{\phi - \theta} + \epsilon_{\phi + \theta} = (\epsilon_1 + \epsilon_2) + (\epsilon_1 - \epsilon_2)\cos 2\phi \cos 2\theta \tag{e}$$

Note from the Mohr's strain circles presented earlier in this chapter that

$$\epsilon_{xx} + \epsilon_{yy} = \epsilon_1 + \epsilon_2 \tag{f}$$

$$\epsilon_{xx} - \epsilon_{yy} = (\epsilon_1 - \epsilon_2)\cos 2\phi \tag{g}$$

Substituting Eqs. (f) and (g) into Eq. (e) gives

$$\epsilon_{\phi-\theta} + \epsilon_{\phi+\theta} = (\epsilon_{xx} + \epsilon_{yy}) + (\epsilon_{xx} - \epsilon_{yy}) \cos 2\theta$$
$$= 2(\epsilon_{xx} \cos^2 \theta + \epsilon_{yy} \sin^2 \theta)$$
$$= 2 \cos^2 \theta(\epsilon_{xx} + \epsilon_{yy} \tan^2 \theta) \qquad (h)$$

If the gage is manufactured so that θ is equal to arctan $\sqrt{v}$, then

$$\tan^2 \theta = v \qquad \cos^2 \theta = \frac{1}{1+v}$$

and Eq. (h) becomes

$$\epsilon_{\phi-\theta} + \epsilon_{\phi+\theta} = \frac{2}{1+v}(\epsilon_{xx} + v\epsilon_{yy}) \qquad (i)$$

Substituting Eq. (i) into Eqs. (2.20) results in

$$\sigma_{xx} = \frac{E}{2(1-v)}(\epsilon_{\phi+\theta} + \epsilon_{\phi-\theta}) \qquad (10.21)$$

where $\frac{1}{2}(\epsilon_{\phi+\theta} + \epsilon_{\phi-\theta})$ is the average strain indicated by the two elements of the gage and is equal to $(\Delta R/R)/S_g$.

The gage reading will give $\frac{1}{2}(\epsilon_{\phi+\theta} + \epsilon_{\phi-\theta})$, and it is only necessary to multiply this reading by $E/(1-v)$ to obtain σ_{xx}. The stress gage will thus give σ_{xx} directly with a single gage. However, it does not give any data regarding σ_{yy} or the principal angle ϕ. Moreover, σ_{xx} may not be the most important stress since it may differ appreciably from σ_1. If the directions of the principal stresses are known, the stress gage may be used more effectively by choosing the x axis to coincide with the principal axis corresponding to σ_1 so that $\sigma_{xx} = \sigma_1$. In fact, when the principal directions are known, a conventional single-element strain gage can be employed as a stress gage.

This adaption is possible if the gage is located along a line which makes an angle θ with respect to the principal axis, as shown in Fig. 10.9. In this case the

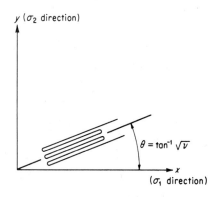

y (σ_2 direction)

$\theta = \tan^{-1} \sqrt{v}$

x
(σ_1 direction)

Figure 10.9 A single-element strain gage employed as a stress gage when the principal directions are known.

strains will be symmetrical about the principal axis; hence it is clear that

$$\epsilon_{\phi-\theta} = \epsilon_{\phi+\theta} = \epsilon_\theta$$

and Eq. (10.21) reduces to

$$\sigma_1 = \sigma_{xx} = \frac{E}{1-v}\epsilon_\theta \tag{10.22}$$

The value ϵ_θ is recorded on the strain gage and converted to σ_{xx} or σ_1 directly by multiplying by $E/(1-v)$. This procedure reduces the number of gages necessary if only the value of σ_1 is to be determined. The saving of a gage is of particular importance in dynamic work when the instrumentation required becomes complex and the number of available channels of recording equipment is limited.

10.6 PLANE-SHEAR GAGES [19]

Consider two strain gages A and B oriented at angles θ_A and θ_B with respect to the x axis, as shown in Fig. 10.10. The strains along the gage axes are given by a modified form of Eqs. (2.18) as

$$\epsilon_A = \frac{\epsilon_{xx} + \epsilon_{yy}}{2} + \frac{\epsilon_{xx} - \epsilon_{yy}}{2}\cos 2\theta_A + \frac{\gamma_{xy}}{2}\sin 2\theta_A$$

$$\epsilon_B = \frac{\epsilon_{xx} + \epsilon_{yy}}{2} + \frac{\epsilon_{xx} - \epsilon_{yy}}{2}\cos 2\theta_B + \frac{\gamma_{xy}}{2}\sin 2\theta_B \tag{10.23}$$

From Eqs. (10.23), the shear strain γ_{xy} is

$$\gamma_{xy} = \frac{2(\epsilon_A - \epsilon_B) - (\epsilon_{xx} - \epsilon_{yy})(\cos 2\theta_A - \cos 2\theta_B)}{\sin 2\theta_A - \sin 2\theta_B} \tag{a}$$

If gages A and B are oriented such that

$$\cos 2\theta_A = \cos 2\theta_B \tag{b}$$

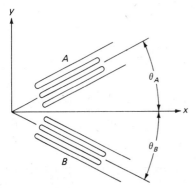

Figure 10.10 Positions of gages A and B for measuring γ_{xy}.

250US

120NC

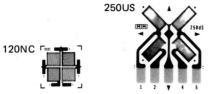

Figure 10.11 Four-element shear-strain gages.

then Eq. (a) reduces to

$$\gamma_{xy} = \frac{(\epsilon_A - \epsilon_B)}{\sin 2\theta_A - \sin 2\theta_B} \tag{10.24}$$

Since the cosine is an even function, $\theta_A = -\theta_B$ satisfies Eq. (b). Thus, the shearing strain γ_{xy} is proportional to the difference between normal strains experienced by gages A and B when they are oriented with respect to the x axis as shown in Fig. 10.10. The angle $\theta_A = -\theta_B$ can be arbitrary; however, for the angle $\theta_A = \pi/4$, Eq. (10.24) reduces simply to

$$\gamma_{xy} = \epsilon_A - \epsilon_B \tag{10.25}$$

Equation (10.25) indicates that the shearing strain γ_{xy} can be measured with a two-element rectangular rosette by orienting the gages at 45 and $-45°$ with respect to the x axis and connecting one gage into arm R_1 and the other into arm R_4 of a Wheatstone bridge. The subtraction $\epsilon_A - \epsilon_B$ will be performed automatically in the bridge, and the output will give γ_{xy} directly.

Four-element shear gages are marketed commercially for this measurement. The four elements provide a complete four-arm bridge with twice the output of the two-element rectangular rosette. Typical shear gages are illustrated in Fig. 10.11.

EXERCISES

10.1 Suppose a state of pure shear stress occurs (say on a circular shaft under pure torsion). Show how a single-element strain gage can be employed to determine the principal stresses σ_1 and σ_2.

10.2 In a thin-walled cylinder loaded with internal pressure ($\sigma_h = 2\sigma_a$), show how a single-element strain gage in the hoop direction can be used to establish the stresses σ_h and σ_a.

10.3 A two-element rectangular rosette was used to determine the two principal stresses at the point shown in Fig. E10.3. If $\sigma_1 = 200$ MPa and $\sigma_2 = 75$ MPa, find σ_{xx}, σ_{yy}, and τ_{xy} when $\phi = 30°$.

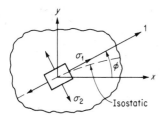

Figure E10.3

10.4 Prove the validity of Eqs. (10.8) by using the Mohr's strain diagram presented in Fig. 10.5.

10.5 Derive an expression for the maximum shear stress at a point in terms of the strains obtained from a three-element rectangular rosette by using Eqs. (10.9).

10.6 The following observations are made with a rectangular rosette mounted on a steel specimen:

Case number	ϵ_A, μm/m	ϵ_B, μm/m	ϵ_C, μm/m
1	500	−300	0
2	900	300	−200
3	−500	200	200
4	800	−100	−900
5	−200	0	200

Determine the principal strains, the principal stresses, and the principal angles ϕ_1 and ϕ_2.

10.7 Prove the validity of Eqs. (10.14) by using the Mohr's strain diagram in Fig. 10.7.

10.8 The following observations were made with a delta rosette mounted on a steel specimen:

Case number	ϵ_A, μm/m	ϵ_B, μm/m	ϵ_C, μm/m
1	400	−200	200
2	800	400	0
3	−600	300	400
4	700	0	−700
5	−300	100	400

Determine the principal strains, the principal stresses, and the principal angles ϕ_1 and ϕ_2.

10.9 Verify the expressions listed in Table 10.1 for the rectangular four-element rosette.

10.10 Verify the expressions listed in Table 10.1 for the tee-delta rosette.

10.11 Determine the error due to cross sensitivity when a strain gage with $K_t = -0.06$ is used on aluminum ($E = 70$ GPa and $v = 0.33$) to measure (a) a state of hydrostatic compression, (b) a state of pure shear, and (c) Poisson's ratio.

10.12 Solve problem 10.11 if the gage is used on steel ($E = 200$ GPa and $v = 0.29$).

10.13 The following apparent strain data were obtained with two-element rectangular rosettes:

Rosette number	ϵ'_{xx}, μm/m	ϵ'_{yy}, μm/m
1	600	300
2	−200	700
3	1,200	400
4	600	−300

Determine the true strains ϵ_{xx} and ϵ_{yy} if $K_t = 0.01$. In each case, determine the error which would have occurred if the cross sensitivity of the gage had been neglected.

10.14 Solve Exercise 10.13 if $K_t = 0.03$.

10.15 Solve Exercise 10.13 if $K_t = -0.03$.

10.16 Determine the included angle between elements in a V-type stress gage designed for use on:
    (a) Glass   (b) Steel   (c) Aluminum

10.17 Design a double-grid gage which can serve as a stress gage. Specify the length of conductor required in the axial and transverse directions for the design shown in Fig. E10.17.

Figure E10.17

10.18 Does the double-grid stress gage shown in Fig. E10.17 have any advantages over the V-configuration stress gage shown in Fig. 10.8?

10.19 Determine $\Delta R/R$ for the circular-arc gage element shown in Fig. E10.19. Neglect transverse-sensitivity effects.

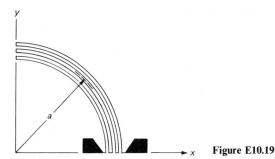

Figure E10.19

10.20 Show that two circular-arc gage elements (see Fig. E10.19) located in adjacent quadrants and positioned in arms R_1 and R_4 of a Wheatstone bridge have an output $\Delta E/V$ proportional to $2\gamma_{xy}/\pi$.

10.21 Determine $\Delta E/V$ for a shear gage consisting of four circular-arc elements arranged in all four quadrants and connected into the four arms of a Wheatstone bridge.

REFERENCES

1. Ades, C. S.: Reduction of Strain Rosettes in the Plastic Range, *Exp. Mech.*, vol. 2, no. 11, pp. 345–349, 1962.
2. Bossart, K. J., and G. A. Brewer: A Graphical Method of Rosette Analysis, *Proc. SESA*, vol. IV, no. 1, pp. 1–8, 1946.
3. Duke, M. E., and E. Wenk, Jr.: The Graphical Solution of 45 Degree Strain Rosette Data and Determination of Error in the Calculated Stress Due to Errors in Measured Strains, *David Taylor Model Basin Rep.* 600, 1949.
4. Grossman, N.: A Nomographic Rosette Computer, *Proc. SESA*, vol. IV, no. 1, pp. 27–35, 1946.
5. Manson, S. S.: Charts for Rapid Analysis of 45° Strain Rosette Data, *NACA*, TN940, May, 1944.
6. McClintock, F. A.: On Determining Principal Strains from Strain Rosettes with Arbitrary Angles, *Proc. SESA*, vol. IX, no. 1, pp. 209–210, 1951.

7. Murray, W. M.: Machine Solution of the Strain Rosette Equation, *Proc. SESA*, vol. II, no. 1, pp. 106–112, 1944.
8. Murray, W. M.: Some Simplifications in Rosette Analysis, *Proc. SESA*, vol. XV, no. 2, pp. 39–52, 1958.
9. Murray, W. M.: An Adjunct to the Strain Rosette, *Proc. SESA*, vol. I, no. 1, pp. 128–133, 1943.
10. Stein, P. K.: A Simplified Method of Obtaining Principal Stress Information from Strain Gage Rosettes, *Proc. SESA*, vol. XV, no. 2, pp. 21–38, 1958.
11. Meier, J. H.: On the Transverse-Strain Sensitivity of Foil Gages, *Exp. Mech.*, vol. 1, no. 7, pp. 39–40, 1961.
12. Meyer, M. L.: A Simple Estimate for the Effect of Cross Sensitivity on Evaluated Strain-Gage Measurement, *Exp. Mech.*, vol. 7, no. 11, pp. 476–480, 1967.
13. Wu, C. T.: Transverse Sensitivity of Bonded Strain Gages, *Exp. Mech.*, vol. 2, no. 11, pp. 338–344, 1962.
14. Errors Due to Transverse Sensitivity in Strain Gages, *Micro-Measurements* TN-137.
15. Williams, S. B.: Geometry in the Design of Stress Measuring Circuits, *Proc. SESA*, vol. XVII, no. 2, pp. 161–178, 1960.
16. Hines, F. F.: The Stress-Strain Gage, *Proc. 1st Int. Congr. Exp. Mech. 1963*, pp. 237–253.
17. Kern, K. E.: The Stress Gage, *Proc. SESA*, vol. IV, no. 1, pp. 124–129, 1946.
18. Lissner, H. R., and C. C. Perry: Conventional Wire Strain Gage Used as a Principal Stress Gage, *Proc. SESA*, vol. XIII, no. 1, pp. 25–32, 1955.
19. Perry, C. C.: Plane-Shear Measurement with Strain Gages, *Exp. Mech.*, vol. 9, no. 1, 1969.

FOUR

OPTICAL METHODS OF STRESS ANALYSIS

ELEVEN

BASIC OPTICS

11.1 THE NATURE OF LIGHT

The phenomenon of light has attracted the attention of man from the earliest times. The ancient Greeks considered light to be an emission of small particles by a luminous body which entered the eye and returned to the body. Empedocles (484–424 B.C.) suggested that light takes time to travel from one point to another; however, Aristotle (384–322 B.C.) later rejected this idea as being too much to assume. The ideas of Aristotle concerning the nature of light persisted for approximately 2000 years.

In the seventeenth century, considerable effort was devoted to a study of the optical effects associated with thin films, lenses, and prisms. Huygens (1629–1695) and Hooke (1635–1703) attempted to explain some of these effects with a wave theory. In the wave theory, a hypothetical substance of zero mass, called the *ether*, was assumed to occupy all space. Initially, light propagation was assumed to be a longitudinal vibratory disturbance moving through the ether. The idea of secondary wavelets, in which each point on a wavefront can be regarded as a new source of waves, was proposed by Huygens to explain refraction. Huygens' concept of secondary wavelets is widely used today to explain, in a simple way, other optical effects such as diffraction and interference. At about the same time, Newton (1642–1727) proposed his corpuscular theory, in which light is visualized as a stream of small but swift particles emanating from shining bodies. The theory was

339

able to explain most of the optical effects observed at the time, and, thanks to Newton's stature, was widely accepted for approximately 100 years.

A revival of interest in the wave theory of light began with the work of Young (1773–1829), who demonstrated that the presence of a refracted ray at an interface between two materials was to be expected from a wave theory while the corpuscular theory of Newton could explain the effect only with difficulty. His two-pinhole experiment, which demonstrated the interference of light, together with the work of Fresnel (1788–1827) on polarized light, which required transverse rather than longitudinal vibrations, firmly established the transverse ether wave theory of light.

The next major step in the evolution of the theory of light was due to Maxwell (1831–1879). His electromagnetic theory predicts the presence of two vector fields in light waves, an electric field and a magnetic field. Since these fields can propagate through space unsupported by any known matter, the need for the hypothetical ether of the previous wave theory was eliminated. The electromagnetic wave theory also unites light with all the other invisible entities of the electromagnetic spectrum, e.g., cosmic rays, gamma rays, x-rays, ultraviolet rays, infrared rays, microwaves, radio waves, and electric-power-transmission waves. The wide range of wavelengths and frequencies available for study in the electromagnetic spectrum has led to rapid development of additional theory and understanding.

Observation of the photoelectric effect by Hertz in 1887, which cannot be explained by a wave theory but is easily explained by a particle theory, led to Einstein's photon theory in 1907. The modern theory of wave mechanics successfully reconciles these two approaches, in which energy can be manifest in either particle or wave forms. For most of the effects to be described in later sections of this text the wave properties of light are important and the particle characteristics of individual photons have little application. For this reason, simple wave theory will be used in most of the discussions which follow.

11.2 WAVE THEORY OF LIGHT [1]

Electromagnetic radiation is predicted by Maxwell's theory to be a transverse wave motion which propagates with an extremely high velocity. Associated with the wave are oscillating electric and magnetic fields which can be described with electric and magnetic vectors **E** and **H**. These vectors are in phase, perpendicular to each other, and at right angles to the direction of propagation. A simple representation of the electric and magnetic vectors associated with an electromagnetic wave at a given instant of time is illustrated in Fig. 11.1. For simplicity and convenience of representation, the wave has been given sinusoidal form. Other wave forms such as the *sawtooth waveform* or the *square waveform* are often encountered in electronics. These complicated waveforms are frequently represented for mathematical analysis by a Fourier series; therefore, the simple sinusoidal representation provides the basic information needed for the analysis of more complicated shapes.

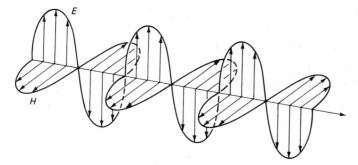

Figure 11.1 The electric and magnetic vectors associated with a plane electromagnetic wave.

All types of electromagnetic radiation propagate with the same velocity in free space (approximately 3×10^8 m/s, or 186,000 mi/s). Characteristics used to differentiate between the various radiations are wavelength and frequency. These two quantities are related to the velocity by the relationship

$$\lambda f = c \tag{11.1}$$

where λ = wavelength
  f = frequency
  c = velocity of propagation

The electromagnetic spectrum has no upper or lower limit of wavelength or frequency. The radiations observed to date have been classified in the broad general categories shown in Fig. 11.2.

Light is usually defined as radiation that can affect the human eye. From Fig. 11.2 it can be seen that the visible range of the spectrum is a small band centered about a wavelength of approximately 550 nm. The limits of the visible spectrum are not well defined because the eye ceases to be sensitive at both long and short wavelengths within the region; however, normal vision is usually

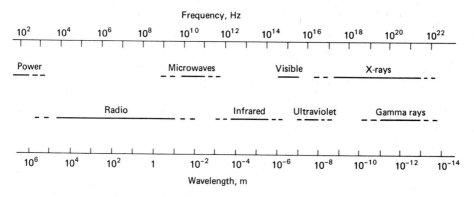

Figure 11.2 The electromagnetic spectrum.

Table 11.1 The visible spectrum

Wavelength range, nm	Color	Wavelength range, nm	Color
400–450	Violet	550–570	Yellow-green
450–480	Blue	570–590	Yellow
480–510	Blue-green	590–630	Orange
510–550	Green	630–700	Red

assumed to be the range from 400 to 700 nm. Within this range the eye interprets the wavelengths as the different colors listed in Table 11.1. Light from a source that emits a continuous spectrum with equal energy for every wavelength is interpreted as *white light*. Light of a single wavelength is known as *monochromatic light.*

Electromagnetic waves can be classified as one-, two-, or three-dimensional according to the number of dimensions in which they propagate energy. Light waves which emanate radially from a small source are three-dimensional. Two quantities associated with a propagating wave which will be useful in discussions involving geometrical and physical optics are wavefronts and rays. For a three-dimensional pulse of light emanating from a source, both the electric vector and the magnetic vector exhibit the periodic variation in magnitude shown in Fig. 11.1 along any radial line. The locus of points on different radial lines from the source exhibiting the same disturbance at a given instant of time, e.g., maximum or minimum values, is a surface known as a *wavefront*. As time passes, the surface moves and indicates how the pulse is propagating. If the medium is optically homogeneous and isotropic, the direction of propagation will be at right angles to the wavefront. A line normal to the wavefront, indicating the direction of propagation of the waves, is called a *ray*. When the waves are propagated out in all directions from a point source, the wavefronts are spheres and the rays are radial lines in all directions from the source. At large distances from the source, the spherical wavefronts have very little curvature, and over a limited region they can be regarded as plane. Plane wavefronts can also be produced by using a lens or mirror to direct a portion of the light from a point source into a parallel beam.

In ordinary light, which is emitted from, say, a tungsten-filament light bulb, the light vector is not restricted in any sense and may be considered to be composed of a number of arbitrary transverse vibrations. Each of the components may have a different wavelength, a different amplitude, a different orientation (plane of vibration), and a different phase with respect to the others. The vector used to represent the light wave can be either the electric vector or the magnetic vector. Both exist simultaneously, as shown in Fig. 11.1, and either or both can be used to describe the optical effects associated with photoelasticity, moiré, and holography. The electric vector has been shown in experiments by Wiener (1890) to be the active agent in interactions between light and a photographic plate; therefore, in all future discussions, attention will be devoted exclusively to vibrations associated with the electric vector.

A. The Wave Equation

Since the disturbance producing light can be represented by a transverse wave motion, it is possible to express the magnitude of the light (electric) vector in terms of the solution of the one-dimensional wave equation:

$$E = f(z - ct) + g(z + ct) \tag{11.2}$$

where
$\quad E$ = magnitude of light vector
$\quad z$ = position along axis of propagation
$\quad t$ = time
$\quad f(z - ct)$ = wave motion in positive z direction
$\quad g(z + ct)$ = wave motion in negative z direction

Most optical effects of interest in experimental stress analysis can be described with a simple sinusoidal or harmonic waveform. Thus, light propagating in the positive z direction away from the source can be represented by Eq. (11.2) as

$$E = f(z - ct) = \frac{K}{z} \cos \frac{2\pi}{\lambda}(z - ct) \tag{11.3}$$

where K is related to the strength of the source and K/z is an attenuation coefficient associated with the expanding spherical wavefront. At distances far from the source, the attenuation is small over short observation distances, and therefore it is frequently neglected. For plane waves, the attenuation does not occur since the beam of light maintains a constant cross section. Equation (11.3) can then be written as

FOR   PLANE WAVES

$$E = a \cos \frac{2\pi}{\lambda}(z - ct) \tag{11.4}$$

where a is a constant known as the *amplitude* of the wave. A graphical representation of the magnitude of the light vector as a function of position along the positive z axis, at two different times, is shown for a plane light wave in Fig. 11.3. The length from peak to peak on the magnitude curve for the light vector is

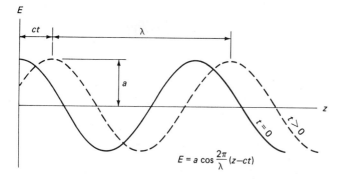

Figure 11.3 Magnitude of the light vector as a function of position along the axis of propagation at two different times.

defined as the _wavelength_ λ. The time required for passage of two successive peaks at some fixed value of z is defined as the _period T_ of the wave and is given by

$$T = \frac{\lambda}{c} \tag{11.5}$$

The _frequency_ of the light vector is defined as the number of oscillations per second. Thus, the frequency is the reciprocal of the period, or

$$f = \frac{1}{T} = \frac{c}{\lambda} \tag{11.6}$$

The terms angular frequency and wave number are frequently used to simplify the argument in a sinusoidal representation of a light wave. The _angular frequency_ ω and the _wave number_ ξ are given by

$$\omega = \frac{2\pi}{T} = 2\pi f \tag{11.7}$$

$$\xi = \frac{2\pi}{\lambda} \tag{11.8}$$

Substituting Eqs. (11.7) and (11.8) into Eq. (11.4) yields

$$E = a \cos{(\xi z - \omega t)} \tag{a}$$

Two waves having the same wavelength and amplitude but a different phase are shown in Fig. 11.4. The two waves can be expressed by

$$E_1 = a \cos{\frac{2\pi}{\lambda}(z + \delta_1 - ct)} \qquad E_2 = a \cos{\frac{2\pi}{\lambda}(z + \delta_2 - ct)} \tag{11.9}$$

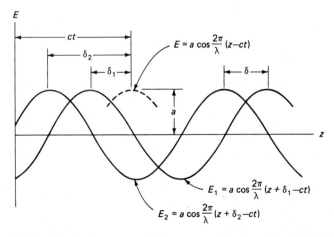

Figure 11.4 Magnitude of the light vector as a function of position along the axis of propagation for two waves with different initial phases.

where δ_1 = initial phase of wave E_1

δ_2 = initial phase of wave E_2

$\delta = \delta_2 - \delta_1$ = the *linear phase difference* between waves

The linear phase difference δ is often referred to as a *retardation* since wave 2 trails wave 1.

The magnitude of the light vector can also be plotted as a function of time at a fixed position along the beam. This representation is useful for many applications since the eye, photographic films, and other light-detecting devices are normally located at fixed positions for observations.

B. Superposition of Waves

In later chapters on photoelasticity and moiré, the phenomena associated with the superposition of two waves having the same frequency but different amplitude and phase will be encountered. At a fixed position z_0 along the light beam, where the observations will be made, the equations for the waves can be expressed as

$$E_1 = a_1 \cos \frac{2\pi}{\lambda}(z_0 + \delta_1 - ct) = a_1 \cos (\phi_1 - \omega t)$$

$$E_2 = a_2 \cos \frac{2\pi}{\lambda}(z_0 + \delta_2 - ct) = a_2 \cos (\phi_2 - \omega t) \qquad (11.10)$$

where ϕ_1 = phase angle associated with wave E_1 at position z_0

ϕ_2 = phase angle associated with wave E_2 at position z_0

a_1 = amplitude of wave E_1

a_2 = amplitude of wave E_2

Consider first the case where the light vectors associated with the two waves oscillate in the same plane. The magnitude of the resulting light vector is simply

$$E = E_1 + E_2 \qquad (b)$$

Substituting Eqs. (11.10) into Eq. (b) yields

$$E = a_1(\cos \omega t \cos \phi_1 + \sin \omega t \sin \phi_1)$$

$$+ a_2(\cos \omega t \cos \phi_2 + \sin \omega t \sin \phi_2)$$

$$= (a_1 \cos \phi_1 + a_2 \cos \phi_2) \cos \omega t + (a_1 \sin \phi_1 + a_2 \sin \phi_2) \sin \omega t$$

$$= a \cos \phi \cos \omega t + a \sin \phi \sin \omega t$$

$$= a \cos (\phi - \omega t) \qquad (c)$$

where

$$a^2 = a_1^2 + a_2^2 + 2a_1 a_2 \cos (\phi_2 - \phi_1) \qquad (11.11)$$

$$\tan \phi = \frac{a_1 \sin \phi_1 + a_2 \sin \phi_2}{a_1 \cos \phi_1 + a_2 \cos \phi_2} \qquad (11.12)$$

Equation (c) indicates that the resulting wave has the same frequency as the original waves but a different amplitude and a different phase angle. The above procedure can easily be extended to the addition of three or more waves.

A special case frequently arises where the amplitudes of the original waves are equal. In this case the amplitude of the resulting wave is given by Eq. (11.11) as

$$a = \sqrt{2a_1^2[1 + \cos(\phi_2 - \phi_1)]} \qquad (d)$$

Since $\phi_2 - \phi_1 = 2\pi\delta/\lambda$,

$$a = \sqrt{2a_1^2\left(1 + \cos\frac{2\pi\delta}{\lambda}\right)} = \sqrt{4a_1^2 \cos^2\frac{\pi\delta}{\lambda}} \qquad (e)$$

In most problems in optics the amplitude of the resulting wave is important, but the time variation is not. This results from the fact that the eye and other sensing instruments respond to the *intensity* of light (intensity is proportional to the square of the amplitude) but cannot detect the rapid time variations (for sodium light the frequency is 5.1×10^{14} Hz). Thus for the special case of two waves of equal amplitude the intensity is given by

$$I \sim a^2 = 4a_1^2 \cos^2\frac{\pi\delta}{\lambda} \qquad (11.13)$$

Equation (11.13) indicates that the intensity of the light wave resulting from superposition of two waves of equal amplitude is a function of the linear phase difference δ between the waves. The intensity of the resultant wave assumes its maximum value when $\delta = n\lambda, n = 0, 1, 2, 3, \ldots$, or when the linear phase difference is an integral number of wavelengths. Under this condition

$$I = 4a_1^2 \qquad (f)$$

which indicates that the intensity of the resultant wave is four times the intensity of one of the individual waves. The intensity of the resultant wave assumes its minimum value when $\delta = [(2n + 1)/2]\lambda, n = 0, 1, 2, 3, \ldots$, or when the linear phase difference is an odd number of half wavelengths. Under these conditions

$$I = 0 \qquad (g)$$

The modification of intensity by superposition of light waves is referred to as an *interference effect.* The effect represented by Eq. (f) is *constructive interference.* The effect represented by Eq. (g) is *destructive interference.* Interference effects have important application in photoelasticity, moiré, and holography.

In previous discussions, the electric vector used to describe the light wave was restricted to a single plane. Light exhibiting this preference for a plane of vibration is known as *plane-* or *linearly polarized light.*

Two other important forms of polarized light arise as a result of the superposition of two linearly polarized light waves having the same frequency but mutually perpendicular planes of vibration, as shown in Fig. 11.5. At a fixed position z_0 along the light beam, the equations for the two waves can be expressed as

$$E_x = a_x \cos\frac{2\pi}{\lambda}(z_0 + \delta_x - ct) = a_x \cos(\phi_x - \omega t)$$

$$E_y = a_y \cos\frac{2\pi}{\lambda}(z_0 + \delta_y - ct) = a_y \cos(\phi_y - \omega t) \qquad (11.14)$$

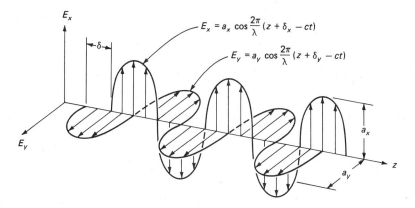

$$E_x = a_x \cos \frac{2\pi}{\lambda}(z + \delta_x - ct)$$

$$E_y = a_y \cos \frac{2\pi}{\lambda}(z + \delta_y - ct)$$

Figure 11.5 Two linearly polarized light waves having the same frequency but mutually perpendicular planes of vibration.

where ϕ_x = phase angle associated with wave in xz plane
$\quad\quad\phi_y$ = phase angle associated with wave in yz plane
$\quad\quad a_x$ = amplitude of wave in xz plane
$\quad\quad a_y$ = amplitude of wave in yz plane
The magnitude of the resulting light vector is given by

$$E = \sqrt{E_x^2 + E_y^2} \tag{h}$$

Considerable insight into the nature of the light resulting from the superposition of two mutually perpendicular waves is provided by a study of the trace of the tip of the resulting electric vector on a plane perpendicular to the axis of propagation at the point z_0. An expression for this trace can be obtained by eliminating time from Eqs. (11.14). Thus

$$\frac{E_x^2}{a_x^2} - 2\frac{E_x E_y}{a_x a_y}\cos(\phi_y - \phi_x) + \frac{E_y^2}{a_y^2} = \sin^2(\phi_y - \phi_x) \tag{i}$$

or since

$$\phi_y - \phi_x = \frac{2\pi}{\lambda}(\delta_y - \delta_x) = \frac{2\pi\delta}{\lambda}$$

$$\frac{E_x^2}{a_x^2} - 2\frac{E_x E_y}{a_x a_y}\cos\frac{2\pi\delta}{\lambda} + \frac{E_y^2}{a_y^2} = \sin^2\frac{2\pi\delta}{\lambda} \tag{11.15}$$

Equation (11.15) is the equation of an ellipse; therefore, light exhibiting this behavior is known as *elliptically polarized light*. At a fixed instant in time, the tips of the electric vectors at different positions along the z axis form an elliptical helix, as shown in Fig. 11.6. During an interval of time t, the helix will translate a distance ct in the positive z direction. As a result, the electric vector at position z_0 will rotate in a counterclockwise direction as the translating helix intersects the perpendicular plane at position z_0. The locus of points representing the trace of

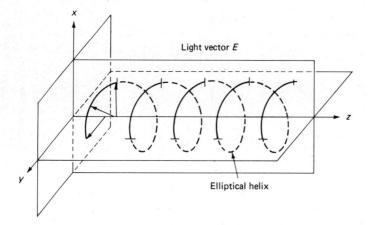

Figure 11.6 The elliptical helix formed by the tips of the light vectors along the axis of propagation at a fixed instant of time.

the tip of the light vector on the perpendicular plane is the ellipse described by Eq. (11.15) and illustrated in Fig. 11.7.

A special case of elliptically polarized light occurs when the amplitudes of the two waves E_x and E_y are equal and $\delta = [(2n + 1)/4]\lambda$, $n = 0, 1, 2, \ldots$, so that Eq. (11.15) reduces to

$$E_x^2 + E_y^2 = a^2 \tag{j}$$

Equation (j) is the equation of a circle; therefore, light exhibiting this behavior is known as *circularly polarized light.* In this case, the tips of the light vectors form a circular helix along the z axis at a given instant of time. For $\delta = \lambda/4, 5\lambda/4, \ldots$, the helix is a left circular helix, and the light vector at position z_0 rotates counterclockwise with time when viewed from a distant position along the z axis. For $\delta = 3\lambda/4, 7\lambda/4, \ldots$, the helix is a right circular helix, and the electric vector at position z_0 rotates clockwise with time.

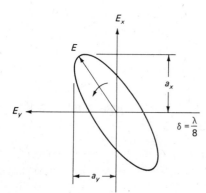

Figure 11.7 Trace of the tip of the light vector on the perpendicular plane at position z_0.

A second special case of elliptically polarized light occurs when the linear phase difference δ between the two waves E_x and E_y is an integral number of half wavelengths ($\delta = n\lambda/2$, $n = 0, 1, 2, \ldots$). For this case, Eq. (11.15) reduces to

$$E_y = \frac{a_y}{a_x} E_x \qquad (k)$$

Equation (k) is the equation of a straight line; therefore, light exhibiting this behavior is known as _plane- or linearly polarized light._ The amplitude of the resulting wave depends upon the amplitudes of the original waves since

$$a = \sqrt{a_x^2 + a_y^2} \qquad (l)$$

The orientation of the plane of vibration depends upon the ratio of the amplitudes of the original waves a_y/a_x and upon the linear phase difference δ between the waves. For $\delta = 0$, λ, 2λ, $\ldots$, the plane of vibration lies in the first and third quadrants. For $\delta = \lambda/2$, $3\lambda/2$, $5\lambda/2$, $\ldots$, the plane of vibration lies in the second and fourth quadrants.

Thus far in the discussions, light has been treated as a wave motion without beginning or end. The light emitted by a conventional light source, e.g., a tungsten-filament light bulb, consists of numerous short pulses originating from a large number of different atoms. Each pulse consists of a finite number of oscillations known as a _wavetrain_. Each wavetrain is thought to be a few meters long with a duration of approximately 10^{-8} s. Since the emissions occur in individual atoms which do not act together in a cooperative manner, the wavetrains may differ from each other in plane of vibration, frequency, amplitude, and phase. Such light is referred to as _incoherent light_. Light sources such as the _laser_, in which the atoms act cooperatively in emitting light, produce _coherent light_, in which the wavetrains are monochromatic, in phase, linearly polarized, and extremely intense. For the interference effects discussed previously, coherent wavetrains are required.

C. The Wave Equation in Complex Notation

A convenient way to represent both the amplitude and phase of a light wave, such as the one represented by Eq. (11.4), for calculations involving a number of optical elements is through the use of complex or exponential notation. Recall the _Euler Identity_

$$e^{i\theta} = \cos\theta + i\sin\theta \qquad (m)$$

where $i^2 = -1$. The sinusoidal wave of Eq. (11.4) is obviously the real part of the complex expression

$$\bar{E} = ae^{i(2\pi/\lambda)(z-ct)} = ae^{i(\phi-\omega t)} \qquad (n)$$

The imaginary part of Eq. (m) could also be used to represent the physical wave; however, it is normally assumed that the real part of a complex quantity is the one having physical significance.

If the amplitude of the wave is also considered to be a complex quantity, then

$$\bar{a} = a_r + ia_i = ae^{i[(2\pi/\lambda)\delta]}$$

where

$$a = \sqrt{a_r^2 + a_i^2} \qquad (o)$$

and

$$\tan\frac{2\pi}{\lambda}\delta = \frac{a_i}{a_r} \qquad (p)$$

A wave with an initial phase δ can be expressed in exponential notation as

$$\bar{E} = \bar{a}e^{i(2\pi/\lambda)(z-ct)} = ae^{i(2\pi/\lambda)(z+\delta-ct)} \qquad (11.16)$$

The physical waves previously represented by Eqs. (11.9) are simply the real part of Eq. (11.16) when represented in exponential notation. Superposition of two or more waves having the same frequency but different amplitude and phase is easily performed with the exponential representation. The real and imaginary parts of the amplitudes of the individual waves are added separately in an algebraic manner. The resultant complex amplitude gives the amplitude and phase of a single wave equivalent to the sum of the individual waves. Extensive use will be made of this representation in Chap. 13, where the theory of photoelasticity is discussed.

The real and imaginary parts of a complex quantity such as the amplitude of a wave may also be written

$$a_r = \tfrac{1}{2}(\bar{a} + \bar{a}^*) \qquad a_i = \tfrac{1}{2}(\bar{a} - \bar{a}^*)$$

where

$$\bar{a}^* = a_r - ia_i$$

is the complex conjugate of the original complex amplitude

$$\bar{a} = a_r + ia_i$$

From Eq. (o),

$$a^2 = a_r^2 + a_i^2$$

Thus

$$a^2 = \bar{a}\bar{a}^* \qquad (11.17)$$

This representation for the square of the amplitude of a complex quantity will be useful in future calculations dealing with the intensity of light.

11.3 REFLECTION AND REFRACTION [2]

In the previous section the electromagnetic wave nature of light was discussed, and wavefronts and rays were defined. The discussions were limited to light propagating in free space. Most optical effects of interest, however, occur as a result of the interaction between a beam of light and some physical material. In free space, light propagates with a velocity c, which is approximately 3×10^8 m/s. In any other medium, the velocity is less than the velocity in free space. The ratio of the velocity in free space to the velocity in a medium is a property of the medium known as the *index of refraction n.* The index of refraction for most gases is only slightly greater than unity (for air, $n = 1.0003$). Values for liquids range from 1.3 to 1.5 (for water, $n = 1.33$) and for solids range from 1.4 to 1.8 (for glass, $n = 1.5$). The index of refraction for a material is not constant but varies slightly with wavelength of the light being transmitted. This dependence of index of refraction on wavelength is referred to as *dispersion.*

Since the frequency of a light wave is independent of the material being traversed, the wavelength is shorter in a material than in free space. Thus a wave propagating in a material will develop a linear phase shift δ with respect to a similar wave propagating in free space. The magnitude of the phase shift, in terms of the index of refraction of the material, can be developed as follows. The time required for passage through a material of thickness h is

$$t = \frac{h}{v} \tag{a}$$

where h is the thickness of the material along the path of light propagation and v is the velocity of light in the material. The distance s traveled during the same time by a wave in free space is

$$s = ct = \frac{ch}{v} \tag{b}$$

Thus the distance δ by which the wave in the material trails the wave in free space is given by

$$\delta = s - h = \frac{ch}{v} - h = h(n - 1) \tag{11.18}$$

since the index of refraction of the material has been defined as $n = c/v$. The retardation δ, as given by Eq. (11.18), is a positive quantity since the index of refraction of a material is always greater than unity. The relative position of one wave with respect to another can be controlled by including the retardation in the phase of the appropriate wave equation.

When a beam of light strikes a surface between two transparent materials with different indices of refraction, it is experimentally observed that, in general, it is divided into a reflected ray and a refracted ray, as shown in Fig. 11.8. The reflected and refracted rays lie in the plane formed by the incident ray and the normal to the surface and known as the *plane of incidence.* The *angle of incidence* α, the *angle of reflection* β, and the *angle of refraction* γ are related as follows:

For reflection:
$$\alpha = \beta \tag{11.19}$$

For refraction:
$$\frac{\sin \alpha}{\sin \gamma} = \frac{n_2}{n_1} = n_{21} \tag{11.20}$$

where n_1 = index of refraction of material 1
  n_2 = index of refraction of material 2
  n_{21} = index of refraction of material 2 with respect to material 1

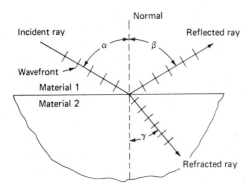

Figure 11.8 Reflection and refraction of a plane light wave at an interface between two transparent materials.

If the light beam originates in the material having the higher index of refraction, n_{21} will be a number less than unity. Under these conditions, some *critical angle of incidence* α_c is reached for which the angle of refraction is 90°. For angles of incidence greater than the critical angle, there is no refracted ray and *total internal reflection occurs*. Total internal reflection cannot occur when the beam of light originates in the medium with the lower index of refraction.

The laws of reflection and refraction give information about the direction of reflected and refracted rays but no information with regard to intensity. Intensity relationships, which can be derived from Maxwell's equations, indicate that the intensity of a reflected beam depends upon both the angle of incidence and the direction of polarization of the incident beam. Consider a completely unpolarized beam of light falling on a surface between two transparent materials, as shown in Fig. 11.9. The electric vector for each wavetrain in the beam can be resolved into two components, one perpendicular to the plane of incidence (the perpendicular component) and the other parallel to the plane of incidence (the parallel component). For completely unpolarized incident light, the two components would have equal intensity. The intensity of the reflected beam can be expressed as

$$I_r = RI_i \tag{11.21}$$

where I_i = intensity of incident beam
I_r = intensity of reflected beam
R = reflection coefficient

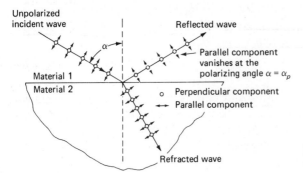

Figure 11.9 Reflection and refraction at the polarizing angle.

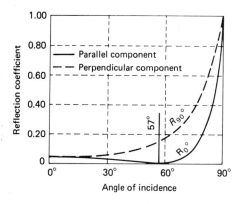

Figure 11.10 Reflection coefficients for an air-glass interface ($n_{21} = 1.5$).

For the perpendicular component

$$R_{90°} = \frac{\sin^2 (\alpha - \gamma)}{\sin^2 (\alpha + \gamma)} \tag{11.22}$$

For the parallel component

$$R_{0°} = \frac{\tan^2 (\alpha - \gamma)}{\tan^2 (\alpha + \gamma)} \tag{11.23}$$

Reflection coefficients for an air-glass interface ($n_{21} = 1.5$) are shown in Fig. 11.10. These data indicate that there is a particular angle of incidence for which the reflection coefficient for the parallel component is zero. This angle is referred to as the *polarizing angle α_p*. Since the parallel component is zero when the angle of incidence is equal to the polarizing angle, the beam reflected from the surface is plane-polarized with the plane of vibration perpendicular to the plane of incidence. Equation (11.23) indicates that the reflection coefficient for the parallel component is zero when $\tan (\alpha + \gamma) = \infty$ or when $\alpha + \gamma = 90°$. Thus from Eq. (11.20),

$$\tan \alpha_p = \frac{n_2}{n_1} = n_{21}$$

It is also experimentally observed that phase changes occur during reflection. A phase change of $\delta = \lambda/2$ occurs when light is incident from the medium with the lower index of refraction. When light is incident from the medium with the higher index of refraction, no phase change occurs upon reflection.

Metal surfaces exhibit relatively large reflection coefficients, as shown in Fig. 11.11. At oblique incidence, the coefficients for light polarized parallel to the plane of incidence are less than the coefficients for the perpendicular component. A change of phase also occurs with metallic reflection. Unfortunately, the phase change varies with both angle of incidence and direction of polarization. As a result, plane-polarized light is changed by oblique reflection to elliptically polarized light.

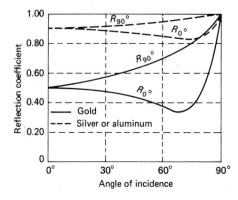

Figure 11.11 Reflection coefficients for several air-metal interfaces.

Previous discussions of reflection and refraction have dealt with materials that are optically isotropic (the index of refraction is the same for all directions in the material); therefore, light propagates with the same velocity in all directions. When a beam of ordinary light is directed at oblique incidence onto the surface of certain crystalline materials such as calcite and quartz, it is observed that, in general, a reflected beam and two refracted beams are produced. This phenomenon is known as *double refraction,* and the material is said to be *birefringent.* One of the refracted rays, known as the *ordinary ray,* propagates with the same velocity along all directions through the crystal and obeys Snell's law of refraction as given by Eq. (11.20). The second refracted ray, known as the *extraordinary ray,* propagates with different velocities along different directions through the crystal and is not governed by Eq. (11.20). Along one direction through the crystal, known as the *optic axis* (the optic axis is a direction and not a specific line), the ordinary and extraordinary rays travel with the same velocity.

If a ray of ordinary light is directed onto the face of a calcite crystal at normal incidence, the ordinary ray passes through the crystal with no deviation while the extraordinary ray is refracted at some angle with respect to the normal to the surface. Since opposite faces of the crystal are parallel, the two rays emerge as parallel beams. The two beams are observed to be plane-polarized in orthogonal directions. The vibrations associated with the ordinary ray are perpendicular to the plane containing the optic axis and the ordinary ray. The vibrations associated with the extraordinary ray lie in the plane containing the optic axis and the extraordinary ray. The ordinary ray and the extraordinary ray lie in the same plane when the plane of incidence coincides with the plane containing the optic axis and the normal to the surface.

For many optical applications, calcite crystals are cut into rectangular blocks with the faces parallel and perpendicular to the optic axis. Light entering the crystal at normal incidence on any of the faces does not deviate from the normal to the surface; therefore, the ordinary and extraordinary rays follow the same path through the crystal. For two of the faces (those perpendicular to the optic axis), the path is along the optic axis, and so the two rays travel with the same velocity. For the other four faces, the path is perpendicular to the optic axis, and so the two rays travel with different velocities. As a result, the waves emerge from the crystal with a linear phase difference δ. Plane-polarized incident light would emerge from such a crystal as elliptically polarized light. Optical elements which convert one form of polarized light into another are referred to as _retarders_ or _wave plates_.

11.4 IMAGE FORMATION BY LENSES AND MIRRORS [2]

In the previous section, reflection and refraction of a plane light wave at a plane interface between two materials was considered. More complicated situations frequently arise in the optical systems used for experimental stress-analysis work. Since lenses and mirrors are widely used in many of these systems, a brief discussion of the significant features of these elements is provided here for future reference.

A. Plane Mirrors

Figure 11.12 shows an object O placed at a distance u in front of a plane mirror. The light from each point on the object (such as point A) is a spherical wave which reflects from the mirror in the manner discussed in Sec. 11.3. When the eye or other light-sensing instrument intercepts the reflected rays, they perceive an image I of the object O at a distance v behind the mirror, which can be determined by extending the reflected rays to the position A' as shown in Fig. 11.12. In this instance, the image I is a _virtual image_ since light rays do not pass through image points such as A'. From the geometry of Fig. 11.12 it is obvious that the magnitudes of u and v are the same. The image is _erect_ and has the same height as the object. One difference between the object and the image not apparent from Fig. 11.12 is that left and right are interchanged (an image of a left hand appears as a right hand in the mirror).

B. Spherical Mirrors

Figure 11.13 shows an object O placed at a distance u in front of a concave spherical mirror. The center of curvature of the mirror is located at C, and the focal point is at F. The light from each point on the object reflects in the manner

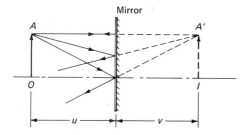

Figure 11.12 Image formation by a plane mirror.

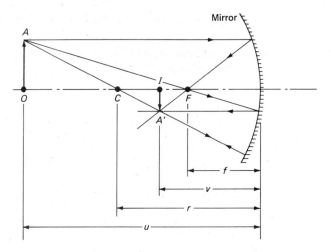

Figure 11.13 Image formation by a concave spherical mirror.

shown in Fig. 11.13 for point A. In this instance the image I is a *real image* since light rays pass through image points such as A'. From the geometry of Fig. 11.13 it can be shown that if all rays from the object make a small angle with respect to the axis of the mirror, the distance to the image satisfies the expression

$$\frac{1}{u} + \frac{1}{v} = \frac{2}{r} = \frac{1}{f} \qquad (11.24)$$

For the case illustrated, both u and v are positive since both the object and the image are real. The distance v will be negative when dealing with a virtual image. With this sign convention, Eq. (11.24) applies to all concave, plane, and convex mirrors. The ratio of the size of the image to the size of the object is known as the magnification M and is given by the expression

$$M = -\frac{v}{u} \qquad (11.25)$$

where the minus sign is used to indicate an inverted image. Equation (11.25) also applies to all concave, plane, and convex mirrors. For example, when Eqs. (11.24) and (11.25) are applied to the plane mirror of Fig. 11.12,

$$v = -u \qquad M = 1$$

This indicates a virtual image which is erect and identical in size to the object.

C. Thin Lenses

At least one and often a series combination of lenses is employed in optical equipment to magnify and focus the image of an object on a photographic plate. For this reason, the passage of light through a single convex lens and a pair of convex lenses in series will be examined in detail. The discussion will be limited to instances where the thin-lens approximation can be applied; i.e., the thickness of the lens can be neglected with respect to other distances such as the focal length f of the lens and the object and image distances u and v.

D. Single-Lens System

The classical optical representation of a single-lens system is shown in Fig. 11.14. The light from each point on the object can be considered as a spherical wave which is reflected and refracted at the air-glass interface, as indicated in Sec. 11.3. In the study of mirrors, the reflected rays are of interest. In the case of lenses, the refracted rays produce the desired optical effects. From the geometry of Fig. 11.14 it can be shown that the object distance u, the image distance v, and the focal length f of the lens are related by

$$\frac{1}{u} + \frac{1}{v} = \frac{1}{f} \tag{a}$$

Similarly the magnification is given by

$$M = -\frac{v}{u} \tag{b}$$

Equations (a) and (b) are identical to Eqs. (11.24) and (11.25) for mirrors. In the development of Eq. (a), the assumption was made that all rays from the object make a small angle with respect to the axis of the lens. The image in Fig. 11.14 is a real image; therefore, the image distance v is positive even though the image is on the opposite side of the lens from the object. With v positive, Eq. (b) indicates that the magnification is negative (thus the image is inverted, as shown).

The situation illustrated in Fig. 11.14 occurs when the object is located beyond the focal point of the lens. The object may also be placed between the focal point and the lens surface, as illustrated in Fig. 11.15. In this case, the distance u is positive and v is negative since the image formed is virtual. Equation (b) then yields a positive magnification, which indicates that the image is erect (as shown).

A similar analysis for a concave lens indicates that Eqs. (11.24) and (11.25)

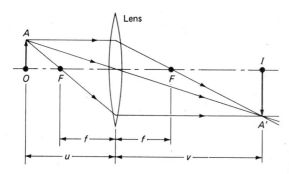

Figure 11.14 Image formation by a single convex lens (object outside the focal point).

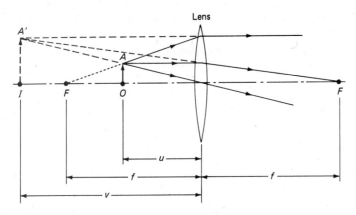

Figure 11.15 Image formation by a single convex lens (object inside the focal point).

apply so long as virtual-image distances are considered negative and real-image distances positive. The proof is left as an exercise for the student.

If Eqs. (11.24) and (11.25) are combined and solved for u and v in terms of M and f,

$$u = \frac{M - 1}{M}f \qquad v = (1 - M)f \tag{11.26}$$

The total length from the model (object) to the focusing plane of a camera (image) in terms of M and f is

$$L = u + v = \left(2 - M - \frac{1}{M}\right)f \tag{11.27}$$

Equations (11.26) and (11.27) show that the distances u, v, and L are directly related to the focal length of the lens. Since the focal length of the camera lens on a

Table 11.2 Influence of the magnification M on the lengths u, v, and L in a camera assembly

$-M$	L/f	u/f	v/f
$\frac{1}{4}$	6.25	5.00	1.25
$\frac{1}{2}$	4.50	3.00	1.50
1	4.00	2.00	2.00
2	4.50	1.50	3.00
4	6.25	1.25	5.00

diffused-light polariscope must be relatively large (see Sec. 11.6), the length L usually equals or exceeds 10 ft. This length, of course, depends upon the magnification capabilities of the camera, as shown in Table 11.2.

E. Two Lenses in Series

The classical optical representation of a series combination of two convex lenses is shown in Fig. 11.16. This series arrangement of convex lenses is often employed in large-field lens polariscopes to decrease their total length. The object is positioned inside the focal point of the long-focal-length field lens F_1 to reduce the length of the system. The slowly diverging rays from lens F_1 are then converged by a shorter-focal-length condenser lens F_2. By applying Eq. (11.24) to both lenses, it is clear that

$$\frac{1}{u_1} + \frac{1}{v_1} = \frac{1}{f_1} \qquad \text{for field lens} \qquad (c)$$

and

$$\frac{1}{u_2} + \frac{1}{v_2} = \frac{1}{f_2} \qquad \text{for condenser lens} \qquad (d)$$

Since the virtual image from the field lens serves as the object for the condenser lens,

$$u_2 = -v_1 + s \qquad (e)$$

The magnification M for the combined lens system is

$$M = M_1 M_2 = \left(-\frac{v_1}{u_1}\right)\left(-\frac{v_2}{u_2}\right) = \frac{v_1 v_2}{u_1 u_2} \qquad (f)$$

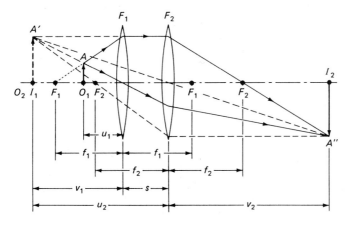

Figure 11.16 Image formation by a series combination of two convex lenses.

By combining Eqs. (c) to (f) and solving for M, u_1, v_2, and s, it is possible to show that

$$M = \frac{f_1 \, f_2}{(u_1 - f_1)(s - f_2) - u_1 \, f_1} \tag{11.28}$$

$$s = \frac{f_2(f_1 - u_1) - u_1 \, f_1 - f_1 \, f_2/M}{f_1 - u_1} \tag{11.29}$$

$$u_1 = \frac{f_1 \, f_2[(1 - M)/M] + sf_1}{s - f_1 - f_2} \tag{11.30}$$

$$v_2 = M\left[s\left(\frac{u_1}{f_1} - 1\right) - u_1\right] \tag{11.31}$$

Equations (11.28) to (11.31) are useful in a number of ways. They can be employed to design a polariscope, i.e., to choose f_1 and f_2, to give a range of magnification M within a certain length $L = u_1 + v_2$. Or if the polariscope already exists, where f_1 and f_2 are fixed and u_1 and s can be varied within fixed limits, the equations can be employed to determine M within these limits. It should be noted that M is a function of f_1, f_2, u_1, and s; thus a given magnification M can be achieved in a number of different ways. Even when f_1 and f_2 are fixed, as is so often the case, a given value of M can still be achieved in a number of different ways by varying both u_1 and s.

All the relationships for the lens combination illustrated in Fig. 11.16, where the object O_1 is placed inside the focal point of lens F_1, are applicable for the case where the object is placed outside the focal point of the lens. This can be illustrated by assuming in Fig. 11.16 that the image I_2 is the object $O_{1'}$. If the rays are then traced through the lens combination in a reverse direction from the ones shown, the object O_1 in the figure becomes the image $I_{2'}$. The magnification in the two cases is different. In practice, the position of the object and the location of the viewing screen will depend on the object being studied and the characteristics of the lenses available for the series combination.

The previous discussion for lenses and mirrors was based on the assumption that all light rays from an object were confined to a region near the axis of the element and made small angles with respect to the axis. In systems where the rays are not confined to a region near the axis, all rays from an object point do not focus at exactly the same image point. This effect is known as a *spherical aberration*. Similarly, when an object point is located such that rays from the point make an appreciable angle with respect to the axis of the element, the image tends to become two mutually perpendicular line images rather than a point image. This defect in optical elements is known as *astigmatism*. Other lens defects which must be considered in the design of optical systems include *distortion, coma, curvature of field,* and *chromatic aberration.* The interested reader should consult an optics textbook for a complete discussion of defects in mirrors and lenses.

11.5 OPTICAL DIFFRACTION AND INTERFERENCE [1, 2]

Consider light from a point source which is being used to illuminate a viewing screen. If an opaque plate with a large hole is placed between the source and the screen, the area illuminated on the screen can be defined by rays drawn from the source to the boundary of the hole and extended to the screen. As the hole in the plate is made smaller and smaller, the area illuminated on the screen decreases until some minimum area of illumination is achieved. Further decreases in the size of the hole cause the area of illumination to increase. This phenomenon, associated with an apparent bending of the light rays, is known as _diffraction_. Diffraction plays an important role in several methods of experimental stress analysis.

Diffraction of waves at a small aperture can be explained by Huygens' principle, which states that "each point on a wavefront may be regarded as a new source of waves." For example, consider a series of plane wavefronts impinging on a long slender rectangular slit such as the one shown in cross section in Fig. 11.17. When the width of the slit is small (less than 0.10 mm), diffraction becomes important. In accordance with Huygens' principle, the light emerging from the slit can be considered to consist of a series of spherical wavefronts which expand and ultimately illuminate a viewing screen placed in the path of the propagating waves. Equation (11.3) can be used to describe these spherical wavefronts. Thus

$$E = \frac{Kb}{r} \cos \frac{2\pi}{\lambda} (r - ct)$$

The amplitude of the wave depends on the strength of the source, the width of the slit, and the radial position of the expanding spherical wavefront.

If the viewing screen is placed at a large distance from the slit, or if a lens is used to focus parallel rays from the slit at a point P, the intensity distribution I_P

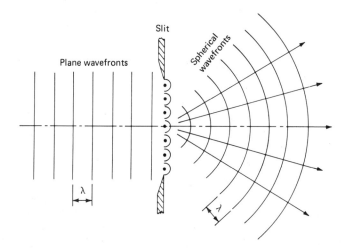

Figure 11.17 Diffraction of plane wavefronts at a narrow rectangular slit.

on the screen as a function of the angle of inclination θ of the rays is easily calculated. For example, consider the emerging parallel rays shown in Fig. 11.18. The contribution to the total amount of light reaching P from an increment of width ds of the slit on the axis of the system can be expressed as

$$dE_P(0) = \frac{K\, ds}{r} \cos \frac{2\pi}{\lambda} (r - ct) \tag{a}$$

Similarly, the contributions from increments of width at a distance s above and below the centerline are

$$dE_P(s) = \frac{K\, ds}{r} \cos \frac{2\pi}{\lambda} (r - ct - s \sin \theta) \tag{b}$$

$$dE_P(-s) = \frac{K\, ds}{r} \cos \frac{2\pi}{\lambda} (r - ct + s \sin \theta) \tag{c}$$

The total amount of light reaching P can be determined by summing the contributions from each increment of width of the slit. This is most easily accomplished by adding Eqs. (b) and (c) and integrating over half the width of the slit. Thus

$$E_P = \int_0^{b/2} dE_P(s) + dE_P(-s)$$

$$= \int_0^{b/2} \frac{K}{r} \left[\cos \frac{2\pi}{\lambda} (r - ct - s \sin \theta) + \cos \frac{2\pi}{\lambda} (r - ct + s \sin \theta) \right] ds$$

$$= \int_0^{b/2} \frac{2K}{r} \left[\cos \frac{2\pi}{\lambda} (r - ct) \cos \frac{2\pi}{\lambda} (s \sin \theta) \right] ds$$

$$= \frac{2K}{r} \frac{\lambda b}{2\pi b \sin \theta} \sin \frac{\pi b \sin \theta}{\lambda} \cos \frac{2\pi}{\lambda} (r - ct)$$

$$= \frac{Kb \sin \beta}{r} \cos \frac{2\pi}{\lambda} (r - ct)$$

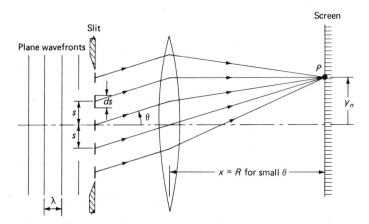

Figure 11.18 Light rays from a narrow rectangular slit which contribute to the intensity at P.

where
$$\beta = \frac{\pi b \sin \theta}{\lambda} \tag{11.32}$$

Since the intensity at P is proportional to the square of the amplitude,

$$I_P = \left(\frac{Kb}{r}\right)^2 \frac{\sin^2 \beta}{\beta^2} \tag{11.33}$$

For maximum or minimum intensity $dI_P/d\beta = 0$. Thus

$$\frac{2 \sin \beta}{\beta^3}(\beta \cos \beta - \sin \beta) = 0$$

Thus either
$$\beta = n\pi \quad \text{or} \quad \beta = \tan \beta \tag{d}$$

The condition $\beta = n\pi$ yields zero intensity except at $\beta = 0$. The roots of $\beta = \tan \beta$ define positions of maximum intensity. The maximums are located at $\beta = 0$, 1.430π, 2.459π, 3.471π, 4.477π, ... If the intensity at $\beta = 0$ $(\theta = 0)$ is arbitrarily taken as 1, the intensities of successive maxima are $\frac{1}{21}$, $\frac{1}{61}$, $\frac{1}{120}$, ...

The dark bands which form in regions of zero intensity are called _diffraction fringes._ For a rectangular slit, the fringes will be parallel straight lines. For a circular pinhole, the fringes will be a series of concentric circles about the axis of the optical system. If the incident light is parallel and a lens is used to produce a focused image on a screen, the phenomenon is referred to as _Fraunhofer diffraction._ If the incident light comes from a point source at a finite distance from the slit or pinhole, the wavefronts are divergent and the phenomenon is known as _Fresnel diffraction._

For the narrow slit, the fringe positions can be determined by substituting Eq. (d) into Eq. (11.32). Thus

$$\frac{\pi b \sin \theta}{\lambda} = n\pi$$

For small values of θ, $\sin \theta = y_n/R$, where R is the distance between the slit or lens and the screen and y_n is the distance from the axis of the optical system to the diffraction fringe of order n. Thus

$$y_n = \frac{n\lambda R}{b} \qquad n = 1, 2, 3, \ldots \tag{11.34}$$

The spacing of diffraction fringes is used as the basis for several strain-measurement methods.

11.6 OPTICAL INSTRUMENTS: THE POLARISCOPE

The polariscope is an optical instrument that utilizes the properties of polarized light in its operation. For experimental stress-analysis work, two types are frequently employed, the plane polariscope and the circular polariscope. The names follow from the type of polarized light used in their operation.

In practice, plane-polarized light is produced with an optical element known as a plane or linear polarizer. Production of circularly polarized light or the more general elliptically polarized light requires the use of a linear polarizer together with an optical element known as a wave plate. A brief discussion of linear polarizers, wave plates, and their series combination follows.

A. Linear or Plane Polarizers [3–5]

When a light wave strikes a plane polarizer, this optical element resolves the wave into two mutually perpendicular components, as shown in Fig. 11.19. The component parallel to the axis of polarization is transmitted while the component perpendicular to the axis of polarization is either absorbed, as in the case of Polaroid, or suffers total internal reflection, as in the case of a calcite crystal such as the Nicol prism.

If the plane polarizer is fixed at some point z_0 along the z axis, the equation for the light vector can be written

$$E = a \cos \frac{2\pi}{\lambda}(z_0 - ct) \tag{a}$$

Since the initial phase of the wave is not important in the developments which follow, Eq. (a) can be reduced through the use of Eqs. (11.6) and (11.7) to

$$E = a \cos 2\pi ft = a \cos \omega t \tag{11.35}$$

where $\omega = 2\pi f$ is known as the circular frequency of the wave. The absorbed and transmitted components of the light vector are

$$E_a = a \cos \omega t \sin \alpha \qquad E_t = a \cos \omega t \cos \alpha$$

where α is the angle between the axis of polarization and the light vector.

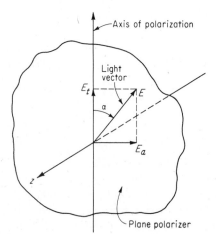

Figure 11.19 Absorbing and transmitting characteristics of a plane polarizer.

In the early days of photoelasticity the production of plane-polarized light was a difficult problem, and as a consequence a number of methods were employed, which included reflected light at a 57° angle of incidence, a glass pile, and the Nicol prism. However, these methods of producing plane-polarized light have been largely displaced with the advent of Polaroid filters, which have the advantage of providing a large field of very well polarized light at a relatively low cost. Most modern polariscopes containing linear polarizers employ Polaroid H sheet,† a transparent material with stained and oriented molecules. In the manufacture of H-type Polaroid films, a thin sheet of polyvinyl alcohol is heated, stretched, and immediately bonded to a supporting sheet of cellulose acetate butyrate. The polyvinyl face of the assembly is then stained by a liquid rich in iodine. The amount of iodine diffused into the sheet determines its quality, and the Polaroid Corporation produces three grades, denoted according to their transmittance of the light as HN-22, HN-32, and HN-38. Since the quality of a polarizer is judged by its transmission ratio, HN-22 (with a transmission ratio of the order of 10^5 at wavelengths normally employed in photoelasticity) is recommended for photoelastic purposes.

B. Wave Plates [3, 6, 7]

A wave plate has previously been defined as an optical element which has the ability to resolve a light vector into two orthogonal components and to transmit the components with different velocities. Such a material has been referred to as doubly refracting or birefringent. The doubly refracting plate illustrated in Fig. 11.20 has two principal axes labeled 1 and 2. The transmission of light along axis 1 proceeds at velocity c_1 and along axis 2 at velocity c_2. Since c_1 is greater than c_2, axis 1 is often called the *fast axis* and axis 2 the *slow axis*.

† Manufactured by the Polaroid Corporation, Cambridge, Mass.

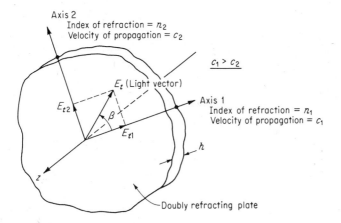

Figure 11.20 A plane-polarized light vector entering a doubly refracting plate.

If this doubly refracting plate is placed in a field of plane-polarized light so that the light vector E_t makes an angle β with axis 1 (the fast axis), then upon entering the plate the light vector is first resolved into two components E_{t1} and E_{t2} along axes 1 and 2, respectively. The magnitudes of the individual components E_{t1} and E_{t2} are given by

$$E_{t1} = E_t \cos \beta = a \cos \alpha \cos \omega t \cos \beta = k \cos \omega t \cos \beta$$

$$E_{t2} = E_t \sin \beta = a \cos \alpha \cos \omega t \sin \beta = k \cos \omega t \sin \beta$$

where $k = a \cos \alpha$. The light components E_{t1} and E_{t2} travel through the plate with different velocities c_1 and c_2, respectively. Because of this velocity difference, the two components will emerge from the plate at different times. In other words, one component is retarded in time relative to the other component. This retardation can be handled most effectively by considering the relative phase shift between the two components. From Eq. (11.18), the linear phase shifts for components E_{t1} and E_{t2} with respect to a wave in air can be expressed as

$$\delta_1 = h(n_1 - n) \qquad \delta_2 = h(n_2 - n)$$

where n is the index of refraction of air.

The relative linear phase shift is then computed simply as

$$\delta = \delta_2 - \delta_1 = h(n_2 - n_1) \tag{11.36}$$

The relative angular phase shift Δ between the two components as they emerge from the plate (recall from Sec. 11.2 that two mutually perpendicular components having the same frequency are equivalent to a rotating vector with angular frequency ω) is given by

$$\Delta = \frac{2\pi}{\lambda} \delta = \frac{2\pi h}{\lambda}(n_2 - n_1) \tag{11.37}$$

The relative phase shift Δ produced by a doubly refracting plate is dependent upon its thickness h, the wavelength of the light λ, and the properties of the plate as described by $n_2 - n_1$. When the doubly refracting plate is designed to give an angular retardation of $\pi/2$, it is called a quarter-wave plate. Doubly refracting plates designed to give angular retardations of π and 2π are known as half- and full-wave plates, respectively. Upon emergence from a general wave plate exhibiting a retardation Δ, the two components of light are described by the equations

$$E'_{t1} = k \cos \beta \cos \omega t \qquad E'_{t2} = k \sin \beta \cos (\omega t - \Delta) \tag{11.38}$$

With this representation, only the relative phase shift between components has been considered. The *identical* additional phase shift suffered by both components, as a result of passage through the wave-plate material (as opposed to free space), has been neglected since it has no effect on the phenomenon being considered.

The magnitude of the light vector which is equivalent to these two components can be expressed as

$$E'_t = \sqrt{(E'_{t1})^2 + (E'_{t2})^2} = k\sqrt{\cos^2 \beta \cos^2 \omega t + \sin^2 \beta \cos^2 (\omega t - \Delta)} \tag{11.39}$$

The angle that the emerging light vector makes with axis 1 (the fast axis) is given by

$$\tan \gamma = \frac{E'_{t2}}{E'_{t1}} = \frac{\cos (\omega t - \Delta)}{\cos \omega t} \tan \beta \tag{11.40}$$

Thus, it is clear that both the amplitude and the rotation of the emerging light vector can be controlled by the wave plate. Controlling factors are the relative phase difference Δ and the orientation angle β. Various combinations of Δ and β and their influence on the type of polarized light produced will be discussed in the next section.

Wave plates employed in a photoelastic polariscope may consist of a single plate of quartz or calcite cut parallel to the optic axis, a single plate of mica, a sheet of oriented cellophane, or a sheet of oriented polyvinyl alcohol. In recent years, as the design of the modern polariscope has tended toward a field of relatively large diameter, most wave plates employed have been fabricated from oriented sheets of polyvinyl alcohol. These wave plates are manufactured by the Polaroid Corporation by warming and unidirectionally stretching the sheet. Since the oriented polyvinyl alcohol sheet is only about 20 μm thick (for a quarter-wave plate), the commercial wave-plate filters are usually laminated between two sheets of cellulose acetate butyrate.

C. Conditioning of Light by a Series Combination of a Linear Polarizer and a Wave Plate

The magnitude and direction of the light vector emerging from a series combination of a linear polarizer and a wave plate are given by Eqs. (11.39) and (11.40). The light emerging from this combination of optical elements is always polarized; however, the type of polarization may be plane, circular, or elliptical. The factors which control the type of polarized light produced by this combination are the relative phase difference Δ imposed by the wave plate and the orientation angle β. Three well-defined cases exist.

Case 1: Plane-polarized light If the angle β is set equal to zero and the relative retardation Δ is not restricted in any sense, the magnitude and direction of the emerging light vector are given by Eqs. (11.39) and (11.40) as

$$E'_t = k \cos \omega t \qquad \gamma = 0$$

Since $\gamma = 0$, the light vector is not rotated as it passes through the wave plate; hence, the light upon emergence remains plane-polarized. The wave plate in this instance does not influence the light except to produce a retardation with respect to a wave in free space which depends on the plate thickness and the index of refraction associated with the fast axis. Similar results are obtained by letting $\beta = \pi/2$. Thus

$$E'_t = k \cos (\omega t - \Delta) \qquad \gamma = \frac{\pi}{2}$$

Case 2: Circularly polarized light If a wave plate is selected so that $\Delta = \pi/2$, that is, a quarter-wave plate, and β is set equal to $\pi/4$, the magnitude and direction of the light vector as it emerges from the plate are given by Eqs. (11.39) and (11.40) as

$$E'_t = \frac{\sqrt{2}}{2} k \sqrt{\cos^2 \omega t + \sin^2 \omega t} = \frac{\sqrt{2}}{2} k \qquad \gamma = \omega t$$

The light vector described by these expressions has a constant magnitude; therefore, the tip of the light vector traces out a circle as it rotates. The vector rotates with a constant angular velocity in a counterclockwise direction when viewed from a distant position along the path of propagation of the light beam. Such light is known as *left circularly polarized light*. *Right circularly polarized light* could be obtained by setting β equal to $3\pi/4$. The light vector would then rotate with a constant angular velocity in the clockwise direction.

Case 3: Elliptically polarized light If a quarter-wave plate ($\Delta = \pi/2$) is selected and β is permitted to be any angle other than $\beta = n\pi/4$ ($n = 0, 1, 2, 3, \ldots$), then by Eqs. (11.39) and (11.40), the magnitude and direction of the emerging light vector are

$$E'_t = k \sqrt{\cos^2 \beta \cos^2 \omega t + \sin^2 \beta \sin^2 \omega t} \qquad \tan \gamma = \tan \beta \tan \omega t$$

The light vector described by these expressions has a magnitude which varies with angular position in such a way that the tip of the light vector traces out an ellipse as it rotates. The shape and orientation of the ellipse and the direction of rotation of the light vector are controlled by the angle β.

Consider now the significance of Eq. (11.37) in the production of circularly polarized light

$$\Delta = \frac{2\pi h}{\lambda} (n_2 - n_1) \tag{11.37}$$

Recall that circularly polarized light requires the use of a quarter-wave plate; therefore, the phase difference Δ must be $\pi/2$. It is clear that the thickness h can be determined to give $\Delta = \pi/2$ once the plate material, $n_2 - n_1$, and the wavelength λ of the light are selected. However, a quarter-wave plate suitable for one wavelength of monochromatic light, i.e., a constant wavelength, will not be suitable for a different wavelength. Also, no quarter-wave plate can be designed for white light since it contains components of different wavelengths.

D. Arrangement of the Optical Elements in a Polariscope [8–12]

Plane polariscope The plane polariscope is the simplest optical system used in photoelasticity; it consists of two linear polarizers and a light source arranged as illustrated in Fig. 11.21a.

The linear polarizer nearest the light source is called the *polarizer*, while the second linear polarizer is known as the *analyzer*. In the plane polariscope the two axes of polarization are always crossed; hence no light is transmitted through the

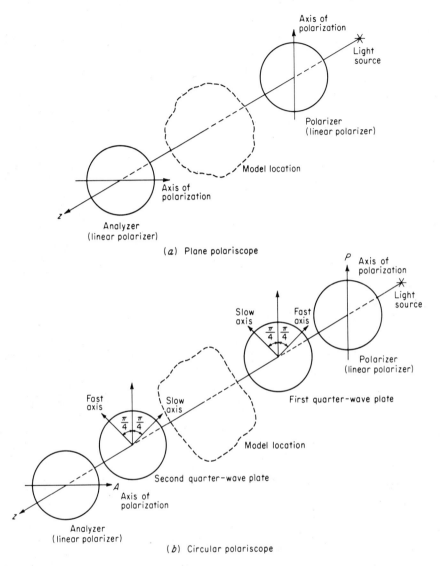

Figure 11.21 Arrangement of the optical elements in a plane polariscope and in a circular polariscope.

analyzer, and this optical system produces a dark field. In operation a photoelastic model is inserted between the two crossed elements and viewed through the analyzer. The behavior of the photoelastic model in a plane polariscope will be covered in Sec. 13.3.

Circular polariscope As the name implies, the circular polariscope employs circularly polarized light; consequently, the photoelastic apparatus contains four optical elements and a light source, which is illustrated in Fig. 11.21*b*.

Table 11.3 Four arrangements of the optical elements in a circular polariscope

Arrangement	Quarter-wave plates	Polarizer and analyzer	Field
A†	Crossed	Crossed	Dark
B	Crossed	Parallel	Light
C	Parallel	Crossed	Light
D	Parallel	Parallel	Dark

† Shown in Fig. 11.21.

The first element following the light source is called the polarizer. It converts the ordinary light into plane-polarized light. The second element is a quarter-wave plate set at an angle $\beta = \pi/4$ to the plane of polarization. This first quarter-wave plate converts the plane-polarized light into circularly polarized light. The second quarter-wave plate is set with its fast axis parallel to the slow axis of the first quarter-wave plate. The purpose of this element is to convert the circularly polarized light into plane-polarized light, which is again vibrating in the vertical plane. The last element is the analyzer, with its axis of polarization in the horizontal plane, and its purpose is to extinguish the light. This series of optical elements constitutes the standard arrangement for a circular polariscope, and it produces a dark field. Actually, four arrangements of the optical elements in the polariscope are possible, depending upon whether the polarizers and quarter-wave plates are crossed or parallel. These four optical arrangements are described in Table 11.3.

Arrangements A and B are normally recommended for light- and dark-field use of the polariscope since a portion of the error introduced by imperfect quarter-wave plates, i.e., both quarter-wave plates differ from $\pi/2$ by an amount ϵ, is canceled out. Since quarter-wave plates are often of poor quality, this fact is important to recall in aligning the polariscope.

E. Construction Details of Diffused-Light and Lens-Type Polariscopes [8–12]

Diffused-light polariscope The arrangement of the optical elements discussed previously is not sufficiently complete or detailed for the visualization of a working polariscope. The degree of complexity of a polariscope varies widely with the investigator and ranges from highly complex lens systems with servomotor drives on the four optical elements to very simple arrangements with no lenses and no provision for rotation of any element.

The diffused-light polariscope described here is one of the simplest and least expensive polariscopes available; yet it can be employed to produce very high quality photoelastic results. This polariscope requires only one lens; however, its field can be made very large since its diameter is dependent upon only the size of the available linear polarizers and quarter-wave plates. Actually, diffused-light polariscopes with field diameters up to 18 in (450 mm) can readily be constructed.

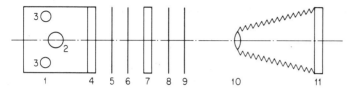

Figure 11.22 Design of a circular diffused-light polariscope with both white and monochromatic light sources: (1) light house (flat-white diffusing paint on interior): (2) monochromatic light source, sodium street lamp 12 in long, 3 in diameter, 10,000 lm); (3) white-light source, 300 W tungsten-filament lamp located on side of light house; (4) diffusing plates, flashed opal glass; (5) polarizer, glass-laminated Polaroid; (6) first quarter-wave plate, glass-laminated orientated polyvinyl alcohol; (7) loading frame; (8) second quarter-wave plate, glass-laminated orientated polyvinyl alcohol; (9) analyzer glass-laminated Polaroid; (10) camera lens, good-quality process lens with about 20- to 24-in focal length; (11) camera for 4 by 5 or 5 by 7 film.

A schematic illustration of the construction details of a diffused-light polariscope is shown in Fig. 11.22. A photograph of an 18-in-diameter (450 mm) diffused-light polariscope is shown in Fig. 11.23.

Lens polariscope In the earlier days of photoelasticity, Nicol prisms (available only in small diameters) were almost exclusively used as the linear polarizing elements. Consequently, it was necessary to employ a lens system to expand the

Figure 11.23 An 18-in-diameter diffused-light polariscope at IIT Research Institute; note that a portion of the polariscope has been suspended from the ceiling in order to open the working area behind the analyzer.

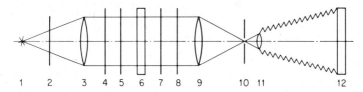

Figure 11.24 Construction details of a circular lens-type polariscope. (1) Light source (usually a small mercury arc), (2) color filter, (3) first field lens, (4) polarizer, (5) first quarter-wave plate, (6) loading frame and model, (7) second quarter-wave plate, (8) analyzer, (9) second field lens, (10) diaphragm stop, (11) camera lens, (12) camera back.

field of view so that reasonably sized models could be studied. However, with the advent of high-quality large-diameter sheets of Polaroid, it is no longer necessary to extend the diameter of the field through the use of a multiple-lens system. Instead, lens polariscopes should be employed only where parallel light over the whole field is a necessity. Instances where parallel light is important include applications where extremely precise definition of the entire model boundary is critical and where partial mirrors are to be employed in the photoelastic bench (for fringe sharpening and fringe multiplication).

Several variations of the lens systems are possible. The arrangement shown in Fig. 11.24 is one of the simplest types that can be employed to obtain parallel light. The polarizer, quarter-wave plates, and analyzer should be placed in the parallel beam between the two field lenses to achieve more complete polarization and to avoid problems associated with internal stresses in the field lenses. A point source of light is required for parallel light, and this point source is usually approached by employing a high-intensity mercury lamp with a very short arc. When a photoelastic model is placed in the field, a slight degree of scattering of the light occurs, which disturbs the parallelism of the light. To improve the parallelism of the light, it is appropriate to regard the model as the source of illumination and to control the light as it emerges from the model. The effects of the light scattered by the model can be minimized by placing a diaphragm stop at the focal point of the second field lens. As the diameter of the stop is reduced, the parallelism of the light is improved; however, the intensity of the light striking the camera back is decreased and film exposure times increase.

Comparison of diffused-light and lens polariscopes In this comparison the following properties of a photoelastic polariscope will be considered: definition of the boundaries, direct view of the model, light intensity, length of the unit, operational characteristics, and cost.

Since a lens polariscope employs a parallel beam of light, the definition of the image of the model boundaries on the camera back is sharper than that obtainable with a diffused-light polariscope. However, two precautions can be taken with a diffused-light polariscope to improve the image of the model boundaries. (1) A long-focal-length lens should be employed; moreover, it should be stopped down as far as possible. Results obtained by employing a lens with a focal length of 24 in (610 mm) stopped down to f: 45 are quite satisfactory in almost all applications. (2) If one region of the model is quite important (say a fillet or a hole), this region should be centered in the field of the polariscope. The width of the boundary shadow in a diffused-light polariscope is a function of the distance of the boundary from the centerline of the polariscope. Thus minimizing this distance for critical regions of the model, minimizes the boundary shadow where it may be detrimental.

For direct viewing of the model, the diffused-light polariscope is much more satisfactory than the lens polariscope. In a diffused-light polariscope the fringe pattern occurring in the model can be viewed by looking directly into the analyzer. Moreover, if the analyzer is advanced until it is quite close to the model, the investigator can load and align the model while viewing the pattern. With a lens polariscope the model fringe pattern cannot be viewed directly through the analyzer; instead, it is necessary to project the image on a ground-glass screen in a darkened room. This procedure is less suitable for rapid testing than that established with a diffusion polariscope.

Light intensity offers little or no difficulty for either polariscope provided they both are properly designed. High-intensity light sources and relatively fast lenses can be employed with both polariscopes to hold exposure times well below 1 min while using relatively slow high-contrast film.

The length of the polariscope is an item to be considered when laboratory space is limited. A lens polariscope is necessarily long, running from 10 to 15 ft (3 to 4.5 m), depending primarily upon the diameter of the field and the distance between the two field lenses. In a diffused-light polariscope the only component which requires any appreciable length is the camera. If the camera is eliminated, the system can be designed to occupy a length of less than 1 ft (0.3 m) and the image can be viewed directly. However, if a camera is required on a diffused-light polariscope, then a long-focal-length lens is necessary and the length of the camera will approach or exceed that of a lens polariscope.

The diffused-light polariscope is, in general, easier to operate than a lens polariscope, as the adjustments required are fewer and less precise. Another very important advantage of the diffused-light polariscope is that the surface finish of the model being tested does not require a high degree of polishing. The lens polariscope, on the other hand, has the advantage of a parallel beam of light which permits the utilization of partial mirrors for fringe sharpening and fringe multiplication. The fact that partial mirrors cannot be employed with a diffused-light polariscope is a serious disadvantage.

Finally, the cost of a large-field diffused-light polariscope is appreciably less than that of a large-field lens polariscope. The diffuser plate in a diffused-light polariscope has in effect replaced the lens system of a lens polariscope. As a consequence of the low cost of the diffuser plate in comparison with the expensive field lenses (i.e., the price of lenses increases roughly as the cube of their diameter), an appreciable savings in the purchase price of the diffused-light polariscope is effected.

11.7 OPTICAL INSTRUMENTS: THE INTERFEROMETER [1, 2, 13]

An interferometer is an optical device which can be used to measure lengths or changes in length with great accuracy by means of interference fringes. The modification of intensity of light by superposition of light waves was defined in Sec. 11.2 as an interference effect. The intensity of the wave resulting from the superposition of two waves of equal amplitude was shown by Eq. (11.13) to be a function of the linear phase difference δ between the waves. A fundamental requirement for the existence of well-defined interference fringes is that the light waves producing the fringes have a sharply defined phase difference which remains constant with time. When light beams from two independent sources are superimposed, interference fringes are not observed since the phase difference between the beams varies in a random way (the beams are incoherent). Two beams from the same source, on the other hand, interfere, since the individual wavetrains in the two beams have the same phase initially (the beams are coherent) and any difference in phase at the point of superposition results solely from differences in optical paths. Here optical-path length is defined as

$$\sum_{i=1}^{i=m} n_i L_i \qquad\qquad (a)$$

where L_i is the mechanical-path length in a material having an index of refraction n_i.

   The concept of optical-path difference and its effect on the production of interference fringes can be illustrated by considering the reflection and refraction of light rays from a transparent plate having a thickness h, as shown in Fig. 11.25. Consider a plane wavefront associated with the light ray A which strikes the plate at an angle of incidence α. Ray B results from reflection at the front surface of the plate and, as discussed in Sec. 11.3, suffers a phase change of $\lambda/2$. A second ray is refracted at the front surface, reflected at the back surface, and refracted from the

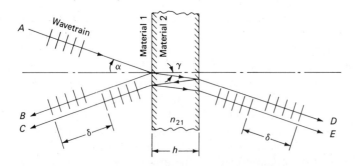

Figure 11.25 Reflection and refraction of light rays from a transparent plate $(n_2 > n_1)$.

front surface before emerging from the front surface of the plate as ray C. The optical-path difference between rays B and C can be computed as

$$\delta = \frac{2h}{\cos\gamma} n_{21} - \frac{2h}{\cos\gamma} \sin\gamma \sin\alpha$$

But from Eq. (11.20)

$$\sin\alpha = n_{21} \sin\gamma$$

therefore $\qquad \delta = \frac{2h}{\cos\gamma} n_{21} - \frac{2h}{\cos\gamma} n_{21} \sin^2\gamma = 2hn_{21} \cos\gamma \qquad (11.41)$

Since ray B suffered a phase change of $\lambda/2$ on reflection, rays B and C will interfere destructively (produce minimum intensity or extinction when brought together) whenever

$$\delta = m\pi \qquad m = 0, 1, 2, 3, \ldots$$

If the light beam illuminates an extended area of the plate, and if the thickness of the plate varies slightly with position, the locus of points experiencing the same order of extinction will combine to form an interference fringe. The fringe spacings will represent thickness variations of approximately 7 μin or 180 nm (in glass with mercury light and a small angle γ).

Rays emerging from the back surface of the plate can also be used to produce interference effects. In Fig. 11.25, a third ray is refracted at both the front and back surfaces of the plate before emerging as ray D. A fourth ray undergoes two internal reflections before being refracted from the back surface of the plate as ray E. The optical-path difference between rays D and E is identical to the difference between rays B and C as given by Eq. (11.41). Since neither ray D nor E suffers a phase change on reflection, the two rays will interfere destructively when brought together whenever

$$\delta = (2m + 1)\frac{\lambda}{2} \qquad m = 0, 1, 2, 3, \ldots$$

The previous discussion serves to illustrate the principles associated with measurements employing interference effects. For the system illustrated in Fig. 11.25, the optical-path difference δ would be many wavelengths ($m \to \infty$). Such a system would involve high-order interference; therefore, it would require extreme coherence and long wavetrains such as those provided by a laser for successful operation. Other systems which utilize low-order interference ($m = 0, 1, 2, 3$) place less stringent requirements on the light source. Two low-order systems used in experimental stress analysis work are the Mach-Zehnder interferometer and the series interferometer.

A. Mach-Zehnder Interferometer

The essential features of a Mach-Zehnder interferometer are illustrated in Fig. 11.26. The light beam from a source is divided into a reference beam and an

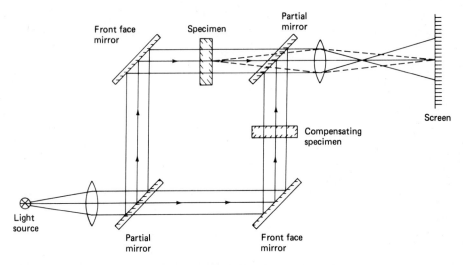

Figure 11.26 Light paths through a Mach-Zehnder interferometer.

active beam with a beam splitter (partial mirror). The beams are recombined after the active beam passes through the specimen of interest. Low-order interference can be obtained by inserting a compensating specimen in the reference beam to adjust for the difference in index of refraction between the specimen material and air.

B. Series Interferometer [13]

The essential features of the series interferometer are illustrated in Fig. 11.27. This instrument, developed by Post [13], is a relatively simple and stable instrument with a large field. It is well suited for photoelasticity applications. The series interferometer utilizes three partial mirrors in series. A fraction of the incident light is reflected and a fraction is transmitted at each of the mirrors. One portion of the light is transmitted directly through the mirrors, as depicted by ray A in Fig. 11.27. Other portions (see rays B and C) are characterized by only two reflections. A majority of the light, however, undergoes a variety of multiple reflections, as illustrated by ray D. When the optical path l_1 is nearly equal to the optical path l_2, rays which traverse paths similar to B and C interfere constructively and destructively to form an interference fringe pattern. This low-order fringe pattern gives the difference between l_1 and l_2 at any point in the field. Superimposed on this pattern is a uniform background intensity due to all the other rays transmitted through the three series mirrors. When a specimen is placed in the field between mirrors 1 and 2 and the optical path l_2 is adjusted to approximately equal l_1, a fringe pattern related to thickness variations in the model is obtained.

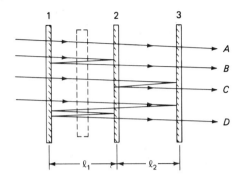

Figure 11.27 Typical forms of reflection as light passes through three partial mirrors in a series interferometer.

EXERCISES

11.1 Verify Eqs. (11.11) and (11.12).

11.2 The wavelength of light from a helium-neon laser is 632.8 nm. Determine:
   (a) The frequency of this light
   (b) The wavelength of this light in a glass plate $(n = 1.522)$
   (c) The velocity of propagation in the glass plate
   (d) The linear phase shift (in terms of wavelength in free space) after the light has passed through a 25-mm-thick glass plate

11.3 A plate of glass having an index of refraction of 1.57 with respect to air is to be used as a polarizer. Determine the polarizing angle and the angle of refraction of the transmitted ray.

11.4 Unpolarized light is directed onto a plane glass surface $(n = 1.57)$ at an angle of incidence of 50°. Determine reflection coefficients associated with the parallel and perpendicular components of the reflected beam.

11.5 A light source is located 15 m below the surface of a body of water $(n = 1.33)$. Determine the maximum distance (measured from a point directly above the source) at which the source will be visible from the air side of the air-water interface.

11.6 A ray of monochromatic light is directed at oblique incidence onto the surface of a glass plate. The ray emerges from the opposite side of the plate in a direction parallel to its initial direction but with a transverse displacement. Develop an expression for this transverse displacement in terms of the plate thickness h, the index of refraction of the glass n, and the angle of incidence α of the light beam.

11.7 The radius of a concave spherical mirror is 1000 mm. An object is located 1500 mm from the mirror. Determine the image location and the magnification. Show the results on a sketch similar to Fig. 11.13.

11.8 Solve Exercise 11.7 if the object is located 750 mm from the mirror.

11.9 Solve Exercise 11.7 if the object is located 200 mm from the mirror.

11.10 A concave mirror will be used to focus the image of an object onto a screen 1.50 m from the object. If a magnification of -2 is required, what radius of curvature must the mirror have?

11.11 Determine the image location and the magnification for an object located 750 mm from a mirror if the mirror is (a) a plane mirror and (b) a convex mirror with a radius of curvature of 1000 mm.

11.12 A convex mirror has a radius of curvature of 2500 mm. Determine the magnification and the image location for an object located 500 mm from the mirror. Show the results on a sketch similar to Fig. 11.13.

11.13 A thin convex lens has a focal length of 600 mm. An object is located 1200 mm to the left of the lens. Determine the image location and the magnification. Show the results on a sketch similar to Fig. 11.14.

11.14 Solve Exercise 11.13 if the object is located 400 mm from the lens.

11.15 An object is placed 500 mm to the left of a thin convex lens having a focal length of 250 mm. A second lens having a focal length of 300 mm is placed 625 mm to the right of the first lens. Determine the location and magnification of the resulting image. Show the results on a sketch similar to Fig. 11.16.

11.16 An object is located 225 mm to the left of a thin convex lens having a focal length of 300 mm. A second lens having a focal length of 250 mm is located 150 mm to the right of the first lens. Determine the location and magnification of the resulting image. Show the results on a sketch similar to Fig. 11.16.

11.17 An object is located 125 mm to the left of a thin convex lens having a focal length of 250 mm. A spherical concave mirror having a radius of curvature of 500 mm is located 625 mm to the right of the lens. Determine the location and magnification of the image resulting from this lens-mirror combination. Show the results on a sketch similar to Fig. 11.16.

11.18 Two thin lenses having focal lengths f_1 and f_2 are placed in contact with one another. Develop an expression for the equivalent focal length of this series combination.

11.19 Verify Eqs. (11.28) through (11.31).

11.20 Parallel light of wavelength 546.1 nm is directed at normal incidence onto a slit having a width of 0.050 mm. A lens having a focal length of 600 mm is positioned behind the slit and is used to focus the light passing through the slit onto a screen. Determine the distance (a) from the center of the diffraction pattern to the first diffraction fringe (minimum intensity) and (b) between the third and fourth diffraction fringes.

11.21 Solve Exercise 11.20 if light from a helium-neon laser is used which has a wavelength of 632.8 nm.

11.22 In a single-slit diffraction pattern the distance from the first minimum on the left to the first minimum on the right is 25 mm. The screen on which the pattern is displayed is 4 m from the slit. The wavelength of the light is 589.3 nm. Determine the slit width.

11.23 If two linear polarizers are arranged in series with an arbitrary angle θ between their planes of polarization, plot the amplitude of the transmitted light as a function of θ as it varies from 0 to 90°.

11.24 What will be the relative angular retardation Δ in a quarter-wave plate designed for operation at $\lambda = 546.1$ nm if it is employed with sodium light, where $\lambda = 589.3$ nm?

11.25 Show that two uniformly imperfect quarter-wave plates can be arranged in series with an appropriate angle between the two plates to produce the effect of a perfect quarter-wave plate. Solve for this angle by assuming that both imperfect quarter-wave plates have an angular retardation of $\Delta = \pi/2 + \epsilon$.

11.26 Determine the magnitude and direction of the light vector emerging from a series combination of a linear polarizer and a half-wave plate oriented at an arbitrary angle θ with respect to the plane of vibration of the linear polarizer.

11.27 Show that the four possible arrangements for a circular polariscope listed in Table 11.3 produce either a light field or a dark field.

11.28 Design the lens system for a 300-mm-diameter lens-type polariscope. Specify a range of magnification M available within the limits of the design parameters $f_1, f_2, u_1,$ and u_2.

11.29 Light having a wavelength of 546.1 nm is directed at normal incidence onto a thin film of transparent material having an index of refraction of 1.63. Ten dark and nine bright fringes are observed over a 25-mm length of the film. Determine the thickness variation over this length.

REFERENCES

1. Born, M., and E. Wolf: "Principles of Optics," Pergamon Press, New York, 1959.
2. Jenkins, F. A., and H. E. White: "Fundamentals of Optics," 4th ed., McGraw-Hill Book Company, New York, 1976.

3. Shurcliff, W. A.: " Polarized Light," Harvard University Press, Cambridge, Mass., 1962.
4. Grabau, M.: Optical Properties of Polaroid for Visible Light, *J. Opt. Soc. Am.*, vol. 27, pp. 420–424, 1937.
5. Land, E. H.: Some Aspects of the Development of Sheet Polarizers, *J. Opt. Soc. Am.*, vol. 41, pp. 957–963, 1951.
6. Jerrard, H. G.: The Calibration of Quarter-wave Plates, *J. Opt. Soc. Am.*, vol. 42, pp. 159–165, 1952.
7. Tuzi, Z., and H. Oosima: On the Artifical Quarter Wave Plate for Photoelasticity Apparatus, *Sci. Pap. Inst. Phys. Chem. Res. (Tokyo)*, vol. 36, pp. 72–81, 1939.
8. Jessop, H. T.: The Optical System in Photo-elastic Observations, *J. Sci. Instrum.*, vol. 25, p. 124, 1948.
9. Mindlin, R. D.: A Reflection Polariscope for Photoelastic Analysis, *Rev. Sci. Instrum.*, vol. 5, pp. 224–228, 1934.
10. Mylonas, C.: The Optical System of Polariscopes, *J. Sci. Instrum.*, vol. 25, pp. 77–81, 1948.
11. Durelli, A. J.: discussion of paper entitled The Photoelastic Laboratory at the Newport News Shipbuilding and Dry Dock Company, *Proc. SESA*, vol. VI, no. 1, pp. 106–110, 1948.
12. Lee, B. R., R. Meadows, Jr., and W. F. Taylor: The Photoelastic Laboratory at the Newport News Shipbuilding and Dry Dock Company, *Proc. SESA*, vol. VI, no. 1, pp. 83–106, 1948.
13. Post, D.: A New Photoelastic Interferometer Suitable for Static and Dynamic Measurements, *Proc. SESA*, vol. XII, no. 1, pp. 191–202, 1954.

TWELVE

MOIRÉ METHODS

12.1 INTRODUCTION [1–10]

The word *moiré* is the French name for a fabric known as *watered silk*, which exhibits patterns of light and dark bands. This moiré effect occurs whenever two similar but not quite identical arrays of equally spaced lines or dots are arranged so that one array can be viewed through the other. Almost everyone has seen the effect in two parallel snow fences or when two layers of window screen are placed in contact.

The first practical application of the moiré effect may have been its use in judging the quality of line rulings used for diffraction gratings or halftone screens. In this application, the moiré fringes provide information on errors in spacing, parallelism, and straightness of the lines in the ruling. All these factors contribute to the quality of the ruling.

Elimination of the moiré effect has always been a major problem associated with screen photography in the printing industry. In multicolor printing, for example, where several screened images must be superimposed, the direction of screening must be carefully controlled to minimize moiré effects.

Considerable insight into the moiré effect can be gained by studying the relationships which exist between the spacings and inclinations of the moiré fringes in a pattern and the geometry of the two interfering line arrays which produce the pattern. This geometrical interpretation of the moiré effect was first published by Tollenaar [1] in 1945. In 1960 Morse, Durelli, and Sciammarella [2] presented a complete analysis of the geometry of moiré fringes in strain analysis, in which the fundamental equations of the moiré method are derived and presented in the form of curves which can be used to give strains and rotations with a minimum of computation.

A second method of analysis of moiré fringe patterns was presented by Weller and Shepard [3] in 1948, describing an application in which moiré fringes were used to measure displacements. Dantu [4, 5] followed the same approach in 1954 and introduced the interpretation of moiré fringes as components of displacements for plane elasticity problems. Sciammarella and Durelli [6] extended this approach into the region of large strains in a paper in 1961.

Moiré fringes have also been used by Theocaris [7, 8] to measure out-of-plane displacements and by Ligtenberg [9] to measure slopes and moment distributions in flat slabs. The use of moiré fringes for measuring displacements in structural models has been outlined in a paper by Durelli and Daniel [10]. The previous discussion serves to illustrate the broad field of application of the moiré method in the determination of displacements and strains.

In the following sections of this chapter, both the *geometrical* and the *displacement-field* approaches to moiré fringe analysis will be outlined. The advantages and limitations of the approaches will be discussed.

12.2 MECHANISM OF FORMATION OF MOIRÉ FRINGES [2–6, 14]

The arrays used to produce moiré fringes may be a series of straight parallel lines, a series of radial lines emanating from a point, a series of concentric circles, or a pattern of dots. In stress-analysis work, arrays consisting of straight parallel lines (ideally, opaque bars with transparent interspaces of equal width) are the most commonly used. Such arrays are frequently referred to as grids, gratings, or grills. In this book, the term *grid* has been used to denote a coarse array (10 lines per inch or less) of perpendicular lines that does not produce a moiré effect (see Sec. 5.4). Comparators or microscopes are normally employed to measure changes in spacing between the intersecting points of a grid network before and after loading, and from these data displacements and strains can be determined. In this book the term *grating* will be used to denote a parallel-line array suitable for moiré work (50 to 1000 lines per inch). When two perpendicular line arrays are used on a specimen, the term *cross-grating* will be employed.

Two gratings must be overlaid to produce moiré fringes. The overlaying can be accomplished by mechanical or optical means. In the following discussions, the two gratings will be referred to as the *model*, or specimen, grating and the *master*, or reference, grating. Quite frequently, the model grating is applied by coating the specimen with a photographic emulsion and contact-printing through the master grating. In this way, the model and master gratings are essentially identical (matched) when the specimen is in the undeformed state. Model arrays can also be applied by bonding, etching, ruling, etc. In one method of analysis, the shadow of the master grating on the model serves as the model array.

A typical moiré fringe pattern, obtained using transmitted light through the

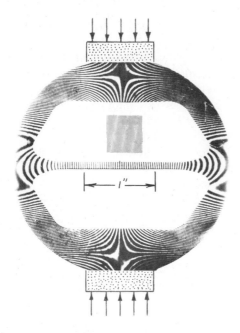

Figure 12.1 Moiré fringe pattern in a special tensile-strength specimen. (*Courtesy of A. J. Durelli.*)

model and master gratings, is shown in Fig. 12.1. In this instance, both the master grating and the model grating before deformation had 1000 lines per inch (39.4 lines per millimeter). The number of lines per unit length is frequently referred to as the *density* of the grating. The master grating was oriented with its lines running vertically, as shown in the inset.

In the discussion which follows, the center-to-center distance between the master grating lines will be referred to as the *pitch* of the grating (reciprocal of the density) and will be designated by the symbol p. The center-to-center distance on the model grating in the deformed state will be designated p'. The direction perpendicular to the lines of the master grating will be referred to as a *primary direction*. The direction parallel to the lines of the master grating will be referred to as a *secondary direction*.

The mechanism of formation of moiré fringes can be illustrated by considering the transmission of a beam of light through model and reference arrays, as shown in Fig. 12.2. If model and master gratings are identical, and if they are aligned such that the opaque bars of one grating coincide exactly with the opaque bars of the other grating, the light will be transmitted as a series of bands having a width equal to one-half the pitch of the gratings. However, owing to diffraction and the resolution capabilities of the eye, this series of bands will appear as a uniform grey field with an intensity equal to approximately one-half the intensity of the incident beam when the pitch of the gratings is small.

If the model is then subjected to a uniform deformation, e.g., the one shown in the central tensile bar of Fig. 12.1, the model grating will exhibit a deformed pitch p', as shown in Fig. 12.3. The transmission of light through the two gratings will

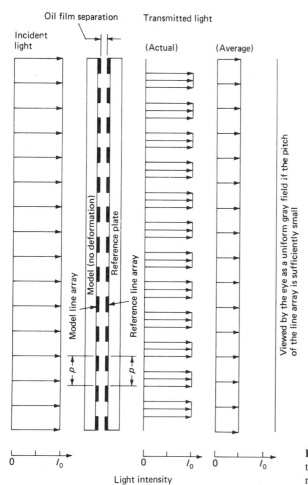

Figure 12.2 Light transmission through matched and aligned model and master gratings.

now occur as a series of bands of different width, the width of the band depending on the overlap of an opaque bar with a transparent interspace. If the intensity of the emerging light is averaged over the pitch length of the master grating, to account for diffraction effects and the resolution capabilities of the eye, the intensity is observed to vary as a staircase function of position. The peaks of the function occur at positions where the transparent interspaces of the two gratings are aligned. A light band is perceived by the eye in these regions. When an opaque bar of one grating is aligned with the transparent interspace of the other grating, the light transmitted is minimum and a dark band known as a *moiré fringe* is formed.

Inspection of the opaque bars in Fig. 12.3 indicates that a moiré fringe is formed, within a given gage length, each time the model grating within the gage length suffers a deformation in the primary direction equal to the pitch p of the

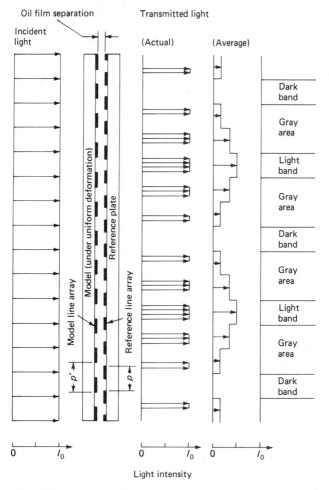

Figure 12.3 Formation of moiré fringes in a uniformly deformed specimen.

master grating. Specimen deformations in the secondary direction do not produce moiré fringes. In the case illustrated in Fig. 12.1, 32 fringes have formed in the 1-in (25 mm) gage length indicated on the specimen. Thus, the change in length of the specimen in this 1-in interval is

$$\Delta l = np = 32(0.001) = 0.032 \text{ in}$$

It should be noted in this illustration that the 1-in gage length represents the final or deformed length of the specimen rather than the original length. Thus, the engineering strain in this interval could be expressed as

$$\epsilon = \frac{\Delta l}{l_0} = \frac{np}{1 - np} = \frac{0.032}{1 - 0.032} = 0.033$$

The results of the previous observations can be generalized to yield expressions for either tensile or compressive strains over an arbitrary gage length for cases involving uniform elongation or contraction but no rotation. Thus

$$\epsilon = \begin{cases} \dfrac{\Delta l}{l_0} = \dfrac{np}{l_g - np} & \text{for tensile strains} \quad (12.1) \\[3mm] -\dfrac{\Delta l}{l_0} = -\dfrac{np}{l_g + np} & \text{for compressive strains} \quad (12.2) \end{cases}$$

where p = pitch of master grating and undeformed model grating
  n = number of moiré fringes in gage length
  l_g = gage length

In the previous discussions, the moiré fringes were formed by elongations or contractions of the specimen in a direction perpendicular to the lines of the master grating. Simple experiments with a pair of identical gratings indicates that moiré fringes can also be formed by pure rotations (no elongations or contractions), as illustrated in Fig. 12.4. In this illustration, two gratings having a line density of 59 lines per inch (2.32 lines per millimeter) have been rotated through an angle θ with respect to one another. Note that the moiré fringes have formed in a direction which bisects the obtuse angle between the lines of the two gratings. The relationship between angle of rotation θ and angle of inclination ϕ of the moiré fringes, both measured in the same direction and with respect to the lines of the master grating, can be expressed as

$$\phi = \frac{\pi}{2} + \frac{\theta}{2} \qquad (12.3)$$

or
$$\theta = 2\phi - \pi \qquad (12.4)$$

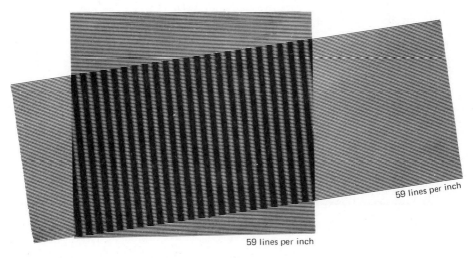

59 lines per inch

59 lines per inch

Figure 12.4 Moiré fringes formed by rotation of one grating with respect to the other.

12.3 THE GEOMETRICAL APPROACH TO MOIRÉ-FRINGE ANALYSIS [2]

In the previous section it was established that moiré fringes are produced either by rotation of the specimen grating with respect to the master grating or by changes in pitch of the specimen grating as a result of load induced deformations. At a general point in a stressed specimen, these two effects occur simultaneously and produce a fringe pattern similar to the one shown in Fig. 12.5. The information readily available from such a pattern is the angle of inclination ϕ of the fringes with respect to the lines of the master grating and the distance between fringes δ. The quantities to be determined at the point of interest are the angle of rotation θ of the specimen grating with respect to the lines of the master grating and the pitch p' of the specimen grating in the deformed state. This approach to moiré-fringe analysis, which is very convenient when information is desired at only a few selected points, has become known as the geometrical approach. Since the geometrical approach gives rotations and strains that are average values between two fringes, such analyses should be limited to homogeneous fields or to very small regions of nonhomogeneous fields. The following presentation will provide only the significant features of the approach. The reader interested in more detail should consult the original work by Morse, Durelli, and Sciammarella [2].

Relationships between the master grating pitch p, the deformed specimen grating pitch p', the angle of rotation θ of the specimen grating with respect to the lines of the master grating, and the data available from a moiré fringe pattern can be obtained from a geometric analysis of the intersections of the lines of the two gratings as shown in Figs. 12.6 and 12.7. In both these figures the opaque bars of the gratings are defined by their centerlines, and light rather than dark bands are referred to as fringes since they are easier to locate accurately.

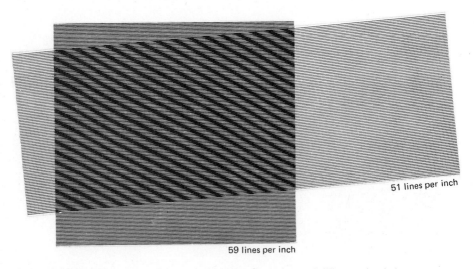

51 lines per inch

59 lines per inch

Figure 12.5 Moiré fringes formed by a combination of rotation and difference in pitch.

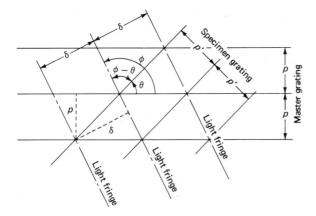

Figure 12.6 Geometry of moiré fringes in terms of fringe spacing.

Consider first the distance δ between fringes. From the geometry of the intersections shown in Fig. 12.6 it is observed that

$$\frac{p}{\sin \theta} = \frac{\delta}{\sin (\phi - \theta)} \qquad (12.5)$$

Solving for the angle of rotation θ in terms of the pitch p of the master grating and the quantities δ and ϕ which can be obtained from the moiré fringe pattern gives

$$\tan \theta = \frac{\sin \phi}{\delta/p + \cos \phi} \qquad (12.6)$$

If gratings with a very fine pitch are used to measure small angles of rotation, then $\phi \approx \pi/2$ and Eq. (12.6) reduces to

$$\theta = \frac{p}{\delta} \qquad (12.7)$$

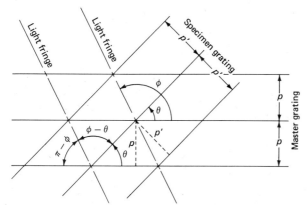

Figure 12.7 Geometry of moiré fringes in terms of angle of inclination.

In a similar manner, it is observed in Fig. 12.7 that

$$\frac{p}{\sin(\pi - \phi)} = \frac{p'}{\sin(\phi - \theta)} \tag{12.8}$$

Thus

$$p' = \frac{p \sin(\phi - \theta)}{\sin \phi} \tag{12.9}$$

or from Eq. (12.5)

$$p' = \frac{\delta \sin \theta}{\sin \phi} \tag{12.10}$$

The angle θ can be eliminated from Eq. (12.10) by use of the trigonometric identity

$$\sin \theta = \frac{\tan \theta}{\sqrt{1 + \tan^2 \theta}}$$

Thus, from Eqs. (12.10) and (12.6), the deformed specimen pitch p' can be expressed in terms of the pitch p of the master grating and the quantities δ and ϕ which can be determined from the moiré fringe pattern

$$p' = \frac{\delta}{\sqrt{1 + (\delta/p)^2 + 2(\delta/p)\cos\phi}} \tag{12.11}$$

In many instances, moiré fringe patterns will be evaluated in regions where rotations are small. In these cases, $\phi \approx 0$ or π, and Eq. (12.11) reduces to

$$p' = \frac{p\delta}{p \pm \delta} \tag{12.12}$$

Once the deformed specimen pitch p' has been determined, the component of normal strain in a direction perpendicular to the lines of the master grating can be computed as

$$\epsilon = \frac{p' - p}{p} \tag{12.13}$$

12.4 THE DISPLACEMENT-FIELD APPROACH TO MOIRÉ-FRINGE ANALYSIS [3–6, 11–13]

Moiré fringe patterns can also be interpreted by relating them to a displacement field. In Sec. 12.2 it was shown that a moiré fringe is formed within a given gage length in a uniformly deformed specimen (the central tensile bar of Fig. 12.1, for example) each time the specimen grating within the gage length is extended (or shortened) by an amount equal to the pitch p of the master grating in a direction perpendicular to the lines of the master grating. Deformations in a direction parallel to the lines of the master grating have no effect on the moiré fringe pattern. The deformation within the gage length can also be expressed in terms of a relative displacement between points at the ends of the gage length.

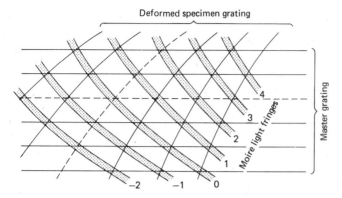

Deformed specimen grating

Master grating

Moire light fringes

4
3
2
1

−2   −1   0

Figure 12.8 Moiré fringes at an arbitrary point in a stressed specimen.

The displacement-field concept can be extended to the general case of extension or contraction combined with rotation. In Fig. 12.8 the lines of a deformed specimen grating are shown superimposed on a master grating. One of the lines of each of the gratings is shown dotted for easy reference, and it will be assumed that these lines coincided in the undeformed state. Both gratings are assumed to have had an initial pitch p. Interference between the master grating and the deformed specimen grating would produce the light moiré fringes shown in the figure. In this pattern it should be noted that the intersection of the two dotted lines has been used to locate the zero-order fringe. This point on the specimen did not displace in the vertical direction as the specimen deformed. Similar intersections of specimen and master grating lines which initially coincided are also located on the zero-order fringe. The intersection of the dotted line on the specimen with the line above the dotted line on the master grating lies on the fringe of order 1. This point on the specimen has moved a distance p in the vertical direction from its original position. Similarly, a point lying on the fringe of order 4 has moved a distance $4p$ in the vertical direction from its original position. Thus, a moiré fringe is a locus of points exhibiting the same component of displacement in a direction perpendicular to the lines of the master grating.

For purposes of analysis, the moiré fringe pattern can be visualized as a displacement surface where the height of a point on the surface above a plane of reference represents the displacement of the point in a direction perpendicular to the lines of the master grating. A simple illustration of this concept is shown in Fig. 12.9. Similar patterns and displacement surfaces can be obtained for any other orientation of the specimen and master gratings.

Once u and v displacement surfaces have been established by using line arrays perpendicular to the x and y axes of a specimen, respectively, the cartesian components of strain can be computed from the derivatives of the displacements (slopes of the displacement surfaces). Two models are normally required for these determinations unless an axis of symmetry exists in the specimen so that mutually

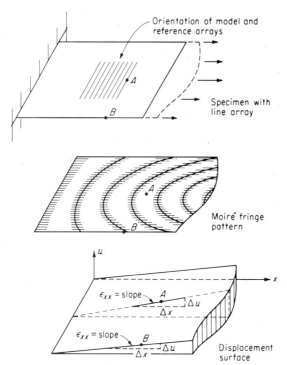

Figure 12.9 Moiré fringe pattern and associated displacement surface for a nonuniform strain field.

perpendicular arrays can be placed on the two halves of the specimen. For the case of large strains, the relationships between displacements and strains are given by Eqs. (2.2) and (2.3) as

$$\epsilon_{xx} = \sqrt{1 + 2\frac{\partial u}{\partial x} + \left(\frac{\partial u}{\partial x}\right)^2 + \left(\frac{\partial v}{\partial x}\right)^2 + \left(\frac{\partial w}{\partial x}\right)^2} - 1$$

$$\epsilon_{yy} = \sqrt{1 + 2\frac{\partial v}{\partial y} + \left(\frac{\partial v}{\partial y}\right)^2 + \left(\frac{\partial w}{\partial y}\right)^2 + \left(\frac{\partial u}{\partial y}\right)^2} - 1$$

$$\gamma_{xy} = \arcsin \frac{\dfrac{\partial u}{\partial y} + \dfrac{\partial v}{\partial x} + \dfrac{\partial u}{\partial x}\dfrac{\partial u}{\partial y} + \dfrac{\partial v}{\partial x}\dfrac{\partial v}{\partial y} + \dfrac{\partial w}{\partial x}\dfrac{\partial w}{\partial y}}{(1 + \epsilon_{xx})(1 + \epsilon_{yy})} \tag{12.14}$$

where u, v, and w are the displacement components in the x, y, and z directions, respectively. When products and powers of derivatives are small enough to be neglected, the previous equations reduce to those commonly employed in the classical linear theory of elasticity:

$$\epsilon_{xx} = \frac{\partial u}{\partial x} \qquad \epsilon_{yy} = \frac{\partial v}{\partial y} \qquad \gamma_{xy} = \frac{\partial v}{\partial x} + \frac{\partial u}{\partial y} \tag{12.15}$$

The displacement gradients $\partial u/\partial x$ and $\partial v/\partial y$ are obtained from the slopes of the two displacement surfaces in a direction perpendicular to the lines of the master gratings. The displacement gradients $\partial u/\partial y$ and $\partial v/\partial x$ are obtained from the slopes of the displacement surfaces in a direction parallel to the lines of the master gratings. The displacement gradients $\partial w/\partial x$ and $\partial w/\partial y$ are usually small and can be neglected in most analyses.

In application of the moiré method, the following technique is used to determine the strains. Two moiré fringe patterns are obtained with specimen and master gratings oriented perpendicular to the x and y axes. Schematic illustrations of the two fringe patterns are shown in Figs. 12.10 and 12.11. Lines along the x and y axes, say AB and CD, are drawn, and displacements u and v along each of these lines are plotted by noting that

$$u, v = np \tag{12.16}$$

where n is the order of the moiré fringe at the point and p is the pitch of the master

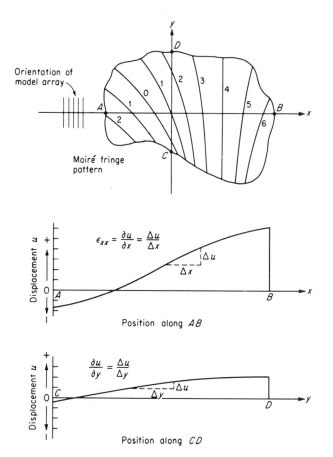

Figure 12.10 Displacement-position plots used to determine $\partial u/\partial x$ and $\partial u/\partial y$.

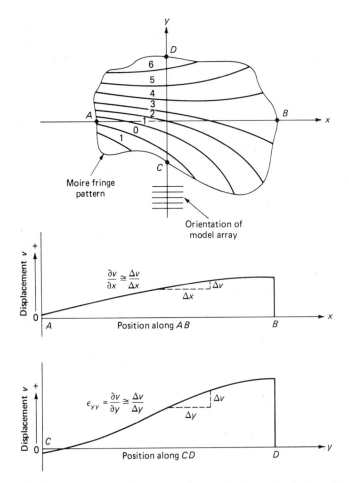

Figure 12.11 Displacement-position plots used to determine $\partial v/\partial x$ and $\partial v/\partial y$.

grating. Tangents drawn to these curves give $\partial u/\partial x$, $\partial u/\partial y$, $\partial v/\partial x$, and $\partial v/\partial y$, as shown at the bottom of Figs. 12.10 and 12.11. Equations (12.14) or (12.15) can then be used to determine the strains ϵ_{xx}, ϵ_{yy}, and γ_{xy}. If lines similar to AB and CD are drawn at all critical areas of the specimen, a complete description of the strain over the entire field of the model can be achieved.

The displacement-field approach to moiré fringe analysis, as previously outlined, is based on an accurate determination of the displacement derivatives $\partial u/\partial x$, $\partial u/\partial y$, $\partial v/\partial x$, and $\partial v/\partial y$. In practice, when the u and v moiré fringe patterns are obtained from separate models, from the two halves of a symmetric model, or with separate u and v master gratings on a crossed model grating (an array of orthogonal lines), the direct derivatives $\partial u/\partial x$ and $\partial v/\partial y$ can usually be obtained with acceptable accuracy but the cross derivatives $\partial u/\partial y$ and $\partial v/\partial x$ cannot. This results from the fact that slight errors in alignment of either the specimen or master gratings with the x or y axes produce a fringe pattern due to rotation misalignment (see Fig. 12.4) in addition to the load-induced pattern. It should be noted in such a rotation pattern that the displacement gradient perpendicular to the lines of the master grating is small while the displacement gradient parallel to the lines of the master grating is large. This sensitivity of the moiré fringe pattern to small rigid-body rotations associated with any misalignment produces large errors in the determinations of the cross derivatives and thus in the shear strains.

One method proposed to eliminate shear-strain error makes use of crossed gratings on both the specimen and master to obtain simultaneous displays of the u and v displacement fields. Since any rotational misalignment is then equal for the two fields, its contribution to the cross derivatives is equal in magnitude but opposite in sign and thus cancels in the shear-strain determination.

In 1948, Weller and Shepard [3] recognized the possibility of displaying two moiré fringe patterns simultaneously with crossed gratings but recommended against its use because of the interweaving between the two families of fringes, as shown in Fig. 12.12, which makes interpretation difficult if not impossible. Post [12] finally resolved this difficulty by using crossed gratings with slightly different pitches on the specimen and master to produce an initial pattern. Proper selection of the grating mismatch provides an initial pattern which can be used to eliminate uncertainties in the assignment of fringe orders throughout the field and

Figure 12.12 Moiré fringe pattern with crossed gratings of identical pitch on the master and the specimen. (*Courtesy of D. Post.*)

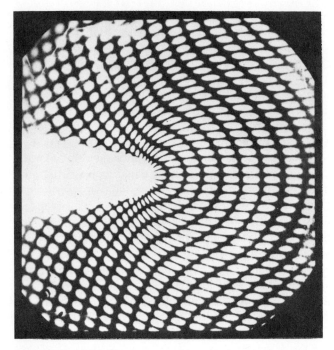

Figure 12.13 Moiré fringe pattern with crossed gratings of different pitch on the master and the specimen. (*Courtesy of D. Post.*)

to provide sufficient data points in any local region for reliable evaluation of fringe gradients. The pattern shown in Fig. 12.12 without mismatch appears as Fig. 12.13 with mismatch. For further information and a detailed discussion of the interpretation of moiré fringe patterns with mismatch, the reader should consult the original paper by Post [12].

The problem of shear-strain errors can also be solved by using the strain-rosette concept commonly employed with electrical-resistance strain gages. With this method, model gratings are employed perpendicular to the x, n (usually $45°$ with respect to the x axis), and y axes. Once the three normal strains ϵ_{xx}, ϵ_n, and ϵ_{yy} are determined at a point, the complete state of strain at the point can be calculated by means of the rosette equations of Chap. 10. In practice, the three moiré patterns can be obtained by using a square array of dots on the specimen, as suggested by Dantu [13]. The master gratings used for the x and y directions have a pitch p, while the grating for the $45°$ direction has a pitch $(\sqrt{2}/2)p$.

12.5 OUT-OF-PLANE DISPLACEMENT MEASUREMENTS [7, 8]

The moiré fringe methods discussed in the previous sections of this chapter have been concerned with the determination of the in-plane displacements u and v, rotations θ_z, and the strains ϵ_{xx}, ϵ_{yy}, and γ_{xy}. In certain plane-stress problems and in a wide variety of problems involving laterally loaded plates, out-of-plane displacements w become important considerations. A moiré method for determining out-of-plane displacements has been developed by Theocaris [7, 8] and applied to a number of these problems. The essential features of the method are as follows.

For out-of-plane displacement measurements, a master grating is employed in front of the specimen, and a collimated beam of light is directed at oblique incidence through the master grating and onto the surface of the specimen, as shown in Fig. 12.14. The shadow of the master grating on the surface of the specimen serves as the specimen grating. When the specimen is viewed at normal incidence, moiré fringes form as a result of interference between the lines of the master and the shadows. Use of a matte surface to ensure distinct shadows improves the quality of the moiré fringe patterns.

From the geometry illustrated in Fig. 12.14 it can be seen that the difference in distance between the master grating and the specimen surface at two adjacent fringe locations can be expressed as

$$d_2 - d_1 = \frac{p}{\tan \alpha}$$

where p is the pitch of the master grating and α is the angle of incidence of the collimated light beam.

In practice, the master grating is located a small distance away from the specimen to accommodate any surface displacements toward the master grating and to serve as a datum plane for the measurement of load-induced, out-of-plane displacements. Any distribution of moiré fringes appearing with the master grating in this initial position will represent irregularities in the surface of the specimen. The presence of any irregularity must be accounted for in the final determination of the out-of-plane displacement.

If a point of zero out-of-plane displacement is known to exist at some point in the specimen, e.g., from theoretical considerations, the master grating can be

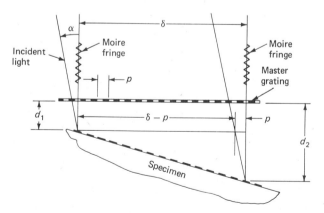

Figure 12.14 Moiré method for measuring out-of-plane displacements.

positioned to locate a moiré fringe (reference, or zero-order, fringe) over this point. At all other fringe locations, the out-of-plane displacement w can then be expressed as

$$w = \frac{np}{\tan \alpha} \tag{12.17}$$

where n is the order of the moiré fringe at the point.

If no point of zero out-of-plane displacement exists in the specimen, it may be necessary to measure the displacement at some convenient point by other means, after which the displacements at all other points can be referred to this reference point.

12.6 OUT-OF-PLANE SLOPE MEASUREMENTS [9]

The determination of stress distributions and deflections in laterally loaded plates is a difficult but important engineering problem. From the theory of elasticity it is known that the stresses at a point in the plate can be expressed in terms of the local curvatures of the plate as

$$\sigma_x = \frac{Ez}{1 - v^2}\left(\frac{1}{\rho_x} + v\,\frac{1}{\rho_y}\right) \qquad \sigma_y = \frac{Ez}{1 - v^2}\left(\frac{1}{\rho_y} + v\,\frac{1}{\rho_x}\right)$$

The deflections are related to the curvatures by the approximate expressions

$$\frac{1}{\rho_x} = -\frac{\partial^2 w}{\partial x^2} \qquad \frac{1}{\rho_y} = -\frac{\partial^2 w}{\partial y^2}$$

In theory, the out-of-plane displacement-measuring technique discussed in the previous section can provide the required curvatures for a solution to the stress problem. In practice, the double differentiations cannot be performed with sufficient accuracy to provide suitable values for the curvatures. To overcome this experimental difficulty, Ligtenberg [9] has developed a moiré method for measuring the partial slopes $\partial w/\partial x$ and $\partial w/\partial y$. A single differentiation then provides reasonably accurate values for the required curvatures.

The essential features of the Ligtenberg method are illustrated in Fig. 12.15. The equipment consists of a fixture for holding and loading the plate, a large cylindrical surface with a coarse line grating, and a camera for recording the moiré fringe patterns. The surface of the plate is made reflecting since the camera views the image of the grating on the surface of the plate. Since the image does not depend on the angle of incidence of the light, a collimated beam is not required for this method. The moiré fringe pattern is formed by superimposing grating images before and after loading. Double-exposure photography is useful for performing the superposition of images.

From the geometry of Fig. 12.15, it can be seen that the location on the grating being viewed by the camera, as a result of reflections from a typical point P

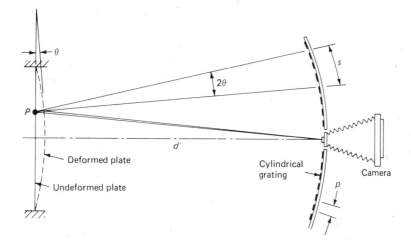

Figure 12.15 Moiré method for measuring out-of-plane slopes.

on the surface of the specimen, shifts as the plate deflects under load. The moiré fringe pattern formed by the superposition of the two images provides a measure of this shift. Observe in Fig. 12.15 that the shift can be expressed in terms of the local slope of the plate as

$$s = 2\theta d$$

where s = magnitude of shift
   θ = local slope of plate in a plane perpendicular to undeformed plate and lines of grating
   d = distance between plate and grating

A moiré fringe will form upon superposition of the two images if the shift s is equal to the pitch p of the grating. Thus the order of the moiré fringe can be expressed as

$$n = \frac{2\theta d}{p}$$

or
$$\theta = \frac{np}{2d} \tag{12.18}$$

The distance d should be large to minimize the effects of out-of-plane displacements w on the shift distance s. Ligtenberg has also shown that the curved grating should have a radius of the order of $3.5d$.

The angle θ given by Eq. (12.18) is the partial slope $\partial w/\partial x$ or $\partial w/\partial y$ depending on the orientation of the grating. Two moiré patterns of these slopes will be needed to solve a plate problem completely. The two patterns can be obtained by rotating the grating 90° after the first pattern is recorded.

12.7 SHARPENING AND MULTIPLICATION OF MOIRÉ FRINGES [14–20]

Application of moiré methods to the study of deformations and strains in the elastic range of material response is usually limited by the lack of sensitivity of the method with respect to other methods, e.g., electrical-resistance strain gages. Both the geometrical and displacement-field approaches to moiré fringe analysis, which were discussed in Sec. 12.3 and 12.4, require accurate determinations of either the fringe spacings or the fringe gradients at the point of interest in the specimen. In a typical moiré fringe pattern obtained with line gratings having a density of 1000 lines per inch (40 lines per millimeter) or less, only a few fringes are normally present; therefore, the spacings or gradients cannot be established with the required accuracy. Since it is not practical to increase the line densities of the gratings much beyond 1000 lines per inch (40 lines per millimeter) to achieve the required sensitivity, moiré-fringe-sharpening and moiré-fringe-multiplication methods have been developed for use with the coarse gratings to solve the sensitivity problem.

Moiré fringe sharpening and moiré fringe multiplication were first discussed by Post [14] in 1967. Application of the method proposed in this initial study has been limited since the special gratings required are difficult to manufacture and are not readily available from a commercial source. Since the method provides additional insight into the mechanism of moiré fringe formation, brief coverage will be provided in this section. The reader interested in more detailed coverage of the method should consult the original paper.

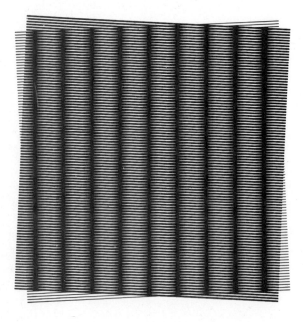

Figure 12.16 Moiré fringe sharpening with complementary gratings. (*Courtesy of D. Post.*)

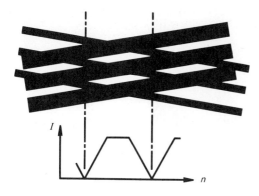

Figure 12.17 Mechanism of formation of sharpened fringes with complementary gratings, $R = 2$ and $\frac{1}{2}$.

In all previous discussions in this chapter, the master grating and the model grating before deformation were identical and were assumed to have opaque bars and clear spaces of equal width. Post has shown that moiré fringe patterns can also be formed when the opaque bars and the clear spaces in the gratings do not have equal widths and that under certain conditions such gratings can be used to produce very desirable effects. For example, consider the case where one grating has wide opaque bars and narrow clear spaces while the other one has narrow opaque bars and wide clear spaces. Such gratings are said to be complementary if the ratio R (opaque width to clear width) of one of the gratings is the reciprocal of the other. A typical example of a sharpened moiré fringe pattern, obtained with complementary gratings, is shown in Fig. 12.16. The mechanism of formation of the sharpened fringes is illustrated in Fig. 12.17, where the intensity of light transmitted through the pair of gratings is plotted as a function of position along a line which is perpendicular to the moiré fringes. This type of intensity distribution is characteristic of complementary gratings.

Moiré fringe multiplication can be accomplished by introducing additional narrow opaque bars in the wide clear spaces of one of the complementary gratings, as shown in Fig. 12.18. The mechanism of formation of the additional fringes is illustrated in Fig. 12.19. Under proper conditions, several additional bars can be inserted in each of the wide clear spaces of one of the gratings. The moiré pattern formed by such a pair of gratings will then have several times as many fringes as the pattern formed by the original complementary pair of gratings.

The methods of moiré fringe sharpening and moiré fringe multiplication discussed in the previous paragraphs can be referred to as geometrical methods since they are based on a simple geometric-optics (or ray-optics) treatment of the passage of light through the superimposed gratings. Other methods of fringe sharpening and fringe multiplication introduced by Holister [15], Post [16, 17], Sciammarella [18], and Chiang [19] utilize the diffraction effects associated with light passage through the gratings.

The diffraction of light from a narrow slit was considered in Sec. 11.5. When light consisting of a series of plane wavefronts is used to illuminate a grating, each of the clear spaces in the grating acts as a slit and, in accordance with Huygens'

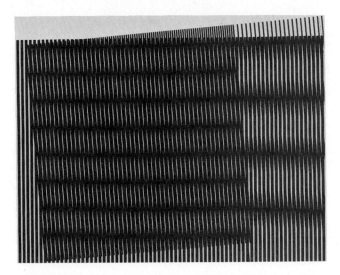

Figure 12.18 Moiré fringe multiplication with special gratings. (*Courtesy of C. Post.*)

principle, emits light in the form of a series of secondary cylindrical wavelets. A surface which is tangent to these wavelets will be a wavefront in the region beyond the grating. A number of wavefronts are possible and are referred to as the different diffraction orders. For example, the zero-order beam is formed by wavelets which have identical phase. This wavefront is parallel to the incident-plane wavefront and propagates along the axis of the optical system. The first-order beams, which are deflected to both sides of the zero-order beam, are formed by adjoining wavelets which are out of phase by one wavelength. Similarly, the second-order beams are formed by adjoining wavelets which are out of phase by two wavelengths, and so on. The diffraction of a plane wavefront by a grating is illustrated in Fig. 12.20. The rays emerging from each of the slits of the grating are associated with the different diffraction orders. The image formed by collecting all the diffraction orders from the grating with a lens is known as the diffraction spectrum of the grating. A photograph of an actual diffraction spectrum of a 300-line-per-inch grating is shown in Fig. 12.21. The dots are images of the point source.

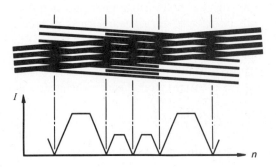

Figure 12.19 Mechanism of formation of additional fringes in moiré fringe multiplication with special gratings.

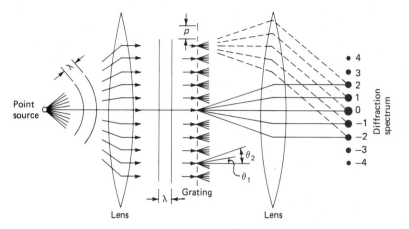

Figure 12.20 Mechanism of formation of the diffraction spectrum of a grating.

Since the wavelets forming the first-order beam are out of phase by one wavelength, the angle θ_1 can be determined from the expression

$$\sin \theta_1 = \theta_1 = \frac{\lambda}{p}$$

since the angles are small. Similarly for the other diffraction orders

$$\sin \theta_n = \theta_n = \frac{n\lambda}{p} \tag{12.19}$$

The distance d between any two dots is then given simply as

$$d = f\theta_1 = \frac{f\lambda}{p} \tag{12.20}$$

where f is the focal length of the decollimating lens.

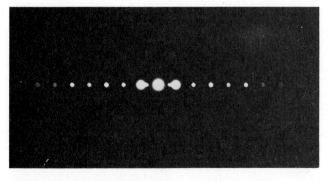

Figure 12.21 The diffraction spectrum of a 300-line-per-inch grating. (*Courtesy of Fu-pen Chiang.*)

One of the optical methods used to produce moiré fringe sharpening and moiré fringe multiplication is based on the simple diffraction phenomenon illustrated in Fig. 12.20. When both a model grating and a master grating are inserted in the collimated light beam between the lenses, the plane wavefronts generated by the first grating are further diffracted by the second grating to produce a diffraction spectrum associated with the superimposed pair. If the complete diffraction spectrum is collected with a camera lens (an ideal situation not realized in practice), the image recorded by the camera will be an exact reproduction of the grating pair and their moiré pattern. If all the diffraction orders are not collected by the camera lens, the image recorded by the camera will be modified. The theory associated with these modifications is beyond the scope of this book, but images recorded with certain individual diffraction orders can be shown to provide both moiré fringe sharpening and moiré fringe multiplication.

An optical system used for moiré fringe sharpening and moiré fringe multiplication is illustrated in Fig. 12.22. Note the presence of the light stop, or aperture, which is used to isolate the particular diffraction order being permitted to enter the camera lens. A selection of photographs recorded using different diffraction orders is shown in Fig. 12.23. The first photograph shows the image recorded if a large number of diffraction orders is permitted to enter the camera. The moiré fringes and the lines of the gratings are visible. This pattern is similar to the ones previously shown, which were obtained with coarse gratings in contact and diffused light. The second photograph shows the image recorded with the zero diffraction order. The moiré fringes in this image have been broadened, but the grating lines have been eliminated. The third photograph shows the image produced by the +1 diffraction order. In this instance, the moiré fringes are sharpened, but the lines of the gratings remain extinguished. It can be shown that at least two diffraction orders must enter the camera before a crude gratinglike image is produced. In the fourth, fifth, and sixth photographs, the images produced by the +2, +3, and +5 diffraction orders, respectively, are shown. In each of these photographs, moiré fringe multiplication ($2\times$, $3\times$, and $5\times$, respectively) is indicated. Higher levels of fringe multiplication can be achieved with the

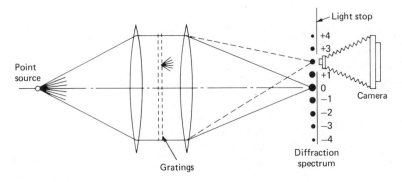

Figure 12.22 Basic optical system for moiré fringe multiplication.

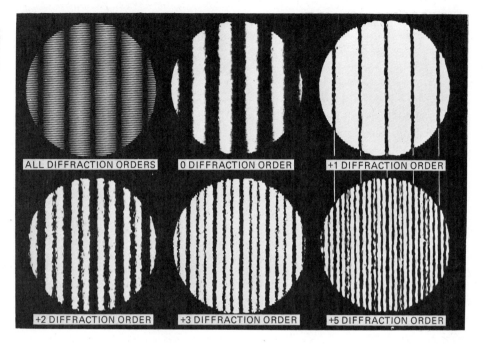

Figure 12.23 Moiré patterns produced by different diffraction orders. (*Courtesy of A. J. Durelli.*)

higher diffraction orders, but the intensity of light associated with these orders is very small.

Other methods of optical fringe multiplication have been reported in the technical literature. In a modified process developed by Sciammarella [18], the need for a master grating is eliminated. Methods of fringe shifting, combined with multiple-exposure photography, to achieve fringe multiplication are discussed by Durelli and Parks [20].

12.8 EXPERIMENTAL PROCEDURE AND TECHNIQUE [21–25]

Master gratings for moiré work with 300, 1000, and 2000 lines per inch (12, 40, and 80 lines per millimeter) are readily available from several commercial suppliers. Because of the inherent cost of the master grating and the risk involved in its use, high-quality duplicates are usually made for routine stress-analysis work. If proper photographic procedures are followed, little difficulty is encountered in producing good-quality duplicates on high-resolution film.

Printing or etching the grating on the model, on the other hand, is often difficult. Work on the development of techniques to apply gratings to the various model materials is constantly in progress, and some interesting procedures have been developed for specific applications. Two factors are extremely important in the production of a satisfactory model array: (1) the photographic emulsion must adhere well to the model material and have a high resolution; (2) the exposure must be carefully controlled to produce an array with proper line width and spacing from the master array.

Since materials, techniques, and procedures associated with moiré work are constantly being improved, no attempt will be made to outline specific procedures. For the latest materials, procedures, and techniques, the interested reader should consult the current technical literature. In particular, the papers by Holister and Luxmoore [21], Chiang [22], and Zandman [23] will in general be helpful. The books on moiré by Durelli and Parks [24] and Theocaris [25] should be consulted for more detail on many of the topics discussed in this chapter.

EXERCISES

12.1 Verify Eq. (12.11).

12.2 In many moiré fringe patterns, the perpendicular distance δ between fringes is difficult to measure accurately while the distance δ_p between fringes in a direction perpendicular to the master grating lines is easy to measure. For this case, develop an expression for the deformed-specimen grating pitch p' in terms of δ_p, the angle of inclination ϕ of the fringes, and the master grating pitch p.

12.3 In many moiré fringe patterns, the perpendicular distance δ between fringes is difficult to measure accurately while the distance δ_s between fringes in a direction parallel to the master grating lines is easy to measure. For this case, develop an expression for the deformed-specimen grating pitch p' in terms of δ_s, the angle of inclination ϕ of the fringes, and the master grating pitch p.

12.4 Use the moiré fringe pattern shown in Fig. E12.4 to prepare a plot of the displacement v as a function of position along the vertical axis of symmetry of the disk. Note that $v = 0$ at the center of the disk. The grating lines are horizontal and have a density of 12 lines per millimeter. The original

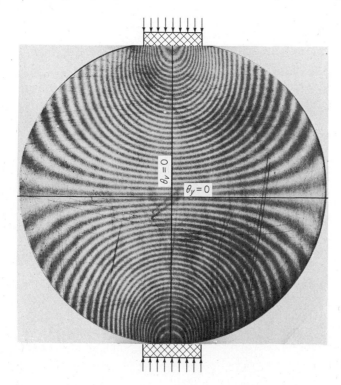

Figure E12.4

diameter of the disk was 100 mm. From the displacement-position curve, determine the distribution of strain ϵ_{yy} along the vertical axis of symmetry of the disk.

12.5 Determine the distribution of strain ϵ_{yy} along the horizontal axis of symmetry of the disk shown in Fig. E12.4.

REFERENCES

1. Tollenaar, D.: Moire: Interferentieverschijnselen bij Rasterdruk, Amsterdam Instituut voor Grafische Techniek, 1945.
2. Morse, S., A. J. Durelli, and C. A. Sciammarella: Geometry of Moiré Fringes in Strain Analysis, *J. Eng. Mec. Div., ASCE*, vol. 86, no. EM4, pp. 105–126, 1960.
3. Weller, R., and B. M. Shepard: Displacement Measurement by Mechanical Interferometry, *Proc. SESA*, vol. VI, no. 1, pp. 35–38, 1948.
4. Dantu, P.: Recherches diverses d'extensometrie et de détermination des contraintes, *Anal. Contraintes, Mem. GAMAC*, tome II, no. 2, pp. 3–14, 1954.
5. Dantu, P.: Utilisation des réseaux pour l'étude des déformations, *Lab. Centr. Ponts Chaussees, Paris, Pub.* 57-6, pp. 26–46, 1957.
6. Sciammarella, C. A., and A. J. Durelli: Moiré Fringes as a Means of Analysing Strains, *J. Eng. Mech. Div., ASCE*, vol. 87, no. EM1, pp. 55–74, 1961.
7. Theocaris, P. S.: Isopachic Patterns by the Moiré Method, *Exp. Mech.*, vol. 4, no. 6, pp. 153–159, 1964.
8. Theocaris, P. S.: Moire Patterns of Isopachics, *J. Sci. Instrum.*, vol. 41, pp. 133–138, 1964.
9. Ligtenberg, F. K.: The Moiré Method: A New Experimental Method for the Determination of Moments in Small Slab Models, *Proc. SESA*, vol. XII, no. 2, pp. 83–98, 1954.
10. Durelli, A. J., and I. M. Daniel: Structural Model Analysis by Means of Moiré Fringes, *J. Struct. Div., ASCE*, vol. 86, no. ST12, pp. 93–102, 1960.
11. Parks, V. J., and A. J. Durelli: Various Forms of the Strain Displacement Relations Applied to Experimental Strain Analysis, *Exp. Mech.*, vol. 4, no. 2, pp. 37–47, 1964.
12. Post, D.: The Moiré Grid-Analyzer Method for Strain Analysis, *Exp. Mech.*, vol. 5, no. 11, pp. 368–377, 1965.
13. Dantu, P.: Extension of the Moiré Method to Thermal Problems, *Exp. Mech.*, vol. 4, no. 3, pp. 64–69, 1964.
14. Post, D.: Sharpening and Multiplication of Moiré Fringes, *Exp. Mech.*, vol. 7, no. 4, pp. 154–159, 1967.
15. Holister, G. S.: Moiré Method of Surface Strain Measurement, *Engineer*, Jan. 27, 1967.
16. Post, D.: Analysis of Moiré Fringe Multiplication Phenomena, *Appl. Opt.*, vol. 6, no. 11, pp. 1939–1942, 1967.
17. Post, D.: New Optical Methods of Moiré Fringe Multiplication, *Exp. Mech.*, vol. 8, no. 2, pp. 63–68, 1968.
18. Sciammarella, C. A.: Moiré-Fringe Multiplication by Means of Filtering and a Wave-front Reconstruction Process, *Exp. Mech.*, vol. 9, no. 4, pp. 179–185, 1969.
19. Chiang, F.: Techniques of Optical Spatial Filtering Applied to the Processing of Moiré-fringe Patterns, *Exp. Mech.*, vol. 9, no. 11, pp. 523–526, 1969.
20. Durelli, A. J., and V. J. Parks: "Moiré Analysis of Strain," chap. 16, Prentice-Hall, Inc., Englewood Cliffs, N.J., 1970.
21. Holister, G. S., and A. R. Luxmoore: The Production of High-Density Moiré Grids, *Exp. Mech.*, vol. 8, no. 5, pp. 210–216, 1968.
22. Chiang, F.: Discussion of the Production of High-Density Moiré Grids, *Exp. Mech.*, vol. 9, no. 6, pp. 286–288, 1969.
23. Zandman, F.: The Transfer-Grid Method: A Practical Moiré Stress-Analysis Tool, *Exp. Mech.*, vol. 7, no. 7, pp. 19A–22A, 1967.
24. Durelli, A. J., and V. J. Parks: "Moiré Analysis of Strain," chaps. 14 and 15, Prentice-Hall, Inc., Englewood Cliffs, N.J., 1970.
25. Theocaris, P. S.: "Moiré Fringes in Strain Analysis," chap. 11, Pergamon Press, New York, 1969.

THIRTEEN

THEORY OF PHOTOELASTICITY

13.1 INTRODUCTION

In Chap. 11, the working optical instrument of photoelasticity, a polariscope, was described in detail. The purpose of this chapter is to discuss the theory of photoelasticity, i.e., what happens in the polariscope when a photoelastic model is placed in the field and loaded. This discussion will be kept as simple as possible, yet it will be sufficiently complete to describe most of the photoelastic effects that can be observed in a transmission polariscope.

13.2 TEMPORARY DOUBLE REFRACTION [1–3]

Many transparent noncrystalline materials that are optically isotropic when free of stress become optically anisotropic and display characteristics similar to crystals when they are stressed. These characteristics persist while loads on the material are maintained but disappear when the loads are removed. This behavior, known as _temporary double refraction_, was first observed by Sir David Brewster in 1816. The method of photoelasticity is based on this physical behavior of transparent noncrystalline materials.

In Chap. 1 it was shown that at least three mutually perpendicular planes, which are free of shear stress, exist at each point of a loaded body. These planes were defined as principal planes and the normal stresses acting on them were defined as the principal stresses σ_1, σ_2, and σ_3. In general, the three principal

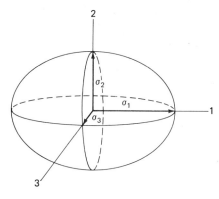

Figure 13.1 The stress ellipsoid.

stresses at a point have different magnitudes. The resultant stress T_n on any plane through the point can be expressed in terms of the three principal stresses at the point and the direction cosines associated with the plane by means of Eqs. (1.15). Normal and shear stresses on the plane for a particular set of coordinate directions can be obtained from Eqs. (1.16).

A geometric representation that provides considerable physical insight into the nature of the state of stress at a point is known as the *stress ellipsoid* or the *ellipsoid of Lamé*. An equation for this stress ellipsoid, shown in Fig. 13.1, is obtained by solving Eqs. (1.15) for the direction cosines and adding the sum of the squares to obtain

$$\frac{T_{nx}^2}{\sigma_1^2} + \frac{T_{ny}^2}{\sigma_2^2} + \frac{T_{nz}^2}{\sigma_3^2} = 1 \tag{13.1}$$

Each radius of this ellipsoid represents the magnitude of the resultant stress T_n on some plane through the point. Since Eqs. (1.15) were expressed in terms of the principal directions, the semiaxes of the ellipsoid represent the magnitudes of the principal stresses at the point. The particular plane associated with an arbitrary radius must be determined by some auxiliary construction such as the stress-director surface [2, p. 65].

The intersection of an arbitrary plane through the origin with the surface of the stress ellipsoid is an ellipse. On planes through the point which have normals along one of the semiaxes of the ellipse, the resultant stress can be resolved into a normal component and a shear component in a direction perpendicular to the plane of the ellipse. The shear stress component in a direction parallel to the plane of the ellipse is zero. The normal stress component on such a plane is known as a *secondary principal stress.* Later it will be shown that the optical response of a doubly refracting material for light passage at normal incidence to the plane of the ellipse is not affected by the presence of this shear stress component or by the normal stress component in a direction perpendicular to the plane of the ellipse. Secondary principal stresses play an important role in three-dimensional photoelasticity studies.

The optical anisotropy (temporary double refraction) which develops in a material as a result of stress can also be represented by an ellipsoid, known in this

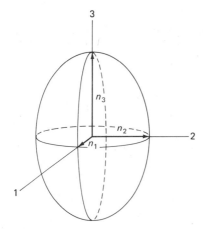

Figure 13.2 The index ellipsoid.

case as the *index ellipsoid*. The semiaxes of the index ellipsoid represent the principal indices of refraction of the material at the point, as shown in Fig. 13.2. Any radius of the ellipsoid represents a direction of light propagation through the point. A plane through the origin, which is perpendicular to the radius, intersects the ellipsoid as an ellipse. The semiaxes of the ellipse represent the indices of refraction associated with light waves having planes of vibration which contain the radius vector and an axis of the ellipse. For a material which is optically isotropic, the three principal indices of refraction are equal, and the index ellipsoid becomes a sphere. The index of refraction is then the same for all directions of light propagation through the material.

The similarities which exist between the stress-ellipsoid representation of the state of stress at a point in a loaded body and the index-ellipsoid representation of the optical properties of a material exhibiting temporary double refraction suggest the presence of a relationship between the two quantities which may form the basis for an experimental determination of stresses (or strains). The relationship is known as the *stress-optic law*.

13.3 THE STRESS-OPTIC LAW [3–7]

The theory which relates changes in the indices of refraction of a material exhibiting temporary double refraction to the state of stress in the material is due to Maxwell, who reported the phenomenon in 1853. Maxwell noted that the changes in the indices of refraction were linearly proportional to the loads (thus to the stresses or strains for a linearly elastic material) and followed the relationships

$$n_1 - n_0 = c_1\sigma_1 + c_2(\sigma_2 + \sigma_3)$$
$$n_2 - n_0 = c_1\sigma_2 + c_2(\sigma_3 + \sigma_1)$$
$$n_3 - n_0 = c_1\sigma_3 + c_2(\sigma_1 + \sigma_2) \tag{13.2}$$

where $\sigma_1, \sigma_2, \sigma_3$ = principal stresses at point

n_0 = index of refraction of material in unstressed state

n_1, n_2, n_3 = indices of refraction of material in stressed state associated with principal stress directions (principal indices of refraction)

c_1, c_2 = constants known as stress-optic coefficients

Equations (13.2) are the fundamental relationships between stress and optical effect and are known as the stress-optic law. These equations indicate that the complete state of stress at a point can be determined by measuring the three principal indices of refraction and establishing the directions of the three principal optical axes. Since the measurements are extremely difficult to make in the three-dimensional case, practical application has been limited to cases of plane stress $(\sigma_3 = 0)$. For plane-stress situations, Eqs. (13.2) reduce to

$$n_1 - n_0 = c_1\sigma_1 + c_2\sigma_2 \qquad n_2 - n_0 = c_1\sigma_2 + c_2\sigma_1 \qquad (13.3)$$

Favre [6] has used a Mach-Zender interferometer and Post [7] has used a series interferometer to make measurements of absolute retardations. Such measurements permitted direct determination of the individual principal stresses at interior points of a loaded two-dimensional model. Measurements with an interferometer can be very precise, but they are difficult and time-consuming. For this reason, absolute-retardation methods are seldom used, while the method of photoelasticity, which utilizes relative retardations, enjoys wide application.

13.4 THE STRESS-OPTIC LAW IN TERMS OF RELATIVE RETARDATION [3, 8–15]

Equations (13.2) provide the changes in index of refraction experienced by a material exhibiting temporary double refraction as the result of an applied state of stress. In the previous section, it was indicated that these absolute changes in index of refraction can be used as the basis for a stress-measurement method. The more widely used method of photoelasticity, however, makes use of relative rather than absolute changes in index of refraction. For example, consider the equations obtained by eliminating n_0 from Eqs. (13.2):

$$n_2 - n_1 = (c_2 - c_1)(\sigma_1 - \sigma_2) = c(\sigma_1 - \sigma_2)$$

$$n_3 - n_2 = (c_2 - c_1)(\sigma_2 - \sigma_3) = c(\sigma_2 - \sigma_3)$$

$$n_1 - n_3 = (c_2 - c_1)(\sigma_3 - \sigma_1) = c(\sigma_3 - \sigma_1) \qquad (13.4)$$

where $c = c_2 - c_1$ is the relative stress-optic coefficient expressed in terms of brewsters (1 brewster = 10^{-13} cm^2/dyn = 10^{-12} m^2/N = 6.895×10^{-9} in^2/lb). Materials exhibiting temporary double refraction may display the properties of either positive or negative crystals. In a positive crystal like quartz, the extraordinary index of refraction n_e is greater than the ordinary index n_o. In a negative crystal like calcite the opposite is true. Photoelastic materials are considered to exhibit positive birefringence when the velocity of propagation of the light wave associated with the principal stress σ_1 is greater than the velocity of the wave associated with the principal stress σ_2. Since the principal stresses are ordered such that $\sigma_1 \geq \sigma_2 \geq \sigma_3$, the principal indices of refraction of a positive doubly refracting material can be ordered such that $n_3 \geq n_2 \geq n_1$. The form of Eqs. (13.4) has been selected to make the relative stress-optic coefficient c a positive constant.

Since a stressed photoelastic model behaves like a temporary wave plate, Eq. (11.37) can be used to relate the relative angular phase shift Δ (or relative retardation) to changes in the indices of refraction in the material resulting from the stresses. For example, consider a slice of material (thickness h) oriented perpendicular to one of the principal-stress directions at the point of interest in the model. If a beam of plane-polarized light is passed through the slice at normal incidence, the relative retardation Δ accumulated along each of the principal-stress directions can be obtained by substituting Eq. (11.37) in turn into each of Eqs. (13.4). Thus

STRESS - OPTIC
LAW
AS USED IN
PHOTO ELASTICITY

$$\Delta_{12} = \frac{2\pi hc}{\lambda}(\sigma_1 - \sigma_2) \qquad \Delta_{23} = \frac{2\pi hc}{\lambda}(\sigma_2 - \sigma_3) \qquad \Delta_{31} = \frac{2\pi hc}{\lambda}(\sigma_3 - \sigma_1)$$

$$(13.5)$$

where Δ_{12} is the magnitude of the relative angular phase shift (relative retardation) developed between components of a light beam propagating in the σ_3 direction. The two components of the beam would have electric vectors oriented in the σ_1 and σ_2 directions. The component associated with the principal stress σ_1 would propagate at a higher velocity than the one associated with the stress σ_2 (since $\sigma_1 \geq \sigma_2$) if the material exhibits positive birefringence. Similar meanings can be attached to the retardations Δ_{23} and Δ_{31}.

Equations (13.5) express the stress-optic law as it is commonly applied in photoelasticity. The relative retardation Δ is linearly proportional to the difference between the two principal stresses having directions perpendicular to the path of propagation of the light beam. The third principal stress, having a direction parallel to the path of propagation of the light beam, has no effect on the relative retardation. Also, the relative retardation Δ is linearly proportional to the model or slice thickness h and inversely proportional to the wavelength λ of the light being used.

The relative stress-optic coefficient c is usually assumed to be a material constant that is independent of the wavelength of the light being used. A study by Vandaele-Dossche and van Geen [13] has shown, however, that this coefficient may depend on wavelength in some cases as the model material passes from the elastic to the plastic state. The dependence of the relative stress-optic coefficient c on the wavelength of the light being used is referred to as *photoelastic dispersion* or *dispersion of birefringence.*

From an analysis of the general three-dimensional state of stress at a point, as represented by the stress ellipsoid, and from an analysis of the change in index of refraction with direction of light propagation in the stressed material, as represented by the index ellipsoid, it can be shown that Eqs. (13.5) apply not only for principal stresses but also for secondary principal stresses. Thus

$$\Delta' = \frac{2\pi hc}{\lambda}(\sigma'_1 - \sigma'_2) \tag{13.6}$$

where σ'_1 and σ'_2 are secondary principal stresses $(\sigma'_1 \geq \sigma'_2)$ at the point of interest in directions perpendicular to the path of propagation of the light beam. The stress-optic law in terms of secondary principal stresses is widely used in three-dimensional photoelasticity work.

For two-dimensional or plane-stress problems, where one of the principal stresses is zero (say $\sigma_3 = 0$), the stress-optic law in terms of the nonzero principal stresses and for light at normal incidence to the plane of the model can be written without the subscripts on the retardation simply as

$$\Delta = \frac{2\pi hc}{\lambda}(\sigma_1 - \sigma_2) \tag{13.7}$$

Here it is understood that σ_1 and σ_2 are the in-plane principal stresses and that σ_1 is greater than σ_2 but not greater than $\sigma_3 = 0$ if both in-plane stresses are compressive.

Since brewsters are not commonly employed in engineering practice, Eq. (13.7) is frequently expressed in the following form for practical work:

$$\sigma_1 - \sigma_2 = \frac{Nf_\sigma}{h} \quad \text{lb/in}^2 \text{ or N/m}^2 \tag{13.8}$$

where

$$N = \frac{\Delta}{2\pi} \quad \text{dimensionless} \tag{13.9}$$

is the relative retardation in terms of a complete cycle of retardation,

MATERIAL FRINGE CONSTANT

$$f_\sigma = \frac{\lambda}{c} \quad \text{lb/in or N/m} \tag{13.10}†$$

is a property of the model material for a given wavelength of light known as the *material fringe value*, and h is the model thickness in inches or meters.

It is immediately apparent from Eq. (13.8) that the stress difference $\sigma_1 - \sigma_2$ in a two-dimensional model can be determined if the relative retardation N can be measured and if the material fringe value f_σ can be established by means of calibration. Actually, the function of the polariscope is to determine the value of N at each point in the model.

If a photoelastic model exhibits a perfectly linear elastic behavior, the difference in the principal strains $\epsilon_1 - \epsilon_2$ can also be measured by establishing the

---

† Often Eq. (13.8) is written as $\tau = (\sigma_1 - \sigma_2)/2 = Nf_\sigma/h$, where f_σ is the material fringe value in terms of shear and is equal to one-half the f_σ value defined here.

fringe order N. The stress-strain relations for a two-dimensional or plane state of stress are given by Eqs. (2.19) as

$$\epsilon_1 = \frac{1}{E}(\sigma_1 - v\sigma_2) \qquad \epsilon_2 = \frac{1}{E}(\sigma_2 - v\sigma_1)$$

Thus
$$\epsilon_1 - \epsilon_2 = \frac{1+v}{E}(\sigma_1 - \sigma_2)$$

Substituting these results into Eq. (13.8) yields

$$\frac{Nf_\sigma}{h} = \frac{E}{1+v}(\epsilon_1 - \epsilon_2) \tag{13.11}$$

If Eq. (13.11) is rewritten as

$$\frac{Nf_\epsilon}{h} = \epsilon_1 - \epsilon_2 \tag{13.12}$$

it is clear that

$$f_\epsilon = \frac{1+v}{E}f_\sigma \tag{13.13}$$

where f_ϵ is the material fringe value in terms of strain.

For a perfectly linear elastic photoelastic model, the determination of N is sufficient to establish both $\sigma_1 - \sigma_2$ and $\epsilon_1 - \epsilon_2$ if any three of the material properties E, v, f_σ, or f_ϵ are known. However, many photoelastic materials exhibit viscoelastic properties, and Eq. (13.13) is not valid. The viscoelastic behavior of photoelastic materials is discussed in Sec. 14.7.

13.5 EFFECTS OF A STRESSED MODEL IN A PLANE POLARISCOPE [3, 8–12]

It has been established (Sec. 13.4) that the principal-stress difference $\sigma_1 - \sigma_2$ can be determined in a two-dimensional model if N is measured at each point in the model. Also, it was stated that the optical axes of the model (a temporary wave plate) coincide with the principal-stress directions. These two facts can be effectively utilized once a method to measure the optical properties of a stressed model has been established.

Consider first the case of a plane-stressed model inserted into the field of a plane polariscope with its normal coincident with the axis of the polariscope, as illustrated in Fig. 13.3. Note that the principal-stress direction at the point under consideration in the model makes an angle α with the axis of polarization of the polarizer.

In Sec. 11.6 it was shown that a plane polarizer resolves an incident light wave into components which vibrate parallel and perpendicular to the axis of the polarizer. The component parallel to the axis is transmitted, and the component

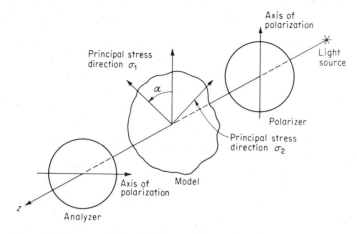

Figure 13.3 A stressed photoelastic model in a plane polariscope.

perpendicular to the axis is internally absorbed. Since the initial phase of the wave is not important in the development which follows, the plane-polarized light beam emerging from the polarizer can be represented by the simple expression

$$E_{py} = k \cos \omega t \qquad\qquad (a)$$

After leaving the polarizer, this plane-polarized light wave enters the model, as shown in Fig. 13.4. Since the stressed model exhibits the optical properties of a wave plate, the impinging light vector is resolved into two components E_1 and E_2 with vibrations parallel to the two in-plane principal stress directions at the point. Thus

$$E_1 = k \cos \alpha \cos \omega t \qquad E_2 = k \sin \alpha \cos \omega t \qquad\qquad (b)$$

Since the two components propagate through the model with different velocities $(c > v_1 > v_2)$, they develop phase shifts Δ_1 and Δ_2 with respect to a wave in air.

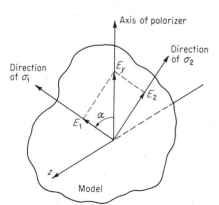

Figure 13.4 Resolution of the light vector as it enters a stressed model in a plane polariscope.

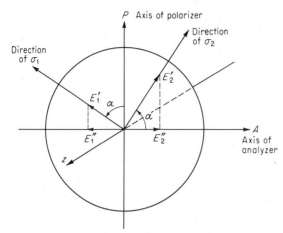

Figure 13.5 Components of the light vectors which are transmitted through the analyzer of a plane polariscope.

Thus, the waves upon emerging from the model can be expressed as

$$E_1' = k \cos \alpha \cos (\omega t - \Delta_1) \qquad E_2' = k \sin \alpha \cos (\omega t - \Delta_2) \qquad (c)$$

where
$$\Delta_1 = \frac{2\pi h}{\lambda} (n_1 - 1) \qquad \Delta_2 = \frac{2\pi h}{\lambda} (n_2 - 1)$$

After leaving the model, the two components continue to propagate without further change and enter the analyzer in the manner shown in Fig. 13.5. The light components E_1' and E_2' are resolved when they enter the analyzer into horizontal components E_1'' and E_2'' and into vertical components. Since the vertical components are internally absorbed in the analyzer, they have not been shown in Fig. 13.5.

The horizontal components transmitted by the analyzer combine to produce an emerging light vector E_{ax}, which is given by

$$E_{ax} = E_2'' - E_1'' = E_2' \cos \alpha - E_1' \sin \alpha \qquad (d)$$

Substituting Eqs. (c) into (d) yields

$$E_{ax} = k \sin \alpha \cos \alpha [\cos (\omega t - \Delta_2) - \cos (\omega t - \Delta_1)]$$

$$= k \sin 2\alpha \sin \frac{\Delta_2 - \Delta_1}{2} \sin \left(\omega t - \frac{\Delta_2 + \Delta_1}{2} \right) \qquad (13.14)$$

It is interesting to note in Eq. (13.14) that the average angular phase shift $(\Delta_2 + \Delta_1)/2$ affects the phase of the light wave emerging from the analyzer but not the amplitude (coefficient of the time-dependent term). Thus, it has no influence on the intensity (intensity is proportional to the square of the amplitude) of the light emerging from the analyzer but serves only to change the phase of the emerging wave with respect to the phase of the initial wave. The relative retardation $\Delta = \Delta_2 - \Delta_1$, however, appears in the amplitude of the wave; therefore, it is one of the factors which controls the intensity of light emerging from the analyzer.

Since the average angular phase shift $(\Delta_2 + \Delta_1)/2$ has no effect on the intensity, it does not contribute to the optical patterns observed in a photoelastic model. In future photoelasticity developments only relative retardations will be considered in order to simplify the analysis.

Since the intensity of light is proportional to the square of the amplitude of the light wave, the light emerging from the analyzer of a plane polariscope is given by

$$I = K \sin^2 2\alpha \sin^2 \frac{\Delta}{2} \qquad (13.15)$$

where

$$\Delta = \Delta_2 - \Delta_1 = \frac{2\pi h}{\lambda}(n_2 - n_1) = \frac{2\pi h c}{\lambda}(\sigma_1 - \sigma_2)$$

Examination of Eq. (13.15) indicates that extinction $(I = 0)$ occurs either when $\sin^2 2\alpha = 0$ or when $\sin^2 (\Delta/2) = 0$. Thus, one of the conditions for extinction is related to the principal-stress directions and the other is related to the principal-stress difference.

A. Effect of Principal-Stress Directions

When $2\alpha = n\pi$, where $n = 0, 1, 2, \ldots$, $\sin^2 2\alpha = 0$ and extinction occurs. In other words, when one of the principal-stress directions coincides with the axis of the polarizer ($\alpha = 0$, $\pi/2$, or any exact multiple of $\pi/2$), the intensity of the light is zero. Since the analysis of the optical effects produced by a stressed model in a plane polariscope was conducted for an arbitrary point in the model, the analysis is valid for all points of the model. When the entire model is viewed in the polariscope, a fringe pattern is observed; the fringes are loci of points where the principal-stress directions (either σ_1 or σ_2) coincide with the axis of the polarizer. The fringe pattern produced by the $\sin^2 2\alpha$ term in Eq. (13.15) is known as an *isoclinic fringe pattern*. Isoclinic fringe patterns are used to determine the principal-stress directions at all points of a photoelastic model. Since isoclinics represent a very important segment of the data obtained from a photoelastic model, the topic of isoclinic-fringe-pattern interpretation will be treated in more detail in Sec. 14.3.

B. Effect of Principal-Stress Difference

When $\Delta/2 = n\pi$, where $n = 0, 1, 2, 3, \ldots$, $\sin^2 (\Delta/2) = 0$ and extinction occurs. In other words, when the principal-stress difference is either zero ($n = 0$) or sufficient to produce an integral number of wavelengths of retardation ($n = 1, 2, 3, \ldots$), the intensity of light emerging from the analyzer is zero. When a complete model is viewed in the polariscope, this second condition for extinction yields a second fringe pattern where the fringes are loci of points exhibiting the same order of extinction ($n = 0, 1, 2, 3, \ldots$). The fringe pattern produced by the $\sin^2 (\Delta/2)$ term in Eq. (13.15) is known as an *isochromatic fringe pattern*. The nature of the optical effect producing the isochromatic fringe pattern requires some additional discussion.

Recall from Eq. (13.7) that the relative retardation Δ may be expressed as

$$\Delta = \frac{2\pi hc}{\lambda}(\sigma_1 - \sigma_2)$$

Thus
$$n = \frac{\Delta}{2\pi} = \frac{hc}{\lambda}(\sigma_1 - \sigma_2) \tag{e}$$

Examination of Eq. (e) indicates that the order of extinction n depends on both the principal-stress difference $\sigma_1 - \sigma_2$ and the wavelength λ of the light being used. Thus, for a given principal-stress difference, the order of extinction n can be an integer only for light of a single wavelength (monochromatic light). When a model is viewed in monochromatic light, the isochromatic fringe pattern appears as a series of dark bands since the intensity of light is zero when $n = 0, 1, 2, 3, \ldots$. When a model is viewed with white light (all wavelengths of the visible spectrum present), the isochromatic fringe pattern appears as a series of colored bands. The intensity of light is zero, and a black fringe appears only when the principal-stress difference is zero and a zero order of extinction occurs for all wavelengths of light. No other region of zero intensity is possible since the principal-stress difference required to produce a given order of extinction is different for each of the wavelengths. Thus, not all the wavelengths can be extinguished simultaneously to produce a condition of zero intensity. The various colored bands form in regions where the principal-stress difference is sufficient to produce extinction of a particular wavelength of the white light. For example, when the principal-stress difference is sufficient to produce extinction of the green wavelengths, the complementary color, red, appears as the isochromatic fringe. At the higher levels of principal-stress difference, where several wavelengths of light can be extinguished simultaneously, e.g., second order red and third order violet, the isochromatic fringes become pale and very difficult to identify; therefore, they are seldom used for stress analysis work.

With monochromatic light, the individual fringes in an isochromatic-fringe pattern remain sharp and clear to very high orders of extinction. Since the wavelength of the light is fixed, Eq. (e) can be written in terms of the material fringe value f_σ and the isochromatic fringe order N as

$$n = N = \frac{h}{f_\sigma}(\sigma_1 - \sigma_2) \tag{f}$$

Hence, the number of fringes appearing in an isochromatic fringe pattern is controlled by the magnitude of the principal-stress difference $\sigma_1 - \sigma_2$, by the thickness h of the model, and by the sensitivity of the photoelastic material, as denoted by the material fringe value f_σ.

In general, the principal-stress difference $\sigma_1 - \sigma_2$ and the principal-stress dir-

Figure 13.6 Superimposed isochromatic and isoclinic fringe patterns for a ring loaded in diametral compression.

ections vary from point to point in a photoelastic model. As a result, the isoclinic fringe pattern and the isochromatic fringe pattern are superimposed, as shown in Fig. 13.6, when the model is viewed in a plane polariscope. Separation of the patterns requires special techniques which will be discussed later.

Theoretically, the isoclinic and isochromatic fringes should be lines of zero width; however, the photograph in Fig. 13.6 shows the fringes as bands with considerable width. Also, direct visual examination of the fringe pattern in a polariscope will show again that the fringes are bands and not lines. In both instances, the width of the fringes is due to the recording characteristics of the eye and the photographic film and not to inaccuracies in the previous development. If the intensity of light emerging from the analyzer is measured with a suitable photoelectric cell, a minimum intensity is recorded at some point near the center of the fringe which coincides with the exact extinction line.

C. Frequency Response of a Polariscope

The circular frequency ω for light in the visible spectrum is approximately 10^{15} rad/s. As a result, neither the eye nor any type of existing high-speed photographic equipment can detect the periodic extinction associated with the ωt term of the expression for the light wave. To date, no dynamic-stress problem has been attempted where the frequency response of the polariscope has proved inadequate. Instead, the problems encountered are associated with loading the model, recording the fringe pattern, and selecting a suitable model material.

13.6 EFFECTS OF A STRESSED MODEL IN A CIRCULAR POLARISCOPE (DARK FIELD, ARRANGEMENT A)
[3, 8–12, 16, 17]

When a stressed photoelastic model is placed in the field of a circular polariscope with its normal coincident with the z axis of the polariscope, the optical effects differ somewhat from those obtained in a plane polariscope. The use of a circular polariscope eliminates the isoclinic fringe pattern while it maintains the isochromatic fringe pattern, and as a result the circular polariscope is more widely used than the plane polariscope. To illustrate this effect, consider the stressed model in the circular polariscope (arrangement A) shown in Fig. 13.7.

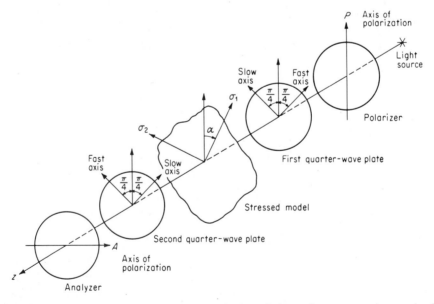

Figure 13.7 A stressed photoelastic model in a circular polariscope (arrangement A, crossed polarizer and analyzer, crossed quarter-wave plates).

The plane-polarized light beam emerging from the polarizer can be represented by the same simple expression used for the plane polariscope, namely,

$$E_{py} = k \cos \omega t \qquad (a)$$

As the light enters the first quarter-wave plate, it is resolved into components E_f and E_s with vibrations parallel to the fast and slow axes, respectively. Since the axes of the quarter-wave plate are oriented at 45° with respect to the axis of the polarizer,

$$E_f = \frac{\sqrt{2}}{2} k \cos \omega t \qquad E_s = \frac{\sqrt{2}}{2} k \cos \omega t$$

As the components propagate through the plate, they develop a relative angular phase shift $\Delta = \pi/2$; therefore, the components emerging from the plate can be expressed as

$$E'_f = \frac{\sqrt{2}}{2} k \cos \omega t \qquad E'_s = \frac{\sqrt{2}}{2} k \cos \left(\omega t - \frac{\pi}{2} \right) = \frac{\sqrt{2}}{2} k \sin \omega t \qquad (b)$$

It has previously been shown that these two plane-polarized beams represent circularly polarized light with the light vector rotating counterclockwise at any point along the axis of propagation of the light between the quarter-wave plate and the model.

After leaving the quarter-wave plate, the components enter the model in the manner illustrated in Fig. 13.8. Since the stressed model exhibits the character-

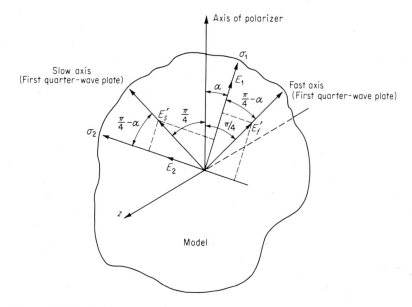

Figure 13.8 Resolution of the light components as they enter the stressed model.

istics of a temporary wave plate, the components E'_f and E'_s are resolved into components E_1 and E_2, which have directions of vibration parallel to the principal-stress directions in the model. Thus

$$E_1 = E'_f \cos\left(\frac{\pi}{4} - \alpha\right) + E'_s \sin\left(\frac{\pi}{4} - \alpha\right)$$

$$E_2 = E'_s \cos\left(\frac{\pi}{4} - \alpha\right) - E'_f \sin\left(\frac{\pi}{4} - \alpha\right) \tag{c}$$

Substituting Eqs. (b) into (c) yields

$$E_1 = \frac{\sqrt{2}}{2} k \left[\cos \omega t \cos\left(\frac{\pi}{4} - \alpha\right) + \sin \omega t \sin\left(\frac{\pi}{4} - \alpha\right)\right]$$

$$= \frac{\sqrt{2}}{2} k \cos\left(\omega t + \alpha - \frac{\pi}{4}\right)$$

$$E_2 = \frac{\sqrt{2}}{2} k \left[\sin \omega t \cos\left(\frac{\pi}{4} - \alpha\right) - \cos \omega t \sin\left(\frac{\pi}{4} - \alpha\right)\right]$$

$$= \frac{\sqrt{2}}{2} k \sin\left(\omega t + \alpha - \frac{\pi}{4}\right)$$

The two components E_1 and E_2 propagate through the model with different velocities. The additional relative retardation Δ accumulated during passage through the model is given by Eq. (13.7) as

$$\Delta = \frac{2\pi hc}{\lambda}(\sigma_1 - \sigma_2)$$

Thus the waves upon emerging from the plate can be expressed as

$$E'_1 = \frac{\sqrt{2}}{2} k \cos\left(\omega t + \alpha - \frac{\pi}{4}\right)$$

$$E'_2 = \frac{\sqrt{2}}{2} k \sin\left(\omega t + \alpha - \frac{\pi}{4} - \Delta\right) \tag{d}$$

The light emerging from the model propagates to the second quarter-wave plate and enters it according to the diagram shown in Fig. 13.9. The components associated with the fast and slow axes of the second quarter-wave plate are

$$E_f = E'_1 \sin\left(\frac{\pi}{4} - \alpha\right) + E'_2 \cos\left(\frac{\pi}{4} - \alpha\right)$$

$$E_s = E'_1 \cos\left(\frac{\pi}{4} - \alpha\right) - E'_2 \sin\left(\frac{\pi}{4} - \alpha\right) \tag{e}$$

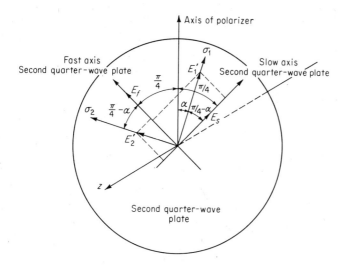

Figure 13.9 Resolution of the light components as they enter the second quarter-wave plate.

Substituting Eqs. (*d*) into (*e*) yields

$$E_f = \frac{\sqrt{2}}{2} k \left[\cos\left(\omega t + \alpha - \frac{\pi}{4}\right) \sin\left(\frac{\pi}{4} - \alpha\right) + \sin\left(\omega t + \alpha - \frac{\pi}{4} - \Delta\right) \cos\left(\frac{\pi}{4} - \alpha\right) \right]$$

$$E_s = \frac{\sqrt{2}}{2} k \left[\cos\left(\omega t + \alpha - \frac{\pi}{4}\right) \cos\left(\frac{\pi}{4} - \alpha\right) - \sin\left(\omega t + \alpha - \frac{\pi}{4} - \Delta\right) \sin\left(\frac{\pi}{4} - \alpha\right) \right]$$

As the light passes through the second quarter-wave plate, a relative phase shift of $\Delta = \pi/2$ develops between the fast and slow components. Thus the waves emerging from the plate can be expressed as

$$E'_f = \frac{\sqrt{2}}{2} k \left[\cos\left(\omega t + \alpha - \frac{\pi}{4}\right) \sin\left(\frac{\pi}{4} - \alpha\right) + \sin\left(\omega t + \alpha - \frac{\pi}{4} - \Delta\right) \cos\left(\frac{\pi}{4} - \alpha\right) \right]$$

$$E'_s = \frac{\sqrt{2}}{2} k \left[\sin\left(\omega t + \alpha - \frac{\pi}{4}\right) \cos\left(\frac{\pi}{4} - \alpha\right) + \cos\left(\omega t + \alpha - \frac{\pi}{4} - \Delta\right) \sin\left(\frac{\pi}{4} - \alpha\right) \right]$$

$$(13.16)$$

Finally, the light enters the analyzer, as shown in Fig. 13.10. The vertical components of E'_f and E'_s are absorbed in the analyzer while the horizontal components are transmitted to give

$$E_{ax} = \frac{\sqrt{2}}{2} (E'_s - E'_f) \qquad\qquad (f)$$

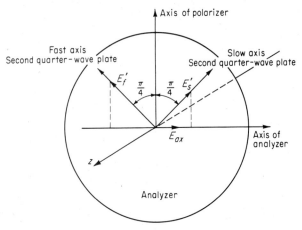

Figure 13.10 Components of the light vectors which are transmitted through the analyzer (dark field).

Substituting Eqs. (13.16) into Eq. (f) gives an expression for the light emerging from the analyzer of a circular polariscope (arrangement A). Thus

$$E_{ax} = \frac{k}{2}\left[\sin\left(\omega t + \alpha - \frac{\pi}{4}\right)\cos\left(\frac{\pi}{4} - \alpha\right) + \cos\left(\omega t + \alpha - \frac{\pi}{4} - \Delta\right)\sin\left(\frac{\pi}{4} - \alpha\right)\right.$$
$$\left. - \cos\left(\omega t + \alpha - \frac{\pi}{4}\right)\sin\left(\frac{\pi}{4} - \alpha\right) - \sin\left(\omega t + \alpha - \frac{\pi}{4} - \Delta\right)\cos\left(\frac{\pi}{4} - \alpha\right)\right]$$

$$= k\sin\frac{\Delta}{2}\sin\left(\omega t + 2\alpha - \frac{\Delta}{2}\right) \tag{13.17}$$

Since the intensity of light is proportional to the square of the amplitude of the light wave, the light emerging from the analyzer of a circular polariscope (arrangement A) is given by

$$I = K\sin^2\frac{\Delta}{2} \tag{13.18}$$

Inspection of Eq. (13.18) indicates that the intensity of the light beam emerging from the circular polariscope is a function only of the principal-stress difference since the angle α does not appear in the amplitude of the wave. This indicates that isoclinics have been eliminated from the fringe pattern observed with the circular polariscope. From the $\sin^2 (\Delta/2)$ term in Eq. (13.18) it is clear that extinction will occur when $\Delta/2 = n\pi$, where $n = 0, 1, 2, 3, \ldots$. This type of extinction is identical with that previously described for the plane polariscope and referred to as an isochromatic fringe pattern. An example of this fringe pattern is shown in Fig. 13.11.

Figure 13.11 Dark-field isochromatic fringe pattern of a ring loaded in diametral compression.

13.7 EFFECTS OF A STRESSED MODEL IN A CIRCULAR POLARISCOPE (LIGHT FIELD, ARRANGEMENT B)
[3, 8–12, 16, 17]

A circular polariscope is usually employed with both the dark- and light-field arrangements (*A* and *B*). The circular polariscope can be converted from dark field (arrangement *A*) to light field (arrangement *B*) simply by rotating the analyzer through 90°. The advantage of employing both light- and dark-field arrangements is that twice as many data are obtained for the whole-field determination of $\sigma_1 - \sigma_2$. Recall from Secs. 13.5 and 13.6 that the order of the fringes N coincides with n for the plane polariscope and for the dark-field circular polariscope; therefore, the fringes are counted in the sequence 0, 1, 2, 3, ... With the light-field arrangement of the circular polariscope, N and n do not coincide. Instead,

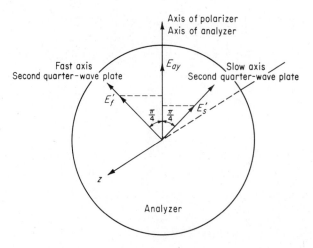

Figure 13.12 Components of the light vectors which are transmitted through the analyzer (light field).

$N = \frac{1}{2} + n$. Hence, with the light-field arrangement, the orders of the fringes are counted $\frac{1}{2}$, $1\frac{1}{2}$, $2\frac{1}{2}$, $3\frac{1}{2}$, ...

To establish the effect of a stressed model in a light-field circular polariscope, it is only necessary to consider the light components emerging from the second quarter-wave plate as represented by Eqs. (13.16) and the analyzer in its new position as indicated in Fig. 13.12. Since the axis of the analyzer is oriented in the vertical direction, the horizontal components of E'_f and E'_s will be absorbed while the vertical components will be transmitted as the light propagates through the analyzer. Thus, the emerging light vector, which lies in a vertical plane, can be expressed as

$$E_{ay} = \frac{\sqrt{2}}{2}(E'_s + E'_f) \qquad (a)$$

Substituting Eqs. (13.16) into Eq. (a) yields

$$E_{ay} = \frac{k}{2}\left[\sin\left(\omega t + \alpha - \frac{\pi}{4}\right)\cos\left(\frac{\pi}{4} - \alpha\right) + \cos\left(\omega t + \alpha - \frac{\pi}{4} - \Delta\right)\sin\left(\frac{\pi}{4} - \alpha\right)\right.$$

$$\left. + \cos\left(\omega t + \alpha - \frac{\pi}{4}\right)\sin\left(\frac{\pi}{4} - \alpha\right) + \sin\left(\omega t + \alpha - \frac{\pi}{4} - \Delta\right)\cos\left(\frac{\pi}{4} - \alpha\right)\right]$$

$$= k\cos\frac{\Delta}{2}\sin\left(\omega t - \frac{\Delta}{2}\right) \qquad (13.19)$$

Since the intensity of light is proportional to the square of the amplitude of the light wave, the light emerging from the analyzer of a circular polariscope (arrangement B) is given by

$$I = K\cos^2\frac{\Delta}{2} \qquad (13.20)$$

Equation (13.20) shows that extinction $(I = 0)$ will occur when

$$\frac{\Delta}{2} = \frac{1 + 2n}{2}\pi \qquad \text{for } n = 0, 1, 2, 3, \ldots$$

or from Eq. (13.9) when

$$N = \frac{\Delta}{2\pi} = \frac{1}{2} + n$$

which implies that the order of the first fringe observed in a light-field polariscope is $\frac{1}{2}$, which corresponds to $n = 0$. An example of a light-field isochromatic fringe pattern is presented in Fig. 13.13.

By using the circular polariscope with both light- and dark-field arrangements, it is possible to obtain two photographs of the resulting isochromatic fringe patterns. The data thus obtained will give a whole-field representation of the order

Figure 13.13 Light-field isochromatic fringe pattern of a ring loaded in diametral compression.

of the fringes to the nearest $\frac{1}{2}$ order. Interpolation between fringes often permits an estimate of the order of the fringes to ± 0.1, which results in accuracies for the principal-stress difference of $\pm 0.1 f_\sigma / h$. If more accurate determinations are necessary, the refined fringe-analysis techniques described in the following section can be used.

13.8 EFFECTS OF A STRESSED MODEL IN A CIRCULAR POLARISCOPE (ARBITRARY ANALYZER POSITION, TARDY COMPENSATION) [18–24]

The analysis presented in the previous two sections for the dark- and light-field arrangements of the circular polariscope can be carried one step further to include rotation of the analyzer through some arbitrary angle. The purpose of such a rotation is to provide a means for determining fractional fringe orders.

In the previous three sections, the optical effects produced by a stressed model in both plane and circular polariscopes were studied by using a trigonometric representation of the light wave. For more complicated situations, the derivations with this representation quickly become an exercise in the manipulation of trigonometric identities and very little of the physical significance of the problem is retained. Under such circumstances, an exponential representation of the light wave can be used to simplify the derivations; therefore, it will be used in all future developments.

Consider first the passage of light through the optical elements of a circular polariscope (arrangement A). With the exponential representation, the light wave emerging from the polarizer can be expressed as

$$E_{py} = k e^{i\omega t} \tag{a}$$

As the light enters the first quarter-wave plate, it is resolved into components with vibrations parallel to the fast and slow axes of the plate. The components develop a phase shift $\Delta = \pi/2$ as they propagate through the plate and emerge as

$$E'_f = \frac{\sqrt{2}}{2} k e^{i\omega t} \qquad E'_s = -i \frac{\sqrt{2}}{2} k e^{i\omega t} \tag{b}$$

As the light enters the model, it is resolved into components with vibrations parallel to the principal-stress directions. The components develop an additional phase shift Δ which depends on the principal-stress difference and emerge from the model as

$$E'_1 = \frac{\sqrt{2}}{2} k e^{i(\omega t + \alpha - \pi/4)} \qquad E'_2 = -i \frac{\sqrt{2}}{2} k e^{i(\omega t + \alpha - \pi/4 - \Delta)} \tag{c}$$

As the light enters the second quarter-wave plate, it is again resolved into components with vibrations parallel to the fast and slow axes of the plate. The components experience a phase shift of $\Delta = \pi/2$ as they propagate through the plate and emerge as

$$E'_f = \frac{\sqrt{2}}{2} k \left[\sin\left(\frac{\pi}{4} - \alpha\right) - ie^{-i\Delta} \cos\left(\frac{\pi}{4} - \alpha\right) \right] e^{i(\omega t + \alpha - \pi/4)}$$

$$E'_s = \frac{\sqrt{2}}{2} k \left[e^{-i\Delta} \sin\left(\frac{\pi}{4} - \alpha\right) - i \cos\left(\frac{\pi}{4} - \alpha\right) \right] e^{i(\omega t + \alpha - \pi/4)} \tag{13.21}$$

Finally, as the light passes through the analyzer, the vertical components of E'_f and E'_s are absorbed while the horizontal components are transmitted. Thus

$$E_{ax} = \frac{k}{2} \left[(e^{-i\Delta} - 1) \sin\left(\frac{\pi}{4} - \alpha\right) + i(e^{-i\Delta} - 1) \cos\left(\frac{\pi}{4} - \alpha\right) \right] e^{i(\omega t + \alpha - \pi/4)}$$

$$= \frac{k}{2} (e^{-i\Delta} - 1) e^{i(\omega t + 2\alpha)} \tag{13.22}$$

Recall from Eq. (11.17) that the square of the amplitude of a wave in exponential notation is the product of the amplitude and its complex conjugate. Thus

$$I \propto E_{ax} E_{ax}^* = K \sin^2 \frac{\Delta}{2} \tag{13.23}$$

This expression for the intensity of the light emerging from the analyzer of a circular polariscope (arrangement A) is identical with that previously determined using a trigonometric representation of the light wave and presented in Eq. (13.18).

To establish the effect of a stressed model in a circular polariscope with the analyzer oriented at an arbitrary angle γ with respect to its dark-field position, it is only necessary to consider the light components emerging from the second quarter-wave plate, as represented by Eqs. (13.21), and their transmission through the analyzer when it is positioned as shown in Fig. 13.14. Thus

$$E_{a\gamma} = E'_s \cos\left(\frac{\pi}{4} + \gamma\right) - E'_f \sin\left(\frac{\pi}{4} + \gamma\right) \tag{d}$$

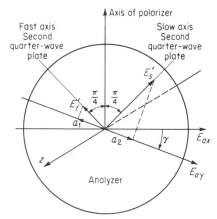

Figure 13.14 Rotation of the analyzer to obtain extinction in the Tardy method of compensation.

Substituting Eqs. (13.21) into Eq. (d) and combining terms yields

$$E_{a\gamma} = \frac{\sqrt{2}}{2} k \left\{ \sin\left(\frac{\pi}{4} - \alpha\right) \left[e^{-i\Delta} \cos\left(\frac{\pi}{4} + \gamma\right) - \sin\left(\frac{\pi}{4} + \gamma\right) \right] \right.$$
$$\left. + i \cos\left(\frac{\pi}{4} - \alpha\right) \left[e^{-i\Delta} \sin\left(\frac{\pi}{4} + \gamma\right) - \cos\left(\frac{\pi}{4} + \gamma\right) \right] \right\} e^{i(\omega t + \alpha - \pi/4)}$$

(13.24)

The intensity of the light emerging from the analyzer is given by Eq. (11.17) as

$$I \propto E_{a\gamma} E_{a\gamma}^* \tag{e}$$

Substituting Eq. (13.24) and its complex conjugate into Eq. (e) and combining terms through the use of suitable trigonometric identities yields

$$I = K(1 - \cos 2\gamma \cos \Delta - \cos 2\alpha \sin 2\gamma \sin \Delta) \tag{13.25}$$

For a given angle of analyzer rotation γ, values of α and Δ required for maximum intensity or minimum intensity are obtained from

$$\frac{\partial I}{\partial \alpha} = K(2 \sin 2\alpha \sin 2\gamma \sin \Delta) = 0 \tag{f}$$

$$\frac{\partial I}{\partial \Delta} = K(\cos 2\gamma \sin \Delta - \cos 2\alpha \sin 2\gamma \cos \Delta) = 0 \tag{g}$$

Values of α and Δ satisfying Eqs. (f) and (g) simultaneously are

$$\alpha = \begin{cases} \dfrac{(2n + 1)\pi}{4} \\[2ex] \dfrac{n\pi}{2} \end{cases} \quad \text{and} \quad \Delta = \begin{cases} n\pi & n = 0, 1, 2, 3, \ldots \\ & \quad \text{maximum intensity} \\ 2\gamma \pm 2n\pi & n = 0, 1, 2, 3, \ldots \end{cases}$$

minimum intensity

The above conditions for extinction ($I = 0$) indicate that a principal-stress direction must be parallel to the axis of the polarizer ($\alpha = 0, \pi/2, \ldots$). The fringe order at the point is then

$$N = \frac{\Delta}{2\pi} = n \pm \frac{\gamma}{\pi} \tag{13.26}$$

Rotation of the analyzer through an angle γ (Tardy method of compensation) is widely used to determine fractional fringe orders at all points of a photoelastic model. A plane polariscope is first employed so that isoclinics can be used to establish the directions of the principal stresses at the point of interest, as shown in Fig. 13.15. The axis of the polarizer is then aligned with a principal-stress direction ($\alpha = 0$ or $\pi/2$), and the other elements of the polariscope are oriented to

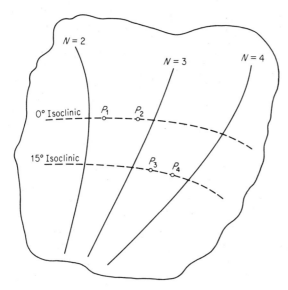

Figure 13.15 Locations of the points of interest relative to the dark-field isochromatic fringe pattern.

produce a standard dark-field circular polariscope. The analyzer is then rotated until extinction occurs at the point of interest, as indicated by Eq. (13.26). To illustrate the procedure, consider the hypothetical dark-field fringe pattern and points of interest shown in Fig. 13.15. At point P_1, which lies between fringes of orders 2 and 3, the value assigned to n is 2. As the analyzer is rotated through an angle γ, the second-order fringe will move toward point P_1 until extinction is obtained. The fringe order at P_1 is then given by $N = 2 + \gamma/\pi$. For point P_2 the value of n is also taken as 2, and the analyzer is rotated through an angle γ_1 until the second-order fringe produces extinction, giving a value for the fringe order of $N = 2 + \gamma_1/\pi$. In this instance n could also be taken as 3, and the analyzer rotated in the opposite direction through an angle $-\gamma_2$ until the third-order fringe produced extinction at point P_2. In this instance the fringe order would be given by $N = 3 - \gamma_2/\pi$, which should check the value of $N = 2 + \gamma_1/\pi$ obtained previously.

The Tardy method of compensation can be quickly and effectively employed to determine fractional fringe orders at arbitrary points in a model, provided isoclinic parameters are used to obtain the directions of the principal stresses. The accuracy of the method depends upon the quality of the quarter-wave plates employed in the polariscope; however, fringe orders obtained with this method and accurate to two decimal points are often quoted in the literature.

13.9 PHOTOELASTIC PHOTOGRAPHY [25, 26]

In most photoelastic analyses, photographs are taken of the isochromatic and isoclinic fringe patterns to establish a permanent record of the test. For this reason it is important that the basic principles of photography be established and that the differences between landscape and photoelastic photography be understood.

A sheet of photographic film is prepared with a coating containing certain silver halides. When this coating is exposed to light, the silver halides undergo a latent change that is permanently distinguishable on the film after a photographic development process. The change is a darkening produced by the formation of metallic silver. The amount of darkening is called the density. The density of a portion of a piece of exposed and developed film is simply a measure of the ability of the deposited silver to prevent the transmission of light. The density of a given type of film is a function of the exposure (light intensity times time) presented to the film. The characteristics of a density-exposure function were first established by Hurter and Driffield in the manner shown in Fig. 13.16.

The curve presented in Fig. 13.16 defines four important characteristics of a photographic film; namely, the fog density D_0, the exposure inertia E_0, the slope of the curve which establishes the gamma number of the film, and the maximum density D_1 which occurs with an exposure E_1. Each of these characteristics is important and should be considered in selecting a film for a particular photoelastic analysis. In a photoelastic photograph, zero exposures occur whenever the intensity goes to zero, i.e., when $N = 0, 1, 2, \ldots$, in a dark-field polariscope; however, the film coating records values of the ranges of exposure above the inertia value E_0. It is this "dead" exposure which produces the fringe width on a negative when in theory the fringe is a line. The slope of the density-versus-log-exposure curve given by γ determines the latitude of the film. The usual landscape film incorporates an emulsion with $\gamma \approx 1$. This relatively low value of γ gives a wide range of exposure values over which the film emulsion will be effective in producing a satisfactory negative. This feature is of course important in landscape photography, where the proper exposure time cannot be precisely established. For photoelastic photographs, film emulsions with a high γ (3 to 6) are often employed

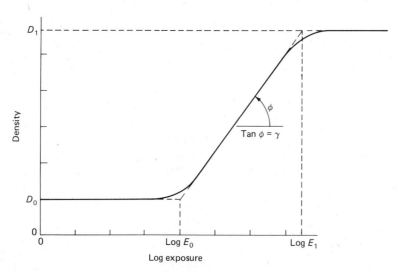

Figure 13.16 Hurter-Driffield graph relating exposure and density.

since this type of film gives a high-contrast negative; i.e., blacks are very black, whites are very white, with very little gray. This is a desirable situation since the fringes tend to sharpen and become better defined. The exposure time is of course more critical, but this time can be accurately established in preliminary exposures for any given polariscope. The fog density D_0 is less important since it implies that there is a thin coating uniformly distributed over the film which absorbs light. This of course detracts from the sharpness of a negative, but since it is a relative factor it is not objectionable.

For the linear portion of the density-versus-log-exposure curve, the density D can be expressed as

$$D = \begin{cases} D_0 + \gamma(\log E - \log E_0) & \text{for } E_0 \leq E \leq E_1 \\ D_0 & \text{for } E \leq E_0 \\ D_1 & \text{for } E \geq E_1 \end{cases} \quad (13.27)$$

where $D = \log I_i/I_e$

I_i = intensity of light incident upon developed negative

I_e = intensity of light emerging from developed negative

D_0 = fog density = $\log I_i/I_{e'}$

D_1 = maximum density = $\log I_i/I_{e''}$

$I_{e'}$ = intensity of light emerging from the unexposed part of developed negative

$I_{e''}$ = intensity of light emerging from an overexposed part of developed negative

$E = It$

I = intensity of light incident on film

t = time of exposure

By employing the above definitions, Eq. (13.27) can be rewritten as

$$\log \frac{I_i}{I_e} = \log \frac{I_i}{I_{e'}} + \gamma \log \frac{E}{E_0} \quad (a)$$

which can be reduced to

$$\rho = \frac{I_e}{I_{e'}} = \left(\frac{E}{E_0}\right)^{-\gamma} \quad \text{for } E_1 \geq E \geq E_0 \quad (13.28)$$

where ρ is called the brightness ratio = $I_e/I_{e'}$. From this definition it follows that $\rho = 1$ corresponds to the brightest area on the negative, while $\rho = 0$ corresponds to an opaque area. Note that $\rho = 1$ when $E \leq E_0$ since $I_e = I_{e'}$ and that $\rho \to 0$ when $E \geq E_1$ since $I_e = I_{e''}$.

Now recall Eq. (13.18), where the light intensity produced by inserting a stressed model in a circular polariscope is given by

$$I = K \sin^2 \frac{\Delta}{2} \quad (13.18)$$

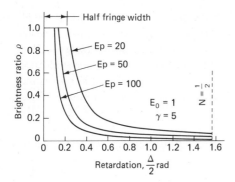

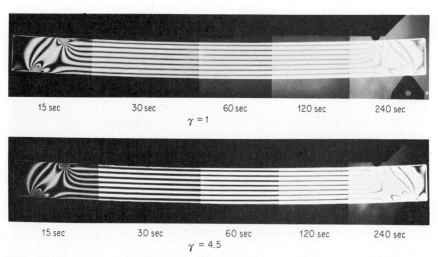

Figure **13.17** Brightness ratio as a function of retardation.

The exposure for the dark-field negative will be

$$E = It = Kt \sin^2 \frac{\Delta}{2} = E_p \sin^2 \frac{\Delta}{2} \qquad (13.29)$$

where $E_p = Kt$ is the uniform exposure produced by the polariscope. Combining Eqs. (13.28) and (13.29) gives

$$\rho \begin{cases} = \dfrac{1}{(E_p/E_0)^\gamma \sin^{2\gamma}(\Delta/2)} & \text{for } E_1 \geq E_p \sin^2 \dfrac{\Delta}{2} \geq E_0 \\[3mm] = 1 & \text{for } E_p \sin^2 \dfrac{\Delta}{2} \leq E_0 \\[3mm] \to 0 & \text{for } E_p \sin^2 \dfrac{\Delta}{2} \geq E_1 \end{cases} \qquad (13.30)$$

Figure **13.18** Influence of exposure time and gamma number on the appearance of the isochromatic fringe pattern for a beam in pure bending.

The results of Eqs. (13.30) describe the brightness ratio ρ for a dark-field photoelastic photograph. An example of the variation of ρ with $\Delta/2$ is given in Fig. 13.17 for three different exposures ($E_p/E_0 = 20$, 50, and 100). It should be noted that increasing the exposure ratio E_p/E_0 reduces the fringe width and improves the contrast. This sharpening of the fringes by increasing the exposure time is illustrated in Fig. 13.18 for both high-contrast and landscape film.

13.10 FRINGE MULTIPLICATION BY PHOTOGRAPHIC METHODS [27, 28]

By employing normal photoelastic procedures, two photographs (one light-field, the other dark-field) are obtained which permit the determination of the order of the fringes in the following sequence: $N = 0, \frac{1}{2}, 1, 1\frac{1}{2}, 2, 2\frac{1}{2}, \dots$. In certain photoelastic applications it is desirable to improve the accuracy of the determination of fractional fringe orders which are between those previously listed. This objective can be accomplished in a number of ways. In this section a photographic technique is described which gives, through superposition of ordinary light- and dark-field isochromatic fringe patterns, a new fringe pattern (mixed-field). This mixed-field pattern has fringes at the $N/4$ and $3N/4$ positions. Use of the mixed-field fringe pattern coupled with ordinary light- and dark-field fringe patterns permits the orders of the fringes to be determined in the $0, \frac{1}{4}, \frac{1}{2}, \frac{3}{4}, 1, \frac{5}{4}, \dots$ sequence, and thus represents a factor of 2 increase in the number of countable fringes.

The proof of this photographic method can easily be established by drawing from the results established in Secs. 13.6, 13.7, and 13.9. From Eqs. (13.30), which describe the brightness of a negative obtained with a dark-field polariscope,

$$\rho_d \begin{cases} = \dfrac{1}{(E_{pd}/E_0)^\gamma \sin^{2\gamma}(\Delta/2)} & \text{for } E_1 \geq E_{pd}\sin^2\dfrac{\Delta}{2} \geq E_0 \quad (a) \\[3ex] = 1 & \text{for } E_{pd}\sin^2\dfrac{\Delta}{2} \leq E_0 \quad (b) \\[3ex] \to 0 & \text{for } E_{pd}\sin^2\dfrac{\Delta}{2} \geq E_1 \quad (c) \end{cases}$$

Similarly by combining Eqs. (13.28) and (13.29) with Eq. (13.20), the brightness ratio for a negative in a light-field polariscope is

$$\rho_l \begin{cases} = \dfrac{1}{(E_{pl}/E_0)^\gamma \cos^{2\gamma}(\Delta/2)} & \text{for } E_1 \geq E_{pl}\cos^2\dfrac{\Delta}{2} \geq E_0 \quad (d) \\[3ex] = 1 & \text{for } E_{pl}\cos^2\dfrac{\Delta}{2} \leq E_0 \quad (e) \\[3ex] \to 0 & \text{for } E_{pl}\cos^2\dfrac{\Delta}{2} \geq E_1 \quad (f) \end{cases}$$

The effect of superimposing light and dark field negatives is obtained by multiplying the expressions for the brightness ratios. Thus, the brightness ratio ρ_m for the mixed field associated with superimposed light- and dark-field negatives is

$$\rho_m = \rho_d \rho_l \tag{g}$$

Equation (g) leads to four nonzero expressions for ρ_m

$$\rho_m = \begin{cases} 1 & \text{from Eqs. (b) and (e)} \\[2ex] \dfrac{1}{(E_{pl}/E_0)^\gamma \cos^{2\gamma}(\Delta/2)} & \text{from Eqs. (b) and (d)} \\[2ex] \dfrac{1}{(E_{pd}/E_0)^\gamma \sin^{2\gamma}(\Delta/2)} & \text{from Eqs. (a) and (e)} \\[2ex] \dfrac{1}{(E_{pd}/E_0)^\gamma (E_{pl}/E_0)^\gamma \sin^{2\gamma}(\Delta/2)\cos^{2\gamma}(\Delta/2)} & \text{from Eqs. (a) and (d)} \end{cases}$$

Of these four solutions for ρ_m, the solution $\rho_m = 1$ locates the fringe position on the superimposed negative. Note that $\rho_m = 1$ only in regions where both

$$\cos^2 \frac{\Delta}{2} \le \frac{E_0}{E_{pl}} \quad \text{and} \quad \sin^2 \frac{\Delta}{2} \le \frac{E_0}{E_{pd}} \tag{h}$$

With equal light- and dark-field exposures

$$E_{pd} = E_{pl} = ME_0 \tag{i}†$$

where M is the exposure multiple. Combining Eqs. (h) and (i) yields the condition $M \le 2$ to obtain any region on the superimposed negative where $\rho_m = 1$. With $M = 2$, Eqs. (h) become

$$\cos^2 \frac{\Delta}{2} \le \frac{1}{2} \quad \sin^2 \frac{\Delta}{2} \le \frac{1}{2} \tag{13.31}$$

It is evident then that

$$\frac{\Delta}{2} = \frac{(2n+1)\pi}{4} \quad \text{where } n = 0, 1, 2, 3, \ldots$$

and

$$N = \frac{\Delta}{2\pi} = \frac{2n+1}{4} \tag{13.32}$$

With $E_{pd} = E_{pl}$ and $M = 2$, the superimposed negatives yield a fringe pattern where the $\frac{1}{4}$-order fringes are displayed with $N = \frac{1}{4}, \frac{3}{4}, \frac{5}{4}, \ldots$. A schematic illustration of the brightness ratio ρ_m as a function of retardation $\Delta/2$ is shown in Fig. 13.19.

---

† A more general treatment of exposure ratio leads to a whole-field compensation technique [32–34].

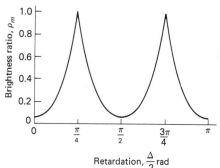

Figure 13.19 Mixed-field brightness ratio as a function of retardation.

The photographic method of fringe multiplication (doubling) is illustrated in the two examples cases presented below.

Beam subjected to a constant moment The model of the beam was machined from $\frac{1}{4}$-in-thick CR-39, and a constant moment was applied to the central portion of the beam by employing the four-point loading technique. Lines were scribed on the model in order to permit easier alignment of the two negatives during superposition. The load was increased so that in the region of constant moment there were about four fringes. So that the photographic method of superposition of negatives could be carried out, the exposure was kept low. With Kodak contrast-process panchromatic films, an aperture setting f: 45, and an exposure time of 15 s, light- and dark-field photographs were taken. The contrast-process film used for this and all the subsequent experiments was developed for 6 min in Kodak D-11 developer. On superposition of light- and dark-field negatives, it was found that the fringes were distinct and of uniform intensity. If the exposure time was greater than the optimum value ($M = 2$), the uniform exposure of the two negatives became prohibitively high and light would not pass through the two negatives. Employing exposures under the optimum value produced films so thin that the fringes were not well defined upon superposition of the negatives. The fringe patterns obtained for the light field, dark field, and superimposed or mixed fields are shown in Fig. 13.20.

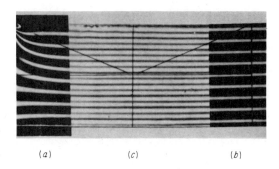

(a) (c) (b)

Figure 13.20 (a) Dark-, (b) light-, and (c) mixed-field isochromatic fringe patterns of a beam subjected to a constant binding moment.

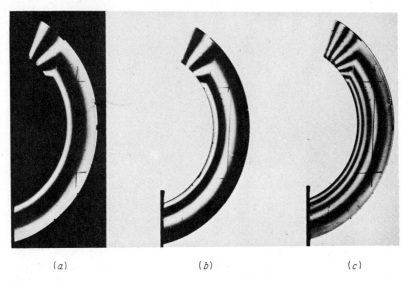

Figure 13.21 (a) Dark-, (b) light-, and (c) mixed-field isochromatic fringe patterns from a three-dimensional slice.

Slice from a three-dimensional model The model was a segment of a hoop slice taken from a thick-walled pressure vessel subjected to internal pressure. One end of the segment was machined into the shape of a wedge in order to locate the zero-order fringe. The slice was polished and mounted in the field of the polariscope, and light- and dark-field photographs were taken. The dark-, light-, and mixed-field photographs shown in Fig. 13.21 illustrate the improvement possible in the fringe-order determination by employing this fringe-doubling technique.

13.11 FRINGE SHARPENING WITH PARTIAL MIRRORS [29]

The bandwidth of the isochromatic fringe can be reduced by a novel technique due to Post [29], which employs partial mirrors† in a circular-lens polariscope. The partial mirrors are inserted into the field of the polariscope on both sides of the model and parallel to it, as illustrated in Fig. 13.22.

The effect of the partial mirrors is to cause the light to propagate back and forth through the model in the manner illustrated in Fig. 13.23. As the light is reflected back and forth between the two mirrors, a portion of it is transmitted at each reflection point. Hence, the intensity of the ray passing back and forth through the model is progressively decreasing. For instance, ray 1 is the most intense, ray 3 less intense, etc. The rays shown in Fig. 13.23 are drawn obliquely

---

† A partial mirror, sometimes called a *beam splitter*, is simply an optical element that transmits a portion of the incident light and reflects the remainder so that $T + R = 1$, where T is the coefficient of transmission and R is the coefficient of reflection.

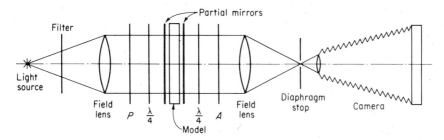

Figure 13.22 Post's modification of the lens polariscope, with partial mirrors for fringe sharpening.

only to present the effects of the partial mirrors; in practice, normal incidence is employed and all the rays coincide and enter and emerge from the same point in the model.

The effect of the partial mirrors on the intensity of the light as it passes through a stressed model can be obtained by modifying Eq. (13.18), which is valid if no partial mirrors are employed in the polariscope:

$$I = K \sin^2 \frac{\Delta}{2} \tag{13.18}$$

Consider ray 1 (see Fig. 13.23) and reduce the intensity due to the light lost by reflection from the partial mirrors at points A and B. Thus

$$I_1 = K(1 - R)^2 \sin^2 \frac{\Delta}{2} = KT^2 \sin^2 \frac{\Delta}{2} \tag{a}$$

where R and T are the reflection and transmission coefficients of the partial mirror.

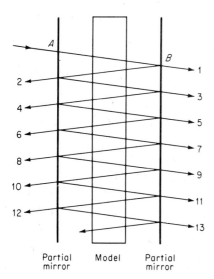

Partial mirror   Model   Partial mirror

Figure 13.23 Light reflection and transmission between two partial mirrors.

The intensity of the light associated with ray 3 can be written as

$$I_3 = KT^2R^2 \sin^2 \frac{3\Delta}{2} \tag{b}$$

In this instance ray 3 has undergone two reflections and two transmissions; hence the T and R terms in Eq. (b) are squared. Also, the light has passed through the model three times, and the argument of the sine function has been multiplied by 3 to account for this fact. By following this same procedure, the intensity of the kth ray may be written as

$$I_k = KT^2R^{k-1} \sin^2 \frac{k\Delta}{2} \qquad k = 1, 3, 5, 7, \ldots \tag{13.33}$$

These intensities I_1, I_3, I_5, $\ldots$, I_k add arithmetically; hence the resultant intensity of the superimposed rays is given by the series expansion

$$I = KT^2 \sum_{k=1}^{\infty} R^{k-1} \sin^2 \frac{k\Delta}{2} \tag{13.34}$$

If this relationship is expanded and plotted as a function of $\Delta/2\pi = N$, the intensity-versus-fringe-order plot shown in Fig. 13.24 is obtained. When this plot is compared with the conventional intensity-versus-$\Delta/2$ plot, also shown in Fig. 13.24, it is clear why the fringes are sharpened. The eye will begin to record a fringe at some minimum intensity I_0; hence the sharpened intensity function produces a much narrower fringe than the conventional intensity function. An example of the isochromatic fringe pattern produced with a pair of partial mirrors, each with a reflection coefficient $R = 0.9$ and a transmission coefficient $T = 0.1$, is shown in Fig. 13.25.

An examination of this figure shows that the sharp light and dark fringes observed are separated by wide gray bands. The dark fringes correspond to the sharp valleys on the intensity-versus-fringe-order plot shown in Fig. 13.24, and the

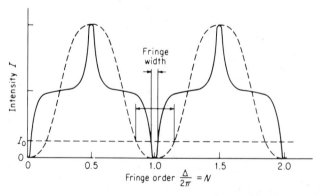

Figure 13.24 Intensity as a function of fringe order in a standard circular polariscope with partial mirrors where $R = 0.85$. (*Courtesy of D. Post.*)

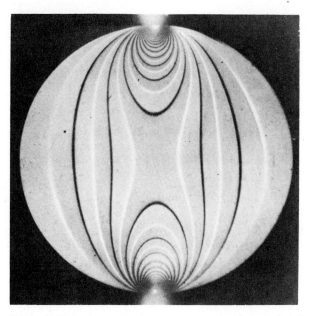

Figure 13.25 Isochromatic fringe pattern produced by a dark-field circular polariscope equipped with partial mirrors. (*Courtesy of D. Post.*)

light fringes correspond to the sharp peaks on this same graph. The wide gray bands are produced by the midrange of intensity also shown in this figure. The dark fringes are ordered in a 0, 1, 2, 3, ... sequence, and the light fringes are ordered in the $\frac{1}{2}$, $1\frac{1}{2}$, $2\frac{1}{2}$, $3\frac{1}{2}$, ... sequence. Thus, the data normally obtained from two conventional light- and dark-field photographs are contained in one photograph if partial mirrors are employed in the standard circular lens polariscope.

13.12 FRINGE MULTIPLICATION WITH PARTIAL MIRRORS [29–31]

Post has also shown that partial mirrors can be quite usefully employed to multiply the number of fringes which can be observed in a photoelastic model. As pointed out in Sec. 13.8, fringe multiplication is quite important since the standard methods of compensation used to evaluate the fractional fringe orders are time-consuming and in certain cases somewhat inaccurate. Fringe multiplication is, in a sense, a whole-field compensation technique where the fractional orders of the fringes can be determined simultaneously at all points on the model.

When partial mirrors are used in fringe multiplication, they are again inserted into a lens polariscope on both sides of the model; however, in this application one of the mirrors is inclined slightly, as illustrated in Fig. 13.26. The effect of the inclined partial mirror on the light passing back and forth through the model is

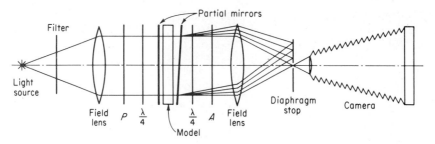

Figure 13.26 Partial mirrors as employed in a circular-lens polariscope for fringe multiplication.

shown in Fig. 13.27. From this figure it is clear that each ray of light emerges from the mirror system at an angle which depends on the number of times the light ray has traversed the model. For instance, rays 1, 3, 5, and 7, which have traversed the model the same number of times as their ray number, emerge at angles 0, 2ϕ, 4ϕ, and 6ϕ. Although the rays do not pass through the same point, the inclination angle ϕ used in the illustration was greatly exaggerated. The length of the line over which the photoelastic effect is averaged depends upon the angle of inclination ϕ, the ray number, and the separation distance between the mirrors. In practice, multiplication by factors of 5 to 7 can be achieved without introducing objectionable errors due to the averaging process which is inherent in this method. Typical patterns obtained from a two-dimensional model and from a three-dimensional slice are illustrated in Figs. 13.28 and 13.29.

The fact that different rays of light are inclined at different angles with respect to the axis of the polariscope permits each ray to be isolated. The rays are all

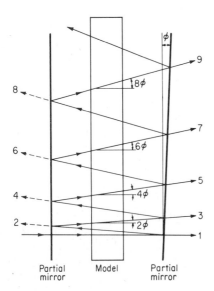

Figure 13.27 Light reflection and transmission between two slightly inclined partial mirrors.

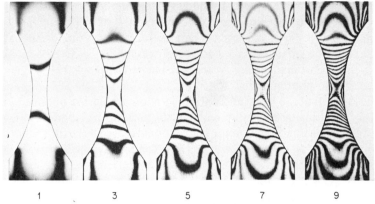

1   3   5   7   9

Figure 13.28 Isochromatic fringe patterns of a tensile specimen obtained by using a lens polariscope equipped with partial mirrors. (*Polarizing Instrument Company.*)

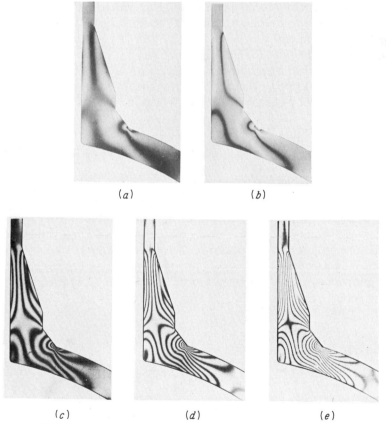

(a)   (b)

(c)   (d)   (e)

Figure 13.29 Isochromatic fringe patterns of a three-dimensional slice obtained by using a lens polariscope equipped with partial mirrors: (a) normal fringe pattern; (b) sharpened fringe pattern; (c) after fringe multiplication 3X; (d) after fringe multiplication 5X; (e) after fringe multiplication 7X. (*Courtesy of C. E. Taylor.*)

collected by the field lens but focused at different points in the focal plane of the field lens (see Fig. 13.26). Any one of these rays can be observed by placing the eye or a camera lens at the proper image point. A diaphragm stop is useful in eliminating all images except the particular one under observation.

In practice, the isochromatic fringe patterns associated with rays 1, 3, 5, 7, etc., can be observed and photographed for both light- and dark-field settings. Suppose, for instance, that the light- and dark-field photographs are obtained for rays 1, 3, and 5. The fringe patterns as recorded on the two photographs associated with ray 1 may be interpreted in the conventional sense where the orders of the fringes are sequenced as $0, \frac{1}{2}, 1, \frac{3}{2}, 2, \frac{5}{2}, \ldots$. However, for ray 3, where the light has passed through the model three times, the orders of the fringes are sequenced as $0, \frac{1}{6}, \frac{1}{3}, \frac{1}{2}, \frac{2}{3}, \frac{5}{6}, 1, \ldots$. Finally, for ray 5, where the light has traversed the model five times, the orders of the fringes are sequenced as $0, \frac{1}{10}, \frac{1}{5}, \frac{3}{10}, \frac{2}{5}, \frac{1}{2}, \ldots$. Thus the superposition of the results obtained from these three rays is sufficient to determine the fringe order to the nearest one-tenth of an order over the entire model. The fringe-multiplication technique, therefore, can be interpreted as a whole-field compensation method where fractional orders of the fringes can be determined with a high degree of accuracy.

The intensity relationship for the mth ray, where $m = 1, 2, 3, 4, 5$, as shown in Fig. 13.27, can be established by modifying Eq. (13.18) to account for the loss in intensity and the added thickness effects as the light traverses the model m times. It is apparent that the intensity of each ray can be written as

$$I_1 = KT^2 \sin^2 \frac{\Delta}{2} \qquad I_3 = KT^2 R^2 \sin^2 \frac{3\Delta}{2}$$

$$I_5 = KT^2 R^4 \sin^2 \frac{5\Delta}{2} \qquad I_m = KT^2 R^{m-1} \sin^2 \frac{m\Delta}{2} \tag{13.35}$$

Thus, fringe multiplication, by Post's partial-mirror method, is accompanied by a considerable loss of light intensity. The intensity of the multiplied fringe pattern as compared with the ordinary fringe pattern is decreased by the term $T^2 R^{m-1}$, which is always much less than 1. The loss in intensity for a particular ray can be minimized by properly selecting the mirror coefficients R and T. Assuming that the mirrors are perfect,

$$T + R = 1 \tag{13.36}$$

Substituting Eq. (13.36) into Eqs. (13.35) yields

$$I_m = K(R - 1)^2 R^{m-1} \sin^2 \frac{m\Delta}{2} \qquad m = 1, 2, 3, 4, \ldots \tag{a}$$

Differentiating this expression with respect to R yields

$$\frac{dI_m}{dR} = K \sin^2 \frac{m\Delta}{2} \{(R - 1)[(R - 1)(m - 1)R^{m-2} + 2R^{m-1}]\} \tag{b}$$

Table 13.1 Optimum mirror properties for fringe multiplication

Multiplication factor m	R	T	Intensity coefficient $T^2 R^{m-1}$
1	0	1	1
3	0.500	0.500	0.0625
5	0.667	0.333	0.0219
7	0.750	0.250	0.0111
9	0.800	0.200	0.0067

Setting Eq. (b) equal to zero and solving for R gives the reflection coefficient for the mirrors, which minimize the intensity lost as follows:

$$R = \frac{m-1}{m+1} \tag{13.37}$$

Since R is a function of m (the number of light traverses through the model), it is not possible to optimize the mirrors for all rays simultaneously. The optimum coefficients of reflection and transmission for each value of m are presented in Table 13.1. A multiplication factor of 5 is usually sufficient for most applications; hence, if only one set of mirrors is available, this set should be selected with $R = 0.667$ and $T = 0.333$ to optimize the minimum intensity condition.

EXERCISES

13.1 If a particular point in a photoelastic model is examined in a polariscope with a mercury light source ($\lambda = 548.1$ nm) and a fringe order of 3.00 is established, what fringe order would be observed if a sodium light source ($\lambda = 589.3$ nm) were used in place of the mercury source?

13.2 The stress fringe value f_σ for a material was determined to be 17.5 kN/m when sodium light ($\lambda = 589.3$ nm) was used in its determination. What would the stress fringe value for the same material be if mercury light ($\lambda = 548.1$ nm) were used in place of the sodium light?

13.3 Derive the equations for light passing through a stressed model in a plane polariscope with the polarizer and analyzer in parallel positions. Under what conditions does extinction ($I = 0$) occur?

13.4 Derive the equations for light passing through a stressed model in a plane polariscope (polarizer and analyzer crossed). Use an exponential representation for the light wave.

13.5 Derive the equations for light passing through a stressed model in a circular polariscope (arrangement C). Use a trigonometric representation for the light wave.

13.6 Derive the equations for light passing through a stressed model in a circular polariscope (arrangement D). Use an exponential representation for the light wave.

13.7 Determine the optical effects produced by light passing through a stressed model in a circular polariscope (arrangements A and B) with imperfect quarter-wave plates. Assume $\Delta = \pi/2 + \epsilon$ for the imperfect plates.

13.8 Determine the optical effects produced by light passing through a stressed model in a circular polariscope with imperfect quarter-wave plates. Use arrangements C and D for the polariscope and assume $\Delta = \pi/2 + \epsilon$ for the imperfect quarter-wave plates.

13.9 With Tardy compensation, the fractional fringe order can be determined at an arbitrary point in a stressed model by aligning a principal-stress direction at the point with the axis of the polarizer and rotating the analyzer for extinction. What effect would be produced by rotating the polarizer instead of the analyzer?

13.10 Investigate the error introduced in the fractional fringe-order determination by improper setting of the isoclinic angle ($\alpha = 0 + \epsilon$ or $\pi/2 + \epsilon$) in Tardy compensation. Numerically evaluate for $\epsilon = \pm 5°$.

13.11 Investigate the error introduced in fractional fringe-order determinations by imperfect quarter-wave plates ($\Delta = \pi/2 + \epsilon$) in Tardy compensation. Make a numerical evaluation for the case of quarter-wave plates for mercury light ($\lambda = 548.1$ nm) being used with sodium light ($\lambda = 589.3$ nm).

13.12 The following procedure has been proposed for establishing the fractional fringe order at a point in a stressed model. Establish the validity of the procedure.

1. Begin with a circular polariscope (arrangement A).
2. Remove the first quarter-wave plate.
3. Position the model so that the principal stress directions at the point of interest are oriented $\pm 45°$ with respect to the axis of the polarizer.
4. Align the fast axis of the second quarter-wave plate with the axis of the polarizer.
5. Rotate the analyzer for extinction.

13.13 Plot the results of Eq. (13.30), that is, $1/\rho$ as a function of $\Delta/2$, for the following values of E/E_0 and γ:

Case	γ	$\dfrac{E_\rho}{E_0}$	Case	γ	$\dfrac{E_\rho}{E_0}$
1	1	1	4	1	3
2	3	1	5	3	3
3	6	1	6	6	3

13.14 During the course of a photoelastic investigation you instruct your technician to photograph the isoclinic patterns in the model under study at 5° intervals. To save film, he decides to record two sets of patterns on the same film by using the following procedure. He first exposes the film for half the normal time with the plane polariscope set at the 0° position. He then rotates both the polarizer and the analyzer 45° counterclockwise but does not move the model and exposes the film for the remaining half of the time. On other sheets of film, he records the 5 and 50° isoclinics, the 10 and 55° isoclinics, etc., using the same procedure. When he develops the film, what does he see?

13.15 Plot the results of Eq. (13.34), that is, I/K versus $\Delta/2$, if $T = 0.1$, $R = 0.9$. Consider the first 10 terms ($k = 1$ to 10).

13.16 Discuss the influence of a uniform preexposure of a given photographic film which would be sufficient to overcome the inertia exposure E_0. Consider in this discussion the width of the fringes and the exposure time required in the polariscope.

REFERENCES

1. Jenkins, F. A., and H. E. White: "Fundamentals of Optics," 4th ed., McGraw-Hill Book Company, New York, 1976.
2. Durelli, A. J., E. A. Phillips, and C. H. Tsao: "Introduction to the Theoretical and Experimental Analysis of Stress and Strain," chap. 3, McGraw-Hill Book Company, New York, 1958.

3. Coker, E. G., and L. N. G. Filon: "A Treatise on Photoelasticity," Cambridge University Press, New York, 1931.
4. Maxwell, J. C.: On the Equilibrium of Elastic Solids, *Trans. R. Soc. Edinburgh*, vol. XX, part 1, pp. 87–120, 1853.
5. Neumann, F. E.: Die Gesetze der Doppelbrechung des Lichts in comprimierten oder ungleichformig erwärmten unkrystallinischen Korpern, *Abh. K. Acad. Wiss. Berlin*, pt. II, pp. 1–254, 1841.
6. Favre, H.: Sur une nouvelle méthode optique de détermination des tensions intérieures, *Rev. Opt.*, vol. 8, pp. 193–213, 241–261, 289–307, 1929.
7. Post, D.: Photoelastic Evaluation of Individual Principal Stresses by Large Field Absolute Retardation Measurements, *Proc. SESA*, vol. XIII, no. 2, pp. 119–132, 1956.
8. Frocht, M. M.: "Photoelasticity," John Wiley & Sons, Inc., New York, vol. 1, 1941, vol. 2, 1948.
9. Jessop, H. T., and F. C. Harris: "Photoelasticity: Principles and Methods," Dover Publications, Inc., New York, 1950.
10. Durelli, A. J., and W. F. Riley: "Introduction to Photomechanics," Prentice-Hall, Englewood Cliffs, N.J., 1965.
11. Kuske, A., and G. Robertson: "Photoelastic Stress Analysis," John Wiley & Sons, Inc., New York, 1974.
12. Mindlin, R. D.: A Review of the Photoelastic Method of Stress Analysis, *J. Appl. Phys.*, vol. 10, pp. 222–241, 273–294, 1939.
13. Vandaele-Dossche, M., and R. van Geen: La biréfringence mécanique en lumière ultra-violette et ses applications, *Bull. Class Sci. Acad. R. Belg.*, vol. 50, no. 2, 1964.
14. Monch, E., and R. Lorek: A Study of the Accuracy and Limits of Application of Plane Photoplastic Experiments, *Proc. Int. Symp. Photoelasticity*, Pergamon Press, New York, 1963.
15. Pindera, J. T., and G. Cloud: On Dispersion of Birefringence of Photoelastic Materials, *Exp. Mech.*, vol. 6, no. 9, pp. 470–480, 1966.
16. Mindlin, R. D.: Distortion of the Photoelastic Fringe Pattern in an Optically Unbalanced Polariscope, *J. Appl. Mech.*, vol. 4, pp. A170–172, 1937.
17. Mindlin, R. D.: Analysis of Doubly Refracting Materials with Circularly and Elliptically Polarized Light, *J. Opt. Soc. Am.*, vol. 27, pp. 288–291, 1937.
18. Tardy, M. H. L.: Méthode pratique d'examen de mesure de la biréfringence des verres d'optique, *Rev. Opt.*, vol. 8, pp. 59–69, 1929.
19. Jessop, H. T.: On the Tardy and Senarmont Methods of Measuring Fractional Relative Retardations, *Br. J. Appl. Phys.*, vol. 4, pp. 138–141, 1953.
20. Flynn, P. D.: Theorems for Senarmont Compensation, *Exp. Mech.*, vol. 10, no. 8, pp. 343–345, 1970.
21. Friedel, G.: Sur un procède de mesure des biréfringences, *Bull. Soc. Fr. Mineral*, vol. 16, pp. 19–33, 1893.
22. Jerrard, H. G.: Optical Compensators for Measurement of Elliptical Polarization, *J. Opt. Soc. Am.*, vol. 38, pp. 35–59, 1948.
23. Chakrabarti, S. K., and K. E. Machin: Accuracy of Compensation Methods in Photoelastic Fringe-Order Measurements, *Exp. Mech.*, vol. 9, no. 9, pp. 429–431, 1969.
24. Sathikh, S. M., and G. W. Bigg: On the Accuracy of Goniometric Compensation Methods in Photoelastic Fringe-Order Measurements, *Exp. Mech.*, vol. 12, no. 1, pp. 47–49, 1972.
25. Flanagan, J. H.: Photoelastic Photography, *Proc. SESA*, vol. XV, no. 2, pp. 1–10, 1958.
26. Koch, W. M., and P. A. Szego: Use of Double-Exposure Photography in Photoelasticity, *J. Appl. Mech.*, vol. 21, p. 198, 1954.
27. Dally, J. W., and F. J. Ahimaz: Photographic Method to Sharpen and Double Isochromatic Fringes, *Exp. Mech.*, vol. 2, no. 6, pp. 170–175, 1962.
28. Das Talukder, N. K., and P. Ghosh: On Fringe Multiplication by Superimposition of Negatives, *Exp. Mech.*, vol. 15, no. 6, pp. 237–239, 1975.
29. Post, D.: Isochromatic Fringe Sharpening and Fringe Multiplication in Photoelasticity, *Proc. SESA*, vol. XII, no. 2, pp. 143–156, 1955.
30. Post, D.: Fringe Multiplication in Three-dimensional Photoelasticity, *J. Strain Anal.*, vol. 1, no. 5, pp. 380–388, 1966.

31. Post, D.: Photoelastic-Fringe Multiplication: For Tenfold Increase in Sensitivity, *Exp. Mech.*, vol. 10, no. 8, pp. 305–312, 1970.

32. Meyer, M. L., C. L. Mehrotra, and N. K. Das Talukder: On a Photographic Method of Determining Fractional Fringe Orders, *CANCAM 73, Montreal, Canada, May–June, 1973.*

33. Das Talukder, N. K., and S. S. Das: Fractional-Fringe Photography by Field-Intensity Variation, *Exp. Mech.*, vol. 15, no. 7, pp. 275–278, 1975.

34. Mehrotra, C. L., and M. L. Meyer: Equidensometry Applied to Photoelasticity, *Exp. Mech.*, vol. 16, no. 10, pp. 383–388, 1976.

FOURTEEN

TWO-DIMENSIONAL PHOTOELASTICITY

14.1 INTRODUCTION [1–5]

In a conventional two-dimensional photoelastic analysis a suitable model is fabricated, loaded, and placed in a polariscope, and the fringe pattern is examined and photographed. The next step in the photoelastic investigation is the interpretation of the fringe patterns, which in reality represent the raw test data. The purpose of this chapter is to discuss the interpretation of the isochromatic and isoclinic fringe patterns, compensation techniques, separation techniques, and scaling of the stresses between the model and prototype.

14.2 ISOCHROMATIC FRINGE PATTERNS [1–6]

The isochromatic fringe pattern obtained from a two-dimensional model gives lines along which the principal-stress difference $\sigma_1 - \sigma_2$ is equal to a constant. A typical example of a light-field isochromatic fringe pattern, which will be utilized to describe the analysis, is presented in Fig. 14.1. This photoelastic model represents a chain link subjected to tensile loads applied axially through roller pins. First, it is necessary to determine the fringe order at each point of interest on the model. In this example the assignment of the fringe order is relatively simple since seven rather obvious $\frac{1}{2}$-order fringes can quickly be identified. The two ovallike fringes located on the flanks of the teeth (labeled A) are of the $\frac{1}{2}$ order since the flank, because of its geometry, cannot support high stresses. The four fringes

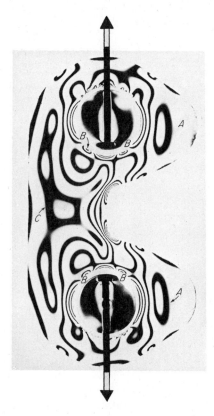

Figure 14.1 Light-field isochromatic fringe pattern of a chain link subjected to axial tensile loads through the roller pins.

located at point B on the pinholes can be identified if the model is viewed with white light since zero-order fringes appear black while higher-order fringes are colored. Finally, the irregularly shaped fringe labeled C near the center of the link is of the $\frac{1}{2}$ order. Since the roller pins are located off the centerline of the plate, a combined state of tension and bending stresses is induced in the link plate with the neutral axis shifted to the right of the centerline of the plate.

With these $\frac{1}{2}$-order fringes established, it is a simple matter to determine the fringe order at any point in the model by progressively counting the fringes outward from the $\frac{1}{2}$-order location. For instance, the order of the fringe at the flank fillets is $7\frac{1}{2}$. When the fringe order at any point on the model has been established, it is possible to compute $\sigma_1 - \sigma_2$ from Eq. (13.8),

$$\sigma_1 - \sigma_2 = \frac{Nf_\sigma}{h}$$

where σ_1 and σ_2 are the principal stresses in the plane of the model. The maximum shear stress is given by

$$\tau_{\max} = \frac{1}{2}(\sigma_1 - \sigma_2) = \frac{Nf_\sigma}{2h} \tag{14.1}$$

if σ_1 & σ_2 OF OPPOSITE SIGN

provided σ_1 and σ_2 are of opposite sign and $\sigma_3 = 0$; otherwise

$$\tau_{max} = \begin{cases} \frac{1}{2}(\sigma_1 - \sigma_3) = \frac{1}{2}\sigma_1 & \text{if } \sigma_1 \text{ and } \sigma_2 \text{ are positive} \\ \frac{1}{2}(\sigma_3 - \sigma_2) = \frac{1}{2}\sigma_2 & \text{if } \sigma_1 \text{ and } \sigma_2 \text{ are negative} \end{cases} \qquad (14.2)$$

The difference between Eq. (14.1) and Eqs. (14.2) is presented graphically in Fig. 14.2, where Mohr stress circles for two cases are given. When $\sigma_1 > 0$ and $\sigma_2 < \sigma_3 = 0$, the maximum shear stress is one-half the value of $\sigma_1 - \sigma_2$ and can be determined directly from the isochromatic fringe pattern according to Eq. (14.1). However, when $\sigma_1 > \sigma_2 > \sigma_3 = 0$, the maximum shear stress does not lie in the plane of the model, and Eq. (14.1) gives τ_p (see Fig. 14.2b) and not τ_{max}. To establish τ_{max} in this case it is necessary to determine σ_1 individually and not the difference $\sigma_1 - \sigma_2$. This is an important point since the maximum-shear theory of failure is often used in the design of machine components.

On the free boundary of the model, either σ_1 or σ_2 is equal to zero; hence, the stress tangential to the boundary can be determined directly from

$$\sigma_1, \sigma_2 = \frac{Nf_\sigma}{h} \qquad (14.3)$$

The sign can usually be determined by inspection, particularly in the critical areas where the boundary stresses are a maximum. By referring to Fig. 14.1, it is apparent that the stresses in the flank fillets are tensile while the stresses along the back of the plate are compressive. On the free surface of the pinholes, the stresses

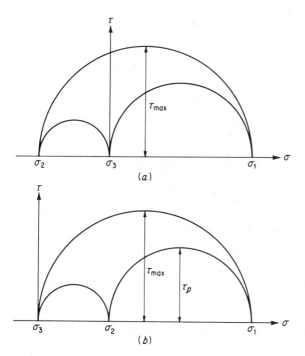

Figure 14.2 Mohr's circle for the state of stress at a point (a) $\sigma_1 > 0$, $\sigma_2 < \sigma_3 = 0$; (b) $\sigma_1 > \sigma_2 > \sigma_3 = 0$.

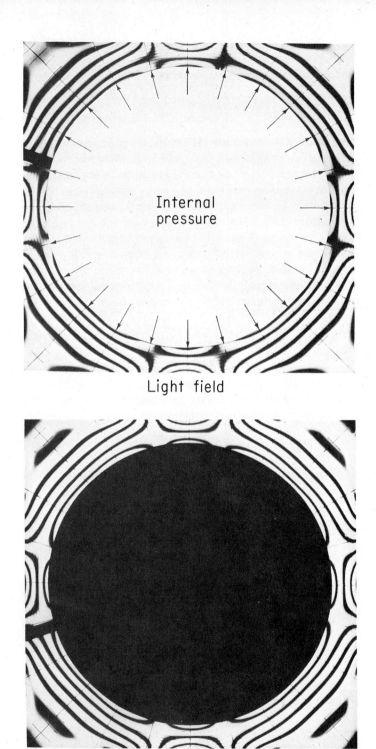

Internal pressure

Light field

Dark field

Figure 14.3 Photoelastic fringe patterns showing the distribution of stress through a section of square tubing with a circular bore.

on the horizontal diameter are tensile and the stresses on the vertical diameter are compressive (see Sec. 3.14). The stresses on that portion of the boundary of the hole where the pin is in contact cannot be determined by applying Eq. (14.3) since the boundary is not free and, in general, σ_1 or σ_2 will not approach zero. In this case neither σ_1 nor σ_2 is known, a priori, and it is necessary to separate the stresses, i.e., individually determine the values of σ_1 and σ_2, at this region of contact on the pinhole. Methods to employ in separating the stresses are presented in Sec. 14.6.

As another example of the interpretation of isochromatic fringe patterns, consider the photograph presented in Fig. 14.3. In this instance a photoelastic model of square conduit with a circular borehole was analyzed. A uniformly distributed load was applied to the circular hole of the model. The stresses along the outside edges of the model can be determined directly from Eq. (14.3) since these edges represent free boundaries of the model. Along the boundary of the circular hole the surface is not free; however, in this case the stress σ_2 acting normal to the boundary is known to be equal to p, the applied load. Hence Eq. (13.8) becomes

$$\sigma_1 - \sigma_2 = \sigma_1 + p = \frac{Nf_\sigma}{h} \quad \text{or} \quad \sigma_1 = \frac{Nf_\sigma}{h} - p \qquad (14.4)$$

where $\sigma_2 = -p$, since the applied pressure p is considered as a positive quantity.

In conclusion, it is clear that the isochromatic fringe pattern, once identified, can be interpreted in the following manner:

1. $\sigma_1 - \sigma_2$ can be determined at any point in the model from Eq. (13.8).
2. If $\sigma_1 > 0$ and $\sigma_2 < 0$, $\sigma_1 - \sigma_2$ can be related to the maximum-shear stress through Eq. (14.1).
3. If $\sigma_1 > \sigma_2 > 0$ or if $0 > \sigma_1 > \sigma_2$, $\sigma_1 - \sigma_2$ cannot be related to the maximum-shear stress and it is necessary to determine σ_1 and σ_2 individually and relate τ_{max} to σ_1 or σ_2 by Eq. (14.2).
4. If the boundary can be considered free (that is, σ_1 or $\sigma_2 = 0$), the other principal stress can be determined directly from Eq. (14.3).
5. If the boundary is not free but the applied normal load is known, the tangential boundary stress can be interpreted by applying Eqs. (14.4).
6. If the boundary is not free and the applied load is not known, separation techniques discussed in Sec. 14.6 must be applied to determine the boundary stresses.

14.3 ISOCLINIC FRINGE PATTERNS [1–6]

The isoclinic fringe pattern obtained in the plane polariscope is employed primarily to give the direction of the principal stresses at any point in the model. In practice this may be accomplished in one of two ways. The first procedure is to obtain a number of isoclinic patterns at different polariscope settings and to

Figure 14.4 Isoclinic fringe patterns for a circular ring subjected to a diametral compressive load.

combine these fringe patterns to give one composite picture showing the isoclinic parameters over the entire field of the model. The second procedure is to isolate the points of interest and then to determine individually the polariscope setting, i.e., the isoclinic parameter, associated with each of these points.

An example of a series of isoclinic fringe patterns is shown in Fig. 14.4, where a thick-walled ring has been loaded in diametral compression. The data presented in this series of photographs are combined to give the composite isoclinic pattern illustrated in Fig. 14.5. Several rules can be followed in sketching the composite isoclinic patterns from the individual isoclinic fringe patterns. These rules are described below:

1. Isoclinics of all parameters must pass through isotropic or singular points.
2. An isoclinic of one parameter must coincide with an axis of symmetry in the model if an axis of symmetry exists.
3. The parameter of an isoclinic intersecting a free boundary is determined by the slope of the boundary at the point of intersection.
4. Isoclinics of all parameters pass through points of concentrated load.

An inspection of Figs. 14.4 and 14.5 will show that isoclinics of all parameters pass through the isotropic points as labeled by A through J. At isotropic points, $\sigma_1 = \sigma_2$ and all directions are principal; hence, isoclinics of all parameters must pass through these points. Again from an inspection of Fig. 14.5 it is clear that the parameter of the isoclinic can be established by the slope of the boundary at its point of intersection. The reasoning for this behavior of isoclinics at free boundaries is based on the fact that the boundaries are isostatics or stress trajectories. That is, the tangential stress at the boundary is principal, and as such the isoclinic parameter must identify the slope of the boundary at the point of intersection. The axes of symmetry cannot support shear stresses; hence, these axes are principal directions and the isoclinics must identify them. The horizontal and vertical axes of the ring shown in Fig. 14.4 are axes of symmetry and are included in the 0° isoclinic family. Finally, at the points where a concentrated load is applied, the stress system in the local neighborhood of the load is principal in the r and θ directions, where the point of load application is the center of this local circle. It is clear then that the principal-stress directions will vary from 0 to 180° in this local region, and isoclinics of all parameters will converge at the point of load application, as illustrated in Fig. 14.5.

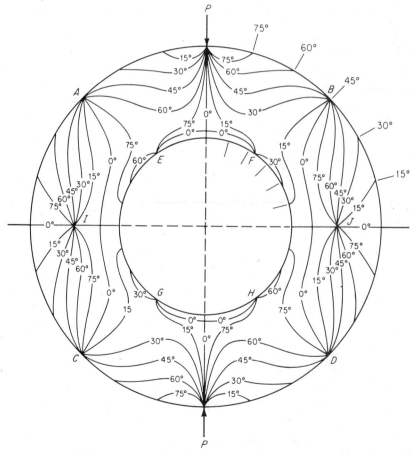

Figure 14.5 Composite isoclinic pattern for a circular ring subjected to a diametral compressive load.

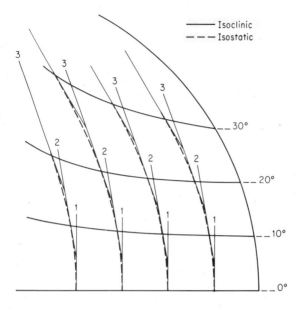

Figure 14.6 Construction technique for converting isoclinics to isostatics.

The isoclinics, lines along which the principal stresses have a constant inclination, give the principal-stress directions in a form which is not appreciated by other factions in the engineering field. As a consequence, it is normal procedure to present the principal-stress directions in the form of an isostatic or stress trajectory diagram where the principal stresses are tangent or normal to the isostatic lines at each point. The isostatic diagram can be constructed directly from the composite isoclinic pattern by utilizing the procedure outlined below and shown graphically in Fig. 14.6. In this construction technique, the stress trajectories are initiated on the 0° isoclinic from arbitrarily spaced points. Lines labeled 1 in Fig. 14.6 and oriented 0° from the normal are drawn through each of these arbitrary points until they intersect the 10° isoclinic line. The lines (1) are bisected, and a new set of lines (2) is drawn, inclined at 10° to the vertical to the next isoclinic parameter. Again these lines are bisected, and another set of construction lines (3) is drawn oriented at an angle of 20° to the vertical. This procedure is repeated until the entire field is covered. The stress trajectories are then sketched by using lines 1, 2, 3, etc., as guides. The stress trajectories are drawn tangent to the construction lines at each isoclinic intersection, as illustrated in Fig. 14.6.

The isoclinic parameters are also employed to determine the shear stresses on an arbitrary plane defined by an Oxy coordinate system. By recalling the fact that the isoclinic parameter gives the direction between the x axis of the coordinate system and the direction of σ_1 or σ_2, and referring to the Mohr's circle representation given in Fig. 1.12 and Eqs. (1.16), it is clear that

$$\tau_{xy} = -\frac{\sigma_1 - \sigma_2}{2} \sin 2\theta_2 = -\frac{Nf_\sigma}{2h} \sin 2\theta_2 \qquad (14.5)$$

where θ_2 is the angle between the x axis and the direction of σ_2 as given by the isoclinic parameter. Also

$$\tau_{xy} = \frac{\sigma_1 - \sigma_2}{2} \sin 2\theta_1 = \frac{Nf_\sigma}{2h} \sin 2\theta_1 \qquad (14.6)$$

where θ_1 is the angle between the x axis and the direction of σ_1 as given by the isoclinic parameter.

The combined isochromatic and isoclinic data represented in Eqs. (14.5) and (14.6) permit the determination of τ_{xy}. This value of τ_{xy} is used in the application of the shear-difference method (see Sec. 14.6A) for individually determining the values of σ_1 and σ_2.

14.4 COMPENSATION TECHNIQUES [7–13]

The isochromatic fringe order can be determined to the nearest $\frac{1}{2}$ order by employing both the light- and the dark-field fringe patterns. Further improvements on the accuracy of the fringe-order determination can be made by employing mixed-field patterns or by using Post's method of fringe multiplication. However, in certain instances even greater accuracies are required, and point-per-point compensation techniques are employed to establish the fringe order N. Two of these methods of compensation will be discussed here: the Babinet-Soleil method and the Tardy method.

A. The Babinet-Soleil Method of Compensation [7, 8]

In employing the Babinet-Soleil method of compensation, one inserts into the field of the polariscope, together with the model, another source of birefringence. As illustrated in Fig. 14.7a, the light passes through both the model and the Babinet-Soleil compensator. The optical effect of the superposition of these two sources of birefringence is presented in Fig. 14.7b. At a general point the state of stress on a principal element can be expressed by σ_1 and σ_2, and the optical effect, i.e., the fringe order, is proportional to $\sigma_1 - \sigma_2$. However, when the compensator is placed into the field with its axis parallel to the σ_2 direction, the optical response of the combined system can be varied by controlling the effective birefringence of the compensator. If, in particular, the birefringence of the compensator is set equal to that of the model by letting $\sigma^* = \sigma_1 - \sigma_2$ (see Fig. 14.7b), then the combined fringe order goes to zero.

The Babinet-Soleil compensator is simply a small variable wave plate which can be inserted into the field of the polariscope, oriented along either the σ_1 or the σ_2 direction, and adjusted to cancel the optical response of the model. The construction details of the Babinet-Soleil compensator are presented in Fig. 14.7a. This instrument contains a quartz plate of uniform thickness and two quartz wedges. The optical axes of the quartz crystals employed in the plate and the wedges are mutually orthogonal. Since quartz is a permanently doubly refracting

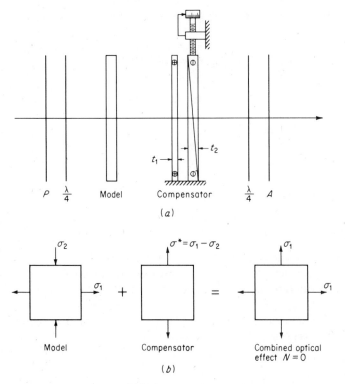

$$t_1 \longrightarrow \quad \longleftarrow t_2$$

$P \quad \dfrac{\lambda}{4} \qquad \text{Model} \qquad \text{Compensator} \qquad \dfrac{\lambda}{4} \quad A$

(a)

$$\sigma_2 \qquad\qquad \sigma^* = \sigma_1 - \sigma_2 \qquad\qquad \sigma_1$$

Model $\qquad\qquad$ + $\qquad\qquad$ Compensator $\qquad\qquad$ = $\qquad\qquad$ Combined optical
effect $N = 0$

(b)

Figure 14.7 The Babinet-Soleil compensator: (a) light passing through the model and compensator; (b) superposition of retardation exhibited by model and compensator.

material, the birefringence exhibited by the compensator can be controlled by adjusting the thickness of the two wedges by turning a calibrated micrometer screw. The resultant retardation produced by this instrument depends upon the thickness of the quartz components t_1 and t_2. When $t_1 = t_2$, the retardation imposed is zero; however, for $t_2 \gtrless t_1$, both positive and negative retardation can be imposed on the system.

In practice, a point is selected on the model where the fringe order is to be determined precisely. Next, isoclinic parameters are established for this point to give the direction of either σ_1 or σ_2. The compensator is then aligned with the principal-stress direction and adjusted to cancel out the model retardation. The reading of the screw micrometer is proportional to the fringe order at the point in question. In this manner the determination of N can be made with at least two- and possibly three-decimal-point precision.

B. The Tardy Method of Compensation [9–13]

The Tardy method of compensation is very commonly employed to determine the order of the fringe at any arbitrary point on the model. Actually, the Tardy method is often preferred over the Babinet-Soleil method since no auxiliary equipment is necessary and the analyzer of the polariscope serves as the compensating device. To employ the Tardy method, the polarizer of the polariscope is aligned with the principal direction of σ_1 at the point in question, and all other elements of the polariscope are rotated relative to the polarizer so that a standard dark-field polariscope exists. The analyzer is then rotated to produce extinction at the point. The interpretation of analyzer rotation in terms of fractional fringe order was discussed previously in Sec. 13.8.

14.5 CALIBRATION METHODS [1]

In most photoelastic analyses the stress distribution in a complex model is sought as a function of the load. To determine this stress distribution accurately requires the careful calibration of the material fringe value f_σ. Although the values of f_σ found in the technical literature are reasonably accurate, the material fringe values of photoelastic materials vary with the supplier, the batch of resin, temperature, and age. For this reason it is always necessary to calibrate each sheet of photoelastic material at the time of the test. Two methods are presented here which are recommended as both simple and accurate means of determining the material fringe value.

In any calibration technique one must select a body for which the theoretical stress distribution is accurately known. Preferably the model should also be easy to machine and simple to load. The calibration model is loaded in increments, and the fringe order and the loads are noted. From these data the material fringe value f_σ can be determined.

Consider first a tensile specimen having a width w and a thickness h, which is often employed as a calibration member. The axial stress induced in the necked region of the tensile specimen by the load P can be expressed as

$$\sigma_1 = \frac{P}{wh} \quad \text{and} \quad \sigma_2 = 0 \tag{14.7}$$

Substituting Eqs. (14.7) into Eq. (13.8) gives

$$\frac{P}{wh} = \frac{Nf_\sigma}{h} \quad \text{or} \quad f_\sigma = \frac{P}{wN} \tag{14.8}$$

TENSILE SPECIMEN

This equation shows that the value of f_σ obtained from the tensile specimen is totally independent of its thickness h. In practice, a curve of the load P is plotted as a function of N (see Fig. 14.8) for five or six different points. The slope of the straight line drawn through these points is used for the value of P/N in Eqs. (14.8) to average out small errors in the reading of P and N.

The circular disk loaded in diametral compression is also employed as a calibration model. The circular disk is somewhat easier to machine and to load than the tensile specimen; moreover, if required, several calibration points can be obtained from a single load with this type of specimen.

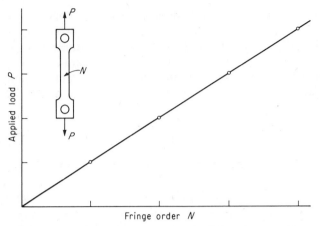

Figure 14.8 Typical calibration curve obtained by using a tensile specimen.

The stress distribution along the horizontal diameter (that is, $y = 0$) is given by

CIRCULAR DISC

$$\sigma_{xx} = \sigma_1 = \frac{2P}{\pi hD}\left(\frac{D^2 - 4x^2}{D^2 + 4x^2}\right)^2$$

$$\sigma_{yy} = \sigma_2 = -\frac{2P}{\pi hD}\left[\frac{4D^4}{(D^2 + 4x^2)^2} - 1\right]$$

$$\tau_{xy} = 0 \tag{14.9}$$

where D = diameter of disk

$\qquad x$ = distance along the horizontal diameter measured from center of disk

$\qquad h$ = thickness of disk

The difference in the principal stresses $\sigma_1 - \sigma_2$ is

$$\sigma_1 - \sigma_2 = \frac{8P}{\pi hD}\frac{D^4 - 4D^2x^2}{(D^2 + 4x^2)^2} = \frac{Nf_\sigma}{h} \tag{14.10}$$

or

$$f_\sigma = \frac{8P}{\pi DN}\frac{D^4 - 4D^2x^2}{(D^2 + 4x^2)^2} \tag{14.11}$$

Equation (14.11) can be employed to calibrate photoelastic materials if a single load P is applied to the disk. In this case the fringe order N is determined as a function of x along the horizontal diameter. These values of N and x are then substituted into Eq. (14.11) to give several values of f_σ, which in turn are averaged to reduce errors in the reading of the fringe order.

More often, however, the center point of the disk, that is, $x = y = 0$, is used for the calibration point, and several values of load are applied to the model. In this instance, Eq. (14.11) reduces to

$$f_\sigma = \frac{8P}{\pi DN} \tag{14.12}$$

Again it is noted that the value of f_σ is independent of the model thickness h. The value of P/N substituted into this equation is determined by plotting several points of P versus N and establishing the slope of this straight line.

14.6 SEPARATION METHODS [1–5, 14–21]

In the analysis of isochromatic patterns, it was shown that the principal-stress difference $\sigma_1 - \sigma_2$ could be determined directly and that the maximum shear stress could be determined provided the two principal stresses are of opposite sign. Also, at free boundaries, the principal stress normal to the boundary is zero; therefore, the isochromatic data yield directly the value of the other principal stress. At interior regions of the model, individual values for the two principal stresses cannot be obtained directly from the isochromatic and isoclinic patterns without using supplementary data or employing numerical methods. The separation methods to be discussed here will be limited to those which are commonly used and will include methods based on: (1) the equilibrium equations, (2) the compatability equations, (3) Hooke's law, and (4) photoelasticity measurements at oblique incidence.

A. Methods Based on the Equilibrium Equations [1–5]

The two methods described in this section are based solely on the equations of equilibrium and as a result are independent of the elastic constants of the photoelastic model material. The first method, commonly referred to as the *shear-difference method*, is based on a graphical integration of the equations of equilibrium as expressed by Eqs. (1.3). The second method, commonly known as *Filon's method*, involves a graphical integration of a form of the equations of equilibrium known as the Lamé-Maxwell equations. Since both these methods are based on graphical integration techniques, they suffer from the limitation that errors are accumulated as the integration proceeds. Thus, extreme care must be exercised to ensure a high degree of accuracy in the original experimental data (isochromatics and isoclinics).

The shear-difference method [1, 2, 5] The equations of equilibrium (1.3) when applied to the plane-stress problem in the absence of body forces reduce to

$$\frac{\partial \sigma_x}{\partial x} + \frac{\partial \tau_{yx}}{\partial y} = 0 \qquad \frac{\partial \sigma_y}{\partial y} + \frac{\partial \tau_{xy}}{\partial x} = 0 \qquad (14.13)$$

where σ_x, σ_y, and τ_{xy} are the normal and shear components of stress at an arbitrary point in the plane stress model under study. Solutions of the equilibrium equations can be obtained in the form

$$\sigma_x = (\sigma_x)_0 - \int \frac{\partial \tau_{yx}}{\partial y} dx \qquad \sigma_y = (\sigma_y)_0 - \int \frac{\partial \tau_{xy}}{\partial x} dy \qquad (14.14)$$

which can be closely approximated by the finite-difference expressions

$$\sigma_x = (\sigma_x)_0 - \sum \frac{\Delta \tau_{yx}}{\Delta y} \Delta x \qquad \sigma_y = (\sigma_y)_0 - \sum \frac{\Delta \tau_{xy}}{\Delta x} \Delta y \qquad (14.15)$$

In the above expressions, the terms $(\sigma_x)_0$ and $(\sigma_y)_0$ represent known stresses at points which have been selected as starting points for the integration process. Usually, these points are selected on free boundaries where the nonzero stress can be computed directly from the isochromatic data. The term τ_{xy} can be computed at any interior point of the model by using Eq. (14.6):

$$\tau_{xy} = \tfrac{1}{2}(\sigma_1 - \sigma_2) \sin 2\theta_1$$

When using this expression, care must be taken to maintain the proper algebraic sign for τ_{xy}. The expression as presented gives the sign of the shear stress in accordance with the theory-of-elasticity sign convention outlined in Chap. 1. The sign should be verified by inspection whenever possible. The term $\Delta \tau_{xy}$ is

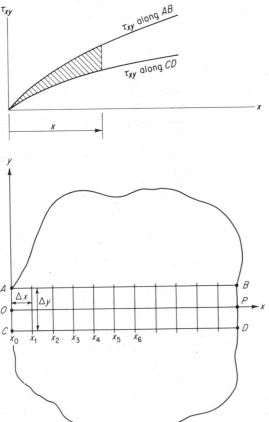

Figure 14.9 Grid system often employed in the application of the shear-difference method.

determined from a plot of the shear-stress distributions along two auxiliary lines (parallel to the line of interest and symmetrically located with respect to it), as illustrated in Fig. 14.9. From the figure it can be noted that the area (shown shaded) between the τ_{xy} curves represents the quantity $\sum \Delta \tau_{xy} \, \Delta x$. Thus, the accumulated area between the origin and a point at a distance x from the origin along the line of interest can be used to compute the difference between $(\sigma_x)_0$ and $(\sigma_x)_x$ simply by dividing by the distance Δy between the two auxiliary lines. Once σ_x is known at a given point, the value for σ_y can be computed from the expression

$$\sigma_y = \sigma_x - (\sigma_1 - \sigma_2) \cos 2\theta_1 \qquad (14.16)$$

Equations (1.8), (1.9), and (13.8) can then be combined to give the two principal stresses as follows:

$$\sigma_1 = \frac{1}{2}(\sigma_x + \sigma_y) + \frac{1}{2}(\sigma_1 - \sigma_2) = \frac{1}{2}(\sigma_x + \sigma_y) + \frac{Nf_\sigma}{2h}$$

$$\sigma_2 = \frac{1}{2}(\sigma_x + \sigma_y) - \frac{1}{2}(\sigma_1 - \sigma_2) = \frac{1}{2}(\sigma_x + \sigma_y) - \frac{Nf_\sigma}{2h} \qquad (14.17)$$

With the principal stresses and their orientation known at every point along the line, the state of stress is completely specified. The procedure can be repeated for any line of interest in the specimen.

Filon's method [1–4] Equilibrium of a small curvilinear rectangle bounded by four isostatics (stress trajectories) leads to a form of the equilibrium equations, commonly known as the Lamé-Maxwell equations, which can be expressed as

$$\frac{\partial \sigma_1}{\partial s_1} = -\frac{\sigma_1 - \sigma_2}{\rho_2} \qquad \frac{\partial \sigma_2}{\partial s_2} = -\frac{\sigma_1 - \sigma_2}{\rho_1} \qquad (14.18)$$

where s_1 and s_2 are orthogonal curvilinear coordinates measured along the σ_1 and σ_2 isostatics, respectively, and ρ_1 and ρ_2 are the respective radii of curvature of these isostatics. When the isostatics are accurately known, this form of the equilibrium equations is easier to use than the cartesian form. Integrating along one of the isostatics yields

$$\sigma_1 = (\sigma_1)_0 - \int \frac{\sigma_1 - \sigma_2}{\rho_2} \, ds_1 \qquad \sigma_2 = (\sigma_2)_0 - \int \frac{\sigma_1 - \sigma_2}{\rho_1} \, ds_2 \qquad (14.19)$$

Integration along an isostatic line can be performed by starting at a point of known stress, usually on a free boundary, and evaluating the integral from a plot of $(\sigma_1 - \sigma_2)/\rho_2$ versus s_1 or $(\sigma_1 - \sigma_2)/\rho_1$ versus s_2. The change in principal stress between the origin and the point of interest along the isostatic is the accumulated area under the curve. This method is certainly the most convenient when the problem of interest involves rotational symmetry. In this case, since the isostatics are concentric circles and radial lines, the isochromatics alone are sufficient for the integration. For an integration along a radial line, the radius of curvature needed for the integration is simply the distance from the axis of symmetry to the point under consideration. The appropriate plot for evaluating the integral would be $(\sigma_1 - \sigma_2)/r$ versus r.

In the more general case of integration along an isostatic in an arbitrary stress field, the radius-of-curvature data are difficult to obtain with sufficient accuracy from photoelastic observations of the isoclinics. Brittle-coating isostatics can also provide the radii of curvatures, but this procedure is not recommended since more accurate methods are discussed in the subsections which follow.

B. Methods Based on the Compatibility Equations [14–17]

The compatibility or continuity equation for plane stress or plane strain in terms of cartesian stress components and with constant or zero body forces can be expressed in terms of the first invariant of stress as

$$\frac{\partial^2}{\partial x^2} (\sigma_{xx} + \sigma_{yy}) + \frac{\partial^2}{\partial y^2} (\sigma_{xx} + \sigma_{yy}) = 0 \tag{14.20}$$

Equations of this form are known as *Laplace's equation*, and any function which satisfies this equation is said to be a *harmonic function*. In photoelasticity, interest in Laplace's equation arises from the fact that the value of the function is uniquely determined at all interior points of a region if the boundary values are known. It was shown previously that the photoelastic isochromatics provide an accurate means for determining both the principal-stress difference $\sigma_1 - \sigma_2$ at all interior points of a two-dimensional model and, in many instances, complete boundary-stress information. Knowledge of the principal-stress sum $\sigma_1 + \sigma_2$ throughout the interior, together with the principal-stress difference $\sigma_1 - \sigma_2$ provides an effective means for evaluating the individual principal stresses.

Rigorous mathematical solution of Laplace's equation is possible only in cases where the boundary is relatively simple. Approximate solutions, which are sufficiently accurate for all practical work, can be obtained by numerical and experimental means. The Laplace equation serves as the governing equation in many other fields of engineering. Included are electrostatic fields in regions enclosed by boundaries at known potential, steady-state temperature distributions, and shapes of uniformly stretched films or membranes. Since the behavior of these different physical systems can be expressed in the same mathematical form, the one permitting the easier form of measurement can be used to study behavior in any of the other systems. These types of measurement methods are referred to as *analogy methods*.

In this section, three methods will be discussed for solving the Laplace equation. They include an analytic method, a numerical method, and an analogy method. The most important limitation of the methods described in this section is the requirement for complete knowledge of the boundary-stress distribution.

The analytic separation method [14] Solution of Laplace's equation by the method of separation of variables yields a sequence of harmonic functions which can be added together in a linear combination to give a series representation H of the first

Table 14.1 Solution of $\nabla^2 H = 0$ in various coordinate systems

Coordinate system	Sequence of harmonic functions		

Cartesian

$$\frac{\partial^2 H}{\partial x^2} + \frac{\partial^2 H}{\partial y^2} = 0$$

1	sinh kx sin ky	sinh ky sin kx
x	sinh kx cos ky	sinh ky cos kx
y	cosh kx sin ky	cosh ky sin kx
xy	cosh kx cos ky	cosh ky cos kx

Polar

$$\frac{\partial^2 H}{\partial r^2} + \frac{1}{r}\frac{\partial H}{\partial r} + \frac{1}{r^2}\frac{\partial^2 H}{\partial \theta^2} = 0$$

$x = r\cos\theta \qquad y = r\sin\theta$

1	$r^n \cos n\theta$	$r^n \sin n\theta$
$\ln r$	$r^{-n}\cos n\theta$	$r^{-n}\sin n\theta$

Bipolar

$$\frac{\partial^2 H}{\partial \alpha^2} + \frac{\partial^2 H}{\partial \beta^2} = 0$$

$x = \dfrac{C\sin\beta}{\cosh\alpha - \cos\beta} \qquad y = \dfrac{C\sinh\alpha}{\cosh\alpha - \cos\beta}$

1	sinh $n\alpha$ sin $n\beta$	cosh $n\alpha$ sin $n\beta$
α	sinh $n\alpha$ cos $n\beta$	cosh $n\alpha$ cos $n\beta$

Elliptic

$$\frac{\partial^2 H}{\partial u^2} + \frac{\partial^2 H}{\partial v^2} = 0$$

$x = c\cosh u\cos v \qquad y = c\sinh u\sin v$

1	sinh nu sin nv	cosh nu sin nv
u	sinh nu cos nv	cosh nu cos nv

Spherical

$$\frac{\partial^2 H}{\partial r^2} + \frac{2}{r}\frac{\partial H}{\partial r} + \frac{1}{r^2}\frac{\partial^2 H}{\partial \phi^2} + \frac{1}{r^2 \tan\phi}\frac{\partial H}{\partial \phi} + \frac{1}{r^2 \sin^2\phi}\frac{\partial^2 H}{\partial \theta^2} = 0$$

$x = r\sin\phi\cos\theta \qquad y = r\sin\phi\sin\theta \qquad z = r\cos\phi$

$r^m P_m^m \cos\phi\sin n\theta$

$r^{-m-1}P_m^m \cos\phi\sin n\theta \qquad$ where P is a Legendre function

$r^m P_m^m \cos\phi\cos n\theta$

$r^{-m-1}P_m^m \cos\phi\cos n\theta$

stress invariant I. Solutions for H referred to several coordinate systems are presented in Table 14.1.

If the region of the model conforms to a regular coordinate system, determination of the proper coefficient for each function is considerably simplified. The sequence of functions when evaluated on the boundaries of the region reduces to a Fourier series. The unknown coefficients in these cases are the Fourier coefficients obtained by integrating the prescribed boundary values.

If the region of the model does not conform to a particular coordinate system, Fourier analysis cannot be employed to determine the coefficients which satisfy the prescribed boundary conditions. Instead, the method of least squares is used for the determination of the coefficients. A finite number of harmonic functions is first selected to appear in the series solution. Coefficients are then chosen such that the mean-square difference between the prescribed boundary values and the evaluation of the series along the boundary is minimized. If N harmonic functions F_1, $F_2, \ldots, F_N$ are selected and the associated unknown coefficients are denoted as C_1, $C_2, \ldots, C_N$, the series solution for the first stress invariant is

$$H = \sum_{n=1}^{N} C_n F_n \tag{14.21}$$

If $I(s)$ is used to represent the distribution of the first stress invariant along a boundary of total length L, H must be selected such that

$$\int_0^L \left[I(s) - \sum_{n=1}^{N} C_n F_n \right]^2 ds = \text{minimum} \tag{14.22}$$

The N unknown coefficients of this series can be evaluated by taking the partial derivative of the integral with respect to each of the coefficients and setting the resulting expressions equal to zero. Thus

$$\frac{\partial}{\partial C_k} \int_0^L \left[I(s) - \sum_{n=1}^{N} C_n F_n \right]^2 ds = 0 \qquad k = 1, 2, \ldots, N$$

which can be reduced to

$$\sum_{n=1}^{N} C_n \int_0^L F_n F_k \, ds = \int_0^L I(s) F_k \, ds \qquad k = 1, 2, \ldots, N \tag{14.23}$$

Equation (14.23) yields N simultaneous equations in terms of the N unknown coefficients. Solution of this set of equations gives the coefficients which provide the best match of boundary values possible with the initial selection of N harmonic functions. By increasing the number of functions in the series for H, the fit can be made as accurate as the original photoelastic determination of $I(s)$.

The four-point influence method [15] Numerical methods can be used very effectively to solve Laplace's equation. The method to be described here utilizes an iteration procedure by which estimated values of the harmonic function at points of interest of a network are systematically improved by making use of the fact that the value of the function at any point depends upon the values of the function at

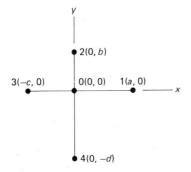

Figure 14.10 Network points in the general case.

neighborhood points. The basic relationship between values of the function at different points can be expressed by the four-point influence equation as

$$\Phi_0 = C_1\Phi_1 + C_2\Phi_2 + C_3\Phi_3 + C_4\Phi_4 \qquad (14.24)$$

where neighborhood points in the most general case are located as shown in Fig. 14.10 and the constants C_1, C_2, C_3, and C_4 have the following values:

$$C_1 = \frac{bcd}{(bd + ac)(a + c)} \qquad C_2 = \frac{acd}{(bd + ac)(b + d)}$$

$$C_3 = \frac{abd}{(bd + ac)(c + a)} \qquad C_4 = \frac{abc}{(bd + ac)(d + b)} \qquad (14.25)$$

In regions where a square array or network of points can be used, the computations are simplified since $C_1 = C_2 = C_3 = C_4 = \frac{1}{4}$.

Once the grid network has been established and the constants associated with each point evaluated, the known boundary values can be assigned to all points of intersection of the network with the boundary and estimated values or zero can be assigned to interior points. The value of each interior point is then improved by traversing the network in a definite sequence and using Eq. (14.24). Each time the network is traversed, the values are improved. The process is continued until the values become stationary or until further corrections do not alter the values more than a predetermined amount. At this stage in the procedure, further accuracy can be obtained only by using a finer-grid network.

In general, the labor involved in the process can be reduced by initially selecting a coarse network to establish reasonable values for the interior points. A fine network can then be introduced in selected regions (areas of high stress gradient) to improve the accuracy of the determinations in these regions.

The method suffers from the limitation that complete boundary-stress data must be available. Also, the method cannot be used to evaluate stress distributions along selected lines of interest without performing the evaluation for the complete model. Only symmetry considerations can be used to effect a reduction in the number of network points to be evaluated. The method has the advantage that isochromatic data alone are sufficient for the determinations. The method also has the advantage that errors made during the iteration process have no influence on the final results. Errors simply increase the number of iterations required to obtain stationary values at interior points.

The electrical-analogy method [16, 17] The electrical-analogy method makes use of the fact that the voltage distribution in a uniformly conducting region enclosed by boundaries at known potential is governed by the Laplace equation in the same manner as the principal-stress sum in a plane-stress situation. Since means are readily available for applying and measuring voltage distributions, electrical measurements provide an excellent means for determining principal-stress sums in the interior of any complicated two-dimensional model.

An electrical model of the same geometry as the photoelastic model is prepared from a uniformly conducting material. Teledeltos paper, which consists of a uniform layer of graphite particles over a thin paper carrier, is a very suitable material from which to fabricate the electrical model. The Teledeltos paper, produced in widths up to 36 in (914 mm) and in lengths of several hundred feet (approximately 100 meters) is available with two different resistance values: 20 and 80 kΩ/in^2 (30 and 120 MΩ/m^2). Voltages applied to the boundary of the electrical model are proportional to the value of I_1 determined on the boundary of the photoelastic model. The values of I_1 on the boundary of the photoelastic model can be established from the isochromatic fringe pattern by using

$$I_1 = \sigma_1 + \sigma_2 = \sigma_1 - \sigma_2 = \frac{N f_\sigma}{h} \tag{14.26}$$

since on a free boundary B either σ_1 or $\sigma_2 = 0$. The voltages applied to the boundary of the electrical model are of both a positive and a negative value, so that care should be exercised in applying the correct sign to the magnitude of the values obtained from Eq. (14.26). A typical circuit diagram illustrating the method

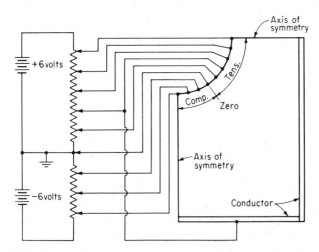

Figure 14.11 Electrical circuit for applying voltages to a Teledeltos paper model.

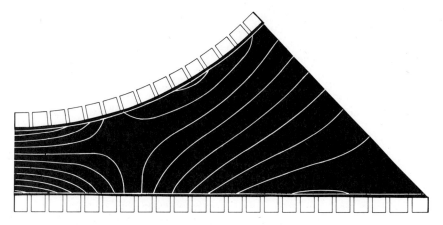

Figure 14.12 Isopachic pattern for a section of square tubing with a pressurized circular bore obtained using Teledeltos paper.

of applying voltages to the boundary of a Teledeltos paper model is illustrated in Fig. 14.11.† The graphite surface of the model is then probed with a high-impedance voltmeter to establish lines of constant voltage on the interior regions of the model. These lines of constant voltage are analogous to the isopachic lines where $\sigma_1 + \sigma_2$ = constant. An example of the constant-voltage lines obtained for a section of square tubing with a pressurized circular bore is illustrated in Fig. 14.12. Isochromatic patterns for the same model were previously shown in Fig. 14.3.

The electrical-analogy method for separating principal stresses is one of the simplest, most rapid, and most accurate of the numerous techniques available. The equipment and skills required in applying the method are quite modest, and excellent results can quickly be achieved.

C. Methods Based on Hooke's Law [18, 19]

Separation methods based on Hooke's law make use of the fact that the strain in a direction perpendicular to the surface of a plane-stress model can be expressed as

$$\epsilon_{zz} = \frac{\Delta h}{h} = -\frac{v}{E}(\sigma_{xx} + \sigma_{yy}) = -\frac{v}{E}(\sigma_1 + \sigma_2)$$

or

$$\sigma_1 + \sigma_2 = -\frac{E}{vh}\Delta h \tag{14.27}$$

From Eq. (14.27) it is obvious that the sum of the principal stresses can be determined if the change in thickness of the model, as a result of the applied loads, can be measured accurately at the point of interest. The procedure requires a

---

† It should be noted here that this method can also be employed in heat transfer to solve the equation $\nabla^2 T = 0$. The constant-voltage lines in this case are analogous to isothermal lines.

device with high sensitivity since the thickness changes are seldom more than a few thousandths of an inch. Instruments which have been developed to make these measurements include lateral extensometers and interferometers.

Lateral extensometers [18] Lateral extensometers can be described as extremely sensitive micrometers or calipers. Quite often, the increased sensitivity is obtained by using delicate mechanical- or optical-lever systems. Specialized designs also exist which permit the use of electrical-resistance strain gages or linear differential transformers as the sensing elements. Extensometers are difficult to position and use if the geometry of the model is at all complicated. This method is recommended only for special situations where limited point data are desired.

Optical interferometers [19] Optical interference between rays of light reflected from a model surface and from an auxiliary optical flat placed close to the model surface provides an accurate means for determining the air gap between the two

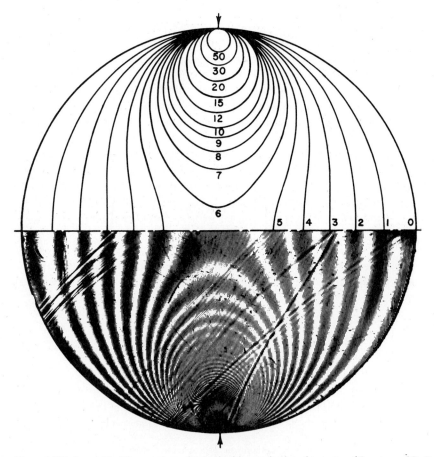

Figure 14.13 Isopachic fringe pattern obtained with a series interferometer. (*Courtesy of D. Post.*)

surfaces. Since the interference fringes which develop are loci of points of constant air-film thickness, they can be used to establish changes in model thickness and hence the sum of the principal stresses. Satisfactory application of this method requires an adequate means for locating and fixing the position of the optical flat with respect to the model and a model surface which is nearly optically flat so that an undesirable initial fringe pattern can be avoided. This second condition is often difficult to achieve.

Other interferometer systems which use transmitted rather than reflected light in their operation have been adapted for use in the determination of principal-stress sums. One such instrument, the series interferometer, was designed for this purpose and as a result is a simple and stable unit with a large-diameter field. The series interferometer and its operation were discussed in Sec. 11.7. A typical iso-pachic fringe pattern obtained with this instrument is shown in Fig. 14.13. Development of more elaborate systems, together with extensive use of the laser with its coherent and monochromatic light, has made optical interferometry a valuable experimental stress-analysis method for other applications.

D. Oblique-Incidence Methods [20, 21]

The equation developed for the stress optic law [Eq. (13.8)] was based on the light's passing through the model at normal incidence. However, if the model is rotated in the polariscope so that the light passes through the model at some other angle, an oblique-incidence fringe pattern can be observed. This oblique-incidence fringe pattern provides additional data which can be employed to separate the principal stresses.

Consider first the case where the principal stress directions are known and rotate the model about the σ_1 axis by an amount θ, as shown in Fig. 14.14. The light passes through the plane of the model obliquely and traverses a distance of $h/\cos\theta$ through the model. The fringe pattern produced is related to the secondary principal stresses lying in the plane normal to the axis of the light, which in

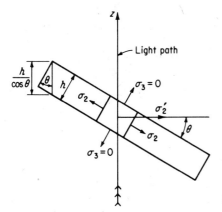

Figure 14.14 Rotation of the model about the σ_1 axis.

this case is the plane containing σ_1 and σ_2'. Thus

$$\sigma_1 - \sigma_2' = \frac{f_\sigma N_\theta}{h/(\cos \theta)} \tag{14.28}$$

where N_θ is the fringe order associated with the oblique-incidence fringe pattern. By applying Eqs. (1.16), it is clear that

$$\sigma_2' = \sigma_2 \cos^2 \theta$$

Hence

$$\sigma_1 - \sigma_2 \cos^2 \theta = \frac{f_\sigma N_\theta}{h/(\cos \theta)} \tag{14.29}$$

By combining Eq. (14.29) (oblique incidence) with Eq. (13.8) (normal incidence), it can be shown that

$$\sigma_1 = \frac{f_\sigma}{h} \frac{\cos \theta}{\sin^2 \theta} (N_\theta - N_0 \cos \theta) \qquad \sigma_2 = \frac{f_\sigma}{h} \frac{1}{\sin^2 \theta} (N_\theta \cos \theta - N_0) \tag{14.30}$$

where N_0 is the fringe order associated with the normal-incidence pattern.

By employing Eqs. (14.30) together with isochromatic fringe patterns from one normal- and one oblique-incidence photograph, it is possible to separate the principal stresses. This approach is often used to separate stresses along a line of symmetry where one rotation of the model about the line of symmetry provides sufficient data to separate the stresses along the entire length of the line.

Next consider the case where the principal-stress directions are not known and rotation of the model is made about arbitrary axes such as Oy and Ox.

In this instance three fringe patterns are obtained: the normal-incidence pattern, the oblique-incidence pattern corresponding to a rotation about Oy, and an oblique-incidence pattern corresponding to a rotation about Ox. The normal-incidence pattern gives, by Eqs. (13.8) and (1.12),

$$N_0 = \frac{h}{f_\sigma} (\sigma_1 - \sigma_2) = \frac{h}{f_\sigma} \sqrt{(\sigma_{xx} - \sigma_{yy})^2 + 4\tau_{xy}^2} \tag{a}$$

The oblique-incidence fringe pattern associated with a model rotation about Oy gives

$$N_{\theta y} = \frac{h}{\cos \theta_y} \frac{\sigma_1' - \sigma_2'}{f_\sigma} \tag{b}$$

where σ_1' and σ_2' are secondary principal stresses in the $y'x'$ plane which is normal to the incident light along the z' axis. From Eqs. (1.12) it is clear that

$$\sigma_1' - \sigma_2' = \sqrt{(\sigma_{x'x'} - \sigma_{y'y'})^2 + 4\tau_{x'y'}^2} \tag{c}$$

and from Eqs. (1.6)

$$\sigma_{x'x'} = \sigma_{xx} \cos^2 \theta_y \qquad \sigma_{y'y'} = \sigma_{yy} \qquad \tau_{x'y'} = \tau_{xy} \cos \theta_y \tag{d}$$

Combining Eqs. (b) to (d) yields

$$N_{\theta y} = \frac{h}{f_\sigma \cos \theta_y} \sqrt{(\sigma_{xx} \cos^2 \theta_y - \sigma_{yy})^2 + 4\tau_{xy}^2 \cos^2 \theta_y} \qquad (e)$$

In a similar manner it can be shown that the oblique-incidence pattern associated with a rotation about Ox yields

$$N_{\theta x} = \frac{h}{f_\sigma \cos \theta_x} \sqrt{(\sigma_{xx} - \sigma_{yy} \cos^2 \theta_x)^2 + 4\tau_{xy}^2 \cos^2 \theta_x} \qquad (f)$$

If the rotation is controlled so that $\theta_x = \theta_y = \theta$, Eqs. (a), (e), and (f) reduce to

$$\frac{N_0^2 f_\sigma^2}{h^2} = \sigma_{xx}^2 - 2\sigma_{xx}\sigma_{yy} + \sigma_{yy}^2 + 4\tau_{xy}^2$$

$$\frac{N_{\theta y}^2 \cos^2 \theta f_\sigma^2}{h^2} = \sigma_{xx}^2 \cos^4 \theta - 2\sigma_{xx}\sigma_{yy} \cos^2 \theta + \sigma_{yy}^2 + 4\tau_{xy}^2 \cos^2 \theta$$

$$\frac{N_{\theta x}^2 \cos^2 \theta f_\sigma^2}{h^2} = \sigma_{xx}^2 - 2\sigma_{xx}\sigma_{yy} \cos^2 \theta + \sigma_{yy}^2 \cos^4 \theta + 4\tau_{xy}^2 \cos^2 \theta \qquad (g)$$

Solving these equations for σ_{xx} or σ_{yy} yields

$$\left(\frac{\sigma_{xx} h}{f_\sigma}\right)^2 = \frac{\cot^2 \theta}{1 - \cos^4 \theta}[N_{\theta x}^2 + N_{\theta y}^2 \cos^2 \theta - N_0^2(1 + \cos^2 \theta)]$$

$$\left(\frac{\sigma_{yy} h}{f_\sigma}\right)^2 = \frac{\cot^2 \theta}{1 - \cos^4 \theta}[N_{\theta y}^2 + N_{\theta x}^2 \cos^2 \theta - N_0^2(1 + \cos^2 \theta)] \qquad (14.31)$$

Solutions of Eqs. (14.31) for σ_{xx} and σ_{yy} are sufficient to permit separation of the principal stresses. Addition of σ_{xx} and σ_{yy} gives the value of the first invariant of stress $I_1 = \sigma_1 + \sigma_2$. All other quantities associated with this two-dimensional state of stress can be obtained from the known quantities $\sigma_1 - \sigma_2, \sigma_1 + \sigma_2, \sigma_{xx}$, and σ_{yy} by employing the relations presented in Secs. 1.5 to 1.7.

14.7 SCALING MODEL-TO-PROTOTYPE STRESSES [22–25]

In the analysis of a photoelastic model fabricated from a polymeric material, the question of the applicability of the results is often raised since the prototype is usually a metallic material. Obviously, the elastic constants of the photoelastic model are greatly different from those of the metallic prototype. However, the stress distribution obtained for a plane-stress or plane-strain problem by a photoelastic analysis is usually independent of the elastic constants, and the results can be applied to a prototype constructed from any material. This statement can be established most readily by reference to the stress equation of compatibility for the plane-stress case. Thus by Eqs. (3.15) and (3.17c) it is clear that

$$\nabla^2(\sigma_{xx} + \sigma_{yy}) = -(v + 1)\left(\frac{\partial F_x}{\partial x} + \frac{\partial F_y}{\partial y}\right) \qquad (14.32)$$

This stress equation of compatibility is independent of the modulus of elasticity E and thus shows that the value of the model modulus does not influence the stress distribution. The influence of the other elastic constant v (Poisson's ratio) depends on the nature of the body-force distribution. If $\partial F_x/\partial x + \partial F_y/\partial y = 0$, the stress distribution is independent of Poisson's ratio. This statement implies that there will be no influence due to Poisson's ratio when

1. $F_x = F_y = 0$ (the absence of body forces)
2. $F_x = C_1, F_y = C_2$ (the uniform body-force field, i.e., gravitational)
3. $F_x = C_1 x, F_y = -C_1 y$ (a linear body-force field in x and y)

There are two exceptions to this general law of similarity of stress distributions in two-dimensional parts. First, if the two-dimensional photoelastic model is multiply connected, Eq. (14.32) does not apply. In this case the multiply connected body has a hole or series of holes, and the influence of Poisson's ratio will depend upon the nature of the loading on the boundary of the hole. If the resultant force acting on the boundary of the hole is zero, the stress distribution will again be independent of Poisson's ratio. However, if the resultant force applied to the boundary of the hole is not zero, the value of Poisson's ratio will influence the distribution of the stresses. Fortunately, in specific examples of this type where the effect of Poisson's ratio has been evaluated, its influence on the maximum principal stress is usually less than about 7 percent.

The second exception to the laws of similitude is the case where the photoelastic model undergoes appreciable distortion under the action of the applied load. Local distortions are a source of error in notches, for example, since curvatures are modified and the stress-concentration factors are decreased. These model distortions can be minimized by selecting a model material with a high figure of merit and reducing the applied load to the lowest value consistent with adequate model response.

Since the photoelastic model may differ from the prototype in respect to scale, thickness, and applied load, as well as the elastic constants, it is necessary to extend this treatment to include the scaling relationships. A great deal has been written concerning scaling relationships employing dimensionless ratios and the Buckingham π theory; however, in most photoelastic applications, scaling the stresses from the model to the prototype is a relatively simple matter where the pertinent dimensionless ratios can be written directly. For instance, in the case of a two-dimensional model with applied loads P, the dimensionless ratio for stresses is $\sigma h l/P$ and for displacements $\delta E h/P$. Thus the prototype stresses can be written as

$$\sigma_p = \sigma_m \frac{P_p}{P_m} \frac{h_m}{h_p} \frac{l_m}{l_p} \qquad (14.33)$$

and the prototype displacements as

$$\delta_p = \delta_m \frac{P_p}{P_m} \frac{E_m}{E_p} \frac{h_m}{h_p} \qquad (14.34)$$

where σ = stress at given point
  δ = displacement at given point
  P = applied load
  h = thickness
  l = typical length dimension
and subscripts p and m refer to the prototype and the model, respectively.

In conclusion, it is clear that scaling between model and prototype can be accomplished in most two-dimensional problems encountered by the photoelastician. The modulus of elasticity is never a consideration in determining the stress distribution unless the loading is such that model deformations change the load distribution, e.g., contact stresses. Also, Poisson's ratio need not be considered when the body is simply connected and the body-force field is either absent or uniform, i.e., dead-weight loading.

14.8 MATERIALS FOR TWO-DIMENSIONAL PHOTOELASTICITY

One of the most important factors in a photoelastic analysis is the selection of the proper material for the photoelastic model. Unfortunately, a perfectly ideal photoelastic material does not exist, and the investigator must select from the list of available materials the one which most closely fits his needs. The quantity of photoelastic plastic used each year is not sufficient to entice a chemical company into the development and subsequent production of a polymeric material especially designed for photoelastic applications. As a consequence, the photoelastician must select a model material which is commerically available for some purpose other than photoelasticity.

The following list gives properties which an ideal photoelastic material should exhibit. These criteria are discussed individually below.

1. The material must be transparent to the light employed in the polariscope.
2. The material should be quite sensitive to either stress or strain, as indicated by a low material fringe value in terms of either stress f_σ or strain f_ϵ.
3. The material should exhibit linear characteristics with respect to
   a. Stress-strain properties
   b. Stress–fringe-order properties
   c. Strain–fringe-order properties
4. The material should have both mechanical and optical isotropy and homogeneity.
5. The material should not exhibit time-dependent properties such as creep.
6. The material should exhibit a high modulus of elasticity and a high proportional limit.
7. The material sensitivity, that is, f_σ or f_ϵ, should not change markedly with small variations in temperature.
8. The material should not exhibit time-edge effects.

9. The material should be capable of being machined by conventional means.
10. The material should be free of residual stresses.
11. The material should not be prohibitively expensive.

A. Transparency

In almost all normal applications the materials selected for photoelastic models are transparent plastics. These plastics must be transparent to visible light, but they need not be crystal clear. This transparency requirement is not difficult to meet since most polymeric materials are colored or made opaque by the addition of fillers. The raw materials in the basic polymer, although not crystal clear, are usually transparent.

In certain special applications which require a study of the stresses in normally opaque materials, e.g., germanium or silicon, an infrared polariscope has been used. A few materials are transparent in either the ultraviolet region or the infrared region of the radiant-energy spectrum. Polariscopes can be constructed to operate in either of these regions if advantages can be gained by employing light with very short or very long wavelengths. However, for stress-analysis purposes the visible-light polariscopes are quite adequate.

B. Sensitivity

A highly sensitive photoelastic material is often desirable since it increases the number of fringes which can be observed in the model. If the value of f_σ for a model material is low, a satisfactory fringe pattern can be achieved in the model with relatively low loads. This feature reduces the complexity of the loading fixture and limits the distortion of the model. In the case of birefringent coatings, which will be discussed later, a material with a low value of f_ϵ is essential to reduce errors introduced by the coating thicknesses.

Photoelastic materials are available with values of f_σ which range from less than 0.2 to over 2000 lb/in (0.035 to 350 kN/m). The situation regarding values of f_ϵ is not so satisfactory since materials with a sufficiently low value of f_ϵ are not yet available (f_ϵ usually ranges between 0.0002 and 0.02 in or from 0.005 to 0.50 mm). A material with a value of $f_\epsilon = 0.00002$ in (0.0005 mm) would greatly enhance the applicability of the birefringent coating method of photoelasticity.

C. Linearity

Photoelastic models are normally employed to predict the stresses which occur in a metallic prototype. Since model-to-prototype scaling must be used to establish prototype stresses, the model material must exhibit linear stress-strain, optical-stress, and optical-strain properties. Very few data are available in the open literature on optical-strain relationships; however, since the photoelastic method is usually employed to determine stress differences, this lack of data on strain behavior is not considered serious. Typical stress-strain curves and stress–fringe-order

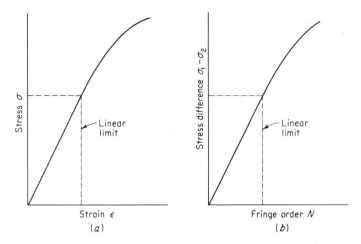

Figure 14.15 Typical (*a*) stress-strain and (*b*) stress-optical response curves for a polymeric photoelastic material.

curves are presented in Fig. 14.15 to show the characteristic behavior of polymeric photoelastic materials. Most polymeric materials exhibit linear stress-strain and stress–fringe-order curves for the initial portion of the curve. However, at higher levels of stress the material may exhibit nonlinear effects. For this reason the higher stress levels are to be avoided in photoelastic tests associated with these nonlinear materials.

D. Isotropy and Homogeneity

Most photoelastic materials are prepared from liquid polymers by casting between two glass plates which form the mold. When the photoelastic materials are prepared by a casting process, the molecular chains of the polymer are randomly oriented and the materials are essentially isotropic and homogeneous. However, certain plastics are rolled or stretched during the production process. In both of these production processes the molecular chains are oriented in the direction of rolling or stretching. These materials will exhibit anisotropic properties (both mechanical and optical) and should therefore be avoided in any photoelastic application except those where anisotropic material properties are required.

E. Creep

Unfortunately, most photoelastic materials of a polymeric base creep both mechanically and optically over the time associated with a photoelastic analysis. Because of the effects of mechanical and optical creep, polymeric materials cannot be truly characterized as elastic materials but must be considered viscoelastic.

One of the first attempts to formulate a mathematical theory of photo-viscoelasticity was made by Mindlin [26] by considering a generalized viscoelastic model consisting of m elastic elements with a shear modulus G_k ($k = 1, 2, 3, \ldots, m$) and m viscous elements with a viscosity coefficient η_k ($k = 1, 2, 3, \ldots, m$) (see Fig. 14.16). Assuming that the photoelastic effect results from only the deformation of the elastic elements of the model, Mindlin showed that the relative retardation, expressed as $n_1 - n_2$, could be related to the stress and strain as

$$(n_1 - n_2) \cos 2\theta_n = R[(\sigma_1 - \sigma_2) \cos 2\theta_\sigma] + 2S[(\epsilon_1 - \epsilon_2) \cos 2\theta_\epsilon] \quad (14.35)$$

where $n_1 - n_2$ = relative retardation

$\theta_n, \theta_\sigma, \theta_\epsilon$ = angles between principal optical, principal stress, principal strain, and x axis, respectively

R, S = linear operators of types relating stress and strain in viscoelastic theory

The operators R and S depend upon the viscoelastic model.

For the three-element elastic model shown in Fig. 14.16, which exhibits instantaneous and delayed elasticity but no flow ($\eta_3 = \infty$), the operators R and S can be expressed as

$$R = \frac{n_0^3}{4G_1}(c_1 - c_2) \qquad S = \frac{n_0^3}{2}c_2 \quad (14.36)$$

These equations show that the birefringence is due to both stress and strain but is independent of stress rate and strain rate. Coker and Filon [3, p. 272] have found that the material xylonite follows this particular viscoelastic model.

It is evident from these results that the optical response exhibited by a photoelastic model depends upon the viscoelastic properties of the model material. In

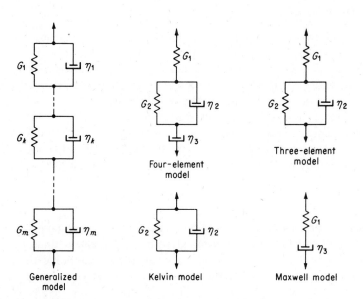

Generalized model

Four-element model

Three-element model

Kelvin model

Maxwell model

Figure 14.16 Generalized model and elementary viscoelastic models.

the most general case, the retardation can be a function of stress, strain, strain rate, and stress rate. Fortunately, most of the polymers used in photoelasticity are linearly viscoelastic. With linearly viscoelastic materials, the stress and strain, which vary with both position and time, can be represented by the product of two functions: one in space coordinates only and the other in a time coordinate only:

$$\sigma^*(x, y, t) = \sigma(x, y)f(t) \qquad \epsilon^*(x, y, t) = \epsilon(x, y)g(t) \qquad (14.37)$$

When Eqs. (14.37) hold, it can be shown that

$$\theta_n = \theta_\sigma = \theta_\epsilon \qquad n_2 - n_1 = C(t)(\sigma_1 - \sigma_2) \qquad n_2 - n_1 = c(t)(\epsilon_1 - \epsilon_2)$$
$$(14.38)$$

where
$$C(t) = R[f(t)] + \frac{1}{G_0} S[g(t)] \qquad c(t) = G_0 R[f(t)] + S[g(t)]$$

and
$$G_0 = \frac{\sigma_1 - \sigma_2}{2(\epsilon_1 - \epsilon_2)}$$

This series of equations implies that Eqs. (13.8) and (13.12) can be rewritten in the following forms:

$$\sigma_1 - \sigma_2 = \frac{N}{h} f_\sigma(t) \qquad \epsilon_1 - \epsilon_2 = \frac{N}{h} f_\epsilon(t) \qquad (14.39)$$

where f_σ and f_ϵ are written as functions of time rather than as constants.

The results of Eqs. (14.39) are significant since they show that viscoelastic model materials can be employed to perform elastic-stress analyses. Because of the viscoelastic nature of the model materials, the stress and strain material fringe values are functions of time; however, over the short time intervals needed to photograph a fringe pattern they can be considered as constants. A typical plot of the variation in material fringe value with time is shown in Fig. 14.17. For most

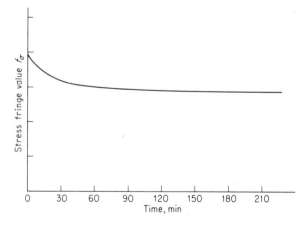

Figure 14.17 Typical curve showing the time-after-loading dependence of the stress fringe value in a viscoelastic polymeric material.

model materials, the stress fringe value f_σ decreases rapidly with time immediately after loading but then tends to stabilize after about 1 h. In practice, the load is maintained on the model until a stable fringe pattern is achieved. The pattern is then photographed, and the stable material fringe value associated with the time of the photograph is used for the analysis. It should be noted that photoelastic materials vary from supplier to supplier and from batch to batch; hence, each sheet of material must be calibrated at the time of the photoelastic analysis to determine $f_\sigma(t)$. Also, in certain photoelastic materials, the polymerization process continues, and f_σ changes with time on a scale of months. Hence, a piece of material stored for a year will, in general, exhibit a higher value of f_σ than the fresh material.

F. Modulus of Elasticity and Proportional Limit

The modulus of elasticity is important in the selection of a photoelastic material because the modulus controls the distortion of the model due to the applied stresses. If a model distorts appreciably, the geometry of its boundary will change and the photoelastic solution is no longer accurate. Errors of considerable magnitude are produced by model distortion, where small changes in the boundary contour are influential in determining the stress distribution. For example, a strip with a very sharp notch theoretically will have a very high stress concentration at the root of the notch. If the model distorts under load, the sharpness of the notch is decreased and the experimentally determined stress concentration is reduced. Another case which exists where model distortion influences the photoelastically determined results is the contact problem. In this instance the stress distribution is a function of the penetration and bearing area of the loading point. Obviously, the modulus of elasticity is influential in controlling both of these quantities.

The factor which can be used to judge the various photoelastic materials in regard to their ability to resist distortion is $1/f_\epsilon$ or $E/f_\sigma(1 + v)$. The best photoelastic materials to resist distortion will exhibit high values of $1/f_\epsilon$ or low values for the material fringe value in terms of strain. Since Poisson's ratio for most rigid polymeric photoelastic materials varies over a limited range between 0.36 and 0.42, the ratio $Q = E/f_\sigma$ is sometimes used to evaluate the merits of the materials. This factor E/f_σ is known as the figure of merit.

The proportional limit σ_{pl} of a photoelastic material is important in two respects. First, a material with a high proportional limit can be loaded to a higher level without endangering the safety of the model (it should be noted here that the polymers normally employed in photoelastic work exhibit a brittle fracture and as such fail disastrously when the ultimate load is reached). Second, a material with a high proportional limit can produce a higher-order fringe pattern which tends to improve the accuracy of the stress determinations. A sensitivity index for a model material can be defined as

$$S = \frac{\sigma_{pl}}{f_\sigma} \tag{14.40}$$

Superior model materials exhibit high values for both the sensitivity index S and the figure of merit Q.

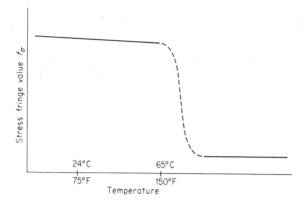

G. Temperature Sensitivity [27]

If the material fringe value in terms of stress changes markedly with temperature, errors can be introduced in a photoelastic analysis by minor temperature variations during the period involved in obtaining a photoelastic pattern. A typical curve showing the general characteristics of the change in material fringe value with temperature is shown in Fig. 14.18. For most polymeric plastics there is a linear region of this curve where f_σ decreases slightly with temperature. For one commonly used epoxy, the change in f_σ is only 0.07 lb/(in)(°F) [0.022 kN/(m)(°C)] in this linear region. However, at temperatures in excess of 150°F (65°C), the value of f_σ begins to drop sharply with increase in temperature, as shown in Fig. 14.18. For conventional two-dimensional photoelastic studies conducted at room temperature (75°F or 24°C), the slope of the curve in the linear region is the important characteristic. For most of the commonly used materials the slope is modest in this region, so that variations in f_σ can be neglected if temperature variations are limited to the range of ±5°F (±3°C).

H. Time-Edge Effect [28, 29]

When a photoelastic model is machined from a sheet of plastic and examined under a no-load condition as a function of time, it is noted that a stress is induced on the boundary which produces a fringe or a series of fringes that parallel the boundary of the model. An example of a model with a severe case of time-edge effect is illustrated in Fig. 14.19.

The influence of these time-edge effects on a photoelastic analysis is quite important. The fringe pattern observed is due to the superposition of two states of stress, the first associated with the load and the second a result of the time-edge stresses. Since the time-edge stresses are the most predominant on the boundary, the errors introduced by the edge stresses may be quite large in the determination of the extremely important boundary stresses.

It has been established that the time-edge effect is caused by diffusion of water vapor from the air into the plastic or from the plastic into the air.

For many photoelastic plastics, the diffusion process is so slow at room

Figure 14.19 Time-edge stresses in a photoelastic model of a turbine-blade dovetail joint (note the distortion of the fringe pattern near the boundary due to time-edge stresses).

temperature that it requires many years to reach an equilibrium state. For this reason, a freshly machined edge of a model usually will be in a condition to accept water from the air (its central region has not been saturated), and time-edge stresses will begin to develop. The rate at which the time-edge stresses develop for a particular model will depend upon the relative humidity of the air and the temperature. Photoelastic tests conducted at relative humidities of greater than 80 percent are often difficult, for the time-edge stresses become objectionably large in less than 2 to 3 h. For most photoelastic plastics the proper procedure to avoid time-edge stresses is to select relatively dry days (relative humidity less than 40 to 50 percent) and to photograph the model as soon as possible after completing the machining process.

The epoxy resins are somewhat different from most other photoelastic materials in that their diffusion rate is sufficiently high that a saturated condition can be established after about 2 to 3 months. If a two-dimensional model is machined from a sheet of material that has been maintained at a constant humidity for several months so that it is in a state of equilibrium (concentration uniform through the thickness of the sheet), and if the model is tested under these same humidity conditions, time-edge stresses will not develop.

I. Machinability

Photoelastic materials must be machinable in order to form the complex models employed in photoelastic analyses. Ideally, it should be possible to turn, mill, route, drill, and grind these plastics. Although machinability properties may appear to be a trivial requirement, it is often extremely difficult to machine a high-quality photoelastic model properly. The action of a cutting tool on the plastic often produces heat coupled with relatively high cutting forces. As a consequence, boundary stresses due to machining can be introduced permanently into the model, making it unsuitable for a quantitative photoelastic analysis.

In machining photoelastic models, care must be taken to avoid high cutting forces and the generation of excessive amounts of heat. These requirements can best be accomplished by using sharp carbide-tipped tools, air cooling, and light cuts coupled with a relatively high cutting speed. For two-dimensional applications, complex models may be routed from almost any thermosetting plastic. In this machining method a router motor (20,000 to 40,000 r/min) is used to drive a carbide rotary file. The photoelastic model is mounted to a metal template which describes the exact shape of the final model. The plastic is rough-cut with a jigsaw to within about $\frac{1}{8}$ in (3 mm) of the template boundary. The final machining operation is accomplished with the router, as illustrated in Fig. 14.20. The metal template is guided by an oversize-diameter pin which is coaxial with the rotary file. The rate of feed along the boundary of the model is carefully controlled by moving the model along the pin by hand. Successive cuts are taken by reducing the diameter of the stationary pin until it finally coincides with the cutter diameter. By using this technique, satisfactory two-dimensional models can be produced in less than 1 h by skilled operators.

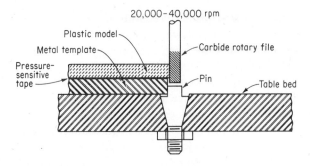

Figure 14.20 Machining a two-dimensional photoelastic model with a router.

J. Residual Stresses

Residual stresses are sometimes introduced into photoelastic plastics during casting and curing operations and almost always by rolling or extrusion processes. They can be observed simply by inserting the sheet of material in a polariscope and noting the order of the fringes in the sheet. The presence of residual stresses in photoelastic models is extremely detrimental since they are superimposed on the true stress distribution produced by loading the model. Since it is difficult if not impossible to subtract out the contribution due to the residual-stress distribution, the presence of residual stresses in the model material often introduces serious errors into a photoelastic analysis.

In certain cases it is possible to reduce the level of the residual stresses by thermally cycling the sheet above its softening point on a flat plate or in an oil bath. However, it is almost impossible to remove all residual stresses completely from a sheet of material once they have been introduced by the manufacturing process. Often it is more expeditious for the photoelastician to cast his own resin into plates, paying particular attention to the curing cycle and mold-release agent, than to purchase materials which must be stress-relieved.

K. Cost of Material

Normally the cost of the model material in a photoelastic analysis represents a very small percentage of the total cost of the investigation. For this reason, the cost of the materials should not be overemphasized, and the most suitable material should be selected on the basis of other parameters, regardless of the apparent difference in cost of the material on a pound basis. Very few two-dimensional photoelastic models require more than 1 or 2 lb of sheet plastic; therefore, prices of $20 to $40 per pound should be considered reasonable if the surface finish is adequate and the sheet is free of residual stress.

14.9 PROPERTIES OF COMMONLY EMPLOYED PHOTOELASTIC MATERIALS [30–39]

A brief examination of the photoelastic literature will show that most polymeric materials exhibit temporary double refraction and that numerous materials have been employed in photoelastic analyses. The list will include several types of glass, celluloid, gelatin, the glyptal resins, natural and synthetic rubber, fused silica, the phenolformaldehydes, polycarbonate, allyl diglycol (CR-39), and several compositions of the epoxies and the polyesters. Today, most elastic-stress analyses are conducted by employing one of the following materials:

1. Columbia resin CR-39
2. Homalite 100
3. Polycarbonate
4. Epoxy resin
5. Urethane rubber

A. Columbia Resin CR-39 [30, 31]

CR-39 is an allyl diglycol carbonate which is produced by reacting phosgene with diethylene glycol to obtain a chloroformate, which is then esterified with allyl alcohol to yield a monomer. The monomer is polymerized by heating in the presence of a catalyst (benzoyl peroxide), and the resultant sheet is a crystal-clear product. The casting operation is performed in such a manner that the final product in sheet form has surfaces of optical quality. Actually this is the predominant advantage of CR-39, namely, that it is available in large sheets 48 by 60 in (1.2 by 1.5 m) of a number of different thicknesses with a surface finish which is nearly optically perfect. The material is quite brittle and is difficult to turn or to mill; however, it can be readily machined by routing, and a high-quality photoelastic model can usually be produced in less than 1 h.

The material has a relatively low fringe value, indicating sensitivity, but its sensitivity index S is impaired by the low stress value at which it becomes nonlinear. CR-39 exhibits appreciable creep; therefore, the material fringe value must be determined as a function of time. The material is less prone to time-edge effects than the epoxies but more susceptible to time-edge effect than the polyesters (Homalite 100) or polycarbonate.

B. Homalite 100

Homalite 100 is a polyester resin which is cast between two plates of glass to form very large sheets. The surfaces of the commercially available sheets are of optical quality, and the material is free of residual stresses. Models can be machined by routing; however, since the material is extremely brittle, edge chipping can be a problem.

Homalite 100 does not exhibit appreciable creep; therefore, the material fringe value can be treated as a constant for loading times in excess of 5 to 10 min. Since moisture absorption is very slow in this material, time-edge effects do not become apparent for several days even under very humid test conditions. The material exhibits both a low figure of merit Q and a low sensitivity index S. High fringe orders cannot be achieved without fracturing the model.

C. Polycarbonate

Polycarbonate is an unusually tough and ductile polymer which yields and flows prior to fracture. It is known by the trade name Lexan in the United States and as Makrolan in Europe. Polycarbonate exhibits both a high figure of merit Q and a high sensitivity index S. It is relatively free of time-edge effects and exhibits very little creep at room temperature.

Polycarbonate is a thermoplastic and is produced in sheet form by an extrusion process. It is available in large sheets with reasonably good surface characteristics. Unfortunately, the extrusion process usually produces some residual birefringence in the sheets. Annealing for an extended period of time at or near the softening temperature is required to eliminate the residual birefringence. The polycarbonate material is also difficult to machine. Any significant heat produced by the cutting tool will cause the material to soften and deform under the tool. Routing can be performed only under water, and side milling is practically impossible. Band sawing and hand filing are often required to produce satisfactory model boundaries. Since the material exhibits both yield and flow characteristics, it can also be employed for photoplastic studies. The birefringence introduced in the plastic state is permanent and is locked into the material on a molecular scale. This behavior makes the material suitable for three-dimensional photoplasticity studies.

D. Epoxy Resin [32]

Epoxy resins were first introduced in photoelastic applications in the mid-1950s, when they were employed predominantly as materials for three-dimensional photoelasticity. However, a brief review of their properties indicates that they are also quite suitable for use in a wide variety of two-dimensional applications. The commercial epoxy resins are condensation products of epichlorohydrin and a polyhydric phenol. The basic monomer can be polymerized by using acid anhydrides, polyamides, or polyamines. In general, curing with the acid anhydrides requires higher temperatures than curing with the polyamides or polyamines.

A wide variety of epoxy materials can be cast into sheet form. The type of the basic monomer, the curing agent, and the percentage of the curing agent relative to the basic monomer can be varied to give an almost infinite number of epoxy materials. One particular epoxy, ERL-2774 with 50 parts per hundred by weight of phthalic anhydride, is a material which can be readily adapted for photoelastic work. This epoxy can be cast into sheets which are light amber in appearance; however, optical-quality surfaces are difficult to produce, and surface finishing by fly cutting or milling is frequently required. The epoxies are usually characterized as brittle materials, but they are easier to machine than the polyesters or CR-39. Most of the epoxies exhibit better optical sensitivity than Homolite 100 or CR-39; they are less sensitive than polycarbonate.

Although the material is susceptible to time-edge effects, the rate of diffusion of water into epoxy is sufficiently high to permit a saturation condition to be achieved in about 2 months. If the sheets are stored until saturated at, say, a 50 percent relative humidity, the model can be cut from the conditioned sheet and little or no time-edge effect will be noted unless the humidity conditions change. Finally, the material creeps approximately the same amount as polycarbonate or Homolite 100 but much less than CR-39.

E. Urethane Rubber [33]

Urethane rubber is an unusual photoelastic material in that it exhibits a very low modulus of elasticity (three orders of magnitude lower than that of the other materials listed) and a very high sensitivity, as indicated by an f_σ value of less than 1 lb/in (0.175 kN/m). The material can be cast between glass plates to produce an amber-colored sheet with optical-quality surfaces. Except for its very low figure of merit, the material ranks relatively well in comparison with the other materials listed. Its strain sensitivity is so low that time-edge effects are negligible; moreover, in spite of its low modulus, the material exhibits little mechanical or optical creep. The material can readily be machined on a high-speed router, but it must be frozen at liquid-nitrogen temperatures before its surfaces can be turned or milled.

The material is particularly suited for demonstration models. Loads applied by hand are sufficient to produce well-defined fringe patterns, and the absence of time-edge effects permits the models to be stored for years. Also, the material is so sensitive to stress that it can be used to study body-force problems if fringe-multiplication techniques are employed with thick models. Finally, urethane rubber can be used for models in dynamic photoelasticity, where its low modulus of elasticity has the effect of lowering the velocity of the stress wave to less than 300 ft/s (90 m/s) as compared with 6000 ft/s (1830 m/s) in CR-39. The low-velocity stress waves in urethane-rubber models are easy to photograph with moderate-speed framing cameras (10,000 frames per second), which are common, while the high-speed stress waves in CR-39 require high-speed cameras (200,000 frames per second or more) to produce satisfactory fringe patterns for analysis.

F. Conclusions Pertaining to Material Selection

A summary of the mechanical and optical properties of the five photoelastic materials is presented in Table 14.2. It is clear by comparing the figure of merit Q and the sensitivity index S that polycarbonate and the epoxies exhibit superior properties. Unfortunately, the polycarbonate material is difficult to machine, and the epoxy resin materials require special precautions to minimize time-edge effects.

Homalite 100 and CR-39 can be used in applications where high precision and low model distortion are not required. Both have the advantage of being available in large sheets with optical-quality surfaces. Homolite 100 is often preferred since it exhibits less creep and exhibits much less time-edge effect than CR-39. The sensitivity index S of CR-39, however, is significantly less than that of Homalite 100.

Table 14.2 Summary of the optical and mechanical properties of several photoelastic materials

Property	CR-39	Homalite 100	Polycarbonate	Epoxy resin ERL-2774 †	‡	Urethane rubber§
Time-edge effect	Poor	Excellent	Excellent	Good	Good	Excellent
Creep	Poor	Excellent	Excellent	Good	Good	Excellent
Machinability	Poor	Good	Poor	Good	Good	Poor
Modulus of elasticity E:						
lb/in^2	250,000	560,000	360,000	475,000	475,000	450
MPa	1725	3860	2480	3275	3275	3
Poisson's ratio v	0.42	0.35	0.38	0.38	0.36	0.46
Proportional limit σ_{pl}:						
lb/in^2	3000	7000	5000	8000	8000	20
MPa	20.7	48.3	34.5	55.2	55.2	0.14
Stress fringe value f_σ ¶:						
lb/in	88	135	40	58	64	1
kN/m	15.4	23.6	7.0	10.2	11.2	0.18
Strain fringe value f_ϵ ¶:						
in	0.00050	0.00033	0.00015	0.00017	0.00018	0.00324
mm	0.0127	0.0084	0.0038	0.0043	0.0046	0.082
Figure of merit Q:						
1/in	2840	4150	9000	8200	7400	450
1/mm	112	163	354	321	292	17
Sensitivity index S:						
1/in	34	52	125	138	125	20
1/mm	1.34	2.05	4.92	5.43	4.92	0.78

† With 50 parts per hundred phthalic anhydride.

‡ With 42 parts per hundred phthalic anhydride and 20 parts per hundred hexahydrophthalic anhydride.

§ 100 parts by weight Hysol 2085 with 24 parts by weight Hysol 3562.

¶ For green light ($\lambda = 546.1$ nm).

Urethane rubber is extremely useful in special-purpose applications such as demonstration models for instructional purposes. Because of its low material stress fringe value it is also useful for modeling where body forces due to gravity produce the loads. Finally, urethane rubber can be used to great advantage in dynamic photoelastic studies, where its low modulus of elasticity results in low-velocity stress waves which are easy to photograph.

EXERCISES

14.1 Plot the fringe orders as a function of position across the horizontal centerline of the chain-link model shown in Fig. 14.1.

14.2 Determine the fringe orders associated with the tensile and compressive stress concentrations at the pinholes of the chain-link shown in Fig. 14.1.

14.3 When $\sigma_1 > 0$ and $\sigma_2 < 0$, show the plane upon which the maximum shear stress acts.

14.4 When $\sigma_1 > 0$ and $\sigma_2 > 0$, show the plane upon which the maximum shear stress acts.

14.5 Determine the stress distribution on the boundary of the pinhole of the chain link shown in Fig. 14.1 if $f_\sigma = 16$ kN/m and $h = 5$ mm.

14.6 Plot the fringe orders as a function of position across the horizontal centerline of the pressurized square conduit shown in Fig. 14.3.

14.7 Plot the fringe orders as a function of position across the diagonal of the pressurized square conduit shown in Fig. 14.3.

14.8 Plot the fringe orders as a function of position along the outer edge of the pressurized square conduit shown in Fig. 14.3.

14.9 Determine the stress distribution on the boundary of the circular hole of the pressurized square conduit shown in Fig. 14.3 if $f_\sigma = 15$ kN/m, $h = 5$ mm, and $p = 1.50$ MPa.

14.10 Determine the maximum tensile and compressive stresses on the inner boundary of the thick-ring specimen shown in Fig. 13.11 if $f_\sigma = 16$ kN/m and $h = 5$ mm. The specimen is subjected to concentrated compressive loads at the ends of the vertical diameter.

14.11 Given a fringe order of 6, a model thickness of 10 mm, a material fringe value of 17.5 kN/m, and an isoclinic parameter of 30° defining the angle between the x axis and σ_1, determine the shear stress τ_{xy} and show the direction of the shear stress on the face of a small element.

14.12 Explain how a simple tension or compression specimen could be used as a compensator.

14.13 Describe how the Tardy method of compensation could be employed to determine the order of the fringe at points P_3 and P_4 of Fig. 13.15.

14.14 If the circular disk shown in Fig. 15.3 has an outside diameter of 100 mm and a thickness $h = 10$ mm, determine the material fringe value if a load of 3.0 kN produced the fringe pattern.

14.15 Verify Eqs. (14.30).

14.16 For $\theta = 30$ and $45°$, show that Eqs. (14.30) reduce to

$$\theta = 30°: \sigma_1 = \frac{\sqrt{3}f_\sigma}{h}(2N_\theta - \sqrt{3}N_0) \qquad \sigma_2 = \frac{2f_\sigma}{h}(\sqrt{3}N_\theta - 2N_0)$$

$$\theta = 45°: \sigma_1 = \frac{\sqrt{2}}{2}\frac{f_\sigma}{h}(2N_\theta - \sqrt{2}N_0) \qquad \sigma_2 = \frac{f_\sigma}{h}(\sqrt{2}N_\theta - 2N_0)$$

14.17 Determine σ_{xx}, σ_{yy}, τ_{xy}, σ_1, σ_2, and the direction of σ_1 if

(a) $\theta_x = \theta_y = \theta = \dfrac{\pi}{4}$ (b) $\theta_x = \theta_y = \theta = \dfrac{\pi}{6}$

and N_0, N_{θ_y}, N_{θ_z} are known.

REFERENCES

1. Durelli, A. J., and W. F. Riley: "Introduction to Photomechanics," Prentice-Hall, Englewood Cliffs, N.J., 1965.
2. Frocht, M. M.: "Photoelasticity," John Wiley & Sons, Inc., New York, vol. 1, 1941, vol. 2, 1948.
3. Coker, E. G., and L. N. G. Filon: "A Treatise on Photoelasticity," Cambridge University Press, London, 1931.
4. Jessop, H. T., and F. C. Harris: "Photoelasticity: Principles and Methods," Dover Publications, Inc., New York, 1950.
5. Kuske, A., and G. Robertson: "Photoelastic Stress Analysis," John Wiley & Sons, Inc., New York, 1974.
6. Durelli, A. J., E. A. Phillips, and C. H. Tsao: "Introduction to the Theoretical and Experimental Analysis of Stress and Strain," chap. 8, McGraw-Hill Book Company, New York, 1958.

7. Jerrard, H. G.: Examination and Calibration of a Babinet Compensator, *J. Sci. Instrum.*, vol. 27, pp. 62–66, 1950.

8. Jerrard, H. G.: Examination and Calibration of Soliel Compensators, *J. Sci. Instrum.*, vol. 30, pp. 65–69, 1953.

9. Tardy, M. H. L.: Méthode pratique d'examen de mesure de la biréfringence des verres d'optique, *Rev. Opt.*, vol. 8, pp. 59–69, 1929.

10. Jessop, H. T.: On the Tardy and Senarmont Methods of Measuring Fractional Relative Retardations, *Br. J. Appl. Phys.*, vol. 4, pp. 138–141, 1953.

11. Chakrabarti, S. K., and K. E. Machin: Accuracy of Compensation Methods in Photoelastic Fringe-Order Measurements, *Exp. Mech.*, vol. 9, no. 9, pp. 429–431, 1969.

12. Flynn, P. D.: Theorems for Senarmont Compensation, *Exp. Mech.*, vol. 10, no. 8, pp. 343–345, 1970.

13. Sathikh, S. M., and G. W. Bigg: On the Accuracy of Goniometric Compensation Methods in Photoelastic Fringe-Order Measurements, *Exp. Mech.*, vol. 12, no. 1, pp. 47–49, 1972.

14. Dally, J. W., and E. R. Erisman: An Analytic Separation Method for Photoelasticity, *Exp. Mech.*, vol. 6, no. 10, pp. 493–499, 1966.

15. Shortly, G. H., and R. Weller: The Numerical Solution of Laplace's Equation, *J. Appl. Phys.*, vol. 9, pp. 334–348, 1938.

16. Waner, N. S., and W. W. Soroka: Stress Concentrations for Structural Angles in Torsion by the Conducting Sheet Analogy, *Proc. SESA*, vol. XI, no. 1, pp. 19–26, 1953.

17. Stokey, W. F., and W. F. Hughes: Tests of the Conducting Paper Analogy for Determining Isopachic Lines, *Proc. SESA*, vol. XII, no. 2, pp. 77–82, 1955.

18. Hiltscher, R.: Development of the Lateral Extensometer Method in Two-dimensional Photoelasticity, pp. 43–56, in "Photoelasticity," Pergamon Press, New York, 1963.

19. Post, D.: A New Photoelastic Interferometer Suitable for Static and Dynamic Measurements, *Proc. SESA*, vol. XII, no. 1, pp. 191–202, 1954.

20. Drucker, D. C.: Photoelastic Separation of Principal Stresses by Oblique Incidence, *J. Appl. Mech.*, vol. 10, no. 3, pp. A156–A160, 1943.

21. Drucker, D. C.: The Method of Oblique Incidence in Photoelasticity, *Proc. SESA*, vol. VIII, no. 1, pp. 51–66, 1950.

22. Clutterbuck, M.: The Dependence of Stress Distribution on Elastic Constants, *Br. J. Appl. Phys.*, vol. 9, 323–329, 1958.

23. Young, D. F.: Basic Principles and Concepts of Model Analysis, *Exp. Mech.*, vol. 11, no. 7, pp. 325–336, 1971.

24. Sanford, R. J.: The Validity of Three-dimensional Photoelastic Analysis of Non-homogeneous Elastic Field Problems, *Br. J. Appl. Phys.*, vol. 17, pp. 99–108, 1966.

25. Dundurs, J.: Dependence of Stress on Poisson's Ratio in Plane Elasticity, *Int. J. Solids Struct.*, vol. 3, pp. 1013–1021, 1967.

26. Mindlin, R. D.: A Mathematical Theory of Photoviscoelasticity, *J. Appl. Phys.*, vol. 20, pp. 206–216, 1949.

27. Lee, G. H., and C. W. Armstrong: Effect of Temperature on Physical and Optical Properties of Photoelastic Material, *J. Appl. Mech.*, vol. 5, pp. A11–A12, 1938.

28. Leaf, W.: The Time-Edge Effect: Its Cause and Prevention, *Proc. 15th Semi-annu. East. Photoelasticity Conf.*, 1942, pp. 20–22.

29. Frocht, M. M.: On the Removal of Time Stresses in Three-dimensional Photoelasticity, *Proc. SESA*, vol. V, no. 2, pp. 9–13, 1948.

30. Coolidge, D. J., Jr.: An Investigation of the Mechanical and Stress-Optical Properties of Columbia Resin, CR-39, *Proc. SESA*, vol. VI, no. 1, pp. 74–82, 1948.

31. Clark, A. B. J.: Static and Dynamic Calibration of a Photoelastic Model Material, CR-39, *Proc. SESA*, vol. XIV, no. 1, pp. 195–204, 1956.

32. Leven, M. M.: Epoxy Resins for Photoelastic Use, pp. 145–165 in "Photoelasticity," Pergamon Press, New York, 1963.

33. Dally, J. W., W. F. Riley, and A. J. Durelli: A Photoelastic Approach to Transient Stress Problems Employing Low Modulus Materials, *J. Appl. Mech.*, vol. 26, no. 4, pp. 613–620, 1959.

34. Mylonas, C.: Mechanical and Optical Properties of Catalin, *Proc. 7th Int. Congr. Appl. Mech. 1948*, vol. 4, p. 165.
35. Taylor, C. E., E. O. Stitz, and R. O. Belsheim: A Casting Material for Three-dimensional Photoelasticity, *Proc. SESA*, vol. VII, no. 2, pp. 155–172, 1950.
36. Leven, M. M.: A New Material for Three-dimensional Photoelasticity, *Proc. SESA*, vol. VI, no. 1, pp. 19–28, 1948.
37. Kolsky, H.: Stress-Birefringence in Polystyrene, *Nature*, vol. 166, pp. 235–236, 1950.
38. Sugarman, B., G. O. Moxley, and I. A. Marshall: A Castable Resin for Photoelastic Work, *Br. J. Appl. Phys.*, vol. 3, no. 7, pp. 233–237, 1952.
39. Spooner, H., and L. D. McConnell: An Ethoxylene Resin for Photoelastic Work, *Br. J. Appl. Phys.*, vol. 4, pp. 181–184, 1953.
40. Frocht, M. M., and Hui Pih: A New Cementable Material for Two- and Three-dimensional Photoelastic Research, *Proc. SESA*, vol. XII, no. 1, pp. 55–64, 1954.
41. D'Agostino, J., D. C. Drucker, C. K. Liu, and C. Mylonas: Epoxy Adhesives and Casting Resins as Photoelastic Plastics, *Proc. SESA*, vol. XII, no. 2, pp. 123–128, 1955.
42. Bayley, H. G.: Gelatin as a Photoelastic Material, *Nature*, vol. 183, pp. 1757–1758, 1959.

THREE-DIMENSIONAL PHOTOELASTICITY

15.1 INTRODUCTION

The coverage of photoelasticity in Chap. 14 was limited to two-dimensional techniques for determining stresses in plane models. Many stress-analysis problems exist, however, which are three-dimensional in character. These problems cannot be effectively approached by using two-dimensional techniques. For example, it is impossible to relate the integrated optical effects in a complicated three-dimensional model to the stresses in the model. It is possible, however, to construct and load a three-dimensional model and to analyze interior planes of the model photoelastically by using either *frozen-stress* or *scattered-light methods*. With the frozen-stress method, model deformations and the associated optical response are locked into a loaded three-dimensional model. Once the stress-freezing process is completed, the model can be sliced and photoelastically analyzed to obtain interior-stress information. With the scattered-light method, interior-stress information can be obtained without stress freezing or slicing of the model. Both these methods will be covered in detail in subsequent sections of this chapter. The frozen-stress method will be covered first.

15.2 LOCKING IN MODEL DEFORMATIONS [1]

The frozen-stress method for three-dimensional photoelasticity was initiated by Oppel [1] in Germany in 1936. The basis of the method is the process by which deformations are permanently locked in the model. The four different techniques for locking deformations in the loaded model include stress-freezing, creep, curing and gamma-ray irradiation methods. In all these methods, the deformations are locked into the model on a molecular scale, thus permitting the models to be sliced without relieving the locked-in deformations. Since the stress-freezing process is by far the most effective and most popular technique for locking the deformations in the model, it will be treated in detail. The creeping, curing, and gamma-ray irradiation techniques will be treated more briefly since they are rarely used in actual practice.

A. The Stress-freezing Method [2, 3]

The stress-freezing method of locking in the model deformations due to an applied load is based on the diphase behavior of many polymeric materials when they are heated. The polymeric materials are composed of long-chain hydrocarbon molecules, as illustrated in Fig. 15.1. Some of the molecular chains are well bonded into a three-dimensional network of primary bonds. However, a large number of molecules are less solidly bonded together into shorter secondary chains. When the polymer is at room temperature, both sets of molecular bonds, the primary and the secondary, act to resist deformation due to applied load. However, as the temperature of the polymer is increased, the secondary bonds break down and the primary bonds in effect carry the entire applied load. Since the secondary bonds constitute a very large portion of the polymer, the deflections which the primary bonds undergo are quite large yet elastic in character. If the temperature of the polymer is lowered to room temperature while the load is maintained on the model, the secondary bonds will re-form between the highly elongated primary bonds and serve to lock them into their extended positions. When the load is removed, the primary bonds relax to a modest degree, but the

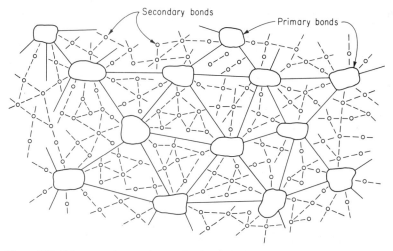

Figure 15.1 Primary and secondary molecular chains in a diphase polymer.

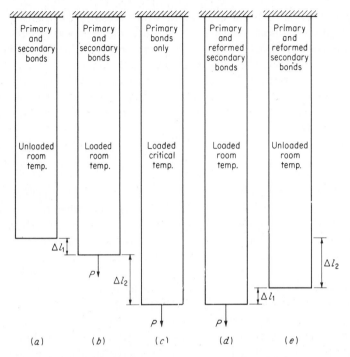

Figure 15.2 Stress freezing of a simple uniaxial tensile specimen.

main portion of their deformation is not recovered. The elastic deformation of the primary bonds is permanently locked into the model by the re-formed secondary bonds. Moreover, these deformations are locked in on a molecular scale; thus the deformation and accompanying birefringence are maintained in any small section cut from the original model.

The diphase behavior of polymeric materials described above constitutes the basis of the stress-freezing process so often employed in three-dimensional photoelasticity. The stress-freezing process can best be illustrated by considering the simple tensile specimen shown in Fig. 15.2. The tensile specimen shown in Fig. 15.2a is first loaded at room temperature with an axially applied force P, and a displacement Δl_1 is produced, as given in Fig. 15.2b. Next, the temperature is increased until the secondary bonds break down and the tensile specimen elongates by an additional amount Δl_2 (see Fig. 15.2c). The temperature is then reduced while the load is maintained until the secondary bonds re-form. If thermal expansions and contractions are neglected, the tensile specimen will not change length during the cooling cycle. Finally, the load is removed and the specimen contracts by an amount Δl_1 while retaining a permanently locked-in deformation Δl_2, as illustrated in Fig. 15.2e.

At this stage a photoelastic model with its locked-in deformations and attendant fringe pattern can be carefully cut or sliced without disturbing the character of either the deformation or the fringe pattern. This fact is due to the molecular nature of the locking process, where the extended primary bonds are locked into place by the re-formed secondary bonds. The cutting or slicing process may relieve one molecular layer on each face of a slice cut from a model, but this relieved layer is so thin relative to the thickness of the slice that the effect cannot be observed. An example of a cut across a locked-in fringe pattern is shown in Fig. 15.3.

The temperature to which the model is heated to break down the secondary bonds is called the *critical temperature*. Actually, this notation is somewhat unfortunate since the temperature required to break down the secondary bonds is not at all critical. Instead, the process of breaking down the secondary bonds depends upon both the temperature and the time under load. If a load is applied to, say, a tensile specimen and maintained at temperatures somewhat below the critical temperature, the deflection or fringe-order response will vary with time under load, as illustrated in Fig. 15.4. At temperatures greater than 95 percent of the critical temperature, the maximum fringe order or deflection is obtained in about 1 min. At the so-called critical temperature the response of the model is almost

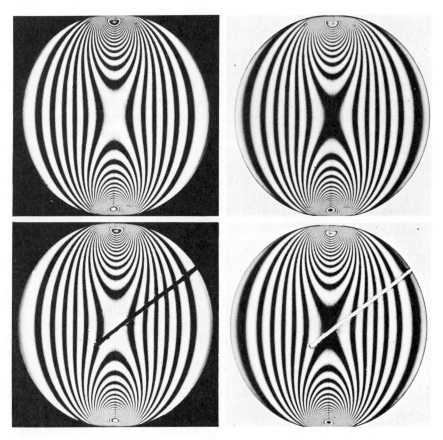

Figure 15.3 Illustration of the fact that careful cutting does not disturb the locked-in fringe pattern.

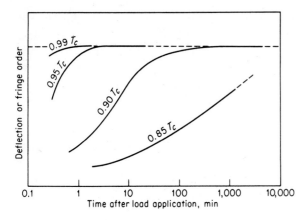

Figure 15.4 Deflection or fringe order as a function of time of load application for temperatures below the critical temperature.

immediate, with the maximum fringe order or deflection attained in less than 0.1 min. For temperatures between 85 and 90 percent of the critical temperature, maximum response can be achieved, but the load must be maintained for several hours.

The elastic behavior of the primary system of bonds at critical temperature can be readily established by conducting loading and unloading tests at the critical temperature. A typical curve showing deflection or maximum fringe order as a function of time for a tensile specimen subjected to a loading and unloading cycle is presented in Fig. 15.5. At or near critical temperature (within 95 percent of the critical value), the maximum response of the model is achieved within a matter of a few minutes; and upon unloading, the model relaxes in about the same period of time. While some creep effects are obviously present during the first minute or so, they may be entirely eliminated by maintaining the load at these temperatures for $\frac{1}{2}$ h or so before beginning to lower the temperature.

The stress-freezing process is used today to lock in the fringe pattern in most three-dimensional photoelastic analyses. The method is extremely simple to apply, as temperature control to ± 5 percent of the critical temperature is usually sufficient. The model response is entirely elastic in character, and the model can be

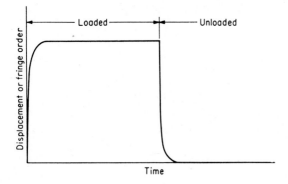

Figure 15.5 Loading-unloading curve for a typical photoelastic plastic at critical temperature.

sliced into plane sections for analysis without disturbing the fringe pattern. The procedure for the stress-freezing process is described below.

1. Place the model in the stress-freezing oven.
2. Heat the model relatively rapidly until the critical temperature is attained.
3. Apply the required loads.
4. Soak the model at least 2 to 4 h until a uniform temperature throughout the model is obtained.
5. Cool the model slowly enough for temperature gradients to be minimized.
6. Remove the load and slice the model.

B. The Creep Method [4]

The creep method for three-dimensional photoelasticity is based upon the fact that many plastics which creep exhibit a fringe pattern which remains directly proportional to the applied load even after considerable creep (see Sec. 14.8E). To employ the creep method, a model (usually manufactured from a phenolformaldehyde type of plastic) is placed under load at room temperature, and the load is maintained for several hours or even days. The model creeps under this maintained load, and the deformation and the order of the fringes increase with time, as illustrated in Fig. 15.6. When the load is removed, the model exhibits an instantaneous recovery and a portion of the response is lost. However, the delayed response due to creep of the model is not recovered instantaneously, and a fringe pattern proportional to the elastic stress distribution remains. This pattern is not permanent, so the fringe order decreases slowly as the model recovers with time after unloading (see Fig. 15.6). Sufficient time is available, however, to remove one or more slices from simple models and to analyze them before the recovery is complete.

   The creep method can be improved in regard to the time available for analysis after unloading by employing materials such as epoxies which creep at very low rates at room temperature. In this case (Fig. 15.7) the epoxy is heated to approxi-

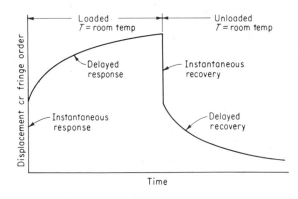

Figure 15.6 Model response and recovery as a function of time after loading and unloading (creep method).

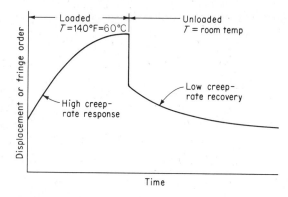

Figure 15.7 Model response and recovery as a function of time after loading and unloading (modified creep method).

mately 140°F (not sufficient to begin to break down the secondary bonds), where the creep rate of the material is greatly accelerated and the response of the model to the sustained load becomes appreciable in about 40 h. If the model is cooled to room temperature before unloading, the creep rate governing the recovery of the model is drastically reduced. This procedure permits the recovery time to be extended over a period of several months, and the model can be sliced and analyzed at the investigator's convenience.

C. The Curing Method [5]

In the application of the curing method, the model response is permanently locked in by making use of the stage-curing phenomena typical of thermosetting resins. To illustrate, consider an epoxy-resin model which is cast from the basic epoxy monomer, an amine-type hardener, and a compatible plasticizer. Initially, in the stage A phase of the curing process the mixture of these constituents will be liquid; however, after a few hours the polymerization process is initiated, and a semicured resin which has the consistency of a soft rubber eraser is obtained (stage B).

The load is applied to the model while it is in the stage B state where some of the molecular chains have been formed. By maintaining the load on the model for a sufficient period of time, the curing process can be completed, and the resin advances to stage C, where it is hard and almost completely polymerized. During this period, new molecular bonds form about the originally deformed bonds and lock them into their deformed state. When the load is removed, the fringe pattern due to the locked-in deformation remains and can be used to determine the elastic stress distribution in a three-dimensional model.

D. The Gamma-Ray Irradiation Method [6]

The gamma-ray irradiation method for stress freezing is similar in some respects to the curing method. In the work reported by Miyazono [6], a partially cured epoxy resin model was loaded and irradiated with a ^{60}Co gamma-ray source of 16 kCu. A total dose of 2×10^8 R was sufficient to complete the cure of the resin and provide complete retention of the fringe pattern associated with the load. The advantages of completing the cure of the epoxy resin by irradiation rather than thermal cycling are associated with the lower values of Poisson's ratio and the higher values of the modulus of elasticity of the model material during stress freezing.

15.3 MATERIALS FOR THREE-DIMENSIONAL PHOTOELASTICITY

Since three-dimensional photoelasticity became a reality late in the 1930s with the discovery of the stress-freezing process, a number of different model materials have been introduced and used. These materials include Catalin 61-893, Fosterite, Kriston, Castolite, and epoxy resins. Of these five materials, only the epoxy-based resins will be discussed here. Fosterite and Kriston are no longer commercially available, and the superiority of the epoxies precludes both Catalin 61-893 and Castolite from practical consideration.

In the discussion of two-dimensional photoelastic materials, 11 requirements of an ideal material were listed. Two additional requirements should be added to this list when considering three-dimensional photoelastic materials: (1) that the material be castable in large sizes and (2) that the material be cementable. In three-dimensional analyses the model often is rather large and intricate, and the need for large castings of the epoxy resin becomes very real. Also, in complex models it is very often desirable to cement parts together much as steel components are welded together to form a complex structure. A complex multicomponent model is shown in Fig. 15.8.

In reality, the epoxies should be considered as a large class of resins since they cannot be characterized by a single molecular structure. Basically they consist of condensation products of epichlorohydrin and a polyhydric phenol. The base monomer is commercially available from several sources in both the United States and Europe. The monomer is usually polymerized by employing either an amine or an acid anhydride curing agent, and a very large number of these curing agents are commercially available. It is clear, then, that a very large number of polymerized epoxy resins can be produced by varying the type of monomer and the type of hardening agent as well as the percentage of each of these constituents.

Leven and Sampson have investigated the problem of compounding epoxy resins for the specific purpose of optimizing their three-dimensional photoelastic properties. From the results of their investigations, it appears that the choice of the basic monomer does not greatly influence the photoelastic properties of the polymerized resin. ERL-2774, a product of Union Carbide, is recommended as the monomer since it has several processing advantages over other available monomers: (1) this monomer is a liquid in which the hardening agent can easily be dissolved and then mixed; (2) the viscosity of this monomer is sufficiently low to permit easy pouring and the release of bubbles produced by stirring; (3) this monomer is one of the least exothermic of the base epoxies, and so the heat generated during curing can be minimized.

The selection of the hardening agent, on the other hand, is extremely important since it influences the photoelastic properties of the resin to an appreciable extent. Of the two general types of hardening agent, the anhydrides are superior to the amines. The amines are not suitable for large castings because of their extremely exothermic reactions. The heat generated in even modest-size castings of amine-cured epoxies is often sufficient to destroy the casting. Moreover, the locked-in fringe pattern in amine-cured epoxies tends to relax with time, and a decrease in the maximum fringe order of 10 to 20 percent in 1 year is not uncommon.

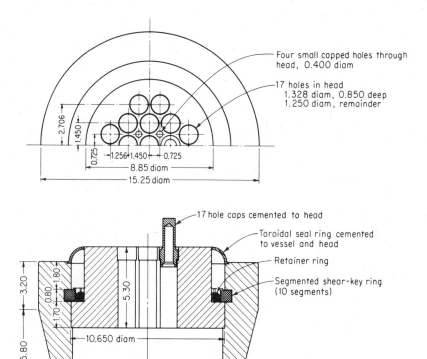

(a)

Figure 15.8 (a) The essential features of a complex three-dimensional photoelasticity model.

Figure 15.8 (*b*) Epoxy-resin components used in fabricating the model. (*Courtesy of M. M. Leven, Westinghouse Electric Corporation.*)

The anhydrides are recommended for the curing of large castings of epoxy resins because of their low exothermic reaction and the polymerized resins' low susceptibility to time-edge stresses. Of the numerous anhydrides available for curing epoxy resins, Leven [10] has recommended the following composition:

ERL-2774, 100 parts by weight
Phthalic anhydride, 42 parts by weight
Hexahydrophthalic anhydride, 20 parts by weight

The pertinent three-dimensional photoelastic properties of this particular epoxy resin are:

Critical temperature: 324 to 347°F (162 to 175°C)
Effective modulus: 5300 to 6500 lb/in^2 (37 to 45 MPa)
Effective material fringe value: 2.48 to 2.84 lb/in (435 to 500 N/m)
Figure of merit: 2100 to 2450 in^{-1} (83 to 96 mm^{-1})

These properties are compared with the photoelastic properties of other resin materials in Table 15.1, which clearly shows the superiority of the epoxy resins over all other materials except Kriston, which is no longer commercially available.

In order to obtain large epoxy castings free of residual stresses, it is almost imperative that the photoelastician prepare the material under closely controlled

Table 15.1 Properties of three-dimensional photoelastic materials at their critical temperature

Material	T_{cr}		$f_{effective}$		$E_{effective}$		$Q = E/f$	
	°F	°C	lb/in	N/m	lb/in²	MPa	in⁻¹	mm⁻¹
Catalin 61–893	230	110	3.20	560	1,100	7.6	344	13.6
Fosterite	189	87	3.85	674	2,320	16.0	603	23.7
Kriston	273	134	6.25	1094	13,800	95.2	2208	87.0
Castrolite	244	118	8.30	1453	4,060	28.0	489	19.3
Epoxy ERL-2774, 50 parts per hundred phthalic anhydride	320	160	2.48	434	5,210	35.9	2100	82.7
Epoxy (recommended)	338	170	2.68	469	6,450	44.5	2407	94.9
PLM-4B‡	248	120	2.20	385	2,500	17.2	1136	44.7

† For green light ($\lambda = 546.1$ nm).
‡ Available from Photolastic, Inc., Malvern, Pa.

conditions. The casting procedure recommended by Leven is summarized in cookbook form below.

1. Carefully clean mold in which the epoxy is to be cast and coat generously with a mold-release agent.
2. Weigh out proper proportions of each of the three constituents according to the formulation previously presented.
3. Heat the ERL-2774 to about 248°F (120°C) in a large container.
4. Heat the hexahydrophthalic anhydride until it melts and filter it into the ERL-2774.
5. Add the phthalic anhydride in its solid form to the mixture; heat and stir thoroughly until all ingredients have dissolved.
6. Cast into the prepared mold and place in an oven at 198°F (92°C). The resin will advance from the A stage to the B stage at this temperature in about 2 days.
7. Continue cure of the B-stage resin until it reaches the consistency of a rubber eraser (about one more day at 212°F or 100°C).
8. Cool slowly to room temperature.
9. Strip casting from mold, replace in oven, and heat to 212°F (100°C) at a rate of 5°F/h (3°C/h).
10. Increase temperature from 212 to 302°F (100 to 150°C) at a rate of 3°F/h (1.5°C/h).
11. Maintain casting at a temperature of 302°F (150°C) for about 8 to 10 days, then cool to room temperature at a rate of about 2°F/h (1°C/h).

The rate of temperature increase employed in reaching the 302°F (150°C) curing temperature and the rate of temperature decrease in cooling to room temperature must be carefully controlled. Since the plastics are, in general, poor conductors of heat, rapid heating may induce thermal stresses, which crack the casting. Also, the increase in temperature must be slow enough to produce a gradual polymerization, and the temperature decrease must be slow enough to prevent the locking in of any appreciable thermal stress in the casting. The heating and cooling rates recommended in the casting procedure are for a casting 12 in (300 mm) in diameter. For larger- or smaller-diameter castings these rates should be reduced or increased in proportion to the square of the diameter. However, Leven does not recommend rates in excess of $3\frac{1}{2}°$F/h ($2°$C/h) for any casting. As a consequence of the relatively low temperature rates, the curing cycle usually requires 2 to 3 weeks, but the casting obtained is relatively free of residual stresses and is ideally suited for three-dimensional photoelastic applications.

The anhydride-cured epoxy resins exhibit a time-edge-stress behavior which is unique in comparison with the behavior of other plastics. As discussed in Sec. 14.8H, time-edge effect is due to the diffusion of water vapor from the air into the plastic. The rate at which the water vapor diffuses into the plastic depends upon the diffusion constant and the humidity of the air, i.e., the concentration of water vapor in the air. For most plastics the diffusion process is so slow that a state of equilibrium is not reached in a period of years; however, for the anhydride-cured epoxy resins the rate of diffusion is sufficiently rapid to produce saturation in about 2 months. This ability of the anhydride-cured epoxy resins to saturate can be used to control the time-edge stresses in a slice taken from a three-dimensional model. Consider that the stress-freezing process drives off most of the water vapor stored in the model. When the model is sliced in preparation for the photoelastic examination, the water vapor again begins to diffuse into the plastic. The concentration gradient produces time-edge stresses, which increase to a maximum after about 5 days and then decrease to zero in about 2 months as the slice becomes saturated, i.e., gradients of concentration go to zero. Any changes in the humidity conditions would again upset the state of equilibrium in the water vapor and create a new set of time-edge stresses. The procedure employed in controlling the time-edge stresses in slices is to store them under constant humidity conditions for 2 months before analysis. If this length of time cannot be tolerated, the slices can be heated to $130°$F ($55°$C) for about 2 days to drive off the absorbed water vapor. However, in this case, the slices must be examined immediately upon their removal from the oven.

15.4 MACHINING, CEMENTING, AND SLICING THREE-DIMENSIONAL MODELS [11, 12]

Cast epoxy resin can be sawed, turned, milled, bored, drilled, ground, or fly cut. Since all epoxy materials are extremely abrasive, carbide-tipped tools should be used whenever available. High-speed tools may also be used, but they tend to dull rapidly and must be sharpened very frequently. Sharp cutting tools are essential for the production of a quality model.

In general, high cutting speeds together with low feed rates should be employed in the production of epoxy models. Finish cuts should range between 0.005 and 0.020 in (0.125 and 0.500 mm), depending on the rigidity of the model and the cutter. All tools should be ground with sufficient clearance to avoid rubbing the machined epoxy surface. Tool pressure should be held to the minimum required for satisfactory machining. Approximately 0.25 in (6 mm) of material should be removed from all cast surfaces to avoid problems with residual stresses.

Epoxy castings are frequently band sawed into blocks of convenient size before machining. Saw blades should have teeth with sufficient set to avoid rubbing. For rough cutting, a $\frac{1}{2}$-in-wide blade with four to eight teeth per inch has proved to be satisfactory. The blade should be used only for epoxy and should be changed frequently. A medium blade speed and slow feed should be employed to minimize pressure on the epoxy. A single-point tool should be used for machining wherever possible because it applies less pressure and results in less heat generation. Two-flute end mills with adequate relief and rake angles should be used for milling. Drilling must be done carefully to avoid excessive heat generation. An indication of proper cutting in any machining operation is the presence of long ribbons of material.

Frequently, complex models must be fabricated by cementing together a number of premachined parts. A base epoxy resin such as ERL 2774 or Epi-Rez 510 cured with triethylenetetraamine hardener makes an excellent room-cure adhesive. A small amount of Cab-O-Sil silica filler can be added to thicken the cement and prevent excessive flow from the joints. The surfaces to be cemented should be roughened with 280-grit silicon-carbide paper, then washed and thoroughly dried. All surfaces of the joints should be wetted with cement, then joined together with 2 to 3 lb/in^2 (14 to 21 kPa) pressure. Extreme care must be exercised to avoid any air bubbles or voids.

The analysis of three-dimensional photoelasticity models is performed on slices cut from the model after stress freezing. Slice preparation involves three steps: layout, cutting the slices, and finishing the surfaces of the slices.

Accurate layout can be accomplished with a glass plate and a surface gage. A sharp scriber should be used to locate center and thickness lines for each slice. Heavy scribe lines should be avoided since they tend to obscure the fringes at the boundary of the slice.

Rough slices can be removed from the model with a band saw. A small cutoff wheel mounted in a flexible power unit is often useful in cutting slices from thin models.

The surfaces of the slices should be finished by fly cutting. The single-point cutter should have a radiused tip and should be adjusted to touch the surface of the slice only on the cutting pass. Double-faced tape can be used to attach the slices to a flat surface for finishing. Since there is less tendency for chipping to occur on the edge of the slice where the cutter enters, the most important points should be located on this side. The slice should not be polished since the same effect can be achieved by immersing the slice in a fluid having the same index of refraction as the model material.

For additional details concerning the fabrication and slicing of three-dimensional models the reader should consult the excellent paper by Johnson [11].

15.5 SLICING THE MODEL AND INTERPRETATION OF THE RESULTING FRINGE PATTERNS [13–17]

If the stress pattern frozen into a three-dimensional photoelastic model is observed in the polariscope, the resulting fringe pattern cannot, in general, be interpreted. The conditioned light passing through the thickness of the model integrates the stress difference $\sigma'_1 - \sigma'_2$ over the length of the path of the light so that little can be concluded regarding the state of stress at any point.

To circumvent this difficulty, the three-dimensional model is usually sliced to remove planes of interest which can then be examined individually to determine the state of stress existing in that particular plane or slice. In studies of this type it must be assumed that the slice is sufficiently thin in relation to the size of the model to ensure that the stresses do not change in either magnitude or direction through the thickness of the slice.

The particular slicing plan employed in sectioning a three-dimensional photoelastic model will, of course, depend upon the geometry of the model and the information being sought in the analysis. There are, however, some general principles which can be followed in preparing a slicing plan; these are outlined below.

A. Surface Slices

The free surfaces of a three-dimensional model are principal surfaces since both the stress normal to the surface and the shearing stresses acting on the surface are zero. As an example of a surface slice, consider the flat head on a thick-walled pressure vessel, illustrated schematically in Fig. 15.9. In this example a surface slice of thickness h is removed from the head and examined in normal incidence, i.e., direction of light coincident with the z axis, in the polariscope. The resultant fringe pattern can be interpreted to give

$$\sigma_1 - \sigma_2 = \frac{N_z f_\sigma}{h} \tag{15.1}$$

which can be written as

$$\sigma_{\theta\theta} - \sigma_{rr} = \frac{N_z f_\sigma}{h} \tag{15.2}$$

since, because of symmetry, it is clear that $\sigma_{\theta\theta} - \sigma_{rr} = \sigma_1 - \sigma_2$. The application of either Eq. (15.1) or Eq. (15.2) tacitly assumes that the value of $\sigma_{\theta\theta} - \sigma_{rr}$ is constant through the thickness of the slice and that the directions of the principal stresses do not change along the z axis. To determine the accuracy of these assumptions or to correct for the errors introduced due to changes in the stress distribution with thickness of the slice, the shaving method is often employed. The shaving method

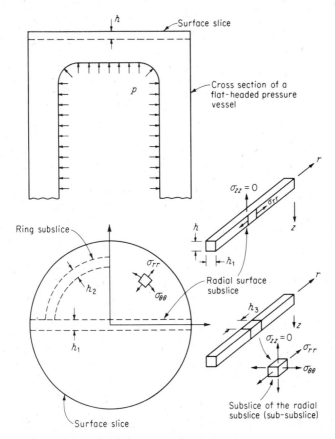

Figure 15.9 Surface slicing on a flat-headed, thick-walled pressure vessel.

consists essentially in removing a thick slice, say $h = \frac{1}{2}$ in, determining the fringe pattern associated with this slice, and then progressively decreasing the thickness of the slice and establishing the fringe pattern corresponding to each thickness. The results of the analysis are then plotted as shown in Fig. 15.10 and extrapolated to $h = 0$ to establish the surface stresses.

It is often advantageous to subslice the surface slice to obtain the individual values of σ_{rr} or $\sigma_{\theta\theta}$. Two such subslices, a radial subslice and a ring subslice, are illustrated in Fig. 15.9. The radial subslice is examined in the polariscope with the light passing through the subslice in the y or θ direction. The resulting fringe pattern gives

$$\sigma_{rr} - \sigma_{zz} = \frac{N_\theta f_\sigma}{h_1} \tag{15.3}$$

Because $$\sigma_{zz} = 0 \text{ at } z = 0,$$

$$\sigma_{rr} = \frac{N_\theta f_\sigma}{h_1} \tag{15.4}$$

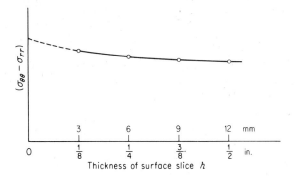

Figure 15.10 Application of the shaving method to establish the surface stress distribution.

By combining Eqs. (15.2) and (15.4) it is clear that

$$\sigma_{\theta\theta} = f_\sigma \left(\frac{N_z}{h} + \frac{N_\theta}{h_1} \right) \tag{15.5}$$

It should be noted in the analysis of the radial subslice that the influence of the $\sigma_{\theta\theta}$ stress, which is coincident with the direction of light, is not effective. Only the stresses which lie in the plane of the slice normal to the direction of light (that is, σ_{zz} or σ_{rr}) influence the fringe pattern. A value of $\sigma_{\theta\theta}$ will occur in this slice; however, this stress will not influence the nature of the fringe pattern.

Next, consider the ring subslice cut from the surface slice, as shown in Fig. 15.9. If this slice is viewed in the polariscope so that the light propagates through the subslice in the radial direction, the resulting fringe pattern will represent

$$\sigma_{\theta\theta} = \sigma_{\theta\theta} - \sigma_{zz} = \frac{N_r f_\sigma}{h_2} \tag{15.6}$$

By combining Eqs. (15.2) and (15.6) it is clear that

$$\sigma_{rr} = f_\sigma \left(\frac{N_r}{h_2} - \frac{N_z}{h} \right) \tag{15.7}$$

Again it is apparent that the radial stress σ_{rr} does not influence the fringe pattern obtained with the ring subslice as long as the light passes through the selected point of the slice in the radial direction.

An alternative procedure which can be employed in the sectioning of the model is the sub-subslice technique. This technique is illustrated in Fig. 15.9, which shows a small cube removed by sectioning the radial subslice. The cube has dimensions h in the z direction, h_1 in the θ direction, and h_3 in the r direction. By viewing the cube in the polariscope in all three possible directions, the following three relationships can be obtained.

With the light in the z direction

$$\sigma_{\theta\theta} - \sigma_{rr} = \frac{N_z f_\sigma}{h} \tag{15.2}$$

With the light in the radial direction

$$\sigma_{\theta\theta} = \sigma_{\theta\theta} - \sigma_{zz} = \frac{N_r f_\sigma}{h_3} \tag{15.6}$$

With the light in the θ direction

$$\sigma_{rr} = \sigma_{rr} - \sigma_{zz} = \frac{N_\theta f_\sigma}{h_1} \tag{15.4}$$

The above example shows how the surface stresses can be established by employing various slicing techniques with a three-dimensional model. Since more than one technique is often available to the investigator, it is advisable to employ at least two different methods and cross-check the results obtained. In this manner the results can be averaged and the errors minimized.

B. Principal Slices Other Than Surface Slices

Often the three-dimensional model will contain planes of symmetry or other planes which are known to be principal. The flat-headed, thick-walled pressure vessel previously discussed in regard to surface slices can also be used as an example to illustrate this topic. The meridional plane presented in Fig. 15.11 represents a plane of symmetry which is also known to be a principal plane. Moreover, in the cylindrical portion of the pressure vessel, the transverse or hoop planes are also known to be principal planes where the theoretical solution presented in Sec. 3.13 will apply. Thus, it is reasonable to section the three-dimensional model of the pressure vessel to obtain a meridional slice and several hoop or transverse slices along the axis of the cylinder. If these slices are examined in the polariscope at normal incidence, the resulting fringe patterns will provide the following data. For the meridional slice

$$\sigma_1 - \sigma_2 = \frac{N f_\sigma}{h} \tag{13.8}$$

where $\sigma_1 - \sigma_2 = \sigma_a - \sigma_{rr}$ in the cylindrical portion of the model and $\sigma_1 - \sigma_2 = \sigma_{rr} - \sigma_{zz}$ in the head portion of the model. In the transition region between the cylinder and the head, the principal directions in the plane are not known since they differ from the axial or radial directions. Isoclinic data can be taken to establish the extent of the transition region and the directions of the principal stresses in this region.

On the external boundaries of the meridional slice

$$\sigma_a = \frac{N f_\sigma}{h} \quad \text{on the vertical boundary (Fig. 15.11)}$$

$$\sigma_{rr} = \frac{N f_\sigma}{h} \quad \text{on the horizontal boundary (Fig. 15.11)}$$

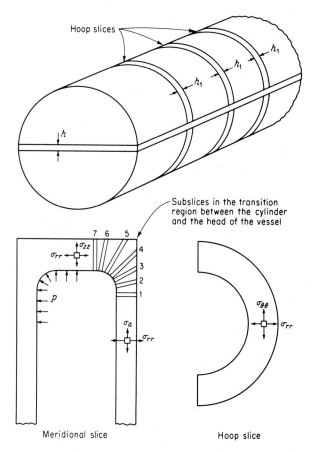

Figure 15.11 Slicing a thick-walled pressure vessel along the planes of principal stress.

On the internal boundaries of the meridional slice

$$\sigma_a = \frac{Nf_\sigma}{h} - p \qquad \text{on the vertical boundary (Fig. 15.11)}$$

$$\sigma_{rr} = \frac{Nf_\sigma}{h} - p \qquad \text{on the horizontal boundary (Fig. 15.11)}$$

At all interior points on the meridional plane, the value of the maximum shear τ_{max} is given by

$$\tau_{max} = \frac{Nf_\sigma}{2h}$$

since one of the two principal stresses is always negative.

The hoop slices in the cylindrical portion of the pressure vessel provide fringe patterns which can be related to the radial and circumferential stresses as shown below:

$$\sigma_{\theta\theta} - \sigma_{rr} = \frac{Nf_{\sigma}}{h_1} \tag{13.8}$$

On the external boundary $\sigma_{rr} = 0$, and

$$\sigma_{\theta\theta} = \frac{Nf_{\sigma}}{h_1}$$

On the internal boundary $\sigma_{rr} = -p$; therefore

$$\sigma_{\theta\theta} = \frac{Nf_{\sigma}}{h_1} - p$$

and at interior points in the hoop slice the maximum shear is given by

$$\tau_{\max} = \frac{\sigma_{\theta\theta} - \sigma_{rr}}{2} = \frac{Nf_{\sigma}}{2h_1}$$

Thus, it is clear that the meridional and hoop slices provide sufficient data to give the individual values of the principal stresses on the surfaces of the pressure vessel and the maximum shear stresses on the two pertinent planes. In the transitional region between the head and cylinder, it is frequently advisable to subslice the meridional slice (see Fig. 15.11) to obtain the value of circumferential stress $\sigma_{\theta\theta}$ on the boundary. Although the subslices are not principal over their entire length, they are principal at the region near the interior boundary and can be employed to obtain $\sigma_{\theta\theta}$ on the interior boundary.

The separation of the principal stresses at interior points in this model is quite involved and requires the use of the shear-difference method in three dimensions, which will be described in Sec. 15.7. If the material from which the prototype is constructed follows the maximum-shear theory of failure, it is not necessary to separate the stresses since the maximum shears are obtained from an examination of the meridional and hoop slices and the pressure required to produce failure by shear can be predicted.

C. The General Slice

In certain instances it is necessary to remove and analyze a slice which does not coincide with a principal plane. In this instance the stresses in the plane of the slice are secondary principal stresses σ_1' and σ_2', which clearly do not coincide with the principal stresses σ_1, σ_2, or σ_3, as shown in Fig. 15.12. The secondary principal stresses in the plane of the slice are given by Eqs. (1.12) as

$$\sigma_1', \sigma_2' = \frac{\sigma_{xx} + \sigma_{yy}}{2} \pm \sqrt{\left(\frac{\sigma_{xx} - \sigma_{yy}}{2}\right)^2 + \tau_{xy}^2} \tag{1.12}$$

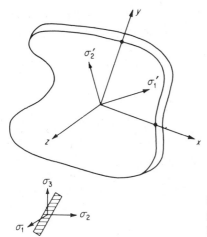

Figure 15.12 A general slice where the secondary principal stresses σ_1' and σ_2' do not coincide with the principal stresses σ_1, σ_2, or σ_3.

Also, the angle which σ_1' makes with the x axis is given by

$$\tan 2\theta = \frac{2\tau_{xy}}{\sigma_{xx} - \sigma_{yy}} \tag{1.14}$$

If this slice is observed at normal incidence, i.e., light transmission parallel to the z axis, the fringe patterns observed, both isoclinics and isochromatics, will be due to the stresses in the xy plane and, further, will not be influenced by the z components of stress. Thus, the isochromatic fringe order can be interpreted by employing

$$\left(\frac{f_\sigma N_z}{h}\right)^2 = [\sigma_1' - \sigma_2']_{xy}^2 = (\sigma_{xx} - \sigma_{yy})^2 + 4\tau_{xy}^2 \tag{15.8}$$

which can easily be derived from Eqs. (13.8) and (1.12).

The isoclinic parameters are related to the stresses in the xy plane by Eqs. (1.14), which show

$$\tan^2 2\theta_z = \frac{4\tau_{xy}^2}{(\sigma_{xx} - \sigma_{yy})^2}$$

$$\cos^2 2\theta_z = \frac{(\sigma_{xx} - \sigma_{yy})^2}{(\sigma_{xx} - \sigma_{yy})^2 + 4\tau_{xy}^2}$$

$$\sin^2 2\theta_z = \frac{4\tau_{xy}^2}{(\sigma_{xx} - \sigma_{yy})^2 + 4\tau_{xy}^2} \tag{15.9}$$

where θ_z is the angle between σ_1' and the x axis with the light passing through the model along the z axis.

The determination of the complete state of stress at an arbitrary interior point on a general slice is extremely involved and requires the use of the shear-difference method in three dimensions. The techniques required for the determination of the complete state of stress (that is, σ_{xx}, σ_{yy}, σ_{zz}, τ_{xy}, τ_{yz}, and τ_{zx}) are presented in Sec. 15.7. The material presented here represents only the interpretation of a normal-incidence examination of a general slice.

15.6 EFFECTIVE STRESSES [17, 18]

The data sought in a three-dimensional analysis will depend to a great degree upon the particular problem being investigated. In certain instances boundary stresses will be sufficient, and the methods presented in Secs. 15.5A and 15.5B can be applied. In other cases a more complete solution is required, and a complete analysis of the general slice will be necessary to obtain the six cartesian components of stress. However, the decision to determine the individual interior stresses should be given a great deal of consideration since the three-dimensional shear-difference method is expensive and time-consuming to use and will lead to inaccurate results unless the method is applied with extreme care and skill.

One alternative to separating the individual stresses in three-dimensional problems is to determine the effective stress σ_e, which is defined below:

$$\sigma_e^2 = \tfrac{1}{2}[(\sigma_1 - \sigma_2)^2 + (\sigma_2 - \sigma_3)^2 + (\sigma_3 - \sigma_1)^2] \qquad (15.10)$$

The effective stress σ_e has found wide acceptance as a criterion (von Mises) for plastic yielding or fatigue failures and as such may have much more significance than the separate shear and normal stresses.

The effective stress can be determined quite simply from a cube cut from a model at any random orientation, as illustrated in Fig. 15.13. The cube is observed in normal incidence on its three mutually orthogonal faces, and the fringe orders N_x, N_y, N_z and the isoclinic parameters θ_x, θ_y, and θ_z are recorded. To illustrate how these six photoelastic measurements can be employed to determine the effective stress σ_e, consider the following expansion of Eq. (15.10):

$$\sigma_e^2 = (\sigma_1^2 + \sigma_2^2 + \sigma_3^2) - (\sigma_1\sigma_2 + \sigma_2\sigma_3 + \sigma_1\sigma_3) \qquad (a)$$

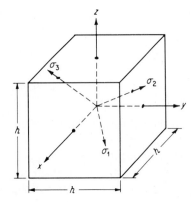

Figure 15.13 Cube removed from any interior point in a three-dimensional photoelastic model. The x, y, z axes are at arbitrary angles relative to the principal coordinate system.

and note from Eqs. (1.9) that

$$\sigma_1^2 + \sigma_2^2 + \sigma_3^2 = I_1^2 - 2I_2 \qquad \sigma_1\sigma_2 + \sigma_2\sigma_3 + \sigma_1\sigma_3 = I_2 \qquad (b)$$

By combining Eqs. (a) and (b) and using Eqs. (1.8), it is clear that

$$\sigma_e^2 = I_1^2 - 3I_2^2 = \sigma_{xx}^2 + \sigma_{yy}^2 + \sigma_{zz}^2 - \sigma_{xx}\sigma_{yy} - \sigma_{yy}\sigma_{zz} - \sigma_{xx}\sigma_{zz}$$

$$+ 3(\tau_{xy}^2 + \tau_{yz}^2 + \tau_{xz}^2) \qquad (15.11)$$

Equation (15.11) can be rearranged into the following form:

$$\sigma_e^2 = \frac{1}{4}\left\{[(\sigma_{yy} - \sigma_{zz})^2 + 4\tau_{yz}^2]\left[2 + \frac{4\tau_{yz}^2}{(\sigma_{yy} - \sigma_{zz})^2 + 4\tau_{yz}^2}\right]\right.$$

$$+ [(\sigma_{xx} - \sigma_{zz})^2 + 4\tau_{xz}^2]\left[2 + \frac{4\tau_{xz}^2}{(\sigma_{xx} - \sigma_{zz})^2 + 4\tau_{xz}^2}\right]$$

$$\left. + [(\sigma_{xx} - \sigma_{yy})^2 + 4\tau_{xy}^2]\left[2 + \frac{4\tau_{xy}^2}{(\sigma_{xx} - \sigma_{yy})^2 + 4\tau_{xy}^2}\right]\right\} \qquad (15.12)$$

Now note from Eq. (15.8) that the three isochromatic observations give

$$\left(\frac{f_\sigma N_x}{h}\right)^2 = [\sigma_1' - \sigma_2']_{yz}^2 = (\sigma_{yy} - \sigma_{zz})^2 + 4\tau_{yz}^2$$

$$\left(\frac{f_\sigma N_y}{h}\right)^2 = [\sigma_1' - \sigma_2']_{xz}^2 = (\sigma_{xx} - \sigma_{zz})^2 + 4\tau_{xz}^2$$

$$\left(\frac{f_\sigma N_z}{h}\right)^2 = [\sigma_1' - \sigma_2']_{xy}^2 = (\sigma_{xx} - \sigma_{yy})^2 + 4\tau_{xy}^2 \qquad (c)$$

and from Eqs. (15.9) that the three isoclinic observations give

$$\sin^2 2\theta_x = \frac{4\tau_{yz}^2}{(\sigma_{yy} - \sigma_{zz})^2 + 4\tau_{yz}^2}$$

$$\sin^2 2\theta_y = \frac{4\tau_{xz}^2}{(\sigma_{xx} - \sigma_{zz})^2 + 4\tau_{xz}^2}$$

$$\sin^2 2\theta_z = \frac{4\tau_{xy}^2}{(\sigma_{xx} - \sigma_{yy})^2 + 4\tau_{xy}^2} \qquad (d)$$

It is evident from an inspection of Eq. (15.12) and Eqs. (c) and (d) that the effective stress may be expressed as

$$\sigma_e^2 = \frac{1}{4}\left[\left(\frac{f_\sigma N_x}{h}\right)^2(2 + \sin^2 2\theta_x) + \left(\frac{f_\sigma N_y}{h}\right)^2(2 + \sin^2 2\theta_y)\right.$$

$$\left. + \left(\frac{f_\sigma N_z}{h}\right)^2(2 + \sin^2 2\theta_z)\right] \qquad (15.13)$$

The value of σ_e obtained through the use of the cube technique together with Eq. (15.13) will permit failure by fatigue or plastic yielding to be predicted for a wide variety of materials frequently employed in the construction of machine components.

15.7 THE SHEAR-DIFFERENCE METHOD IN THREE DIMENSIONS [19, 20]

In certain instances it is necessary to determine the complete state of stress (that is, $\sigma_{xx}, \sigma_{yy}, \sigma_{zz}, \tau_{xy}, \tau_{yz}, \tau_{xz}$) at an arbitrary point in a three-dimensional model. The shear-difference method is the most practical technique available to the photoelastician for completely determining the state of stress at interior points in the model. However, it should be noted that the method involves a stepwise numerical-integration procedure which tends to accumulate error. Hence, the care which must be exercised in collecting the input data is considerable.

The shear-difference method is based on the numerical integration of one of the stress equations of equilibrium, shown below:

$$\frac{\partial \sigma_{xx}}{\partial x} + \frac{\partial \tau_{xy}}{\partial y} + \frac{\partial \tau_{xz}}{\partial z} = 0 \tag{1.3a}$$

If a perfectly general line (OP) is selected in the model in the manner illustrated in Fig. 15.14, Eq. (1.3a) can be integrated to obtain the stress σ_{xx} at the interior point x_1. The integration procedure, which is identical to that presented in Sec. 14.6A, is illustrated below:

$$\int_{x_0}^{x_1} \frac{\partial \sigma_{xx}}{\partial x} \, \partial x + \int_{x_0}^{x_1} \frac{\partial \tau_{xy}}{\partial y} \, \partial x + \int_{x_0}^{x_1} \frac{\partial \tau_{xz}}{\partial z} \, \partial x = 0 \tag{a}$$

If finite but small values of Δx, Δy, and Δz are substituted for the partial differentials, it is possible to write

$$\sigma_{xx}\Big|_{x_1} = \sigma_{xx}\Big|_{x_0} - \frac{\Delta \tau_{xy}}{\Delta y} \Delta x \Big|_{x_0}^{x_1} - \frac{\Delta \tau_{xz}}{\Delta z} \Delta x \Big|_{x_0}^{x_1} \tag{15.14}$$

The value of $\Delta \tau_{xy}/\Delta y$ is obtained by measuring the value of τ_{xy} along lines AB and CD (shown in Fig. 15.14), finding the difference, and dividing by Δy. In a similar fashion, the value of $\Delta \tau_{xz}/\Delta z$ is obtained by determining τ_{xz} along lines EF and GH, finding the difference, and dividing by Δz. If, in particular, $\Delta x = \Delta y = \Delta z$, Eq. (15.14) reduces to

$$\sigma_{xx}\Big|_{x_1} = \sigma_{xx}\Big|_{x_0} - \Delta \tau_{xy}\Big|_{(x_0+x_1)/2} - \Delta \tau_{xz}\Big|_{(x_0+x_1)/2} \tag{15.15a}$$

By continuing this integration in a stepwise procedure, it is possible to write

$$\sigma_{xx}\Big|_{x_2} = \sigma_{xx}\Big|_{x_1} - \Delta \tau_{xy}\Big|_{(x_1+x_2)/2} - \Delta \tau_{xz}\Big|_{(x_1+x_2)/2}$$

$$\cdots\cdots\cdots\cdots\cdots\cdots \tag{15.15b}$$

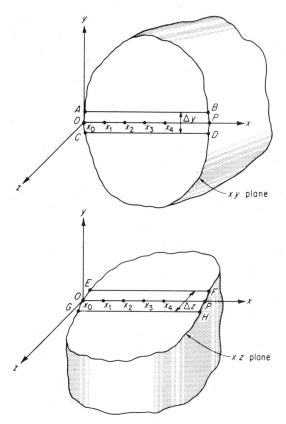

Figure 15.14 An arbitrary line OP and associated auxiliary lines in a general three-dimensional body.

In concept the shear-difference method expressed in terms of Eqs. (15.15) is extremely simple; however, in application the method requires considerable input data obtained along lines OP, AB, CD, EF, and GH. To show the procedure for collecting these data, consider a slice taken from the model which contains the xy plane shown in Fig. 15.14. This slice is then observed in normal incidence, and the isoclinic parameters and isochromatic fringe orders are established along lines OP, AB, and CD. These data can be employed to obtain the shear stresses τ_{xy} along these three lines by employing Eqs. (1.11):

$$\tau_{xy} = \frac{1}{2}(\sigma'_1 - \sigma'_2)_{xy} \sin 2\theta_z = \frac{1}{2}\frac{N_z}{h} f_\sigma \sin 2\theta_z \tag{15.16}$$

where N_z is the isochromatic fringe order observed by passing light through the xy slice in the z direction and θ_z is the angle which σ'_1 makes with the x axis as provided by the isoclinic parameter.

Next a subslice is removed from the xy slice in the manner illustrated in Fig. 15.15 and is observed in the polariscope with the light passing through the

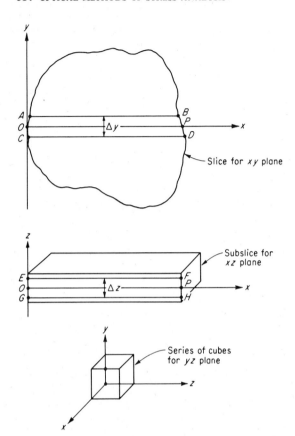

Figure 15.15 The slicing plan normally employed with the shear-difference method in three dimensions.

subslice in the y direction. Isoclinic parameters and isochromatic fringe orders along lines OP, EF, and GH give the shear stress τ_{xz} as

$$\tau_{xz} = \frac{1}{2}(\sigma'_1 - \sigma'_2)_{xz} \sin 2\theta_y = \frac{1}{2}\frac{N_y f_\sigma}{h} \sin 2\theta_y \tag{15.17}$$

At this stage τ_{xy} and τ_{xz} have been established along OP, and sufficient data have been obtained to employ Eqs. (15.15) to arrive at σ_{xx}. The other two normal stresses σ_{yy} and σ_{zz} can be established once σ_{xx} is known by utilizing the following equations, which are apparent from an examination of Mohr's circle (see Fig. 1.12):

$$\sigma_{yy} = \sigma_{xx} - (\sigma'_1 - \sigma'_2)_{xy} \cos 2\theta_z = \sigma_{xx} - \frac{N_x f_\sigma}{h} \cos 2\theta_z$$

$$\sigma_{zz} = \sigma_{xx} - (\sigma'_1 - \sigma'_2)_{xz} \cos 2\theta_y = \sigma_{xx} - \frac{N_y f_\sigma}{h} \cos 2\theta_y \tag{15.18}$$

To evaluate the final cartesian component of stress τ_{yz}, the subslice may be reduced to a series of cubes each containing an evaluation point x_1, x_2, x_3, etc., as its center, as shown in Fig. 15.15. The yz plane of these cubes is examined in the polariscope with the light passing through the cube in the x direction. The isochromatic and isoclinic data are then employed to give τ_{yz}:

$$\tau_{yz} = \frac{1}{2}(\sigma'_1 - \sigma'_2)_{yz} \sin 2\theta_x = \frac{1}{2}\frac{N_x f_\sigma}{h} \sin 2\theta_x \qquad (15.19)$$

To summarize, the shear-difference method requires the determination of isoclinic parameters θ_y and θ_z and isochromatic fringe orders N_y and N_z along selected lines. As indicated in Fig. 15.15, θ_y and N_y are determined along OP, AB, and CD, whereas θ_z and N_z are determined along OP, EF, and GH. Utilizing these data with Eqs. (15.15) to (15.17) will give σ_{xx}, τ_{xy}, and τ_{xz}. The normal stresses σ_{yy} and σ_{zz} are readily determined from existing data by Eqs. (15.18). Finally, a cube technique is presented for determining the final stress component τ_{yz} given in Eq. (15.19). The six cartesian components of stress completely define the state of stress acting at any point along OP. Individual values of the principal stresses and their directions can be obtained by employing Eq. (1.7).

15.8 APPLICATION OF THE FROZEN-STRESS METHOD†

Perhaps one of the best examples of the use of three-dimensional photoelasticity in application to problems related to pressure-vessel design is the analysis of a reactor head. From the point of view of the photoelastician there are two factors which make this type of problem somewhat difficult: (1) the model must be made unusually large in order to scale all the pertinent structural components adequately; (2) the model in its completed state is extremely complex. The size and complexity of the photoelastic model of the reactor head covered in this analysis are shown in Figs. 15.16 and 15.17.

The drawing presented in Fig. 15.16 shows the shape and dimensions of the closure head, the vessel employed for clamping and pressurization, the bolts and springs for loading the flange, and the two sizes of cap used to seal the penetrations in the head. The penetration holes were countersunk to approximately one-half their depth, and end caps were inserted into the countersunk region and cemented on all contacting surfaces.

The model material used for the study was ERL-2774 with 50 parts by weight of phthalic anhydride. Standard procedures for stress-freezing with this material involve heating the model to a critical temperature of about 330°F (165°C). The bolts on the closure head were tightened against the compression springs to give a clamping force equal to 1.57 times the pressure thrust of the head when 5 lb/in² gage (34.5 kPa) pressure was applied to the model. The model was slowly cooled (1.25°F/h or 0.7°C/h) to room temperature with the bolt loads and pressure loads acting on the model. The stresses and strains representing the elastic stress distrib-

---

† The data and figures included in this section have been provided through the courtesy of M. M. Leven, Westinghouse Electric Corp.

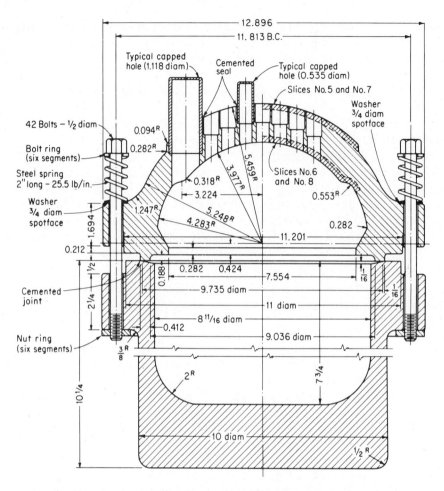

Figure 15.16 Dimensions of the pressure vessel and closure head.

ution due to these forces were permanently locked into the model upon completion of this stress-freezing cycle.

After stress freezing, eight slices were removed from the closure head, as illustrated in Fig. 15.18. Four radial slices, two outer-surface slices, and two inner-surface slices were adequate to provide the necessary photoelastic data.

A. Meridional and Circumferential Stresses in the Transition Region of the Closure Head

The isochromatic fringe patterns obtained from the radial slices, as shown in Fig. 15.19, were interpreted by employing Eqs. (13.8) and (14.4) to give values for the meridional stresses σ_m on the boundary. The circumferential stresses $\sigma_{\theta\theta}$ were obtained by using surface subslices cut from the radial slices, as shown in

Figure 15.17 The assembled closure head ready for testing.

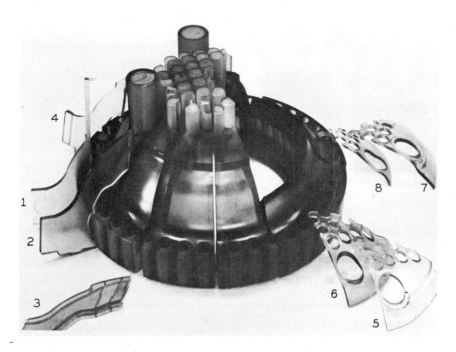

Figure 15.18 The closure head after stress freezing, showing the eight slices removed.

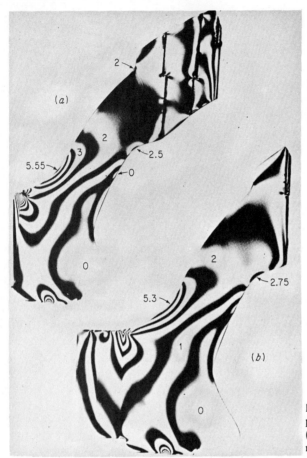

Figure 15.19 Isochromatic fringe patterns for (*a*) radial slice no. 3 (0.254 in thick) and (*b*) radial slice no. 4 (0.248 in thick).

Fig. 15.18. These subslices, when viewed in a direction mutually perpendicular to σ_m and σ_θ, will give

$$\frac{\sigma_m - \sigma_{\theta\theta}}{p} = \frac{N f_\sigma}{hp} \tag{15.20}$$

Since σ_m is known from the examination of the radial slice, the data obtained from the radial subslice, when used with Eq. (15.20), will give $\sigma_{\theta\theta}$ directly.

The distribution of the meridional and circumferential stresses in the transitional region of the closure head is shown in Fig. 15.20. The maximum tensile stress occurs on the outer surface at the knuckle between the head and the flange (see point *A* in Fig. 15.20). The maximum compressive stress occurs at point *D* on the interior surface of the model. The maximum tensile stress at the knuckle fillet *A* is equal to 10.2 times the pressure, or 25,500 lb/in² (175.8 MPa) in the prototype based on a design pressure of 2500 lb/in² (17.2 MPa).

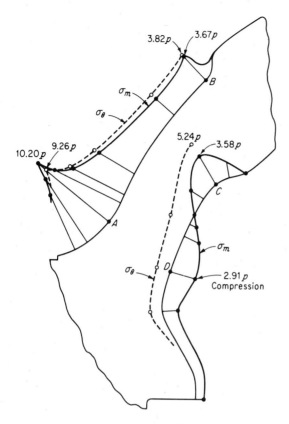

Figure 15.20 Meridional σ_m and circumferential σ_θ stresses along inner and outer surfaces of the closure head, as obtained from radial slice no. 3 and subslices.

B. Stresses in the Region of the Penetrations of the Closure Head

Four surface slices, two from the outer surface and two from the inner surface, were removed from the model as shown in Fig. 15.16 to establish the stresses about the holes in the reactor head. These slices were contoured to give a uniform normal thickness of approximately $\frac{1}{4}$ in (6 mm). Later the slices were thinned to $\frac{3}{16}$ in (4.5 mm) and then to $\frac{1}{8}$ in (3 mm). Since the variation of stress between inner and outer surfaces was very slight (indicating almost pure membrane action), the results obtained for the stresses about the holes were not significantly dependent upon slice thickness.

The direction of propagation of the light in the polariscope was parallel to the axis of the penetrations, so that the thickness of the section traversed by the light was different at all points. Furthermore, at certain ligaments the direction of the tangential stress at the penetration was not perpendicular to the light direction; hence, the component σ_h of the desired tangential stress σ_t was obtained. The effect of the curvature of the slice on the thickness h' and on the horizontal and tangential components of stress is shown in Fig. 15.21. The expressions for determining h', α, β, and σ_t/p are also shown in this figure. In a rectangular array such as this,

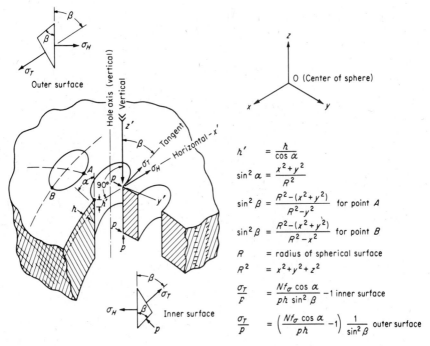

The equations shown in the figure:

$$h' = \frac{h}{\cos \alpha}$$

$$\sin^2 \alpha = \frac{x^2 + y^2}{R^2}$$

$$\sin^2 \beta = \frac{R^2 - (x^2 + y^2)}{R^2 - y^2} \quad \text{for point } A$$

$$\sin^2 \beta = \frac{R^2 - (x^2 + y^2)}{R^2 - x^2} \quad \text{for point } B$$

R = radius of spherical surface

$R^2 = x^2 + y^2 + z^2$

$$\frac{\sigma_T}{P} = \frac{N f_\sigma \cos \alpha}{ph \sin^2 \beta} - 1 \quad \text{inner surface}$$

$$\frac{\sigma_T}{P} = \left(\frac{N f_\sigma \cos \alpha}{ph} - 1\right)\frac{1}{\sin^2 \beta} \quad \text{outer surface}$$

Figure 15.21 Sketch showing the stress $\sigma_h + p$ which vertical incidence of polarized light yields from surface slices.

the maximum stress occurs either at point A, where the ligament is parallel to the yz plane, or at point B, where the ligament is parallel to the xz plane.

A typical example of the fringe pattern obtained from one of the surface slices is shown in Fig. 15.22. It is readily apparent that the maximum fringe order occurred at the refueling penetration V, where a fringe order of 9.8 was established. This proved to be the point of maximum stress (19 times the pressure) in the entire reactor head. At other penetrations on the inner surface, the tangential stresses at the ligaments varied from about 12 to 18.5p. On the outer surface of the head, the tangential stresses at the penetrations varied from 10 to 14 times the pressure. Based on a prototype design pressure of 2500 lb/in² (17.2 MPa), the maximum stress at the refueling penetration V is approximately 43,000 lb/in² (296 MPa).

This example problem shows the applicability of the photoelastic method in solving extremely complex design problems. The whole-field potential of the photoelastic method is quite advantageous in examining the region of the penetrations. It was possible to examine large symmetric portions of the head, select critical locations from among the many possible points of stress concentration, and precisely determine the magnitude of both the model stresses and the prototype stresses. With the degree of complexity offered by this particular reactor-head design, it is doubtful that any analytical solution could be employed to predict the stresses with any degree of confidence. In this example, photoelasticity

Figure 15.22 Isochromatic fringe pattern for inner surface slice no. *6a* of normal thickness $\frac{3}{16}$ in.

was employed to verify the validity of a particular design. The method could also be employed to improve the design. The procedure would involve testing of a number of models after contour changes have been introduced in a systematic manner in regions of high stress.

15.9 THE SCATTERED-LIGHT METHOD [21–27]

The scattered-light method of photoelasticity is based on the scattering character-istics of a wave of light as it passes through a transparent medium. To illustrate this scattering phenomenon, consider a wave of ordinary light propagating in the

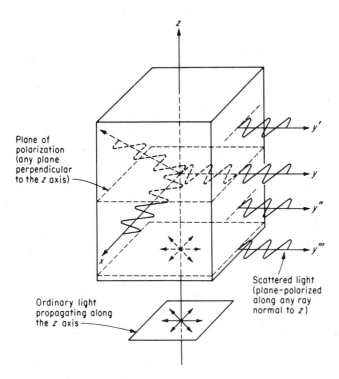

Plane of polarization (any plane perpendicular to the z axis)

Scattered light (plane-polarized along any ray normal to z)

Ordinary light propagating along the z axis

Figure 15.23 Scattered-light phenomenon.

z direction and vibrating in the xy plane, as shown in Fig. 15.23. Normally, this light wave would be viewed in the z direction and would have an intensity proportional to the square of the resultant amplitude of vibrations in the xy plane. However, most materials do not transmit light perfectly, and some scattering of the light occurs. This scattering can be viewed as a secondary set of vibrations which are excited by the main wave and which propagate radially outward from the scattering source. For a main wave of ordinary light propagating in the z direction, the vibrations associated with the scattered light will all lie in planes normal to the z axis. Thus, when the scattered light is viewed along any ray which is normal to the z axis, it will be plane-polarized, as indicated in Fig. 15.23. When the scattered light is viewed along a ray which is not normal to the z axis, the degree of polarization decreases as coincidence with the z axis is approached.

For the purpose of photoelasticity it may be assumed that the photoelastic material has an infinite number of scattering sources uniformly distributed throughout the material. Therefore, the incident light will scatter at every point and will produce a secondary source of plane-polarized light which propagates radially outward from the source. It is possible to use this polarization produced by the scattering of the light within a photoelastic model in place of either the polarizer or the analyzer in a photoelastic polariscope. This utilization of scattered light, which is equivalent to locating a polarizer or an analyzer in the interior of the model, provides an approach to the general three-dimensional problem. Since either the polarizer or the analyzer can be optically positioned at arbitrary planes in the phtoelastic model, stress information can be obtained without stress-freezing or slicing of the model.

A. Scattered Light as an Interior Polarizer or Analyzer

Since the light scattered within a photoelastic model at right angles to the direction of the incident beam is plane-polarized, it can be used as either a polarizer or an analyzer. Consider a beam of unpolarized light which enters through the model at point P, as shown in Fig. 15.24. As the incident light beam passes through the model, each point along the beam acts as a source of plane-polarized light, the direction of vibration being mutually perpendicular to the direction of observation and the incident beam.

Consider first the light scattered from an interior point Q in the model, as defined in Fig. 15.24. If the model is stressed, the polarized light scattered from point Q is resolved into two components along the directions of the secondary principal stresses σ_1' and σ_2'. As the light propagates outward over the distance QR, a phase difference develops between the two components. This phase differ-

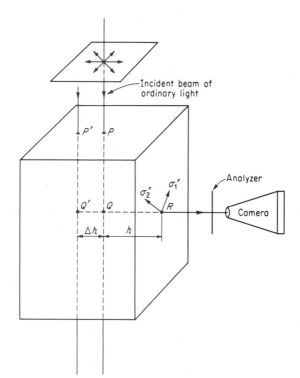

Figure 15.24 Transmission of unpolarized light through a stressed medium.

ence or relative retardation can be measured by inserting an analyzer between point R and the camera. In this instance

$$N_{QR} = \frac{h}{f_\sigma}(\sigma_1' - \sigma_2')_{\text{av } QR} \tag{15.21}$$

where $(\sigma_1' - \sigma_2')_{\text{av}}$ is the average value of $\sigma_1' - \sigma_2'$ over the distance QR if rotational effects are neglected.

If the line which is illuminated is moved from PQ to $P'Q'$, the relative retardation observed at the camera will be given by

$$N_{Q'R} = \frac{h + \Delta h}{f_\sigma}(\sigma_1' - \sigma_2')_{\text{av } Q'R}$$

$$= \frac{1}{f_\sigma}[h(\sigma_1' - \sigma_2')_{\text{av } QR} + \Delta h(\sigma_1' - \sigma_2')_{\text{av } \Delta h}] \tag{15.22}$$

The difference in retardations is due to the additional retardation ΔN acquired by the light in transversing the distance $QQ' = \Delta h$. This value of ΔN can be obtained from Eqs. (15.21) and (15.22) as follows:

$$\Delta N = N_{Q'R} - N_{QR} = \frac{1}{f_\sigma}[\Delta h(\sigma_1' - \sigma_2')_{\text{av } \Delta h}]$$

which may be rewritten as

$$(\sigma_1' - \sigma_2')_{\text{av } \Delta h} = f_\sigma \frac{\Delta N}{\Delta h} \tag{15.23}$$

By determining ΔN and measuring Δh it is possible to measure $\sigma_1' - \sigma_2'$ over a centralized plane of observation of thickness Δh without slicing the model. In practice, a sheet of light is used to illuminate a plane in the model rather than a line as indicated in the previous discussion. The use of a sheet of light permits the determination of a fringe pattern over the whole field of the plane.

In the previously discussed method the scattered light was employed to effectively place a temporary polarizer first at point Q and then at point Q'. It should be noted that the photoelastic effect observed was entirely due to the retardation acquired over the distance QR or $Q'R$. The propagation of the ordinary light over the distance PQ or $P'Q'$ did not influence the fringe pattern.

Scattered light can also be utilized to optically place a temporary analyzer on any plane in the interior of the photoelastic model. This application of scattered light is illustrated in Fig. 15.25, where an incident beam or sheet of plane-polarized light is employed instead of an incident beam or sheet of ordinary light. In this case, the plane-polarized incident beam is resolved as it enters the model into two components along the secondary principal axes associated with σ_1' and σ_2'. These two components travel with different velocities along the secondary principal planes, and in traversing the distance P and Q they acquire a certain relative retardation.

The scattering source located at point P acts as an analyzer since the scattered light is polarized. The resulting image observed along line QC will have an inten-

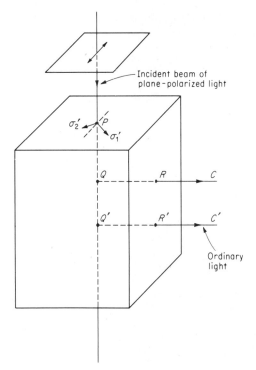

Incident beam of plane-polarized light

Ordinary light

Figure 15.25 Transmission of polarized light through a stressed medium.

sity dependent upon the relative retardation acquired over the path PQ. If the line of observation is moved to $Q'C'$, the relative retardation varies, depending upon the additional retardation acquired over the distance QQ'. Since $QQ' = \Delta h$ and the additional retardation acquired over the distance Δh is ΔN, the ratio $\Delta N/\Delta h$ is the space rate of formation of retardation in successive planes normal to the incident light beam or light sheet. The ratio $\Delta N/\Delta h$ is related to the difference in secondary principal stress $\sigma'_1 - \sigma'_2$ lying in the plane normal to the incident light by

$$(\sigma'_1 - \sigma'_2)_{\text{av } \Delta h} = \frac{\Delta N}{\Delta h} f_\sigma \qquad (15.24)$$

Although Eq. (15.24) is identical to Eq. (15.23), it should be emphasized that the data which must be obtained for Eq. (15.23) require two scattered-light photographs. The data needed for Eq. (15.24) can be obtained from a single scattered-light photograph since the magnitude of the stresses will be proportional to the space gradient of the fringes.

A typical fringe pattern obtained with scattered light incident upon a shaft with a keyway is presented in Fig. 15.26. Inspection of this figure shows that the zero-order fringe is located at the external corners of the keyway where the incident light enters the model. As the incident light propagates through the model, the number of fringes in the scattered-light pattern increases until a fringe order of 10 is obtained near the center of the shaft. It must be emphasized that the fringe

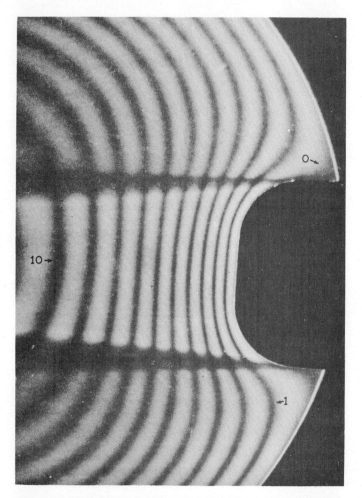

Figure 15.26 Scattered-light fringe pattern for a transverse section of a shaft with a keyway in pure torsion. (*Courtesy of M. M. Leven, Westinghouse Electric Corporation.*)

order obtained in this type of scattered-light fringe pattern is not related to the difference between the principal stresses. Instead, the gradient of the fringes $\Delta N/\Delta h$ gives the magnitude of $\sigma'_1 - \sigma'_2$ in accordance with Eq. (15.24). In this example, the maximum gradient is about the keyway in the shaft; hence the maximum value of $\sigma'_1 - \sigma'_2$ occurs at this location.

Of the two options for use of scattered light, i.e., either as a polarizer or as an analyzer, the scattering is usually used to provide a temporary analyzer. A sheet of light is used to illuminate a given plane in the model, and the light scattered from this plane is photographed to give a whole-field fringe pattern. The gradient of the fringes $\Delta N/\Delta h$ is then used to obtain the magnitude of $\sigma'_1 - \sigma'_2$ in the plane normal to the incident light.

B. A Scattered-Light Method for Determining the Directions of the Secondary Principal Stresses [24]

Since the scattered-light method is most often employed to provide a temporary analyzer in the model, this same scattered-light arrangement will be used to determine the directions of the secondary principal stresses. Recall that plane-polarized incident light is used to illuminate a given plane in the model. This plane-polarized light upon entering the model is resolved into two components, and each propagates with a different velocity along the secondary principal planes. The magnitudes of these two light vectors are

$$E'_{t1} = k \cos \beta \cos \omega t \qquad E'_{t2} = k \sin \beta \cos (\omega t - \Delta) \qquad (11.38)$$

where β is the angle between the plane of vibration of the incident light and the direction of σ'_1.

The intensity of the scattered light is a function of the amplitude of the incident light and as such will vary, depending upon the direction of observation. The magnitude of the scattered-light vector E_s is illustrated in Fig. 15.27 and is, of course, related to the incident-light vectors E'_{t1} and E'_{t2}. Considering the projection of E'_{t1} and E'_{t2} onto a line normal to the direction of observation, the magnitude of the scattered light vector E_s can be expressed as

$$E_s = E'_{t1} \sin \alpha - E'_{t2} \cos \alpha \qquad (a)$$

Combining Eqs. (11.38) and Eq. (a) gives

$$E_s = k[\cos \beta \sin \alpha \cos \omega t - \sin \beta \cos \alpha \cos (\omega t - \Delta)]$$

$$= k(\cos \beta \sin \alpha \cos \omega t - \sin \beta \cos \alpha \cos \omega t \cos \Delta$$

$$- \sin \beta \cos \alpha \sin \omega t \sin \Delta)$$

which can be reduced to

$$E_s = kS \cos (\omega t + \phi) \qquad (b)$$

where

$$S \cos \phi = \cos \beta \sin \alpha - \sin \beta \cos \alpha \cos \Delta$$

$$S \sin \phi = \sin \beta \cos \alpha \sin \Delta \qquad (c)$$

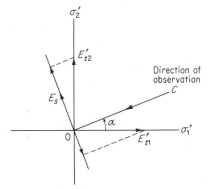

Figure 15.27 The apparent amplitude E_s of the scattered light.

Squaring both sides of Eqs. (*c*) and adding give

$$S^2 = \cos^2 \beta \sin^2 \alpha + \sin^2 \beta \cos^2 \alpha - \tfrac{1}{2} \sin 2\beta \sin 2\alpha \cos \Delta \qquad (d)$$

The intensity of the light scattered in the direction of observation *OC* is proportional to the square of the apparent amplitude *S* and is given by

$$I_s = K(\cos^2 \beta \sin^2 \alpha + \sin^2 \beta \cos^2 \alpha - \tfrac{1}{2} \sin 2\beta \sin 2\alpha \cos \Delta) \qquad (15.25)$$

where *K*, the proportionality constant, includes *k* as well as the scattering properties of the photoelastic material. Thus it is clear that the intensity of the scattered light I_s is controlled by three factors:

1. The relative retardation, as indicated by the $\cos \Delta$ term, which produces the fringe patterns discussed in the previous section
2. The angle between the plane of vibration of the incident beam and the direction of σ_1', as given by $\cos^2 \beta$, $\sin^2 \beta$, or $\sin 2\beta$
3. The angle of observation defined in Fig. 15.27, as given by $\sin^2 \alpha$, $\cos^2 \alpha$, or $\sin 2\alpha$

The results shown in Eq. (15.25) present, in effect, techniques for determining the directions of the secondary principal stresses and also techniques to obtain a clear and sharp scattered-light fringe pattern. These techniques will become obvious upon examination of the five special cases of Eq. (15.25) which are considered below.

Case 1 Let $\beta = 0$, that is, the plane of vibration coincides with σ_1'; $\alpha \neq 0$. Then

$$I_s = K(\sin^2 \alpha)$$

In this case, certain regions on the illuminated plane where $\beta = 0$ will appear bright when the scattered-light pattern is examined. No fringes will be evident in this bright region since the retardation term in Eq. (15.25) ($\sin 2\beta \sin 2\alpha \cos \Delta$) has vanished. When the angle of observation is varied, the intensity of this bright region will change and become a maximum when $\sin \alpha = 1$. In this condition ($\beta = 0$ and $\alpha = \pi/2$) the plane of vibration and the line of observation are at right angles with the former along σ_1' and with the latter along σ_2'. The intensity in this region will go to zero when $\sin \alpha = 0$. In this condition ($\beta = 0$ and $\alpha = 0$) the plane of vibration and the line of observation are parallel and coincident with the direction of σ_1'. These light or dark regions may be considered as scattered-light isoclinic fringes since their formation depends entirely upon the direction of σ_1'.

Case 2 Let $\beta = \pi/2$, that is, the plane of vibration coincides with σ_2'; $\alpha \neq 0$. Then

$$I_s = K(\cos^2 \alpha)$$

Again bright regions will appear on the corresponding scattered-light fringe pattern. The intensity of this pattern will increase to a maximum when α goes to $0°$. In this condition ($\beta = \pi/2$ and $\alpha = 0$) the plane of vibration and the line of observation are crossed with the former along σ'_2 and with the latter along σ'_1. The intensity can be extinguished if α goes to $\pi/2$, where both the plane of vibration and the direction of observation coincide with σ'_2.

Case 3 Let $\beta = \pi/4$; that is, the plane of vibration bisects the angle between σ'_1 and σ'_2. Then

$$I_s = \tfrac{1}{2}K(1 - \sin 2\alpha \cos \Delta)$$

In this case, both the direction of observation and the retardation influence the intensity of the light. To determine directions under these conditions, a compensator is introduced between the polarizer and the model and adjusted to give minimum intensity for any value of α, that is, adjusted to make $\cos \Delta \to 1$. Then the direction of observation is varied until complete extinction is established when $\alpha = \pi/4$. For this condition ($\beta = \pi/4$, $\alpha = \pi/4$, and $\Delta = n\pi$) complete extinction is established; however, this fringe cannot be considered a true isoclinic since its formation depends upon both the retardation and the direction of the principal stresses.

Case 4 Let $\alpha = \pi/4$; that is, the line of observation bisects the angle between σ'_1 and σ'_2. Then

$$I_s = \tfrac{1}{2}K(1 - \sin 2\beta \cos \Delta)$$

This case is similar to case 3 except that the dark fringes obtained correspond to the condition where $\sin 2\beta \cos \Delta = 1$. Again, a compensator is used to control the value of Δ to make $\cos \Delta \to 1$ and the polarizer is rotated to obtain extinction, which occurs when $\beta = \pi/4$.

Case 5 Let $\alpha = \beta = \pi/4$. Then

$$I_s = \tfrac{1}{2}K(1 - \cos \Delta)$$

This case is particularly significant if clear and sharp fringe patterns are to be obtained over a given region of the model where the directions of σ'_1 and σ'_2 remain constant. When $\beta = \alpha = \pi/4$, the intensity relation becomes totally independent of both α and β. This fact implies that only the retardation Δ influences the intensity; hence the quality of the fringe pattern is not compromised because of a uniform background intensity produced by the first two terms in Eq. (15.25). The fringe pattern shown in Fig. 15.26 was obtained with $\alpha = \beta = \pi/4$.

C. Scattered-Light Polariscope [24, 28]

A scattered-light polariscope differs in many respects from the conventional transmitted-light polariscope, and it is therefore usually advisable to construct a new polariscope for scattered-light applications rather than modify a transmitted-light polariscope. A schematic illustration of a simple scattered-light polariscope is presented in Fig. 15.28.

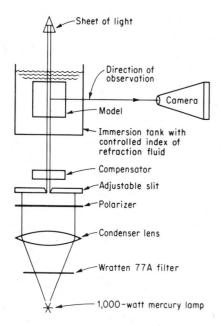

Sheet of light

Direction of observation

Camera

Model

Immersion tank with controlled index of refraction fluid

Compensator

Adjustable slit

Polarizer

Condenser lens

Wratten 77A filter

1,000-watt mercury lamp

Figure 15.28 A schematic illustration of a scattered-light polariscope.

In a scattered-light polariscope the light beam is usually projected in the vertical direction either upward or downward to permit the observation of the scattered-light pattern in the horizontal plane. The light source must be quite intense because of the inefficiency of the scattering process. For this reason a 1000-W mercury-arc lamp or a laser light source is usually employed together with a condenser lens and an adjustable light slit to provide a high-intensity sheet of light which can be used to illuminate any plane in the model.

The model is placed in an immersion tank with an appropriate fluid of the same index of refraction to permit observation from any direction in the horizontal viewing plane. The camera located on this horizontal plane should be capable of rotation about the vertical axis of the polariscope so that the scattered-light pattern can be photographed at any arbitrary angle α. Moreover, the polarizer located forward of the adjustable slit should be mounted so that it can be rotated to vary the angle β. Finally, it is sometimes desirable to locate a compensator or a quarter-wave plate or both in the light path forward of the model. Mountings for these two elements should be provided so that they can also be freely rotated.

REFERENCES

1. Oppel, G.: Polarisationsoptische Untersuchung raumlicher Spannungs- und Dehnungszustande, *Forsch. Geb. Ingenieurw.*, vol. 7, pp. 240–248, 1936.
2. Hetenyi, M.: The Application of the Hardening Resins in Three-dimensional Photoelastic Studies, *J. Appl. Phys.*, vol. 10, pp. 295–300, 1939.
3. Hetenyi, M.: The Fundamentals of Three-dimensional Photoelasticity, *J. Appl. Mech.*, vol. 5, no. 4, pp. 149–155, 1938.

4. Durelli, A. J., and R. L. Lake: Some Unorthodox Procedures in Photoelasticity, *Proc. SESA*, vol. IX, no. 1, pp. 97–122, 1951.
5. Dally, J. W., A. J. Durelli, and W. F. Riley: A New Method to "Lock-in" Elastic Effects for Experimental Stress Analysis, *J. Appl. Mech.*, vol. 25, no. 2, pp. 189–195, 1958.
6. Miyazono, S.: Fixation of Photoelastic Fringe Patterns by Gamma Rays, *J. Appl. Phys.*, vol. 38, no. 5, pp. 2319–2323, 1967.
7. Leven, M. M.: A New Material for Three-dimensional Photoelasticity, *Proc. SESA*, vol. VI, no. 1, pp. 19–28, 1948.
8. Taylor, C. E., E. O. Stitz, and R. O. Belsheim: A Casting Material for Three-dimensional Photoelasticity, *Proc. SESA*, vol. VII, no. 2, pp. 155–172, 1950.
9. Frocht, M. M., and H. Pih: A New Cementable Material for Two- and Three-dimensional Photoelastic Research, *Proc. SESA*, vol. XII, no. 1, pp. 55–64, 1954.
10. Leven, M. M.: Epoxy Resins for Photoelastic Use, pp. 145–165 in "Photoelasticity," Pergamon Press, New York, 1963.
11. Johnson, R. L.: Model Making and Slicing for Three-dimensional Photoelasticity, *Exp. Mech.*, vol. 9, no. 3, pp. 23N–32N, 1969.
12. Nickola, W. E., and M. J. Greaves: Construction of Complex Photoelastic Models Using Thin Molded-Epoxy Sheets, *Exp. Mech.*, vol. 10, no. 4, pp. 23N–30N, 1970.
13. Drucker, D., and R. D. Mindlin: Stress Analysis by Three-dimensional Photoelasticity Methods, *J. Appl. Phys.*, vol. 11, pp. 724–732, 1940.
14. Mindlin, R. D., and L. E. Goodman: The Optical Equations of Three-dimensional Photoelasticity, *J. Appl. Phys.*, vol. 20, pp. 89–95, 1949.
15. Drucker, D. C., and W. B. Woodward: Interpretation of Photoelastic Transmission Patterns for Three-dimensional Models, *J. Appl. Phys.*, vol. 25, no. 4, pp. 510–512, 1954.
16. Leven, M. M.: Quantitative Three-dimensional Photoelasticity, *Proc. SESA*, vol. XII, no. 2, pp. 157–171, 1955.
17. Leven, M. M., and A. M. Wahl: Three-Dimensional Photoelasticity and Its Application in Machine Design, *Trans. ASME*, vol. 80, pp. 1683–1694, 1958.
18. Brock, J. S.: The Determination of Effective Stress and Maximum Shear Stress by Means of Small Cubes Taken from Photoelastic Models, *Proc. SESA*, vol. XVI, no. 1, pp. 1–8, 1958.
19. Frocht, M. M., and R. Guernsey, Jr.: Studies in Three-dimensional Photoelasticity, *Proc. 1st U.S. Natl. Congr. Appl. Mech. 1951*, pp. 301–307.
20. Frocht, M. M., and R. Guernsey, Jr.: Further Work on the General Three-dimensional Photoelastic Problem, *J. Appl. Mech.*, vol. 22, pp. 183–189, 1955.
21. Weller, R., and J. K. Bussey: Photoelastic Analysis of Three-dimensional Stress Systems Using Scattered Light, *J. R. Aeron. Soc.*, vol. 44, pp. 74–88, 1940.
22. Leven, M. M.: Stresses in Keyways by Photoelastic Methods and Comparison with Numerical Solutions, *Proc. SESA*, vol. VII, no. 2, pp. 141–154, 1950.
23. Jessop, H. T.: The Scattered Light Method of Exploration of Stresses in Two- and Three-dimensional Models, *Br. J. Appl. Phys.*, vol. 2, no. 9, pp. 249–260, 1951.
24. Frocht, M. M., and L. S. Srinath: Non-destructive Method for Three-dimensional Photoelasticity, *Proc. 3rd U.S. Natl. Congr. Appl. Mech. 1958*, pp. 329–337.
25. Cheng, Y. F.: Some New Techniques for Scattered-light Photoelasticity, *Exp. Mech.*, vol. 3, no. 11, pp. 275–278, 1963.
26. Srinath, L. S.: Analysis of Scattered-Light Methods in Photoelasticity, *Exp. Mech.*, vol. 9, no. 10, pp. 463–468, 1969.
27. Cheng, Y. F.: An Automatic System for Scattered-Light Photoelasticity, *Exp. Mech.*, vol. 9, no. 9, pp. 407–412, 1969.
28. Swinson, W. F., and C. E. Bowman: Application of Scattered-Light Photoelasticity to Doubly Connected Tapered Torsion Bars, *Exp. Mech.*, vol. 6, no. 6, pp. 297–305, 1966.

SIXTEEN

BIREFRINGENT COATINGS

16.1 INTRODUCTION [1–6]

The method of birefringent coatings represents an extension of the classical procedures of transmission photoelasticity to the determination of surface strains in opaque two- and three-dimensional bodies. The coating is a thin sheet of birefringent material, usually a polymer, which is bonded to the surface of the prototype being analyzed. The coating is mirrored at the interface to provide a reflecting surface for the light. When the prototype is loaded, the displacements on its surface are transmitted to the mirrored side of the coating, where they produce a strain field through the thickness of the coating. The distribution of the strain field over the surface of the prototype, in terms of principal-strain differences, is determined by employing a reflected-light polariscope to record the fringe orders, as illustrated in Fig. 16.1.

The birefringent-coating method has many advantages over other methods of experimental stress analysis. It provides full field data that enable the investigator to visualize the complete distribution of surface strains. The method is nondestructive, and since the coatings can be applied directly to the prototype, the need for models is eliminated. Through proper selection of coating materials, the method can be made applicable over a very wide range of strain. The method is also very useful in converting complex nonlinear stress-analysis problems in the prototype into relatively simple linear elastic problems in the coating. For instance, plastic and viscoelastic response of a prototype can be measured in terms of the elastic response of the birefringent coating. Similarly, the anisotropic re-

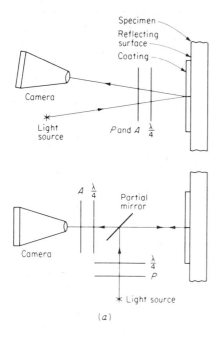

Figure 16.1 Reflection polariscopes commonly employed to record birefringent-coating data.

(a)

sponse of composite materials can be examined in terms of an isotropic response of the coating.

The concept of birefringent coatings was first introduced by Mesnager [1] in France in 1930. The method was reexamined by Oppel [2] in Germany in 1937. Early efforts were not successful due to the lack of suitable adhesives, the low sensitivity of the available polymers, and the reinforcing effects occurring when glass was employed as a coating. When the epoxies became available, in the early 1950s, the bonding and sensitivity problems were alleviated. Significant development of the method in the area of materials and technique were made by Fleury and Zandman [3] in France in 1954 and later by D'Agostino, Drucker, Liu, and Mylonas [4, 5] in the United States and by Kawata [6] in Japan. Activity by many capable researchers in recent years has provided the theory for birefringent coatings, instruments for laboratory analysis, procedures for extending the method to plasticity and anisotropic-elasticity problems, and solutions to a wide variety of industrial problems.

16.2 COATING STRESSES AND STRAINS [7]

When the specimen is loaded, the surface displacements of the specimen at the interface are transmitted to the birefringent coating if the bond is adequate. As the coating responds to these transmitted displacements, stresses and an associated amount of birefringence are induced. Observation of the coating by means of a reflection polariscope gives a fringe pattern which is related to the surface strains of the specimen. If it is assumed that the coating is sufficiently thin, the strains occurring on the surface of the specimen are transmitted to the coating without distortion. Within the framework of this assumption, it is clear that

$$\sigma_3 = \sigma_{zz} = 0 \quad \text{in both coating and specimen}$$

$$\epsilon_1^c(x, y) = \epsilon_1^s(x, y) \qquad \epsilon_2^c(x, y) = \epsilon_2^s(x, y) \tag{16.1}$$

where the coordinate system is defined in Fig. 16.2.

Since the strains expressed in Eqs. (16.1) are identical to those described in Eqs. (4.2), it is possible to employ Eqs. (4.5) directly to express the coating stresses in terms of the specimen stresses.

$$\sigma_1^c = \frac{E^c}{E^s(1 - v^{c2})}[(1 - v^c v^s)\sigma_1^s + (v^c - v^s)\sigma_2^s]$$

$$\sigma_2^c = \frac{E^c}{E^s(1 - v^{c2})}[(1 - v^c v^s)\sigma_2^s + (v^c - v^s)\sigma_1^s] \tag{16.2}$$

Subtracting the two relations given in Eqs. (16.2) yields

$$\sigma_1^c - \sigma_2^c = \frac{E^c}{E^s}\frac{1 + v^s}{1 + v^c}(\sigma_1^s - \sigma_2^s) \tag{16.3}$$

Inspection of Eq. (16.3) shows that the difference in the principal stresses acting in the coating $\sigma_1^c - \sigma_2^c$ is linearly related to the difference in the principal stresses acting on the surface of the specimen $\sigma_1^s - \sigma_2^s$. The elastic constants E^c, E^s, v^c, and v^s influence the magnitude of $\sigma_1^c - \sigma_2^c$. The photoelastic response of the coating is related to $\sigma_1^c - \sigma_2^c$ by employing Eq. (13.8) and noting that the optical path length is $2h^c$. Thus

$$\sigma_1^c - \sigma_2^c = \frac{Nf_\sigma}{2h^c} = \frac{E^c}{E^s}\frac{1 + v^s}{1 + v^c}(\sigma_1^s - \sigma_2^s)$$

and the difference in the principal stresses for the specimen is given by

$$\sigma_1^s - \sigma_2^s = \frac{E^s}{E^c}\frac{1 + v^c}{1 + v^s}\frac{Nf_\sigma}{2h^c} \tag{16.4}$$

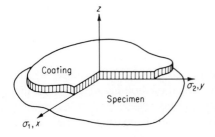

Figure 16.2 Coordinate system related to the specimen and the coating.

It is clear that $\sigma_1^s - \sigma_2^s$ can be determined by the observation of the isochromatic fringe order in the birefringent coating provided E^s, E^c, v^c, v^s, f_σ, and h^c are known. In certain instances it may be preferable to work in terms of strain instead of stress. This transformation is quite simple since it has been assumed that $\epsilon_1^c - \epsilon_2^c = \epsilon_1^s - \epsilon_2^s$; hence

$$\epsilon_1^s - \epsilon_2^s = \frac{Nf_\epsilon}{2h^c} = \frac{N}{2h^c} \frac{1 + v^c}{E^c} f_\sigma \tag{16.5}$$

By using Eq. (16.5), the birefringent coating can be employed as a strain gage to give the difference in principal strains $\epsilon_1^s - \epsilon_2^s$.

The strain-optic relationship presented in Eq. (16.5) is often written in the form

$$\epsilon_1^s - \epsilon_2^s = \epsilon_1^c - \epsilon_2^c = \frac{Nf_\epsilon}{2h^c} = \frac{N}{2h^c} \frac{\lambda}{K} \tag{16.6}$$

where K is the strain coefficient for the coating and λ is the wavelength in microinches of the light being used; that is, 21.5 μin corresponds to 546.1 nm for mercury green light. This alternative form of the strain-optic law is more general since the same strain coefficient K for a coating (provided by the manufacturer) can be employed with light sources having different wavelengths. For a perfectly linear elastic photoelastic material the constants f_σ, f_ϵ, and K are related by the expressions

$$f_\epsilon = \frac{\lambda}{K} = \frac{1 + v^c}{E^c} f_\sigma \qquad f_\sigma = \frac{E^c}{1 + v^c} \frac{\lambda}{K} = \frac{E^c}{1 + v^c} f_\epsilon \tag{16.7}$$

The isochromatic and isoclinic data at a point in a coating can be employed with Eq. (2.18) to establish the shearing strain γ_{xy}^s. Thus

$$\gamma_{xy}^s = \gamma_{xy}^c = (\epsilon_1^c - \epsilon_2^c) \sin 2\phi_1 \tag{16.8}$$

where ϕ_1 is the isoclinic parameter defining the angle between σ_1 and the x axis. From Eqs. (16.5) and (16.8), it is clear that

$$\gamma_{xy}^s = \frac{Nf_\epsilon}{2h^c} \sin 2\phi_1 \tag{16.9}$$

and from the stress-strain relationship expressed by Eq. (2.20)

$$\tau_{xy}^s = \frac{E^s}{2(1 + v^s)} \frac{Nf_\epsilon}{2h^c} \sin 2\phi_1 = \frac{E^s}{E^c} \frac{1 + v^c}{1 + v^s} \frac{Nf_\sigma}{4h^c} \sin 2\phi_1 \tag{16.10}$$

The photoelastic data obtained from the isochromatic and isoclinic fringe patterns are not sufficient, in general, to determine the individual values of the principal stresses and strains. Auxiliary methods, which must be employed to determine σ_1, σ_2, and τ_{max}, will be discussed in Sec. 16.7.

16.3 COATING SENSITIVITY

The ability of a photoelastic coating to respond optically to a stress field in a specimen is controlled by a number of factors. The effects of several important factors can be evaluated by defining a stress sensitivity index S_σ^s as

$$S_\sigma^s = \frac{N}{\sigma_1^s - \sigma_2^s} \tag{16.11}$$

By substituting Eqs. (16.4) and (16.7) into Eq. (16.11) the stress sensitivity index can be expressed in terms of coating properties and specimen properties as

$$S_\sigma^s = \frac{2h^c}{f_\epsilon} \frac{1 + v^s}{E^s} = C_c C_s \tag{16.12}$$

where $C_c = 2h^c/f_\epsilon$ is the coating coefficient of sensitivity and $C_s = (1 + v^s)/E^s$ is the specimen coefficient of sensitivity. The form of the equation for the stress-sensitivity index indicates that optical response can be increased, for a given coating and specimen material, only by increasing the coating thickness. Arbitrary increases in coating thickness are usually not possible, however, because of errors associated with thick coatings.

In applications of coatings to elastic-stress problems, the maximum response of the coating occurs when some point on the specimen yields. If the principal stresses in the specimen are of opposite sign, and if the specimen material follows the Tresca yield criterion, the maximum stresses and strains that the specimen can support before yielding are

$$(\sigma_1^s - \sigma_2^s)_{\max} = S_y \tag{16.13a}$$

and
$$(\epsilon_1^s - \epsilon_2^s)_{\max} = C_s S_y \tag{16.13b}$$

where S_y is the yield strength of the specimen material. The maximum fringe order indicated by the coating just before yielding of the specimen follows from Eqs. (16.11), (16.12), and (16.13a). Thus

$$N_{\max} = C_c C_s S_y = \frac{2h^c}{f_\epsilon} \frac{1 + v^s}{E^s} S_y \tag{16.14}$$

It is evident from Eq. (16.14) that the maximum fringe order is a function of the three parameters S_y, C_s, and C_c. Typical values for the maximum fringe order in a typical coating on a wide range of engineering materials are listed in Table 16.1.

It is evident from Table 16.1 that the maximum optical response exhibited by a photoelastic coating depends to a great extent on the properties of the specimen material. Maximum fringe orders in the table range over two orders of magnitude, with very low responses on concrete and glass and relatively high responses on high-strength steel, aluminum, and fiber-glass materials. The coating thickness and the methods employed to determine fringe orders in any elastic-stress analysis will depend on the specific problem. Where the optical response is low, thick

Table 16.1 Yield strength, specimen coefficient of sensitivity, strain difference, and maximum fringe order at yielding for a typical birefringent coating

Material	S_y		C_s		$(\epsilon_1^s - \epsilon_2^s)_{max}$ μm/m	N_{max},† fringes
	lb/in²	MPa	10^{-8} in²/lb	pm²/N		
Steel:						
HR 1020	35,000	240	4.3	6.2	1,500	1.86
CD 1020	45,000	310	4.3	6.2	1,940	2.40
HT 1040	80,000	550	4.3	6.2	3,450	4.26
HT 4140	130,000	900	4.3	6.2	5,600	6.92
HT 5210	180,000	1240	4.3	6.2	7,700	9.58
Maraging (18 Ni)	250,000	1720	4.9	7.1	12,200	15.20
Aluminum:						
1100 H16	20,000	140	13.3	19.3	2,660	3.30
3004 H34	29,000	200	13.3	19.3	3,860	4.78
2024 T3	50,000	345	13.3	19.3	6,670	8.25
7075 T6	73,000	500	13.3	19.3	9,700	12.00
Magnesium, AM 11	21,000	145	20.8	30.2	4,380	5.42
Cartridge brass	63,000	435	8.7	12.6	5,480	6.80
Phosphor bronze	75,000	515	8.7	12.6	6,520	8.10
Beryllium copper	70,000	480	7.7	11.2	5,400	6.70
Glass	3,000	21	12.5	18.1	375	0.46
Concrete (compression)	4,000	28	42.0	60.9	1,680	2.10
Plastic, nylon 6-6	12,000	83	467	677	56,000	69.00
Glass-reinforced plastic	120,000	830	31.8	46.1	38,200	47.40

† For $C_c = 1.24 \times 10^3$ fringes, which corresponds to $h^c = 0.10$ in = 2.54 mm, $f_\epsilon = 1.62 \times 10^{-4}$ in/fringe = 4.10 × μm/fringe, $K = 0.14$, $\lambda = 22.7$ μin = 578 nm.

coatings or point-by-point compensation methods are required. With higher-strength materials, thin coatings can be employed, and the fringe orders can be photographed or observed directly on the model.

16.4 COATING MATERIALS [8–12]

One of the most important factors in a photoelastic analysis is the selection of the proper material for the photoelastic model. Indeed, major developments in the application of two- and three-dimensional photoelasticity and birefringent coatings occurred only after the introduction of suitable materials. The physical properties which an ideal coating should exhibit include the following:

1. A high optical strain coefficient K to maximize the number of fringes per unit strain
2. A low modulus of elasticity E^c to minimize reinforcing effects
3. A high resistance to both optical and mechanical stress relaxation to maximize stability of the measurement with time

4. A linear strain-optical response to minimize data-reduction problems
5. A good bondability to ensure perfect strain transmission at the interface between coating and specimen
6. A high proportional limit with respect to strain to maximize the range of strain over which the coating can be utilized
7. An adequate pliability to permit utilization on curved surfaces of three-dimensional components

In most instances, the selection of a coating material involves a compromise since no material exhibits all these characteristics. As an example, the first coating material employed by Mesnager was glass, which exhibited a relatively high sensitivity to strain. A glass coating reinforces the component significantly, however, and it cannot be applied to curved surfaces. A wide variety of polymeric materials are available which can be employed as photoelastic coatings. Typical properties of some of the materials are listed in Table 16.2.

An examination of Table 16.2 indicates that the polycarbonate material exhibits a superior combination of properties. Unfortunately, it is available only in sheet form, and techniques have not yet been established to allow contouring of the sheets for three-dimensional work. The epoxy materials also exhibit good sensitivity and are preferred when contouring is necessary. The casting and contouring process with epoxy materials will be discussed in Sec. 16.4. For large strains, which are often encountered in plastic analyses, modified epoxies (a blend of rigid and flexible resins of a copolymer of epoxy and polysulfide) can be

Table 16.2 Properties of different photoelastic materials for coating applications

Material	Modulus E		K	Sensitivity		Strain limit %	Ref.
	lb/in^2	MPa		$1/f_\epsilon$			
				fringes/in[†]	fringes/mm		
Polycarbonate	320,000	2,210	0.16	7,300	287		9, 10
PS-1	360,000	2,480	0.14	6,500	256	10.0	‡
Epoxy with anhydride	475,000	3,280	0.12	5,400	213	2.0	8
Epoxy with amine§	450,000	3,100	0.09	4,000	157		
PS-2	450,000	3,100	0.12	5,400	213	3.0	‡
Polyester¶	560,000	3,860	0.04	1,950	78	1.5	
Modified epoxy	2,800	19	0.02	1,000	39	15.0	6
PS-3	30,000	210	0.02	940	37	30.0	‡
Polyurethane††	500	3.5	0.008	380	15		
PS-4	1,000	7.0	0.005	230	9	150.0	‡
Glass	10,000,000	69,000	0.14	6,500	256	0.10	

† With mercury green light ($\lambda = 21.5\ \mu$in = 546 nm).
‡ Commercially available from Photoelastic, Inc., Malvern, Pa.
§ 100 parts per hundred ERL 2774, 15 parts per hundred TETA.
¶ Homalite 100.
†† 100 parts per hundred Hysol 2085, 24 parts per hundred Hysol 3562.

employed as coatings since they can be compounded to exhibit a linear response over a large range of strain (30 percent). Polyurethane rubber is used in applications involving very large strains where sensitivity is no problem but where the strains encountered range from 30 to 100 percent. Glass is most commonly used in transducer applications, where its excellent stability is an important characteristic.

16.5 APPLICATION OF COATINGS [12]

The successful use of photoelastic coatings depends on a perfect bond between the coating and the component. For this reason, careful attention must be given to component and coating-surface preparation, the adhesive, and the bonding procedure. The surface of the component should be smooth and free of all foreign material. A satisfactory component surface can normally be obtained by sanding and cleaning with suitable chemical agents and solvents. For plane components, a precured sheet of coating can be machined to size, cleaned, and then bonded to the component. Proper matching of the geometry of the coating to that of the test specimen is required, during bonding, with this procedure. Alternatively, the sheet of coating can be bonded to the test specimen and after the adhesive cures, the specimen can be used as a template to machine the coating to the proper shape.

The adhesive normally used with birefringent coatings is an aluminum-filled polymer. Once the adhesive is spread evenly over the surface of the component, the sheet of coating is applied. The coating should be positioned at one end and slowly rotated and pressed into position to work out the air bubbles. Pressure is not normally applied as the adhesive cures since residual stresses could be introduced in the coating by the clamping devices. The final layer of adhesive should be between 0.003 and 0.010 in (0.08 and 0.25 mm) thick, depending primarily on the uniformity of the specimen surface.

The application of photoelastic coatings to curved surfaces, associated with three-dimensional components, is best achieved by using the contoured-sheet method. This method is illustrated in Fig. 16.3, where the essential steps in the operational sequence are depicted. Basically, a sheet of epoxy is cast to the desired thickness on a plate coated with Teflon or silicone rubber. The polymerization process can be made to proceed slowly by selecting proper amine curing agents. During the first stage of polymerization, the coating material slowly transforms from a liquid (stage A) to a rubbery solid (stage B). In stage B, the sheet is soft and pliable. Also, the strain-optic coefficient is very low; therefore, large deformations can be imposed on the sheet without introducing photoelastic response. At this stage, the sheet should be stripped from the surface plate and contoured by hand to fit the curved surfaces of the test specimen.

During the next stage of polymerization, as the coating becomes rigid (stage C) and photoelastically responsive, it should be maintained in contact with the surface of the test specimen. When the polymerization is completed, the rigid contoured shell of coating can be stripped from the specimen, cleaned, and

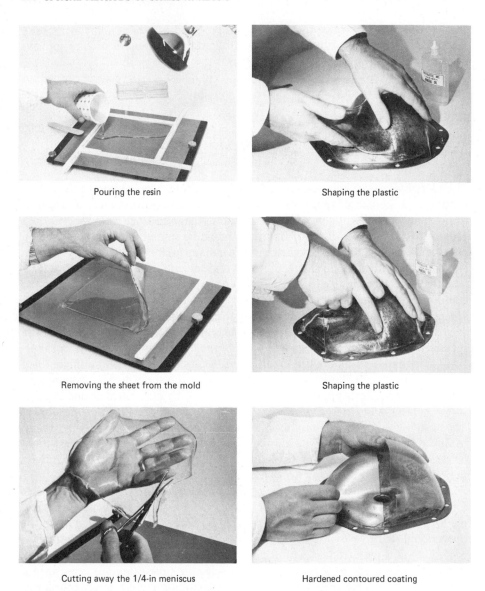

Figure 16.3 Sequence of operations in applying contoured sheets. (*Photoelastic, Inc.*)

checked for thickness variations. The shell can then be trimmed to match adjoining sections of coating and cemented in place.

This contour-sheet method enables the coating to be applied to specimens of almost any shape. It permits the coating to be formed into stress-free shells of reasonably uniform thickness which conform perfectly to the surface to be coated. In actual operation, some experience is required to develop the skills associated with forming the shells used to coat a complicated three-dimensional specimen.

16.6 EFFECTS OF COATING THICKNESS

When a photoelastic coating is used on a specimen, only in very idealized instances are the strains transmitted to the coating without some modification or distortion. More realistically, the coating can be considered as a three-dimensional extension of the specimen which is loaded by means of shear and normal tractions at the interface. These tractions vary so that the displacements experienced by the coating and the specimen at the interface are identical (as dictated by perfect bonding). Thus in the most general case:

1. The average strain in the coating does not equal the strain at the interface.
2. A strain gradient exists through the thickness of the coating.
3. The coating serves to reinforce the specimen.

It is evident that these three thickness effects tend to vanish as the coating thickness approaches zero. However, coatings with finite thickness (usually 0.020 to 0.120 in or 0.50 to 3.00 mm) are required to obtain adequate optical response from the coating for accurate fringe-order determinations. Thus, the question naturally arises regarding the magnitude of the error associated with thickness effects in the application of the coating method to different stress-analysis problems. This issue is complex and has been the subject of some controversy since the mid 1950s. Some aspects of the issue remain unresolved; nevertheless, an understanding has been reached pertaining to many of the factors contributing to thickness effects in photoelastic coatings.

The topic of thickness effects will be treated in this book by beginning with the simple model of the coating considered previously in Sec. 16.2. The model of the coating will be made progressively more complex as additional factors influencing the behavior of the coating are introduced. Where possible, experimental verifications will be used to justify assumptions or to minimize the complexity of the analysis.

A. Reinforcing Effects of Birefringent Coatings [13]

When a metallic specimen is coated with a birefringent coating and subjected to loads, the coating carries a portion of the load and consequently the strain on the part is reduced by a certain amount. It is possible in many cases to calculate the reinforcing effect due to the birefringent coating and establish correction factors which can be employed in a simple fashion to account for the reinforcement. In this section, the reinforcement due to the coating will be computed for plane-stress and flexural problems. In the plane-stress problem, an element from a coated specimen can be isolated as shown in Fig. 16.4. A similar element from an uncoated specimen can also be isolated, and the forces acting in the x direction on both elements can be equated to yield

$$h^s \, dy \, \sigma_{xx}^u = h^s \, dy \, \sigma_{xx}^s + h^c \, dy \, \sigma_{xx}^c \qquad \sigma_{xx}^u = \sigma_{xx}^s + \frac{h^c}{h^s} \sigma_{xx}^c \qquad (16.15a)$$

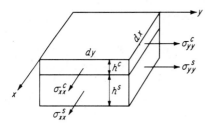

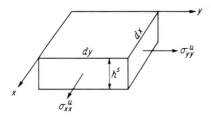

Figure 16.4 Comparison of stresses in coated and un-coated elements from a plane-stress specimen.

The corresponding expression for forces in the y direction is

$$\sigma_{yy}^u = \sigma_{yy}^s + \frac{h^c}{h^s}\sigma_{yy}^c \tag{16.15b}$$

If it is again assumed that

$$\epsilon_{xx}^c = \epsilon_{xx}^s \qquad \epsilon_{yy}^c = \epsilon_{yy}^s \qquad \sigma_{zz}^c = \sigma_{zz}^s = 0$$

both the coating and the specimen are in a state of plane stress, and from Eqs. (2.20) it is apparent that

$$\sigma_{xx} = \frac{E}{1 - v^2}(\epsilon_{xx} + v\epsilon_{yy}) \qquad \sigma_{yy} = \frac{E}{1 - v^2}(\epsilon_{yy} + v\epsilon_{xx}) \tag{16.16}$$

Substitution of Eqs. (16.16) into Eqs. (16.15) gives

$$\frac{E^s}{1 - v^{s2}}(\epsilon_{xx}^u + v^s\epsilon_{yy}^u) = \frac{E^s}{1 - v^{s2}}(\epsilon_{xx}^s + v^s\epsilon_{yy}^s)$$

$$+ \frac{h^c}{h^s}\frac{E^c}{1 - v^{c2}}(\epsilon_{xx}^c + v^c\epsilon_{yy}^c) \tag{a}$$

$$\frac{E^s}{1 - v^{s2}}(\epsilon_{yy}^u + v^s\epsilon_{xx}^u) = \frac{E^s}{1 - v^{s2}}(\epsilon_{yy}^s + v^s\epsilon_{xx}^s)$$

$$+ \frac{h^c}{h^s}\frac{E^c}{1 - v^{c2}}(\epsilon_{yy}^c + v^c\epsilon_{xx}^c) \tag{b}$$

Subtracting Eq. (b) from Eq. (a) and simplifying yields

$$\epsilon_{xx}^u - \epsilon_{yy}^u = \left(1 + \frac{h^c}{h^s}\frac{E^c}{E^s}\frac{1 + v^s}{1 + v^c}\right)(\epsilon_{xx}^c - \epsilon_{yy}^c) \tag{16.17}$$

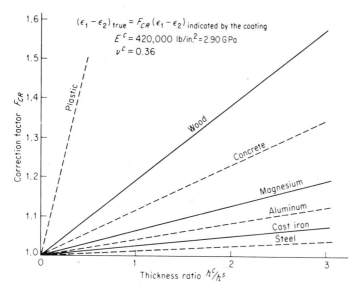

Figure 16.5 Correction factor F_{CR} for a number of different specimen materials.

This equation may be rewritten as

$$\epsilon_{xx}^u - \epsilon_{yy}^u = F_{CR}(\epsilon_{xx}^c - \epsilon_{yy}^c)$$

where

$$F_{CR} = \left(1 + \frac{h^c}{h^s}\frac{E^c}{E^s}\frac{1+v^s}{1+v^c}\right) \tag{16.18}$$

The term F_{CR} represents a correction factor which must be applied to the value of $\epsilon_{xx}^c - \epsilon_{yy}^c$ obtained from the birefringent coating to establish the true value of the principal strain difference in the uncoated specimen. The correction factor F_{CR} adequately accounts for the reinforcement due to the presence of the birefringent coating.

A graph showing F_{CR} as a function of h^c/h^s is presented for a number of different specimen materials in Fig. 16.5. These results are based on values of $E^c = 420,000$ lb/in^2 (2.90 GPa) and $v^c = 0.36$, which are representative of the rigid epoxy and polycarbonate coating materials. Inspection of Fig. 16.5 shows that the correction factor is small for values of h^c/h^s less than 1 provided the specimen material is metallic. If, however, the specimen material is wood, concrete, or plastic, the correction factor becomes quite appreciable.

B. Strain Variations through the Coating Thickness [13]

A second example which illustrates the effect of strain variation through the thickness of the coating together with reinforcing effects is that of a wide plate subjected to a pure bending moment M. Consider an element from the central region of this plate, as indicated in Fig. 16.6. If it is assumed that the strain

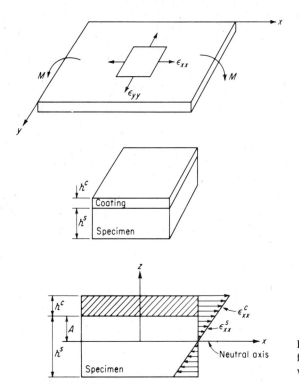

Figure 16.6 Element with a bire-fringent coating from the center of a wide plate in bending.

distribution is linear and that it is transmitted through the specimen-coating interface, then from elementary plate theory the strain in both the specimen and the coating can be expressed as a function of z:

$$\epsilon_{xx}^s = \frac{z}{\rho} \qquad \text{for } (h^s - A) \le z \le A$$

$$\epsilon_{xx}^c = \frac{z}{\rho} \qquad \text{for } A \le z \le (A + h^c)$$

$$\epsilon_{yy}^s = \epsilon_{yy}^c = 0 \qquad \text{for all } z \tag{16.19}$$

where z is measured from the neutral axis, A is the distance from the neutral axis to the interface, and ρ is the radius of curvature. Since σ_{zz} is assumed to vanish for all values of z, Eqs. (16.16) can be employed with Eqs. (16.19) to express the stress σ_{xx} in terms of the strains:

$$\sigma_{xx}^s = \frac{E^s}{1 - v^{s2}} \frac{z}{\rho} \qquad \text{for } h^s - A \le z \le A$$

$$\sigma_{xx}^c = \frac{E^c}{1 - v^{c2}} \frac{z}{\rho} \qquad \text{for } A \le z \le A + h^c \tag{16.20}$$

The position of the neutral axis, as described by the symbol A, can be obtained by considering equilibrium of the plate in the x direction:

$$\int_{A-h^s}^{A} \sigma_{xx}^s \, dz + \int_{A}^{A+h^c} \sigma_{xx}^c \, dz = 0 \tag{16.21}$$

Substituting Eqs. (16.20) into Eq. (16.21), integrating, and solving for A give

$$A = \frac{h^s}{2} \frac{1 - BC^2}{1 + BC} \tag{16.22}$$

where

$$B = \frac{E^c}{E^s} \frac{1 - v^{s2}}{1 - v^{c2}} \qquad C = \frac{h^c}{h^s}$$

The radius of curvature ρ can be computed by considering equilibrium of the moments, where

$$M = \int_{A-h^s}^{A} \sigma_{xx}^s z \, dz + \int_{A}^{A+h^c} \sigma_{xx}^c z \, dz \tag{16.23}$$

By substituting Eqs. (16.20) into Eq. (16.23), integrating, and using Eq. (16.22) to simplify the results, it can be shown that

$$\frac{1}{\rho} = \frac{12M}{H} \frac{1 - v^{s2}}{E^s h^{s3}}$$

where

$$H = 4(1 + BC^3) - \frac{3(1 - BC^2)^2}{1 + BC} \tag{16.24}$$

If the coating is examined in a reflection polariscope, the fringe pattern is proportional to the average of the strain difference $\epsilon_{xx}^c - \epsilon_{yy}^c$ through the coating thickness. The average strain can be computed from Eqs. (16.19), (16.22), and (16.23) as

$$(\epsilon_{xx}^c - \epsilon_{yy}^c)_{av} = \frac{12M}{H} \frac{1 - v^{s2}}{E^s h^{s3}} \frac{1}{h^c} \int_{A}^{A+h^c} z \, dz$$

which yields

$$(\epsilon_{xx}^c - \epsilon_{yy}^c)_{av} = \frac{6M}{H} \frac{1 - v^{s2}}{E^s h^{s2}} \frac{1 + C}{1 + BC} \tag{16.25}$$

Since the true difference of strain on the surface of an uncoated plate is

$$(\epsilon_{xx}^s - \epsilon_{yy}^s)_{true} = 6M \frac{1 - v^{s2}}{E^s h^{s2}} \tag{16.26}$$

it is clear by comparison of Eqs. (16.25) and (16.26) that the coating does not indicate the true difference in the surface strains. It is possible, however, to correct for the error by introducing a bending correction factor. Thus,

$$(\epsilon_{xx}^s - \epsilon_{yy}^s)_{true} = F_{CB}(\epsilon_{xx}^c - \epsilon_{yy}^c)_{av}$$

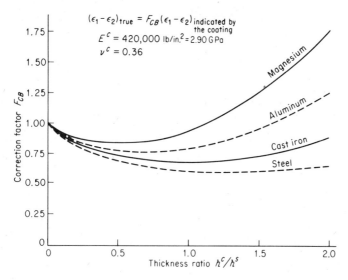

Figure 16.7 Correction factor F_{CB} for a number of different specimen materials.

where the bending correction factor F_{CB} is

$$F_{CB} = \frac{H(1 + BC)}{1 + C} = \frac{1 + BC}{1 + C}\left[4(1 + BC^3) - \frac{3(1 - BC^2)^2}{1 + BC}\right] \quad (16.27)$$

The bending correction factor, which is a function of the dimensionless ratios B and C, is shown in graphical form in Fig. 16.7.

This correction factor accounts for the two thickness effects which occur in this example. The first is a reinforcing effect, which reduces the strain at the interface between the coating and the specimen in comparison with a beam without the coating. The second effect is due to the nonuniform distribution of strain through the coating. The optical response of the coating is related to the average strain, which in this instance represents the strain at the midpoint in the coating. Since the average coating strain is higher than the interface strain, the effects of strain variation through the thickness of the coating tend to offset the reinforcing effect of the coating. In applications of coatings on beams and plates, the coating thickness h^c is usually much less than the specimen thickness h^s. Inspection of Fig. 16.7 indicates that the correction factor for small ratios of h^c/h^s is appreciable and should not be neglected.

C. Poisson's Ratio Mismatch [14]

In plane-stress problems, the errors arising from coating thickness effects, i.e., reinforcement, strain gradient, and curvature, are usually small. However, in almost all cases, a mismatch in Poisson's ratio occurs with v^c usually greater than v^s. This difference in Poisson's ratio produces a distortion of the displacement and strain fields through the thickness of the coating which experimental evidence indicates is particularly pronounced at the boundaries.

The significance of the Poisson's ratio mismatch effect can be examined by assuming that the strains ϵ_1 tangent to the boundary are transmitted without loss or amplification so that $\epsilon_1^c = \epsilon_1^s$. This implies that the distortion of the strain field occurs primarily in the transverse direction. At the interface, the transverse strain in the coating ϵ_2^c is controlled by the specimen. Thus

$$\epsilon_2^c = \epsilon_2^s = -v^s \epsilon_1^s \qquad (a)$$

At the free surface of the coating, the transverse strain can occur in accordance with Poisson's ratio of the coating. Thus

$$\epsilon_2^c = -v^c \epsilon_1^c = -v^c \epsilon_1^s \qquad (b)$$

The average value of ϵ_2^c through the thickness of the coating is bounded by these two limiting values; therefore, the fringe-order response of the coating at the boundary is bounded by

$$\frac{(1 + v^s)\epsilon_1^s}{F_\epsilon} < N < \frac{(1 + v^c)\epsilon_1^s}{F_\epsilon} \qquad (16.28)$$

where $F_\epsilon = f_\epsilon / 2h^c$.

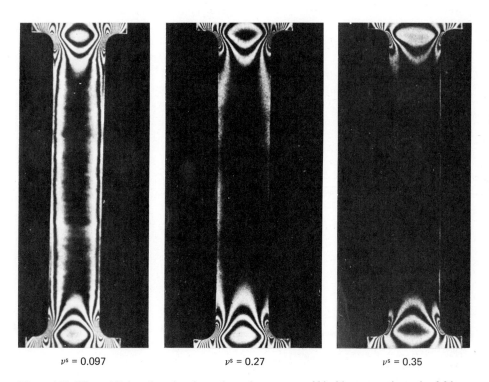

$v^s = 0.097$ $v^s = 0.27$ $v^s = 0.35$

Figure 16.8 Effect of Poisson's ratio mismatch on the response of birefringent coatings, $v^c = 0.36$.

It is clear from this inequality that the magnitude of the distortion is controlled by the mismatch parameter $(1 + v^c)/(1 + v^s)$. For a constant value of $v_c = 0.36$ and variations in v_s between 0 and 0.5, the mismatch parameter ranges from 1.36 to 0.90. Experiments with tensile specimens, illustrated in Fig. 16.8, indicate that the fringe orders on the boundary and in interior regions of the specimen are given by

$$
N = \begin{cases} \dfrac{\epsilon_1^s(1 + v^c)}{F_\epsilon} & \text{on the boundary} \\[3ex] \dfrac{\epsilon_1^s(1 + v^s)}{F_\epsilon} & \text{in the interior} \end{cases} \tag{16.29}
$$

A transition zone exists near the boundary, where the relation between fringe order and strain in a tension specimen can be expressed as

$$
N = [1 + v^s + C_v(v^c - v^s)]\frac{\epsilon_1^s}{F_\epsilon} \tag{16.30}
$$

where C_v is a correction factor accounting for the mismatch. Experiments conducted with tension specimens with a large mismatch parameter (1.24) and with coatings ranging in thickness from 0.022 to 0.130 in (0.55 to 3.25 mm) indicated that the width of the transition zone is about four times the thickness of the coating. The value of C_v decreases from 1 at the boundary to zero at a position $4h^c$ from the boundary, as indicated in Fig. 16.9.

For metallic specimen materials, $v^c - v^s$ is usually less than 0.06, which is relatively small in comparison with $1 + v^s \approx 1.3$ in Eq. (16.30). As a result, the effect of Poisson's ratio mismatch can often be neglected when dealing with metal components.

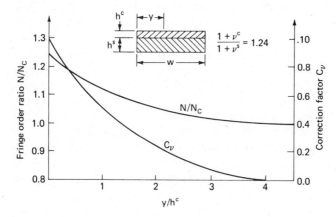

Figure 16.9 Correction factor C_v as a function of h^c/h^s.

D. Influence of Strain Gradient [15–19]

Duffy and his associates [15–17] have approached the problem of the influence of strain gradients and the thickness of the coating by prescribing a displacement field at the interface between the specimen and the coating and then determining the average strain distribution through the thickness of the coating. In the simplest of the three cases treated by Duffy, the displacement field at the interface is only a function of x and is prescribed by the Fourier series

$$u_0 = \frac{A_0}{2} + \sum_{n=1}^{n=\infty} A_n \cos p_n x + \sum_{n=1}^{n=\infty} B_n \sin p_n x$$

$$v_0 = 0$$

$$w_0 = \frac{C_0}{2} + \sum_{n=1}^{n=\infty} C_n \cos p_n x + \sum_{n=1}^{n=\infty} D_n \sin p_n x \qquad (16.31)$$

where A_n, B_n, C_n, D_n = Fourier coefficients needed to describe an arbitrary
strain field
$p_n = 2\pi/\lambda_n$ = wave number
λ_n = wavelength of the displacement field
The coordinate system is defined in Fig. 16.2.

The coating was treated as a three-dimensional elastic body, and an exact solution for the displacement field $u = u(x, z)$, $v = 0$, and $w = w(x, z)$ was obtained. Although the solution of this elasticity problem is beyond the scope of this text, some of the results provide insight into the behavior of a coating. For example, the results indicate that the average strain difference through the thickness of the coating can be expressed as

$$(\epsilon_1^c - \epsilon_2^c)_{av} = \frac{Nf_\epsilon}{2h^c} = \frac{1}{h^c}\left(Fh^c \frac{\partial u_0}{\partial x} + Gw_0\right) \qquad (16.32)$$

where F is a correction factor associated with longitudinal strain at the interface between the specimen and the coating and G is a correction factor associated with curvature of the interface. Values for the correction factors F and G are shown plotted as a function of λ/h^c in Fig. 16.10. Inspection of Fig. 16.10 shows that $F \to 1$ and $G \to 0$ for very thin coatings $(h^c/\lambda \to 0)$. Thus, for very thin coatings, the fringe order N given by Eq. (16.32) becomes $2h^c\epsilon_0/f_\epsilon$, indicating a perfect strain transmission from the specimen to the coating. In the more general case of a coating with finite thickness, $F < 1$ and $G > 0$; therefore, some error will result due to strain variations through the coating.

This theoretical treatment, using sinusoidal displacement fields for u_0, v_0, and w_0, predicts a substantial distortion of the average strain field transmitted to the coating by the specimen. For example, with $h^c/\lambda = 0.1$, the average strain is about 65 percent of the surface strain, indicating an error of 35 percent. The existence of these large errors has not been demonstrated in the laboratory because the very high strain gradients implied by sinusoidal displacement fields are difficult to produce with usual loadings. A more realistic approach is to consider displace-

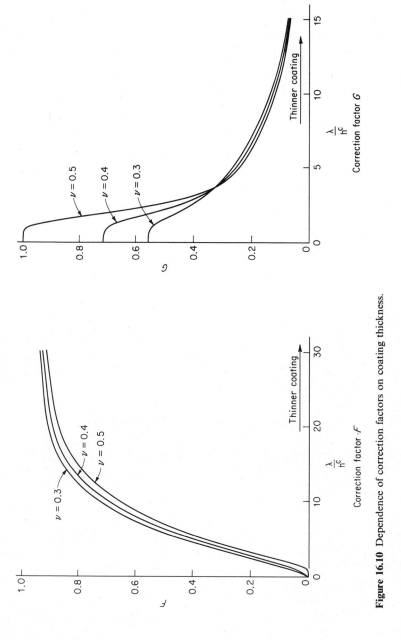

Figure 16.10 Dependence of correction factors on coating thickness.

ment fields as polynomials in a power-series expansion to develop strain fields associated with the usual loadings. Duffy has also considered the power-series representation and has shown that constant strains or strains which vary linearly in x do not produce a variation in strain across the thickness of the coating. Since most strain fields encountered in practice contain significant constant and linear terms, the relative error produced by thickness effects, resulting from the higher-order terms in the strain field, are appreciably reduced.

16.7 FRINGE-ORDER DETERMINATIONS IN COATINGS [20]

The methods employed to determine the order of the fringes exhibited by the coating depends upon the response of the coating and the accuracy required in the analysis. If the response of the coating is large (four or more fringes), monochromatic light can be used to obtain photographs of the light- and dark-field isochromatic fringe patterns. These fringes can usually be interpolated or extrapolated to the nearest 0.2 fringe, giving an accuracy of about 5 percent based on a maximum of four fringes. An example of a high-response fringe pattern which can be analyzed as indicated above is shown in Fig. 16.11.

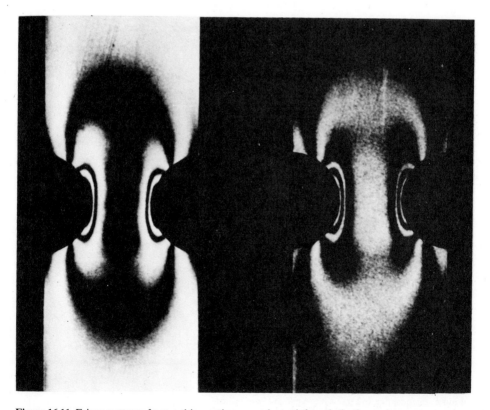

Figure 16.11 Fringe patterns from a thin coating on a glass-reinforced plastic specimen.

For fringe patterns exhibiting between two and four fringes, the use of colored patterns produced with white light is advantageous. The colored pattern is produced by the attenuation and extinction of one or more colors from the white spectrum. The observed colored fringes represent the complementary color produced by the transmitted portion of the white-light spectrum. The sequence of colored fringes produced in a field where the stress is increasing is listed in Table 16.3. The exact color shade will be a function of the energy distribution in the white-light spectrum and the recording characteristics of the colored film being used; however, the sequence listed in the table is adequate for most visual observations.

It is evident from Table 16.3 that the use of white light substantially increases the number of fringes that can be identified. For example, in the interval $0 \leq N \leq 2$, 12 different color bands exist, which can be used to establish fractional fringe orders. Moreover, the polariscope can also be used in the light-field position to yield a second family of colored fringes to effectively double the amount of data

Table 16.3 Sequence of colors produced in a dark-field polariscope with white light

Color	Retardation, nm	Fringe order
Black	0	0
Gray	160	0.28
White	260	0.45
Yellow	350	0.60
Orange	460	0.79
Red	520	0.90
Tint of passage no. 1†	577	1.00
Blue	620	1.06
Blue-green	700	1.20
Green-yellow	800	1.38
Orange	940	1.62
Red	1,050	1.81
Tint of passage no. 2†	1,150	2.00
Green	1,350	2.33
Green-yellow	1,450	2.50
Pink	1,550	2.67
Tint of passage no. 3†	1,730	3.00
Green	1,800	3.10
Pink	2,100	3.60
Tint of passage no. 4†	2,300	4.00
Green	2,400	4.13

† The tint of passage is a sharp dividing zone between red and blue in the first-order fringe, red and green in the second-order fringe, and pink and green in the third-, fourth-, and fifth-order fringes. Beyond five fringes, white-light analysis is not practical since the colors become very pale and difficult to distinguish.

Basic reflection polariscope

Basic polariscope with camera or microscope

Basic polariscope with compensator
and digital readout instrumentation

Figure 16.12 Reflection polariscope. (*Photoelastic, Inc.*).

available for estimating fractional fringe orders. Thus, the fringe order can be established to within 0.1 fringe, giving an accuracy of about 5 percent based on a maximum of 2 fringes.

For precise fringe-order determinations, where the maximum fringe order is less than 2 and accuracies of 5 percent or less are required in the analysis, it is necessary to use compensation techniques. These techniques are point-by-point methods which significantly improve the accuracy of the fringe-order determination but require detailed involvement of the investigator. Compensation methods similar to those described in Sec. 14.4 are employed. When a Babinet-Soleil compensator is properly aligned with a coating free of residual optical effects, the load-induced fringe orders N can be determined to within ± 0.01 fringe, giving an accuracy of about 2 percent based on a maximum of 0.5 fringes. With the Tardy method of compensation, the fringe order can be determined to within ± 0.02 fringe.

Specialized photoelastic equipment for use with birefringent coatings is illustrated in Fig. 16.12.

16.8 STRESS SEPARATION METHODS

Examination of Eq. (16.4) shows that the isochromatic data from a birefringent coating permit the determination of $\sigma_1^s - \sigma_2^s$ but not the individual values of the principal stresses. The separation methods most widely used with photoelastic coatings are oblique incidence and strip coatings.

A. Oblique Incidence [21, 22]

The oblique-incidence method for separation of the principal stresses, where the stress directions at a point are known from previous isoclinic determinations, was discussed previously in Sec. 14.6D. When the method is used with photoelastic coatings, some modifications in the equations are required since it is convenient to work with coating strains rather than coating stresses. A normal-incidence observation at a point in the coating would yield a normal-incidence fringe order N_0 given by Eq. (16.6) as

$$N_0 = \frac{2h^c}{f_\epsilon}(\epsilon_1 - \epsilon_2) \tag{16.33}$$

The superscripts on the strain terms in the equations of this section will be omitted since it is assumed that specimen and coating strains are identical. If an oblique-incidence observation is made by rotating the direction of observation about ϵ_1 by an angle θ_1, the oblique-incidence fringe order $N_{\theta 1}$ is given by Eq. (16.6) as

$$N_{\theta 1} = \frac{2h^c}{f_\epsilon \cos \theta_1}(\epsilon_1 - \epsilon_2') \tag{16.34}$$

where ϵ'_2 is the secondary principal strain in the planes normal to the incident and reflected light beams. The secondary principal strain ϵ'_2 is given by Eq. (2.18) as

$$\epsilon'_2 = \epsilon_2 \cos^2 \theta_1 + \epsilon_3 \sin^2 \theta_1 \tag{16.35}$$

and since the coating is under a state of plane stress, Eq. (2.21) indicates that

$$\epsilon_3 = -\frac{v^c}{1 - v^c}(\epsilon_1 + \epsilon_2) \tag{16.36}$$

Substituting Eqs. (16.35) and (16.36) into Eq. (16.34) yields

$$N_{\theta 1} = \frac{2h^c}{f_\epsilon} \frac{1}{(1 - v^c) \cos \theta_1} [(1 - v^c \cos^2 \theta_1)\epsilon_1 - (\cos^2 \theta_1 - v^c)\epsilon_2] \tag{16.37}$$

Solution of Eqs. (16.33) and (16.37) for ϵ_1 and ϵ_2 gives

$$\epsilon_1 = \frac{f_\epsilon}{2h^c} \frac{1}{(1 + v^c) \sin^2 \theta_1} [(1 - v^c) \cos \theta_1 N_{\theta 1} - (\cos^2 \theta_1 - v^c)N_0]$$

$$\epsilon_2 = \frac{f_\epsilon}{2h^c} \frac{1}{(1 + v^c) \sin^2 \theta_1} [(1 - v^c) \cos \theta_1 N_{\theta 1} - (1 - v^c \cos^2 \theta_1)N_0]$$

$$\tag{16.38}$$

The derivation of Eq. (16.38) implies that N_0 and $N_{\theta 1}$ are always positive numbers. Since the in-plane principal strains are ordered with $\epsilon_1 \geq \epsilon_2$, N_0 will always be positive. From Eq. (16.34), it is clear that $N_{\theta 1}$ will be positive only when $\epsilon_1 > \epsilon'_2$. The sign of $N_{\theta 1}$ can quickly be determined by inserting a compensator perpendicular to ϵ_1. If the optical effect of $\epsilon_1 - \epsilon'_2$ can be canceled by using the compensator as a tension device, then $\epsilon_1 > \epsilon'_2$ and $N_{\theta 1}$ is positive. If the cancellation is accomplished by using the compensator as a compression device, then $\epsilon'_2 > \epsilon_1$ and $N_{\theta 1}$ must be treated as a negative number in computing the principal strains in Eqs. (16.38).

Once the principal strains in the specimen have been determined, the principal stresses in the specimen can be determined by substituting the strains into Eqs. (2.20).

B. Strip Coatings [23]

A photoelastic strip coating is composed of a large number of closely spaced parallel strips, as shown in Fig. 16.13. This coating, because of its discontinuous geometry, does not react isotropically to strain. The strips tend to transmit the strains along the length of the strip but attenuate any transverse or shearing strains. The behavior of a strip coating is similar to the behavior of an electrical-resistance strain gage as expressed by Eq. (6.3). As a result of the strain-transmitting characteristics of a strip coating, its strain-optic law can be expressed as

$$N_s = S_a \epsilon_a + S_t \epsilon_t + S_s \gamma_{at} \tag{16.39}$$

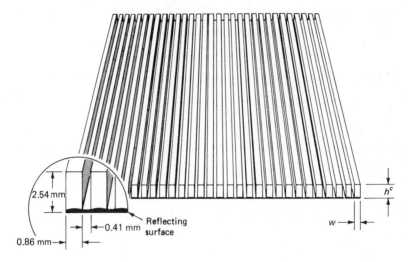

Figure 16.13 Photoelastic strip coatings.

where S_a = sensitivity of strip to axial strains
S_t = sensitivity of strip to transverse strains
S_s = sensitivity of strip to shearing strains
ϵ_a = normal strain in axial direction
ϵ_t = normal strain in transverse direction
γ_{at} = shearing strain associated with axial and transverse directions
N_s = normal-incidence fringe order exhibited by strip

The values of the strain sensitivities depend on the geometry of the strip, i.e., its height h^c and its width w. Strips with a large w/h^c ratio exhibit the optical response of a continuous coating. When the w/h^c ratio is small, the optical response is due primarily to the axial strain since both S_t and S_s are small. O'Regan [23] has shown that $S_t < 0.01S_a$ and $S_s < 0.02S_a$ for a strip coating having a ratio $w/h^c = 0.34$. Thus for strip coatings with a small w/h^c ratio, the terms containing S_t and S_s in Eq. (16.39) can be neglected and the strain-optic law can be written as

$$N_s = S_a\epsilon_a = \frac{2h^c}{f_\epsilon}\epsilon_a \qquad (16.40)$$

When the strips are positioned close together on a specimen, the regions of extinction in adjacent strips blend together to form continuous fringes, as shown in Fig. 16.14. Thus, the strip coatings provide whole-field data for one component of normal strain.

Strip coatings can be employed in three different ways to establish the individual principal stresses in the specimen. First, they can be employed in a manner similar to strain gages, where three different strip coatings are used to determine the strains $\epsilon_{0°}$, $\epsilon_{45°}$, and $\epsilon_{90°}$. Equations (10.9) for the three-element rectangular rosette can then be employed to compute the principal stresses.

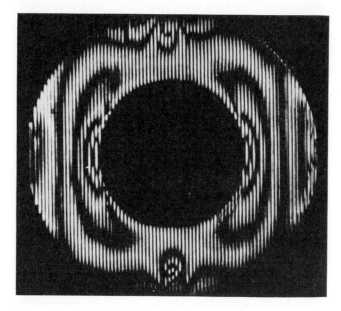

Figure 16.14 Photoelastic fringe pattern from a strip coating. (*Courtesy of R. O'Regan.*)

The second method requires use of two strip coatings to provide the fringe orders N_{sx} and N_{sy} and a continuous coating to provide the fringe order N_0. The principal stresses in the specimen and the principal angle can be expressed in terms of these optical measurements as

$$\sigma_1^s = \frac{E^s f_\epsilon}{2h^c} \left[\frac{N_{sx} + N_{sy}}{2(1 - v^s)} + \frac{N_0}{2(1 + v^s)} \right]$$

$$\sigma_2^s = \frac{E^s f_\epsilon}{2h^c} \left[\frac{N_{sx} + N_{sy}}{2(1 - v^s)} - \frac{N_0}{2(1 + v^s)} \right]$$

$$\cos 2\phi = \frac{N_{sx} - N_{sy}}{N_0} \tag{16.41}$$

The third method involves use of a strip coating to provide N_{sx} and a continuous coating to provide N_0 and ϕ_1, the angle between the principal strain and the x axis. The principal stresses in the specimen can be expressed in terms of these data as

$$\sigma_1^s = \frac{E^s f_\epsilon}{2h^c} \left[\frac{N_{sx}}{1 - v^s} + \frac{N_0(\sin^2 \phi_1 - v^s \cos^2 \phi_1)}{1 - v^{s2}} \right]$$

$$\sigma_2^s = \frac{E^s f_\epsilon}{2h^c} \left[\frac{N_{sx}}{1 - v^s} - \frac{N_0(\cos^2 \phi_1 - v^s \sin^2 \phi_1)}{1 - v^{s2}} \right] \tag{16.42}$$

Current use of strip coatings is limited since they are not marketed commercially. Electrical-resistance strain gages are often used to provide the additional data needed to separate stresses at selected points of a specimen after full-field observations have been made with a continuous birefringent coating.

EXERCISES

16.1 Determine the specimen coefficient of sensitivity for the following materials:
  (a) Mild steel AISI 1010 (b) High-strength steel AISI 4640
  (c) Aluminum 24S (d) Aluminum 2S
  (e) Hastelloy A (f) Inconel X
  (g) Magnesium M1 (h) Red brass
  (i) Admiralty metal (j) Plexiglas
  (k) Titanium

16.2 For a coating with $h^c = 1.50$ mm and $f_\epsilon = 3.80$ μm/fringe, determine the maximum fringe order N which could be developed in the materials listed in Exercise 16.1.

16.3 Specify a coating, i.e., material and thickness, to use for the analysis of the following steel ($E^s = 200$ GPa and $\nu^s = 0.30$) structures:
  (a) A curved beam where $\sigma_{max} = 150$ MPa
  (b) A spherical shell with radius $R = 3.00$ m and wall thickness $t = 15$ mm subjected to a pressure $p = 1500$ kPa
  (c) A cylindrical shell with radius $R = 1500$ mm and wall thickness $t = 15$ mm subjected to a pressure $p = 5.00$ MPa

16.4 Determine the maximum fringe order developed in the coatings specified in Exercise 16.3.

16.5 Discuss the procedure to be used to install a coating on a panel with an elliptical hole.

16.6 Discuss the procedure to be used to install a coating at the intersection between two circular cylinders of the same diameter.

16.7 A coating is placed on an aluminum (2024 T351) tension strip of known dimensions w and h. The tension strip is loaded with a known load P, and the resulting fringe order N is measured with a reflection polariscope. Outline the procedure for determining the material fringe value f_ϵ for the coating from these data. If a sodium ($\lambda = 589.3$ nm) light source is used, determine the strain coefficient K for the coating.

16.8 At an interior point on a steel (AISI 1030) specimen with a polycarbonate coating ($h^c = 4.00$ mm and $f_\epsilon = 5.10$ μm/fringe) a value of $N = 1.85$ and $\phi_1 = 27°$ is measured. Describe the state of stress in the specimen at this point.

16.9 If the coating and measurements of Exercise 16.8 are associated with a point on the free boundary of a panel specimen with in-plane loads, describe the state of stress at the point.

16.10 Determine the correction factor needed to account for reinforcing effects of the coating for a plane-stress specimen fabricated from the materials listed in Exercise 16.1 if:

  (a) $\dfrac{h^c}{h^s} = 0.1$ (b) $\dfrac{h^c}{h^s} = 0.2$ (c) $\dfrac{h^c}{h^s} = 0.5$ (d) $\dfrac{h^c}{h^s} = 1.0$ (e) $\dfrac{h^c}{h^s} = 2.0$

16.11 Verify Eq. (16.17).

16.12 Verify Eq. (16.22).

16.13 Verify Eq. (16.24).

16.14 For the materials listed in Exercise 16.1, determine the correction factor F_{CB} for wide plates in bending if:

  (a) $\dfrac{h^c}{h^s} = 0.1$ (b) $\dfrac{h^c}{h^s} = 0.2$ (c) $\dfrac{h^c}{h^s} = 0.5$ (d) $\dfrac{h^c}{h^s} = 1.0$ (e) $\dfrac{h^c}{h^s} = 2.0$

16.15 A birefringent coating ($E^c = 2.50$ GPa and $v^c = 0.36$) is used on a plane-stress tensile specimen fabricated from glass-reinforced plastic ($E^s = 27.5$ GPa and $v^s = 0.20$) to measure the stress concentration factor resulting from a centrally located hole. If $N_{max} = 4.5$ on the boundary of the hole and $N_0 = 1.00$ at an interior point well removed from the hole, determine the stress concentration factor due to the hole.

16.16 A polycarbonate coating ($h^c = 2.5$ mm) is used on an aluminum 1100 plane-strain specimen. The specimen is loaded (in plane) until certain regions have yielded. If $\epsilon_2 < 0$ in these regions, determine the color of the fringe delineating the elastic plastic boundary. Assume that the observation is made in white light in a dark-field polariscope.

16.17 Simplify Eq. (16.38) for rapid application by selecting $\theta_1 = 45°$.

16.18 Use the results of Exercise 16.17 to determine the principal stresses at a point in a steel specimen if $f_\epsilon = 5.10$ μm/fringe and $h^c = 2.00$ mm when $N_0 = 2.64$ and $N_{\theta 1} = 1.32$.

16.19 Verify Eq. (16.41).

16.20 Verify Eq. (16.42).

REFERENCES

1. Mesnager, M.: Sur la détermination optique des tensions intérieures dans les solides à trois dimensions, *C. R. (Paris)*, vol. 190, p. 1249, 1930.
2. Oppel, G.: Das polarisationsoptische Schichtverfahren zur Messung der Oberflachenspannung am beanspruchten Bauteil ohne Modell, *Z. Ver. Dtsch. Ing.*, vol. 81, pp. 803–804, 1937.
3. Fleury, R., and F. Zandman: Jauge d'efforts photoélastique, *C. R. (Paris)*, vol. 238, p. 1559, 1954.
4. D'Agostino, J., D. C. Drucker, C. K. Liu, and C. Mylonas: An Analysis of Plastic Behavior of Metals with Bonded Birefringent Plastic, *Proc. SESA*, vol. XII, no. 2, pp. 115–122, 1955.
5. D'Agostino, J., D. C. Drucker, C. K. Liu, and C. Mylonas: Epoxy Adhesives and Casting Resins as Photo-elastic Plastics, *Proc. SESA*, vol. XII, no. 2, pp. 123–128, 1955.
6. Kawata, K.: Analysis of Elastoplastic Behavior of Metals by Means of Photoelastic Coating Method, *J. Sci. Res. Instrum., Tokyo*, vol. 52, pp. 17–40, 1958.
7. Durelli, A. J., E. A. Phillips, and C. H. Tsao: "Introduction to the Theoretical and Experimental Analysis of Stress and Strain," McGraw-Hill Book Company, New York, 1958.
8. Leven, M. M.: Epoxy Resins for Photoelastic Use, *Symp. Photoelasticity 1963*, pp. 145–168, Pergamon Press, New York.
9. Ito, K.: New Model Materials for Photoelasticity and Photoplasticity, *Exp. Mech.*, vol. 2, no. 12, pp. 373–376, 1962.
10. Thomas, A. D.: Photoelastic Analysis and Model Fringe Value of Lexan Polycarbonate Resin, *Gen. Elec. Co. Tech. Serv. Mem.*, 1962.
11. McIver, R. W.: Structural Test Applications Utilizing Large Continuous Photoelastic Coatings, *Exp. Mech.*, vol. 5, no. 1, pp. 19A–25A, no. 2, pp. 19A–26A, 1965.
12. Materials for Photoelastic Coatings, *Bull.* P1120, Photoelastic, Inc., Malvern, Penna.
13. Zandman, F., S. S. Redner, and E. I. Riegner: Reinforcing Effect of Birefringent Coatings, *Exp. Mech.*, vol. 2, no. 2, pp. 55–64, 1962.
14. Dally, J. W., and I. Alfirevich: Application of Birefringent Coatings to Glass-fiber-reinforced Plastics, *Exp. Mech.*, vol. 9, no. 3, pp. 97–102, 1969.
15. Duffy, J.: Effects of the Thickness of Birefringent Coatings, *Exp. Mech.*, vol. 1, no. 3, pp. 74–82, 1961.
16. Lee, T. C., C. Mylonas, and J. Duffy: Thickness Effects in Birefringent Coatings with Radial Symmetry, *Exp. Mech.*, vol. 1, no. 10, pp. 134–142, 1961.
17. Duffy, J., and C. Mylonas: An Experimental Study on the Thickness of Birefringent Coatings, *Symp. Photoelasticity 1963*, pp. 27–42, Pergamon Press, New York.
18. Theocaris, P. S., and K. Dafermos: A Critical Review on the Thickness Effect of Birefringent Coatings, *Exp. Mech.*, vol. 4, no. 9, pp. 271–276, 1964.

19. Post, D., and F. Zandman: Accuracy of Birefringent Coating Method for Coatings of Arbitrary Thickness, *Exp. Mech.*, vol. 1, no. 1, pp. 21–32, 1961.
20. Redner, S. S.: Instruction Manual for the 030 Series Reflection Polariscope, Photolastic, Inc., Malvern, Pa.
21. Redner, S. S.: New Oblique Incidence Method for Direct Photoelastic Measurement of Principal Strains, *Exp. Mech.*, vol. 3, no. 3, pp. 67–72, 1963.
22. Hovanesian, J. D.: Sign Determination in Oblique Incidence for Photoelastic Coatings, *Exp. Mech.*, vol. 5, no. 4, p. 128, 1965.
23. O'Regan, R.: New Method for Determining Strain on the Surface of a Body with Photoelastic Coatings, *Exp. Mech.*, vol. 5, no. 8, pp. 241–246, 1965.

INDEX